HARCOURT

Math

Harcourt School Publishers

Orlando • Boston • Dallas • Chicago • San Diego

www.harcourtschool.com

Senior Author

Evan M. Maletsky
Professor of Mathematics
Montclair State University
Upper Montclair, New Jersey

Mathematics Advisor

Richard Askey
Professor of Mathematics
University of Wisconsin
Madison, Wisconsin

Authors

Angela Giglio Andrews
Math Teacher, Scott School
Naperville District #203
Naperville, Illinois

Jennie M. Bennett
Instructional Mathematics Supervisor
Houstion Independent School District
Houston, Texas

Grace M. Burton
Chair, Department of Curricular Studies
Professor, School of Education
University of North Carolina
 at Wilmington
Wilmington, North Carolina

Howard C. Johnson
Dean of the Graduate School
Associate Vice Chancellor for
 Academic Affairs
Professor, Mathematics and
 Mathematics Education
Syracuse University
Syracuse, New York

Lynda A. Luckie
Administrator/Math Specialist
Gwinnett County Public Schools
Lawrenceville, Georgia

Joyce C. McLeod
Visiting Professor
Rollins College
Winter Park, Florida

Vicki Newman
Classroom Teacher
McGaugh Elementary School
Los Alamitos Unified School District
Seal Beach, California

Janet K. Scheer
Executive Director
Create A Vision
Foster City, California

Karen A. Schultz
College of Education
Georgia State University
Atlanta, Georgia

Program Consultants and Specialists

Janet S. Abbott
Mathematics Consultant
California

Lois Harrison-Jones
Education and Management
 Consultant
Dallas, Texas

Elsie Babcock
Director, Mathematics and
 Science Center
Mathematics Consultant
Wayne Regional Educational
 Service Agency
Wayne, Michigan

Arax Miller
Curriculum Coordinator and
 English Department
 Chairperson
Chamlian School
Glendale, California

William J. Driscoll
Professor of Mathematics
Department of Mathematical
 Sciences
Central Connecticut State
 University
New Britain, Connecticut

Rebecca Valbuena
Language Development
 Specialist
Stanton Elementary School
Glendora, California

Understand Numbers and Operations

1 PLACE VALUE AND NUMBER SENSE

- ✅ Check What You Know 1
- **1** Benchmark Numbers 2
- **2** Understand Place Value 4
- **3** Place Value Through Hundred Thousands 6
- **4** Place Value Through Millions 8
  Problem Solving: Thinker's Corner
- **5** **Problem Solving Skill:** Use a Graph 12
  Chapter 1 Review/Test 14
- ⭐ Standardized Test Prep 15
  Intervention: Troubleshooting H2
  Extra Practice H32

2 COMPARE AND ORDER NUMBERS 16

- ✅ Check What You Know 17
- **1** Compare Numbers 18
- **2** Order Numbers 20
  Problem Solving: Thinker's Corner
- **3** **Problem Solving Strategy:** Make a Table 24
- **4** Round Numbers 26
  Problem Solving: Linkup to Reading
  Chapter 2 Review/Test 30
- ⭐ Standardized Test Prep 31
  Intervention: Troubleshooting H3
  Extra Practice H33

Technology Link

Harcourt Math Newsroom Video: *Chapter 1, p. 10*

E-Lab: *Chapter 2, p. 21   Chapter 3, p. 42*

Mighty Math Calculating Crew: *Chapter 4, p. 67*

Multimedia Glossary: *The Learning Site* at
www.harcourtschool.com/mathglossary

3

ADD AND SUBTRACT GREATER NUMBERS 32

- ✔ Check What You Know 33
- 1 Estimate Sums and Differences 34
- 2 Use Mental Math Strategies 36
  - **Problem Solving:** Thinker's Corner
- 3 Add and Subtract 4-Digit Numbers 40
- 4 Subtract Across Zeros 42
- 5 Choose a Method 44
  - **Problem Solving:** Thinker's Corner
- 6 **Problem Solving Skill:**
  - Estimate or Find Exact Answers 48
  - Chapter 3 Review/Test 50
- ★ Standardized Test Prep 51
  - **Intervention:** Troubleshooting H5
  - **Extra Practice** H34

4

a+b⁄c ALGEBRA USE ADDITION AND SUBTRACTION .52

- ✔ Check What You Know 53
- 1 Expressions 54
- 2 Use Parentheses 56
- 3 Match Words and Expressions 58
- 4 Use Variables 60
  - **Problem Solving:** Linkup to Health
- 5 Find a Rule 64
- 6 Equations ✳ HANDS ON Activity 66
  - **Problem Solving:** Thinker's Corner
- 7 **Problem Solving Strategy:** Make a Model 70
  - Chapter 4 Review/Test 72
- ★ Standardized Test Prep 73
  - **Intervention:** Troubleshooting H6
  - **Extra Practice** H35

UNIT WRAPUP

Problem Solving: Math Detective 74
Challenge: Front-End Estimation 75
Study Guide and Review 76
Performance Assessment 78
Technology Linkup E-Lab: Subtracting Across Zeros 79
Problem Solving: On Location in Minnesota 79A

Data, Graphing, and Time

5 COLLECT AND ORGANIZE DATA 80

✔ Check What You Know 81
1 Collect and Organize Data 82
   Problem Solving: Thinker's Corner
2 ☀ **HANDS ON** Find Median and Mode 86
3 Line Plot 88
4 Stem-and-Leaf Plot 90
5 Compare Graphs 92
6 **Problem Solving Strategy:** Make a Graph 94
   Chapter 5 Review/Test 96
⭐ Standardized Test Prep 97
Intervention: Troubleshooting H8
Extra Practice H36

6 ANALYZE AND GRAPH DATA 98

✔ Check What You Know 99
1 ☀ **HANDS ON** Double-Bar Graphs 100
2 Read Line Graphs .102
3 ☀ **HANDS ON** Make Line Graphs 104
4 Choose an Appropriate Graph 106
   Problem Solving: Thinker's Corner
5 **Problem Solving Skill:** Draw Conclusions 110
   Chapter 6 Review/Test 112
⭐ Standardized Test Prep 113
Intervention: Troubleshooting H9
Extra Practice H37

Technology Link

Harcourt Math Newsroom Video: *Chapter 7, p. 127*

E-Lab: *Chapter 6, p. 101*   *Chapter 7, p. 121*

Multimedia Glossary: *The Learning Site* at
www.harcourtschool.com/mathglossary

7 UNDERSTAND TIME .114

✓ Check What You Know 115
1 Before and After the Hour 116
2 A.M. and P.M. 118
3 Elapsed Time 120
   Problem Solving: Linkup to Reading
4 Problem Solving Skill: Sequence Information 124
5 Elapsed Time on a Calendar .126
   Problem Solving: Linkup to Science
   Chapter 7 Review/Test 130
⭐ Standardized Test Prep 131
   Intervention: Troubleshooting H10
   Extra Practice H38

UNIT WRAPUP

Problem Solving: Math Detective .132
Challenge: Time Zones 133
Study Guide and Review 134
Performance Assessment 136
Technology Linkup E-Lab: Finding the Median and Mode . .137
   Problem Solving: On Location in Oregon 137A

UNIT 3
CHAPTERS 8-9

Multiplication and Division Facts

8 PRACTICE MULTIPLICATION AND DIVISION FACTS 138

   ✅ Check What You Know 139
1. Relate Multiplication and Division 140
2. Multiply and Divide Facts Through 5 142
3. Multiply and Divide Facts Through 10 ✴ **HANDS ON** Activity 144
   Problem Solving: Linkup to Art
4. ✴ **HANDS ON** Multiplication Table Through 12 148
5. Multiply 3 Factors 150
6. **Problem Solving Skill:** Choose the Operation 152
   Chapter 8 Review/Test 154
   ⭐ Standardized Test Prep 155
   Intervention: Troubleshooting H12
   Extra Practice H39

Technology Link

Harcourt Math Newsroom Video: *Chapter 8, p. 144*

E-Lab: *Chapter 8, p. 149*

Mighty Math Number Heroes: *Chapter 8, p. 146*

Multimedia Glossary: *The Learning Site* at www.harcourtschool.com/mathglossary

9 $\frac{a+b}{c}$ ALGEBRA USE MULTIPLICATION AND DIVISION FACTS 156

- ✓ Check What You Know 157
- **1** Expressions with Parentheses .158
  Problem Solving: Thinker's Corner
- **2** Match Words and Expressions 162
- **3** Equations ✴HANDS ON Activity 164
  Problem Solving: Linkup to Reading
- **4** Expressions with Variables 168
- **5** Equations with Variables 170
- **6** Find a Rule 172
- **7** **Problem Solving Strategy:** Work Backward 174
  Chapter 9 Review/Test 176
- ⭐ Standardized Test Prep 177
  Intervention: Troubleshooting H7, H13, H16
  Extra Practice H40

UNIT WRAPUP

Problem Solving: Math Detective 178
Challenge: Square Numbers 179
Study Guide and Review 180
Performance Assessment 182
Technology Linkup Mighty Math Calculating Crew:
Multiplication and Division Facts 183
Problem Solving: On Location at the Beach 183A

Multiply by 1- and 2-Digit Numbers

10

MULTIPLY BY 1-DIGIT NUMBERS **184**
- ✓ Check What You Know 185
- **1** Mental Math: Multiplication Patterns 186
- **2** Estimate Products 188
- **3** Multiply 2-Digit Numbers ✹ HANDS ON Activity 190
  Problem Solving: Thinker's Corner
- **4** ✹ HANDS ON Model Multiplication 194
- **5** Multiply 3-Digit Numbers 196
  Problem Solving: Thinker's Corner
- **6** Multiply 4-Digit Numbers .200
  Problem Solving: Linkup to Reading
- **7** **Problem Solving Strategy:** Write an Equation 204
  Chapter 10 Review/Test 206
- ✰ Standardized Test Prep 207
  Intervention: Troubleshooting H13
  Extra Practice H41

11

UNDERSTAND MULTIPLICATION **208**
- ✓ Check What You Know 209
- **1** Mental Math: Patterns with Multiples 210
- **2** Multiply by Multiples of 10 .212
- **3** Estimate Products 214
- **4** ✹ HANDS ON Model Multiplication 216
- **5** **Problem Solving Strategy:** Solve a Simpler Problem . .218
  Chapter 11 Review/Test 220
- ✰ Standardized Test Prep 221
  Intervention: Troubleshooting H13
  Extra Practice H42

Technology Link

Harcourt Math Newsroom Video: *Chapter 11, p. 214*
E-Lab: *Chapter 12, p. 225*
Mighty Math Number Heroes: *Chapter 10, p. 197*
Mighty Math Calculating Crew: *Chapter 10, p. 201*
Multimedia Glossary: *The Learning Site* at
www.harcourtschool.com/mathglossary

12

MULTIPLY BY 2-DIGIT NUMBERS 222

✓ Check What You Know 223
1 Multiply by 2-Digit Numbers 224
   Problem Solving: Linkup to Science
2 More About Multiplying by 2-Digit Numbers 228
3 Choose a Method 230
4 Practice Multiplication 232
5 **Problem Solving Skill:** Multistep Problems 234
   Chapter 12 Review/Test 236
⭐ Standardized Test Prep 237
   Intervention: Troubleshooting H15
   Extra Practice H43

UNIT WRAPUP

Problem Solving: Math Detective 238
Challenge: Mental Multiplication 239
Study Guide and Review 240
Performance Assessment 242
Technology Linkup Calculator: Number Patterns 243
Problem Solving: On Location in New Jersey 243A

Divide by 1- and 2-Digit Divisors

13

UNDERSTAND DIVISION **244**
✔ Check What You Know 245
1 Divide with Remainders ✷ HANDS ON Activity 246
2 ✷ HANDS ON Model Division 248
3 Division Procedures 250
   Problem Solving: Thinker's Corner
4 **Problem Solving Strategy:** Predict and Test 254
5 Mental Math: Division Patterns 256
6 Estimate Quotients 258
   Chapter 13 Review/Test 260
⭐ Standardized Test Prep 261
   Intervention: Troubleshooting H12, H16
   Extra Practice H44

14

DIVIDE BY 1-DIGIT DIVISORS **262**
✔ Check What You Know 263
1 Place the First Digit 264
2 Divide 3-Digit Numbers 266
3 Zeros in Division 268
   Problem Solving: Linkup to Social Studies
4 Choose a Method 272
5 **Problem Solving Skill:** Interpret the Remainder 274
6 Find the Mean 276
   Chapter 14 Review/Test 278
⭐ Standardized Test Prep 279
   Intervention: Troubleshooting H16
   Extra Practice H45

Technology Link

Harcourt Math Newsroom Video: *Chapter 15, p. 283*
E-Lab: *Chapter 14, p. 276   Chapter 16, p. 303*
Mighty Math Calculating Crew: *Chapter 13, p. 250   Chapter 15, p. 289*
Mighty Math Number Heroes: *Chapter 15, p. 286   Chapter 16, p. 308*
Multimedia Glossary: *The Learning Site* at
www.harcourtschool.com/mathglossary

15

DIVIDE BY 2-DIGIT DIVISORS 280

✔ Check What You Know 281
1 Division Patterns to Estimate 282
2 ✦ HANDS ON Model Division 284
3 Division Procedures 286
4 Correcting Quotients 288
   Problem Solving: Linkup to Reading
5 **Problem Solving Skill:** Choose the Operation 292
   Chapter 15 Review/Test 294
⭐ Standardized Test Prep 295
   Intervention: Troubleshooting H17
   Extra Practice H46

16

PATTERNS WITH FACTORS AND MULTIPLES . .296

✔ Check What You Know 297
1 Factors and Multiples 298
2 Factor Numbers ✦ HANDS ON Activity 300
3 Prime and Composite Numbers
   ✦ HANDS ON Activities 302
   Problem Solving: Linkup to History
4 Find Prime Factors 306
5 **Problem Solving Strategy:** Find a Pattern 308
   Chapter 16 Review/Test 310
⭐ Standardized Test Prep 311
   Intervention: Troubleshooting H12, H16
   Extra Practice H47

UNIT WRAPUP

Problem Solving: Math Detective 312
Challenge: Divisibility Rules 313
Study Guide and Review 314
Performance Assessment .316
Technology Linkup Calculator: Remainders 317
Problem Solving: On Location at the Zoo 317A

Geometry

17

PLANE FIGURES **318**

✔ Check What You Know 319

1 Lines, Rays, and Angles ✹ HANDS ON Activity 320
   Problem Solving: Linkup to Art

2 Line Relationships 324

3 Congruent Figures and Motion
   ✹ HANDS ON Activities 326
   Problem Solving: Thinker's Corner

4 Symmetric Figures ✹ HANDS ON Activity 330

5 **Problem Solving Strategy:** Make a Model 332

   Chapter 17 Review/Test 334

⭐ Standardized Test Prep 335

   Intervention: Troubleshooting H25

   Extra Practice H48

Technology Link

Harcourt Math Newsroom Video: *Chapter 17, p. 327*

E-Lab: *Chapter 18, p. 343*

Mighty Math Number Heroes: *Chapter 18, p. 349*

Multimedia Glossary: *The Learning Site* at
www.harcourtschool.com/mathglossary

18

MEASURE AND CLASSIFY PLANE FIGURES . .336

✓ Check What You Know 337
1 ✦ HANDS ON Turns and Degrees 338
2 ✦ HANDS ON Measure Angles .340
3 Circles ✦ HANDS ON Activity .342
4 ✦ HANDS ON Circumference .344
5 Classify Triangles 346
6 Classify Quadrilaterals ✦ HANDS ON Activity 348
   Problem Solving: Linkup to Reading
7 **Problem Solving Strategy:** Draw a Diagram 352
   Chapter 18 Review/Test 354
⭐ Standardized Test Prep 355
   Intervention: Troubleshooting .H30
   Extra Practice H49

UNIT WRAPUP

Problem Solving: Math Detective 356
Challenge: Tessellations 357
Study Guide and Review 358
Performance Assessment 360
Technology Linkup Mighty Math Number Heroes:
   Symmetry 361
Problem Solving: On Location in Chicago 361A

Fractions and Decimals

19 UNDERSTAND FRACTIONS 362

- ✔ Check What You Know 363
- **1** Read and Write Fractions 364
- **2** ✳ **HANDS ON** Equivalent Fractions 366
- **3** Equivalent Fractions 368
  Problem Solving: Linkup to Music
- **4** Compare and Order Fractions ✳ **HANDS ON** Activity . . .372
  Problem Solving: Thinker's Corner
- **5** **Problem Solving Strategy:** Make a Model 376
- **6** Mixed Numbers 378
  Problem Solving: Thinker's Corner
  Chapter 19 Review/Test 382
- ⭐ Standardized Test Prep 383
  Intervention: Troubleshooting .H18
  Extra Practice H50

20 ADD AND SUBTRACT FRACTIONS AND MIXED NUMBERS 384

- ✔ Check What You Know 385
- **1** Add Like Fractions ✳ **HANDS ON** Activity 386
  Problem Solving: Linkup to Art
- **2** ✳ **HANDS ON** Subtract Like Fractions 390
- **3** Add and Subtract Mixed Numbers 392
  Problem Solving: Thinker's Corner
- **4** **Problem Solving Skill:** Choose the Operation 396
- **5** ✳ **HANDS ON** Add Unlike Fractions 398
- **6** ✳ **HANDS ON** Subtract Unlike Fractions 400
  Chapter 20 Review/Test 402
- ⭐ Standardized Test Prep 403
  Intervention: Troubleshooting .H18
  Extra Practice H51

Technology Link

Harcourt Math Newsroom Video: *Chapter 21, p. 407*
E-Lab: *Chapter 19, p. 367   Chapter 20, p. 391   Chapter 21, p. 413*
Mighty Math Number Heroes: *Chapter 19, p. 373*
Mighty Math Calculating Crew: *Chapter 20, p. 393*
    Chapter 21, p. 415   Chapter 22, p. 432
Multimedia Glossary: *The Learning Site* at
www.harcourtschool.com/mathglossary

21

UNDERSTAND DECIMALS 404

✔ *Check What You Know* 405
1 Tenths and Hundredths .406
   Problem Solving: Linkup to Science
2 Thousandths 410
3 ☀ HANDS ON Equivalent Decimals 412
4 Relate Mixed Numbers and Decimals 414
   Problem Solving: Thinker's Corner
5 Compare and Order Decimals 418
   Problem Solving: Thinker's Corner
6 **Problem Solving Strategy:** Use Logical Reasoning . . .422
   Chapter 21 Review/Test .424
⭐ Standardized Test Prep 425
   Intervention: Troubleshooting H20
   Extra Practice H52

22

ADD AND SUBTRACT DECIMALS 426

✔ *Check What You Know* 427
1 Round Decimals 428
2 Estimate Sums and Differences 430
3 Add Decimals ☀ HANDS ON Activity 432
4 Subtract Decimals ☀ HANDS ON Activity 434
5 Add and Subtract Decimals 436
   Problem Solving: Thinker's Corner
6 **Problem Solving Skill:**
   Evaluate Reasonableness of Answers 440
   Chapter 22 Review/Test .442
⭐ Standardized Test Prep 443
   Intervention: Troubleshooting H21
   Extra Practice H53

UNIT WRAPUP

Problem Solving: Math Detective 444
Challenge: Circle Graphs .445
Study Guide and Review 446
Performance Assessment 448
Technology Linkup Mighty Math Number Heroes:
   Add and Subtract Fractions 449
   Problem Solving: On Location at National Parks 449A

Measurement and Geometry

23 CUSTOMARY MEASUREMENT 450

✔ Check What You Know 451
1 Choose the Appropriate Unit 452
2 Measure Fractional Parts ✳ HANDS ON Activity 454
   Problem Solving: Linkup to Reading
3 ALGEBRA Change Linear Units 458
4 ✳ HANDS ON Capacity 460
5 ✳ HANDS ON Weight 462
6 **Problem Solving Strategy:** Compare Strategies 464
   Chapter 23 Review/Test 466
⭐ Standardized Test Prep 467
   Intervention: Troubleshooting H18, H22, H23
   Extra Practice H54

24 METRIC MEASUREMENT 468

✔ Check What You Know 469
1 Linear Measure ✳ HANDS ON Activities 470
   Problem Solving: Thinker's Corner
2 ALGEBRA Change Linear Units 474
3 ✳ HANDS ON Capacity 476
4 ✳ HANDS ON Mass 478
5 **Problem Solving Strategy:** Draw a Diagram 480
   Chapter 24 Review/Test 482
⭐ Standardized Test Prep 483
   Intervention: Troubleshooting H23
   Extra Practice H55

Technology Link

Harcourt Math Newsroom Video: *Chapter 23, p. 454*

E-Lab: *Chapter 24, p. 480   Chapter 25, p. 493*

Mighty Math Calculating Crew: *Chapter 26, pp. 508, 512*

Multimedia Glossary: *The Learning Site* at
www.harcourtschool.com/mathglossary

25

PERIMETER AND AREA OF PLANE FIGURES . .484

- ✓ Check What You Know 485
- **1** Perimeter of Polygons 486
  - **Problem Solving:** Thinker's Corner
- **2** Estimate and Find Perimeter ✷ HANDS ON Activity 490
- **3** Estimate and Find Area 492
  - **Problem Solving:** Linkup to Architecture
- **4** Relate Area and Perimeter ✷ HANDS ON Activities 496
- **5** Relate Formulas and Rules 498
  - **Problem Solving:** Linkup to Reading
- **6** **Problem Solving Strategy:** Find a Pattern 502
  - Chapter 25 Review/Test .504
- ★ Standardized Test Prep 505
  - **Intervention:** Troubleshooting H27
  - **Extra Practice** H56

26

SOLID FIGURES AND VOLUME 506

- ✓ Check What You Know 507
- **1** Faces, Edges, and Vertices ✷ HANDS ON Activity 508
  - **Problem Solving:** Linkup to Social Studies
- **2** Patterns for Solid Figures ✷ HANDS ON Activity 512
- **3** Estimate and Find Volume of Prisms 514
- **4** **Problem Solving Skill:**
  - Too Much/Too Little Information 516
  - Chapter 26 Review/Test .518
- ★ Standardized Test Prep 519
  - **Intervention:** Troubleshooting H29
  - **Extra Practice** H57

UNIT WRAPUP

- **Problem Solving:** Math Detective 520
- **Challenge:** Relate Benchmark Measurements 521
- **Study Guide and Review** 522
- **Performance Assessment** 524
- **Technology Linkup** Mighty Math Calculating Crew:
  - Nets for Solid Figures .525
- **Problem Solving:** On Location with Balloons 525A

Probability, Algebra, and Graphing

27 OUTCOMES 526

✓ Check What You Know 527
1 ✸ **HANDS ON** Record Outcomes 528
2 Tree Diagrams 530
3 **Problem Solving Strategy:** Make an Organized List . .532
4 Predict Outcomes of Experiments
  ✸ **HANDS ON** Activity 534
Problem Solving: Linkup to Reading
Chapter 27 Review/Test 538
★ Standardized Test Prep 539
Intervention: Troubleshooting H31
Extra Practice H58

28 PROBABILITY 540

✓ Check What You Know 541
1 Probability as a Fraction 542
2 ✸ **HANDS ON** More About Probability 544
3 Test for Fairness 546
4 **Problem Solving Skill:** Draw Conclusions 548
Chapter 28 Review/Test 550
★ Standardized Test Prep 551
Intervention: Troubleshooting H18
Extra Practice H59

Technology Link

Harcourt Math Newsroom Video: *Chapter 28, p. 544*

E-Lab: *Chapter 27, p. 528   Chapter 27, p. 535*

Mighty Math Number Heroes: *Chapter 28, p. 542*

Multimedia Glossary: *The Learning Site* at
www.harcourtschool.com/mathglossary

29

a+b/c ALGEBRA EXPLORE NEGATIVE NUMBERS . . .552

- ✓ Check What You Know 553
- **1** Temperature: Fahrenheit .554
- **2** Temperature: Celsius 556
- **3** Negative Numbers 558
  Problem Solving: Linkup to Science
- **4** **Problem Solving Skill:** Make Generalizations 562
  Chapter 29 Review/Test .564
- ☆ Standardized Test Prep .565
  Intervention: Troubleshooting H3, H24
  Extra Practice H60

30

EXPLORE THE COORDINATE GRID 566

- ✓ Check What You Know 567
- **1** Use a Coordinate Grid 568
- **2** Use an Equation 570
  Problem Solving: Linkup to Social Studies
- **3** Graph an Equation ✷ HANDS ON Activities 574
  Problem Solving: Thinker's Corner
- **4** **Problem Solving Skill:** Identify Relationships 578
  Chapter 30 Review/Test .580
- ☆ Standardized Test Prep .581
  Intervention: Troubleshooting H10, H25
  Extra Practice H61

UNIT WRAPUP

Problem Solving: Math Detective 582
Challenge: Find All Possible Ways 583
Study Guide and Review 584
Performance Assessment 586
Technology Linkup E-Lab: Predicting Outcomes 587
  Problem Solving: On Location with
    Monarch Butterflies 587A

STUDENT HANDBOOK

Table of Contents H1
Troubleshooting H2-H31
Extra Practice H32-H61
Sharpen Your Test-Taking Skills H62-H65
Basic Facts Tests H66-H71
Table of Measures H72
Glossary H73-H82
Index H83-H93

Welcome!

The authors of *Harcourt Math* want you to enjoy learning math and to feel confident that you can do it. We invite you to share your math book with family members. Take them on a guided tour through your book!

The Guided Tour

Choose a chapter you are interested in. Show your family some of these things in the chapter that will help you learn.

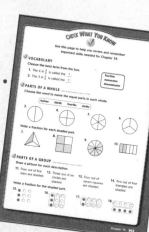

✓ Check What You Know

Do you need to review any skills before you begin the next chapter? If you do, you will find help in the Handbook in the back of your book.

⏱ The Math Lessons

☑ **Quick Review** to check the skills you need for the lesson.

☑ **Learn section** to help you study problems, models, examples, and questions that give you different ways to learn.

☑ **Check** to make sure you understood the lesson.

☑ **Practice and Problem Solving** to practice what you have just learned.

☑ **Mixed Review and Test Prep** to keep your skills sharp and to prepare you for important tests. Look back at the pages shown next to each problem to get help if you need it.

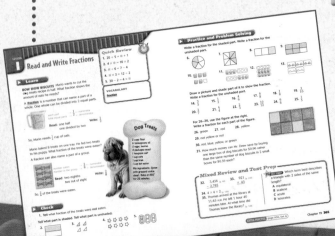

Student Handbook ·············

Now show your family the **Student Handbook** in the back of your book. The sections will help you in many different ways.

- ✅ **Troubleshooting** will help you review and remember skills from last year.

- ✅ **Extra Practice** can be used to make sure that you are ready to move on to the next lesson.

- ✅ **Sharpen Your Test-Taking Skills** will help you feel confident that you can do well on a test.

- ✅ **Basic Facts Tests** will check whether you have memorized all of the basic facts and will show you which facts you still need to practice.

Invite your family members to

- talk with you about what you are learning.

- help you correct errors you have made on completed work.

- help you set a time and find a quiet place to do math homework.

- help you memorize the addition, subtraction, multiplication, and division facts.

- solve problems as you play together, shop together, and do household chores.

- visit **The Learning Site** at www.harcourtschool.com

- have **Fun with Math!**

Have a great year!

The Authors

Be a Good Problem Solver!

You need to organize your thinking. You can use problem solving steps to stay on track.

Use these problem solving steps. They can help you think through a problem.

UNDERSTAND the problem.

What are you asked to find?

What information will you use?

Is there information you will not use? If so, what?

Restate the question in your own words.

List the information given in the problem.

Decide whether you need all the information you are given.

PLAN a strategy to solve.

What strategy can you use to solve the problem?

Think about some problem solving strategies you can use. Then choose one.

SOLVE the problem.

How can you use the strategy to solve the problem?

Follow your plan. Show your solution.

CHECK your answer.

Look back at the problem. Does the answer make sense? Explain.

What other strategy could you use?

Be sure you answered the question that is asked.

Solving the problem by another method is a good way to check your work.

Try It

Here's how you can use the problem solving steps to solve a problem.

Find a Pattern

PROBLEM This figure shows how a larger triangle can be divided into smaller congruent triangles. If the large triangle is extended to 5 rows, how many small triangles will there be in rows 4 and 5?

Row 1 →
Row 2 →
Row 3 →

PROBLEM SOLVING STRATEGIES

Draw a Diagram or Picture
Make a Model or Act it Out
Make an Organized List
▶ **Find a Pattern**
Make a Table or Graph
Predict and Test
Work Backward
Solve a Simpler Problem
Write an Equation
Use Logical Reasoning

UNDERSTAND the problem.

I can count the number of small triangles in rows 1, 2, and 3. I need to find the number of triangles in rows 4 and 5.

PLAN a strategy to solve.

I can *find a pattern* in the number of triangles in rows 1, 2, and 3 to extend the rows.

SOLVE the problem.

Number of the row	1	2	3	4	5
Number of small triangles	1	3	5	7	9

+ 2 + 2 + 2 + 2

The pattern is: In each row, the number of small triangles increases by 2.

$5 + 2 = 7$ and $7 + 2 = 9$

So, the number in row 4 is 7 and the number in row 5 is 9.

CHECK your answer.

I can check my answer by drawing a diagram. I can count the number of triangles in each row. The pattern checks.

Getting Ready!

Remember the Properties of Addition

These rules, called properties, can help you recall basic facts and compute mentally.

ORDER PROPERTY OF ADDITION
Changing the order of the addends does not change the sum.

$$6 + 8 = 14 \qquad\qquad 8 + 6 = 14$$
addends   sum   addends   sum

This means that $6 + 8 = 8 + 6$.

Order Property of Addition
$$6 + 5 = 5 + 6$$
$$25 + 40 = 40 + 25$$
$$70 + 35 = 35 + 70$$

ZERO PROPERTY OF ADDITION
When you add zero to a number, the sum is that number.

$$12 + 0 = 12 \qquad 0 + 25 = 25$$

GROUPING PROPERTY OF ADDITION
When you group addends in different ways, the sums are the same.

$$9 + (15 + 5) = (9 + 15) + 5$$
$$9 + \quad 20 \quad = \quad 24 \quad + 5$$
$$29 \qquad = \qquad 29$$

▶ Practice

Copy and complete. Write the property shown.

1. $5 + 4 = \blacksquare + 5$

2. $135 + 0 = \blacksquare$

3. $9 + \blacksquare = 8 + 9$

4. $13 + (7 + 8) = \blacksquare$
   $(13 + 7) + 8 = \blacksquare$

5. $24 + (16 + 35) = \blacksquare$
   $(24 + 16) + 35 = \blacksquare$

6. $(18 + 4) + 7 = \blacksquare$
   $18 + (4 + 7) = \blacksquare$

7. $0 + 10 = \blacksquare$

8. $7 + 6 = 6 + \blacksquare$

9. $0 + 3{,}240 = \blacksquare$

Practice Addition and Subtraction Facts

Facts Practice

Use addition properties and fact families to review addition and subtraction facts.

You have learned that addition and subtraction are related. They are inverse, or opposite, operations.

A **FACT FAMILY** is a set of related addition and subtraction number sentences that use the same numbers.

8, 6, 14	6, 6, 12
$8 + 6 = 14$ $6 + 8 = 14$	$6 + 6 = 12$
$14 - 6 = 8$ $14 - 8 = 6$	$12 - 6 = 6$

$17 - 9 = \blacksquare$

$8 + 9 = 17$

▶ Practice

Find the sum or difference.

1. $4 + 9$ **2.** $7 + 6$ **3.** $0 + 8$ **4.** $5 + 5$ **5.** $6 + 9$

6. $5 + 7$ **7.** $9 + 9$ **8.** $7 + 8$ **9.** $4 + 6$ **10.** $9 - 8$

11. $12 - 7$ **12.** $15 - 8$ **13.** $11 - 6$ **14.** $17 - 8$ **15.** $16 - 9$

16. $14 - 7$ **17.** $9 - 0$ **18.** $13 - 6$ **19.** $15 - 9$

20. $\begin{array}{r} 8 \\ + 8 \\ \hline \end{array}$ **21.** $\begin{array}{r} 16 \\ - 8 \\ \hline \end{array}$ **22.** $\begin{array}{r} 10 \\ + 7 \\ \hline \end{array}$ **23.** $\begin{array}{r} 9 \\ + 0 \\ \hline \end{array}$

24. $\begin{array}{r} 9 \\ - 9 \\ \hline \end{array}$ **25.** $\begin{array}{r} 11 \\ - 4 \\ \hline \end{array}$ **26.** $\begin{array}{r} 6 \\ + 7 \\ \hline \end{array}$ **27.** $\begin{array}{r} 15 \\ - 6 \\ \hline \end{array}$

28. $\begin{array}{r} 5 \\ + 6 \\ \hline \end{array}$ **29.** $\begin{array}{r} 12 \\ - 7 \\ \hline \end{array}$ **30.** $\begin{array}{r} 8 \\ - 0 \\ \hline \end{array}$ **31.** $\begin{array}{r} 5 \\ + 9 \\ \hline \end{array}$

32. $\begin{array}{r} 7 \\ - 0 \\ \hline \end{array}$ **33.** $\begin{array}{r} 5 \\ + 8 \\ \hline \end{array}$ **34.** $\begin{array}{r} 14 \\ - 8 \\ \hline \end{array}$ **35.** $\begin{array}{r} 7 \\ + 7 \\ \hline \end{array}$

36. $\begin{array}{r} 5 \\ + 9 \\ \hline \end{array}$ **37.** $\begin{array}{r} 18 \\ - 9 \\ \hline \end{array}$ **38.** $\begin{array}{r} 13 \\ - 5 \\ \hline \end{array}$ **39.** $\begin{array}{r} 6 \\ + 6 \\ \hline \end{array}$

40. $\begin{array}{r} 9 \\ + 9 \\ \hline \end{array}$ **41.** $\begin{array}{r} 9 \\ - 1 \\ \hline \end{array}$ **42.** $\begin{array}{r} 17 \\ - 8 \\ \hline \end{array}$ **43.** $\begin{array}{r} 8 \\ + 8 \\ \hline \end{array}$

Place Value and Number Sense

A model of the Statue of Liberty is one of the model landmarks at a park in California. More than 42 million plastic bricks were used to build all the models.

PROBLEM SOLVING Look at the table below. Write each number of bricks in expanded word form.

LANDMARK MODELS	
Landmark	**Number of Bricks**
Capitol, Washington, D.C.	253,000
Statue of Liberty, New York	1,400,000
Mount Rushmore, South Dakota	1,500,000

DATA LINK

Use this page to help you review and remember
important skills needed for Chapter 1.

✓ VOCABULARY

Choose the best term from the box.

1. The number 4,082 is written in ? form.

2. In the number 8,640, the digit 6 is in the hundreds ? position.

3. Each of the symbols 0, 1, 2, 3, 4, 5, 6, 7, 8, and 9 is called a ? .

4. The ? form of 731 is 700 + 30 + 1.

> place-value
> standard
> expanded
> digit
> decimal

✓ BENCHMARK NUMBERS TO 100 (For Intervention, see p. H2.)

5. There are 10 beads in the first jar. Which jars have more than 10 beads?

6. There are 50 beads in the first jar. Which jars have fewer than 50 beads?

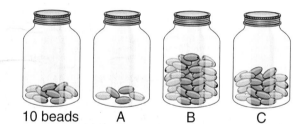

10 beads   A   B   C

50 beads   A   B   C

✓ READ AND WRITE NUMBERS TO THOUSANDS (For Intervention, see p. H2.)

Write each word form in standard form.

7. ninety

8. one hundred ninety-four

9. nine hundred seventy

10. two thousand, seventy

11. six thousand, three hundred forty-nine

12. six thousand, seven

✓ PLACE VALUE TO THOUSANDS (For Intervention, see p. H3.)

Write the value of the blue digit.

13. 5,024

14. 8,653

15. 3,934

16. 2,865

17. 9,752

18. 3,001

19. 9,356

20. 4,090

Benchmark Numbers

VOCABULARY

benchmark

▶ **Learn**

SIZE IT UP!   For a number to have meaning, it should be related to something you already know.

The Washington Monument is 555 feet tall. That is about the same as 25 two-story houses stacked on top of each other.

MATH IDEA   A benchmark is a known number of things that helps you understand the size or amount of a different number of things.

You can use a benchmark when you are estimating a number of items that would take a long time to count.

Examples

A   Use the benchmark to decide which is the most reasonable number of nickels in the full jar.

100   1,000   10,000

500 nickels

Benchmark: 500 nickels

The most reasonable number of nickels in the full jar is 1,000.

B   Use the benchmark to decide which is the most reasonable number of beans in the jar.

100   400   4,000

50 beans

Benchmark: 50 beans

A reasonable number of beans in the jar is 400.

• In Example B, why is 4,000 not a reasonable number?

1. **Tell** whether the number of students in your class is a good benchmark for the number of students in your school. Explain.

Use the benchmark to decide which is the more reasonable number.

2. beads in the jar

20 beads

80 or 800

3. gallons of water in the tank

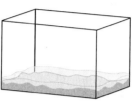 

20 gallons

200 or 2,000

▶ **Practice and Problem Solving**

Use the benchmark to decide which is the more reasonable number.

4. pretzel sticks in the jar

100 pretzels

500 or 5,000

5. beads in the vase

20 beads

200 or 2,000

6. Emily has 6 dolls and Dana has 11. If Laura has 8 more dolls than Emily and Dana combined, how many dolls does Laura have?

7. ✍ **Write About It** Explain when you would use a benchmark number.

Mixed Review and Test Prep

8. $78 + 9$

9. $746 + 24$

10. 5×5

11. $532 - 40$

12. **TEST PREP** Which is the missing addend for $18 + \blacksquare = 26$?

   A 4   **B** 5   **C** 8   **D** 9

Understand Place Value

Quick Review

1. 55 − 21   2. 21 + 46

3. 77 − 20   4. 163 + 36

5. 321 + 270

▶ **Learn**

IT'S DEEP!   The deepest living starfish was collected from a depth of 24,881 feet in the western Pacific Ocean in 1962.

What is the value of the digit 2 in 24,881?

Remember

A digit is one of the ten symbols 0, 1, 2, 3, 4, 5, 6, 7, 8, or 9 used to write numbers.

Ten thousands	Thousands	Hundreds	Tens	Ones
2	4,	8	8	1
2 × 10,000	4 × 1,000	8 × 100	8 × 10	1 × 1
20,000	4,000	800	80	1

Think:
Multiply the digit by its place-value position to find the value of each digit.

So, the value of the digit 2 is 20,000.

MATH IDEA   The value of a digit depends on its place-value position in the number.

Changing a given digit in a number changes the value of the number.

Examples

Ⓐ   58,937 to 59,937   increased by 1,000

Ⓑ   58,937 to 68,937   increased by 10,000

Ⓒ   58,937 to 88,937   increased by 30,000

Ⓓ   58,937 to 57,937   decreased by 1,000

Ⓔ   58,937 to 48,937   decreased by 10,000

Ⓕ   58,937 to 28,937   decreased by 30,000

▲ The Japan Marine Science and Technology Center's *Shinkai 6500* collects ocean data from any depth down to 21,325 feet.

• Which digit in the number 13,872 would be changed to form 19,872? How would the value of 13,872 change?

▶ Check

1. Find the value of the digit 6 in the number 76,308.

Write the value of the digit 4 in each number.

2. 27,345   **3.** 74,960

4. 83,412   **5.** 14,873

Compare the digits to find the value of the change.

6. 8,947 to 3,947   **7.** 82,756 to 82,716   **8.** 14,583 to 16,583

▶ Practice and Problem Solving

Write the value of the digit 8 in each number.

9. 53,489   **10.** 97,806   **11.** 86,239   **12.** 68,391

Compare the digits to find the value of the change.

13. 62,895 to 32,895   **14.** 93,714 to 99,714   **15.** 38,047 to 38,097

16. 49,807 to 49,207   **17.** 51,386 to 11,386   **18.** 29,471 to 29,671

Change the value of the number by the given amount.

19. 5,671 increased by 2,000   **20.** 37,842 decreased by 10,000

21. 63,172 increased by 600   **22.** 24,597 increased by 4,000

23. 71,408 decreased by 20,000   **24.** 52,496 decreased by 70

Complete.

25. $24{,}180 = 20{,}000 + \blacksquare + 100 + 80$ **26.** $5{,}2\blacksquare6 = 5{,}000 + 200 + 30 + 6$

27. NUMBER SENSE In a 4-digit number, the two greatest place-value digits are 2. The sum of the ones and tens digits is 14. What numbers are possible?

28. 📖 **Write About It** If you add a ten thousands digit that is 2 times the ones digit to the number 2,794, what is the new number? Explain.

Mixed Review and Test Prep

For 29–30, find the missing number.

29. $60 - \blacksquare = 20$ **30.** $\blacksquare - 5 = 39$

31. Name a plane figure that has four sides of equal length.

32. $78 + 63$

33. TEST PREP Find the missing number.
318, 324, 330, ■, 342
A 334   **B** 336   **C** 338   **D** 340

Place Value Through Hundred Thousands

▶ **Learn**

EARTH TO MOON The least distance from the Earth to the moon is 225,792 miles.

To understand this number, a column for hundred thousands has to be added to the place-value chart.

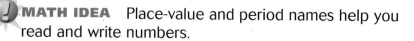

———— PERIOD ————

THOUSANDS			ONES		
Hundreds	Tens	Ones	Hundreds	Tens	Ones
2	2	5,	7	9	2

Each group of three digits is called a **period**. A comma separates each of the periods. The number 225,792 has two periods, *ones* and *thousands*.

MATH IDEA Place-value and period names help you read and write numbers.

Standard Form: 225,792

Word Form: two hundred twenty-five thousand, seven hundred ninety-two

Expanded Form: 200,000 + 20,000 + 5,000 + 700 + 90 + 2

Examples

Standard Form	Word Form	Expanded Form
Ⓐ 40,915	forty thousand, nine hundred fifteen	40,000 + 900 + 10 + 5
Ⓑ 607,304	six hundred seven thousand, three hundred four	600,000 + 7,000 + 300 + 4

Quick Review

Write the place-value name for the digit 3.

1. 49,031   **2.** 35,477

3. 2,386   **4.** 693

5. 83,904

VOCABULARY

period

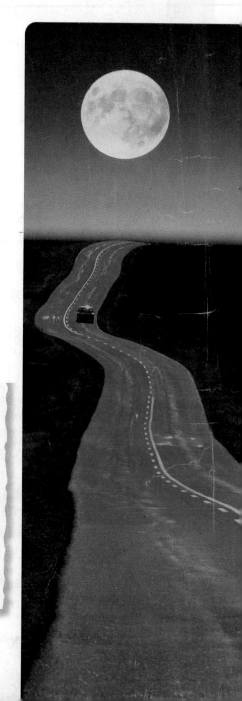

▶ Check

1. **Find** the place value of the digit 9 in 952,700. In 1969, the *Apollo 11* astronauts traveled a total distance of 952,700 miles.

Write each number in two other forms.

2. two hundred five thousand, sixty-one

3. 916,359

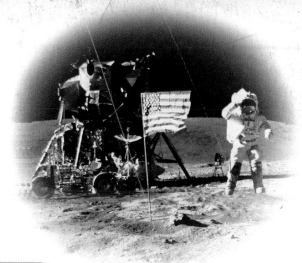

▶ Practice and Problem Solving

Write each number in two other forms.

4. three hundred thousand, ninety-six

5. four hundred sixteen thousand, two hundred ten

6. 40,705

7. 60,000 + 3,000 + 40 + 8

Complete.

8. 52,376 = fifty-two __?__ , three hundred __?__ = ■ + 2,000 + 300 + 70 + ■

9. 90,000 + ■ + 80 = 90,58■ = ninety thousand, five __?__ eighty

Write the value of the blue digit.

10. 534,908   11. 980,571   12. 143,296   13. 278,105

14. 357,841   15. 493,560   16. 782,046   17. 609,428

18. If five hundred thousand, twenty-six is increased by three thousand, what is the new number in standard form?

19. I am 900 more than the greatest possible 4-digit even number that can be made using the digits 1, 4, 2, 5. What number am I?

20. Write the word form of the number that is 1,000 greater than 23,548.

Mixed Review and Test Prep

21. 325 − ■ = 194   22. ■ − 75 = 896

23. Round 615 to the nearest hundred.

24. 35 + ■ = 66 + 35

25. **TEST PREP** Find the sum.

417 + 89 + 123

A 519   C 619

B 529   D 629

EXTRA PRACTICE page H32, Set C

Place Value Through Millions

> ▶ **Learn**

READ ALL ABOUT IT! Newspapers keep people informed of local, national, and world events. The first newspaper was written in Germany in the 1600's.

Look at this story. It contains about 200 words. If there are 5 stories of this size on one page, about how many words are on the page?

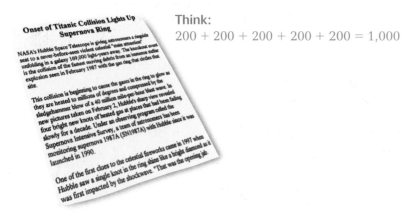

Think:
$200 + 200 + 200 + 200 + 200 = 1{,}000$

So, there are about 1,000 words on a page.

With 1,000 words on a page,
    10 pages would have 10,000 words.
  100 pages would have 100,000 words.
1,000 pages would have 1,000,000 words.

The period to the left of *thousands* is **millions**.

	PERIOD							
MILLIONS			**THOUSANDS**			**ONES**		
Hundreds	Tens	Ones	Hundreds	Tens	Ones	Hundreds	Tens	Ones
		1,	0	0	0,	0	0	0

Write: 1,000,000   **Read:** one million

One million is a large number. If you read 100 words a minute, it would take you almost 7 days non-stop to read 1,000,000 words.

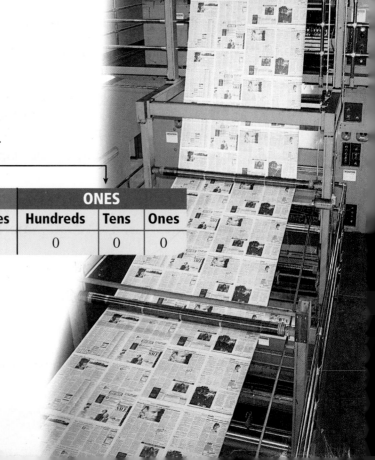

More About Millions

You can use place-value and period names to help you read and write numbers in the millions period.

The world's largest ball of twine is found in Cawker City, Kansas. As of September 1999, it contained 82,979,868 inches of twine.

Look at this number on the place-value chart.

▲ The ball of twine weighs 17,248 pounds. That's more than the weight of 6 cars!

		PERIOD						
MILLIONS			**THOUSANDS**			**ONES**		
Hundreds	Tens	Ones	Hundreds	Tens	Ones	Hundreds	Tens	Ones
	8	2,	9	7	9,	8	6	8

Standard Form: 82,979,868

Word Form: eighty-two million, nine hundred seventy-nine thousand, eight hundred sixty-eight

Expanded Form: 80,000,000 + 2,000,000 + 900,000 + 70,000 + 9,000 + 800 + 60 + 8

Examples

Standard Form	Word Form	Expanded Form
Ⓐ 54,060,900	fifty-four million, sixty thousand, nine hundred	50,000,000 + 4,000,000 + 60,000 + 900
Ⓑ 100,207,054	one hundred million, two hundred seven thousand, fifty-four	100,000,000 + 200,000 + 7,000 + 50 + 4

▶ Check

1. **Tell** how many periods an 8-digit number has.

Write the value of the blue digit.

2. 7,943,120   3. 8,450,203   4. 68,549,227

Write each number in word form.

5. 57,643,120   6. 16,452,003   7. 608,049,227

LESSON CONTINUES ▶

Write the value of the blue digit.

8. 7,534,908

9. 98,745,300

10. 142,980,871

11. 4,371,568

12. 36,420,156

13. 512,604,397

Write each number in word form.

14. 5,769,042

15. 2,831,001

16. 42,168,339

Use place value to find each missing number. Explain.

17. 6,758,324; 6,768,324; ■; 6,788,324

18. 9,537,461; 9,537,561; ■; 9,537,761

19. 2,408,693; 2,409,694; ■; 2,411,696

20. 4,657,839; 4,657,939; ■; 4,658,139

21. Write the word form of the number 10,000,000 greater than 5,670,891.

22. Write 12,097,341 in expanded form.

23. Write 8,000,000 + 100,000 + 70,000 + 3,000 + 900 + 50 + 6 in word form.

Technology Link

To learn more about *millions,* watch the Harcourt Math Newsroom Video *Supernova Blast.*

Complete.

24. 7,523,■46 = 7,000,000 + 500,000 + 20,000 + 3,000 + 800 + 40 + 6

25. 7,903,264 = seven _?_ , nine hundred three _?_ , two hundred sixty-four

26. **? What's the Question?** Mrs. Diaz wrote the number 46,152,780. The answer is 6,000,000. What is the question?

27. ✎ Write a problem using a newspaper article that includes numbers in standard form or word form.

28. **? What's the Error?** In his report, Tarek wrote the number of days that it takes Saturn to orbit the sun as ten million, seven hundred sixty. Describe his error and write the number correctly.

▲ Saturn takes about 10,760 days to orbit the sun.

29. Saturn's rings stretch from about 4,350 miles above the surface to 45,984 miles above the surface. Write these distances in word form.

30. The circulation of the largest newspaper in the world is about 14,976,000. What is the place-value position of the digit 4?

Mixed Review and Test Prep

31. 714 + 836

32. 854 − 138

33. 42 ÷ 7

34. 8 × 4

35. The sum of two numbers is 15. The difference is 3. What are the numbers?

36. What is the place-value position of the digit 6 in 36,280? (p. 4)

37. Write two hundred thousand, eighty-six in standard form. (p. 6)

38. TEST PREP Bobby has 87 basketball cards. He has 13 more than Jeff. How many cards does Jeff have?
A 71   **B** 74   **C** 95   **D** 105

39. TEST PREP Choose the letter that describes the pattern.

1, 7, 13, 19, 25

F Add 6.   **H** Add 4.
G Add 5.   **J** Add 3.

PROBLEM SOLVING ThinKer's CorNer

ROMAN NUMERALS Our *numeration system* uses Arabic numerals, or digits (0, 1, 2, . . .) to write numbers. The Romans used these symbols to name numbers.

I	V	X	L	C	D	M
1	5	10	50	100	500	1,000

a. Add when the symbols are alike or when the symbols' values decrease from left to right. A numeral cannot be added more than three times in a row.

LXIII = 63
50 + 10 + 1 + 1 + 1 = 63

b. Subtract when a symbol's value is less than the value of a symbol to its right.

XIX = 19
IX represents 10 − 1 = 9
10 + 9 = 19

Write the number named.

1. XIV   **2.** XXXIX   **3.** XLI   **4.** XC   **5.** LXVIII

6. REASONING Complete the pattern III, VI, IX, XII, ▪, ▪, XXI.

7. Write 35, 44, and 62 as Roman numerals.

EXTRA PRACTICE page H32, Set D

5 Problem Solving Skill
Use a Graph

Quick Review

1. 210 + 37 2. 364 − 122
3. 97 + 103 4. 527 + 81
5. 2,230 − 1,023

UNDERSTAND PLAN SOLVE CHECK

PRIZED PETS Betty, Marcia, and Ed want to know which pets are the most popular in the United States. Betty thinks cats are the most popular pet. Marcia thinks dogs are the most popular. Ed's choice is fish. Where can you find the information to see who is correct?

You can use a graph that compares the estimated numbers of kinds of pets in the United States.

Remember

A pictograph is a graph that uses pictures to show and compare information.

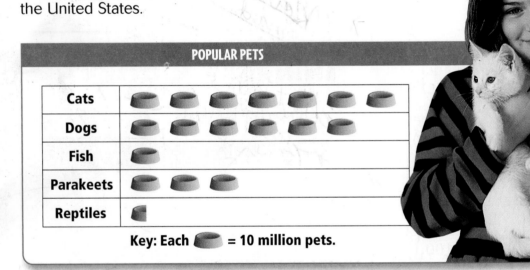

POPULAR PETS

Cats	
Dogs	
Fish	
Parakeets	
Reptiles	

Key: Each ⬭ = 10 million pets.

Look at the pictograph. Cats are shown with the most symbols, 7. Since each symbol stands for 10,000,000 pets, there are about 70 million pet cats.

So, Betty is correct. Cats are the most popular pets in the United States.

Talk About It

• About how many pets are dogs?

• The number of reptiles is shown with one half of a symbol. About how many pets are reptiles?

• **What if** the number of reptiles was 15 million? How many symbols would be used?

▶ Problem Solving Practice

USE DATA The U.S. has over 62,000,000 Internet users. This is more than any other country. For 1–2, use the Internet graph.

1. Find the number of Internet users in Japan. How did you find this number?

2. Which country shown in the graph has the fewest Internet users?

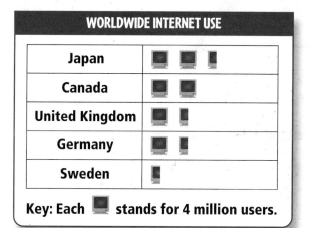

WORLDWIDE INTERNET USE

Japan	
Canada	
United Kingdom	
Germany	
Sweden	

Key: Each ▪ stands for 4 million users.

USE DATA For 3–4, use the cat food graph.

3. During which week were the most bags of cat food sold?
   A Week 1
   B Week 2
   C Week 3
   D Week 4

4. In which weeks were fewer than 500 bags of cat food sold?
   F Weeks 1, 2
   G Weeks 1, 4
   H Weeks 2, 4
   J Weeks 1, 3

Mixed Applications

5. The hour hand on the clock is between the 8 and 9. The minute hand is pointing to 7. What time is it?

6. **ALGEBRA** Tom is 4 years younger than Jan but 2 years older than Sue. If Jan is 15, how old is Sue?

7. In two hours 639 people rode to the top of the Skydeck at Sears Tower. During the first hour 257 people went to the top. How many people rode to the top during the second hour?

8. **REASONING** Yvonne arrived at the party after Irma. Diana arrived before Irma but after Ann. In what order did the students arrive at the party?

9. I am less than 80 and greater than 60. The sum of my digits is 8. I am odd. What number am I?

10. Gloria spent $5.08 for milk and eggs. The eggs cost $1.49. How much did the milk cost?

Review/Test

✅ CHECK VOCABULARY AND CONCEPTS

Choose the best term from the box.

> benchmark
> period
> million

1. You can use the number of students in your class as a ? to help you find the number of students in your grade. (p. 2)

2. Each group of three digits is called a ? . (p. 6)

✅ CHECK SKILLS

10 beads

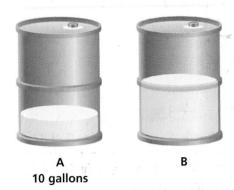

A   B

10 gallons

3. Which is a more reasonable number of beads in the full jar, 70 or 700? (p. 2)

4. Which is a more reasonable number of gallons of water in barrel B, 30 or 300? (p. 2)

Write each number in two other forms. (pp. 6–11)

5. 20,000 + 80 + 3   6. 200,057   7. 4,902,746   8. 48,360,105

Write the value of the blue digit. (pp. 4–11)

9. 8,451   10. 29,710   11. 652,700   12. 4,136,729

✅ CHECK PROBLEM SOLVING (pp. 12–13)

For 13–15, use the graph.

13. Which animals live longer than 20 years?

14. How many years does the grizzly bear live?

15. How many years longer does the Asian elephant live than the grizzly bear?

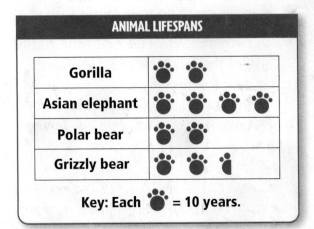

ANIMAL LIFESPANS

Gorilla	🐾 🐾
Asian elephant	🐾 🐾 🐾 🐾
Polar bear	🐾 🐾
Grizzly bear	🐾 🐾 🐾

Key: Each 🐾 = 10 years.

⭐Standardized Test Prep

Understand the problem.
See item **4.**

Use the details in the question. Look for a 3 in the tens place and then compare the numbers to 570.

Also see problem **1,** p. H62.

For 1–8, choose the best answer.

1. How is 504,912 written in expanded form?

  A 500,000 + 40,000 + 900 + 10 + 2

  B 500,000 + 4,000 + 90 + 2

  C 500,000 + 4,000 + 900 + 10 + 2

  D 500,000 + 4,000 + 9,010 + 2

2. In what place is the 7 in 3,704,189?
  F thousands
  G ten thousands
  H hundred thousands
  J millions

3. How is eight million, two hundred thirty thousand, fifty-six written in standard form?
  A 8,203,056   **C** 823,056
  B 8,200,356   **D** 8,230,056

4. Which number is greater than 570 and has a 3 in the tens place?
  F 530 **G** 603 **H** 630 **J** 703

5. What is the value of the 3 in 7,236,054?
  A 300,000   **C** 3,000
  B 30,000   **D** 30

6.   328
   −113
    F 215   **H** 441
    G 221   **J** NOT HERE

For 7–8, use the graph.

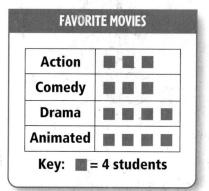

FAVORITE MOVIES

Action	■ ■ ■
Comedy	■ ■ ■
Drama	■ ■ ■ ■
Animated	■ ■ ■ ■

Key: ■ = 4 students

7. How many more students chose dramas than comedies?
  A 1   **C** 5
  B 4   **D** 6

8. How many students chose comedies?
  F 3   **H** 9
  G 5   **J** 12

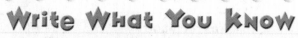

Write What You Know

9. The digits of a 4-digit number have a sum of 11. All the digits are different and the least digit is 1. What is the least number possible? What is the greatest number possible? Explain your answer.

10. Use the graph above. Suppose 8 students who chose action movies changed their vote to drama. Explain how the graph would change. Draw the new graph.

Compare and Order Numbers

The Cascade Mountains begin in Canada and extend through Oregon and Washington into California. Mount Hood is the highest peak in Oregon.

PROBLEM SOLVING List the mountains and their heights from least to greatest.

HIGHEST OREGON CASCADES PEAKS	
Mountain	**Height**
Middle Sister	10,047
Mount Hood	11,235
Mount Jefferson	10,495
North Sister	10,085
South Sister	10,385

CHECK WHAT YOU KNOW

Use this page to help you review and remember
important skills needed for Chapter 2.

VOCABULARY

Choose the best term from the box.

1. The symbol for ? is <.

2. The symbol for ? is >.

3. When you find which of two numbers is greater,
   you ? the numbers.

| compare |
| greater than |
| less than |
| order |

COMPARE NUMBERS TO THOUSANDS (For Intervention, see p. H3.)

Compare. Write < or > for each ●.

4. 476 ● 647

5. 285 ● 99

6. 560 ● 650

7. 4,768 ● 476

8. 5,043 ● 5,034

9. 7,691 ● 7,916

10. 2,389 ● 2,398

11. 1,340 ● 1,444

12. 5,227 ● 4,978

ROUND TO TENS AND HUNDREDS (For Intervention, see p. H4.)

Round each number to the nearest ten.

13. 764

14. 996

15. 857

16. 305

17. 5,892

18. 1,437

19. 3,629

20. 6,431

Round each number to the nearest hundred.

21. 342

22. 681

23. 155

24. 495

25. 4,521

26. 5,306

27. 9,976

28. 1,782

ORDER NUMBERS TO THOUSANDS (For Intervention, see p. H4.)

Write the numbers in order from *least* to *greatest*.

29. 684; 680; 689

30. 540; 504; 603

31. 394; 359; 349

32. 6,809; 6,098; 6,890

33. 3,564; 3,278; 3,782

34. 4,037; 4,370; 3,407

Write the numbers in order from *greatest* to *least*.

35. 724; 472; 747

36. 618; 168; 681

37. 329; 239; 339

38. 8,093; 9,803; 9,380

39. 1,763; 1,637; 7,163

40. 1,527; 1,257; 2,751

Compare Numbers

▶ Learn

RIVER RUN   The Missouri River is 2,315 miles long, and the Mississippi River is 2,348 miles long. Which river is longer?

One Way   Use a number line to compare numbers. Compare 2,315 and 2,348.

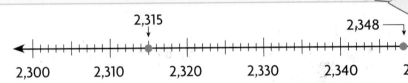

2,315 is to the left of 2,348.
So, 2,315 is less than 2,348.

2,315 < 2,348

2,348 is to the right of 2,315.
So, 2,348 is greater than 2,315.

2,348 > 2,315

So, the Mississippi River is longer.

Another Way   Use place value to compare numbers.
Compare 23,400,836 and 23,317,600.

STEP 1

Start with the first place on the left.
Compare the ten millions.

23,400,836
↓   2 = 2
23,317,600

There are the same number of ten millions.

STEP 2

Compare the millions.

23,400,836
↓   3 = 3
23,317,600

There are the same number of millions.

STEP 3

Compare the hundred thousands.

23,400,836
↓   4 > 3
23,317,600

4 hundred thousands are greater than 3 hundred thousands.

So, 23,400,836 > 23,317,600.

MATH IDEA   To compare numbers, use a number line or use place value.

▶ Check

1. **Explain** how to compare 3,218 and 3,225 using the number line below.

3,200   3,210   3,220   3,230   3,240   3,250

Write the greater number.

2. 2,346 or 2,338

3. 9,531 or 4,631

4. 52,457 or 67,623

5. 4,298,765 or 4,279,465

6. 254,908 or 1,254,980

▶ Practice and Problem Solving

Write the greater number.

7. 4,516 or 4,156

8. 18,927 or 18,937

9. 82,697 or 86,279

Compare. Write <, >, or = for each ●.

10. 2,475 ● 2,475

11. 13,056 ● 13,156

12. 255,136 ● 25,116

13. 5,000,371 ● 500,371

14. 82,245,235 ● 82,245,535

Find all of the digits that can replace each ■.

15. 9■7,536 < 957,549

16. 423,■96,517 < 423,695,815

USE DATA For 17–19, use the table.

17. Which river is longer, Amazon or Nile?

18. Which Asian river has a length greater than 3,500 miles?

19. ✎ **Write a problem** that compares two rivers from the World Rivers table.

WORLD RIVERS		
River	**Continent**	**Length (mi)**
Amazon	South America	4,000
Chang	Asia	3,964
Huang	Asia	3,395
Nile	Africa	4,160

Mixed Review and Test Prep

Write the value of the digit 4. (p. 6)

20. 241,389

21. 759,486

22. Write the number that is 1,000,000 greater than 99,036,871. (p. 8)

23. Write 1,034,506 in word form. (p. 8)

24. **TEST PREP** Which is the standard form of 50,000,000 + 30,000 + 4? (p. 8)
A 15,304,000   C 50,030,004
B 15,034,000   D 50,300,400

EXTRA PRACTICE page H33, Set A

2 Order Numbers

▶ Learn

LOTS OF LAND The map shows the land area, in square miles, of Illinois, Iowa, and Wisconsin. Place the states in order from least to greatest land area.

One Way Use a number line to show the order.

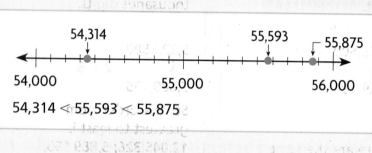

54,314 < 55,593 < 55,875

So, the order is Wisconsin, Illinois, Iowa.

Another Way Use place value to order numbers. Order 437,243; 469,872; and 435,681 from least to greatest.

STEP 1	STEP 2	STEP 3
Start with the first place on the left. Compare the hundred thousands. 437,243 ↓ 469,872 ↓ 435,681 There are the same number of hundred thousands.	Compare the ten thousands. 437,243 ↓ 469,872 3 < 6 ↓ 435,681 Since 3 < 6, 469,872 is the greatest.	Compare the thousands digits in the other two numbers. 437,243 ↓ 5 < 7 435,681 So, the order from least to greatest is 435,681; 437,243; 469,872.

• How would you order 68,195; 681,095; and 61,958 from least to greatest?

Compare Populations

The number of representatives in Congress depends on each state's population. Use the table to place the states in order from greatest to least population. Which state most likely had the fewest representatives in Congress?

1998 STATE POPULATIONS	
State	**Population**
Indiana	5,889,195
Illinois	12,045,326
Wisconsin	5,223,500

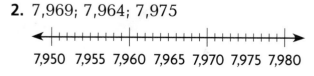

Example Compare.

STEP 1

Start with the first place on the left. Compare the ten millions.

12,045,326
↓   1 > 0
 5,223,500
↓
 5,889,195

Since 1 > 0, the greatest number is 12,045,326.

STEP 2

Compare the millions digits in the other two numbers.

5,223,500
↓
5,889,195

There are the same number of millions.

STEP 3

Compare the hundred thousands digits.

5,223,500
 ↓   2 < 8
5,889,195

So, the order from greatest to least is 12,045,326; 5,889,195; 5,223,500.

So, the states in order from greatest to least population are Illinois, Indiana, and Wisconsin. Of the three states, Wisconsin most likely had the fewest representatives in Congress in 1998.

 MATH IDEA   You can order numbers from least to greatest or greatest to least.

Technology Link

More Practice: Use E-Lab, *Ordering Numbers.*

www.harcourtschool.com/
elab2002

▶ Check

1. **Tell** why you start with the first place on the left when you compare numbers.

Write the numbers in order from *least* to *greatest*.

2. 7,969; 7,964; 7,975

◄─┼┼┼┼┼┼┼┼┼┼┼┼┼┼┼┼┼┼┼┼┼┼┼┼┼┼┼─►
7,950 7,955 7,960 7,965 7,970 7,975 7,980

3. 9,131; 9,155; 9,138

◄─┼┼┼┼┼┼┼┼┼┼┼┼┼┼┼┼┼┼┼┼┼┼┼┼┼┼┼─►
9,130 9,135 9,140 9,145 9,150 9,155 9,160

Write the numbers in order from *greatest* to *least*.

4. 35,000; 35,225; 34,350

5. 870,000; 877,000; 807,000

LESSON CONTINUES

Write the numbers in order from *least* to *greatest*.

6. 12,139; 12,117; 12,109; 12,123

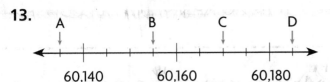

12,100 12,110 12,120 12,130 12,140

7. 36,397; 36,457; 36,384; 36,419

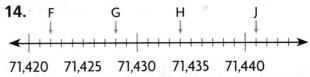

36,380   36,400   36,420   36,440   36,460

8. 190,209; 191,210; 190,201

9. 1,234,410; 1,234,402; 1,434,320

Write the numbers in order from *greatest* to *least*.

10. 16,432; 16,905; 7,906

11. 119,234; 119,819; 1,119,080

12. Order the numbers in the box from greatest to least. Then, underline the numbers greater than 625,000 and less than 650,000.

> 648,279   628,341   642,978
> 682,437   624,879   612,443

For 13–14, write the number represented by each letter. Then, write the *greatest* and *least* numbers.

13.

A   B   C   D

60,140   60,160   60,180

14.   F   G   H   J

71,420   71,425   71,430   71,435   71,440

Name all of the digits that can replace each ■.

15. 358 < 3■3 < 370

16. 1,012 < 1,■20 < 1,200

17. 5,328 < ■,680 < 5,690

18. 82,913 < 8■,086 < 83,096

19. 4,526,109 > 4,526,■17 > 4,526,010

20. 3,942,687 > 3,942,6■3 > 3,942,670

21. I am a number between 149,900 and 150,000. My tens digit is 7 more than my ones digit. The sum of my tens and ones digits is 9. What number am I?

22. **? What's the Error?** Paul ordered the populations of three states from least to greatest. His work is shown. Describe his error. Write the states in the correct order.

23. Jamal and his mom left for the store at 11:40 A.M. They arrived back home at 1:00 P.M. How long were they gone?

24. Write a problem comparing three or more numbers. Use facts from your science book.

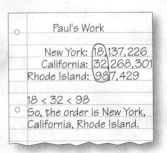

Paul's Work

New York: 18,137,226
California: 32,268,301
Rhode Island: 987,429

18 < 32 < 98
So, the order is New York, California, Rhode Island.

USE DATA For 25–27, use the table.

25. Write the Great Lakes in order from greatest area to least area.

26. Which lakes have areas greater than Lake Erie?

27. Which lakes have areas less than 10,000 square miles?

AREA OF THE GREAT LAKES (in square miles)	
Lake	Area
Erie	9,910
Huron	23,000
Michigan	22,300
Ontario	7,340
Superior	31,700

Mixed Review and Test Prep

Find the sum or difference.

28. $398 + 571$

29. $890 - 231$

30. $700 - 103$

31. $651 + 362$

32. Two 6-digit numbers have three 8's and three 7's. No two digits next to each other are the same. Write the numbers in standard form. (p. 6)

33. **TEST PREP** Find the value of the digit 5 in 54,287. (p. 4)

   A 5
   B 50
   C 5,000
   D 50,000

34. **TEST PREP** Find the value of the digit 3 in 13,720,980. (p. 8)

   F 3,000
   G 30,000
   H 3,000,000
   J 30,000,000

PROBLEM SOLVING THINKER'S CORNER

PICO, CENTRO, NADA In Venezuela, people play a number game called *Pico, Centro, Nada.*

1. Think of a 2-digit number. Have your partner try to guess the number. With each guess, give one of these clues:

   Pico means that one digit is correct, but it's in the wrong place.
   Centro means that one digit is correct, and it's in the correct place.
   Nada means that neither digit is correct.

2. When one number for each player is found, work together to compare the numbers using $<$, $>$, or $=$.

3. When four numbers have been found, list the numbers in order from *least* to *greatest*.

VENEZUELA

EXTRA PRACTICE page H33, Set B

LESSON 3

Problem Solving Strategy
Make a Table

Quick Review

Write the greater number.

1. 4,861 or 40,810
2. 17,092 or 7,920
3. 5,614 or 51,462
4. 32,811 or 37,112
5. 300,020 or 30,200

PROBLEM Mt. Rainier, in Washington, is 14,410 ft tall, Mt. Whitney, in California, is 14,494 ft tall, Mt. Bear, in Alaska, is 14,831 ft tall, and Colorado's Mt. Elbert is 14,433 ft tall. Which mountains are taller than Mt. Elbert?

UNDERSTAND

- What are you asked to find?
- What information will you use?
- Is there any information you will not use? Explain.

PLAN

- What strategy can you use to solve the problem?
  You can *make a table* to organize the information to find the mountains that are taller than Mt. Elbert.

SOLVE

- How can you use the strategy to solve the problem?
  Compare the heights. Place the mountains in the table from the tallest mountain to the shortest. Mt. Bear and Mt. Whitney are taller than Mt. Elbert.

U.S. MOUNTAINS

Name	Height (in feet)
Mt. Bear	14,831
Mt. Whitney	14,494
Mt. Elbert	14,433
Mt. Rainier	14,410

CHECK

- How can you decide if your answer is correct?

Mt. McKinley, in Alaska, is the tallest U.S. mountain peak, at 20,320 ft. It is taller than 30,000 new pencils placed end to end.

1. **What if** University Peak, in Alaska, is added to the table? Its height is 14,470 ft. Make a new table ordering the heights from the tallest mountain to the shortest.

USE DATA For 2–5, make a table to solve.

An author wrote four books about hiking. *Hike for Life* sold 390,457 copies, *Safe Hiking* sold 256,749 copies, *Tricky Trails* sold 354,216 copies, and 391,752 copies of *Miles a Day* were sold.

2. Which book sold the most copies?

3. Which book sold the fewest copies?

4. **What if** *Safe Hiking* had sold 385,485 copies? Which book would have sold the fewest copies?
   A *Miles a Day*
   B *Tricky Trails*
   C *Safe Hiking*
   D *Hike for Life*

5. Which book sold about the same number of copies as *Miles a Day*?
   F *Hiking World*
   G *Tricky Trails*
   H *Safe Hiking*
   J *Hike for Life*

Mixed Strategy Practice

6. Delta County holds its fair every two years. The twenty-sixth fair was in 1948. In what year was the 50th Delta County Fair held?

7. **? What's the Question?** Pat has 5 more posters than Bill. Bill has 8 posters. Roger has 4 fewer posters than Pat. The answer is 9 posters.

USE DATA For 8–10, use the table.

8. If Lenny continues in the same daily pattern, how many pull-ups will he complete on Saturday?

9. Lenny did 84 pull-ups for the week. How many pull-ups did he do Friday through Sunday?

10. How many pull-ups would Lenny complete if he did 8 pull-ups each day for 7 days? 10 pull-ups?

Lenny's Exercise Plan

Day	Number of Pull-Ups
Monday	6
Tuesday	8
Wednesday	10
Thursday	12

Quick Review

Tell whether the number is closer to 100 or 200.

1. 98   **2.** 172

3. 145   **4.** 159

5. 120

▶ **Learn**

NEW NEIGHBORS In 1997, the number of immigrants, or newcomers, who came to the United States was 798,378. A reporter wants to round this number to the nearest ten thousand to make it easier to read. What is this number rounded to the nearest ten thousand?

One Way Use a number line to round greater numbers.

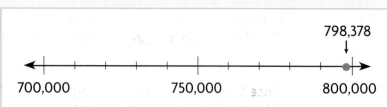

Think: 798,378 is closer to 800,000 than to 790,000.

So, 798,378 rounded to the nearest ten thousand is 800,000.

Another Way Use these rules to round greater numbers.

To round a number:
- Find the place to which you want to round.
- Look at the digit to its right.
- If the digit is *less than 5,* the digit in the rounding place stays the same.
- If the digit is *5 or more,* the digit in the rounding place increases by 1.

Round 798,378 to the nearest hundred thousand.

798,378 is between 700,000 and 800,000.

place to be rounded → 798,378   Look at the ten thousands digit.

Since 9 is greater than 5, the digit 7 increases by 1. So, 798,378 rounds to 800,000.

Round Greater Numbers

Immigrants from all over the world bring their native languages to the United States. The table shows a recent year's data about languages most often used at home, other than English.

LANGUAGES OTHER THAN ENGLISH SPOKEN IN THE U.S.	
Language	**Number of People**
Spanish	17,339,000
French	1,702,000
German	1,547,000
Italian	1,309,000
Chinese	1,249,000

Round the number of people who speak German to the nearest

hundred thousand. 1,547,000 → 1,500,000

ten thousand. 1,547,000 → 1,550,000

Examples

A Round 17,339,000 to the nearest million.

17,339,000 is between 17,000,000 and 18,000,000.

place to be rounded ⟶ 17,339,000 ⟵ Look at the hundred thousands digit.

Since 3 is less than 5, the digit 7 stays the same. So, 17,339,000 rounded to the nearest million is 17,000,000.

B Round 17,339,000 to the nearest ten thousand.

17,339,000 is between 17,330,000 and 17,340,000.

place to be rounded ⟶ 17,339,000 ⟵ Look at the thousands digit.

Since 9 is greater than 5, the digit 3 is increased by 1. So, 17,339,000 rounded to the nearest ten thousand is 17,340,000.

- When rounding, you change some digits to zero. Explain how you know which digits to change.

▶ Check

1. **Explain** how to round 99,999 to the nearest ten thousand.

2. **Find** the least and greatest numbers that round to 600,000.

Round each number to the nearest thousand.

3. 16,822　　　4. 895,104　　　5. 4,286,531　　　6. 9,104,523

Round each number to the place value of the blue digit.

7. 98,749　　　8. 167,403　　　9. 317,482　　　10. 1,247,962

100,000

LESSON CONTINUES

▶ Practice and Problem Solving

Round each number to the nearest thousand.

11. 64,385   **12.** 37,179   **13.** 82,435   **14.** 93,798

15. 399,999   **16.** 8,365,700   **17.** 1,438,607   **18.** 2,513,986

Round each number to the place value of the blue digit.

19. 529,999   **20.** 154,879   **21.** 1,943,672   **22.** 2,837,486

23. 667,841   **24.** 725,639   **25.** 453,602   **26.** 375,926

27. 6,385,837   **28.** 7,384,609   **29.** 9,645,408   **30.** 12,647,813

31. 5,476,301   **32.** 14,358,900   **33.** 41,683,205   **34.** 62,591,073

35. Describe all the numbers that when rounded to the nearest thousand are 312,000.

36. Which number rounds to sixteen million, eight hundred thousand, 16,864,381 or 16,849,268?

37. On the place-value chart, my thousands period is seventy-four. My ones period is 500 + 20 + 3. My millions digit is 3 times my ones digit. What number am I? Round me to the nearest hundred thousand.

USE DATA For 38–41, use the table.

38. Write the state that had a little more than 200,000 immigrants admitted.

39. Which state's number of immigrants has the digit 2 in the hundred thousands place?

40. Is the number of immigrants admitted to New York closer to 123,000 or 124,000?

41. Write the number that is 50,000 greater than the number of immigrants admitted to New York.

IMMIGRANTS ADMITTED, 1997	
State	**Immigrants**
California	203,305
New York	123,716
Florida	82,318

42. **REASONING** Write a number that never repeats a digit and rounds to 7,000,000.

43. **Write About It** Explain how to round 994,685 to the nearest hundred thousand.

44. Round 98,456 to the nearest hundred. Round your answer to the nearest thousand. Round that answer to the nearest 10,000.

45. Rico's mother arrived at the school 15 minutes after the bus left. She arrived at 3:05 P.M. At what time did the bus leave the school?

Mixed Review and Test Prep

46. Write eighteen thousand, five hundred seventy in standard form. (p. 6)

47. What number is 100 greater than the sum of 418 and 975?

48. Use word form to write the number that is 1,000 greater than 99,756. (p. 6)

49. Which number is greater, 568,413 or 568,143? (p. 18)

50. Write 150,210 in word form. (p. 6)

51. Write 5,000,000 + 2,000 + 10 + 4 in standard form. (p. 8)

52. **TEST PREP** Which number below, when subtracted from 800, has a difference greater than 200?
 A 423   **B** 609   **C** 612   **D** 742

53. **TEST PREP** Which number below is between 476,891 and 674,198? (p. 20)
 F 468,981   **H** 676,819
 G 647,918   **J** 746,189

PROBLEM SOLVING LiNKUP ... to Reading

STRATEGY · COMPARE When you **compare** two or more things, you look at how they are alike. When you **contrast** two or more things, you look at how they are different.

Look at the chart below. The chart compares and contrasts the estimated areas of 5 deserts. Can you think of other ways to compare and contrast the deserts?

COMPARE	CONTRAST
The Kalahari and the Sahara are located in Africa.	The Australian Desert is the only desert located in Australia.
All have areas greater than 500,000 sq km.	The Sahara is larger than the other 4 deserts put together.

DESERTS OF THE WORLD

Desert	Location	Area (sq km)
Kalahari	Africa	520,000
Gobi	Asia	1,036,000
Arabian	Asia	1,300,000
Australian	Australia	3,800,000
Sahara	Africa	9,000,000

1. Look at the areas of the Arabian Desert and the Gobi Desert. How do they compare? How do they contrast?

2. Round each area to the nearest million. Compare and contrast the numbers. How does this change the data?

EXTRA PRACTICE page H33, Set C

Review/Test

✓ CHECK VOCABULARY AND CONCEPTS

Choose the best term from the box.

> greatest
> least
> rounded
> 5 or more
> less than 5

1. To compare numbers, use a number line or start with the _?_ place-value position. (p. 18)

2. The number 7,950,614 _?_ to the nearest million is 8,000,000. (p. 26)

3. When rounding, if the digit to the right of the rounding place is _?_, the digit in the rounding place increases by one. (p. 26)

✓ CHECK SKILLS

Compare. Write <, >, or = for each ●. (pp. 18–19)

4. 15,980 ● 15,754

5. 34,980 ● 3,690

6. 780,256 ● 783,130

7. 1,895,006 ● 1,392,950

8. 282,700 ● 2,308,030

9. 562,026 ● 562,026

10. Write 25,908; 25,616; and 25,972 in order from *least* to *greatest*. (pp. 20–23)

11. Write 3,791,808; 3,759,204; and 3,090,910 in order from *greatest* to *least*. (pp. 20–23)

Round each number to the place value of the blue digit. (pp. 26–29)

12. 105,219

13. 983,050

14. 4,591,203

15. 7,948,033

16. 6,839,032

17. 871,094

18. 3,217,849

19. 7,981,936

20. 17,340,206

21. 34,761,502

22. 29,457,863

23. 58,732,141

✓ CHECK PROBLEM SOLVING

Solve. (pp. 24–25, 26–29)

24. City A's population is 342,653. City B has 451,321 people, and City C has 353,308 people. Make a table that lists the cities in order from least to greatest population. Which cities' populations are greater than 345,000?

25. In a recent year, a statewide festival's total attendance was 746,982. Round 746,982 to the nearest hundred thousand, ten thousand, and thousand. Which of these rounded amounts is closest to the actual attendance?

Standardized Test Prep

 Look for important words.
See item **2.**

Not is an important word. **Not** true means that when you compare two numbers with the symbol used, the statement is false.

Also see problem **2,** p. H62.

For 1–9, choose the best answer.

1. Which is true?
   - **A** 36,176 < 35,716
   - **B** 36,671 < 36,176
   - **C** 36,716 > 36,671
   - **D** 36,176 < 36,167

2. Which is **not** true?
   - **F** 2,196 > 1,962
   - **G** 2,196 < 1,962
   - **H** 1,962 < 2,196
   - **J** 3,119 < 3,911

3. 583 − 197 = ■
   - **A** 386
   - **B** 414
   - **C** 486
   - **D** NOT HERE

4. Which numbers are in order from *greatest* to *least*?
   - **F** 3,749,386; 3,497,368; 3,749,602
   - **G** 3,749,602; 3,749,386; 3,497,368
   - **H** 3,497,369; 3,749,386; 3,749,602
   - **J** 3,749,602; 3,497,368; 3,749,386

5. Which numbers are in order from *least* to *greatest*?
   - **A** 16,487; 15,652; 15,607
   - **B** 15,652; 16,487; 15,607
   - **C** 15,607; 16,487; 15,652
   - **D** 15,607; 15,652; 16,487

6. How is two hundred thirteen thousand, seven hundred four written in standard form?
   - **F** 213,704
   - **G** 213,074
   - **H** 213,740
   - **J** 230,074

7. What digit is in the millions place in 6,820,149?
   - **A** 6
   - **B** 8
   - **C** 0
   - **D** 2

8. 386 + 284 = ■
   - **F** 560
   - **G** 670
   - **H** 726
   - **J** NOT HERE

9. What is 4,609,112 rounded to the nearest million?
   - **A** 4,600,000
   - **B** 4,000,000
   - **C** 4,609,000
   - **D** 5,000,000

Write What You Know

10. Explain how to round a number to the nearest thousand. Then write a number that has at least 4 digits and round it to the nearest thousand.

11. A newspaper reported that 18,000 people attended a parade. Tell whether you think that number is an estimate or exact total. Explain.

Add and Subtract Greater Numbers

Insects are part of a group of animals called arthropods (ARTH•ruh•pahdz). Arthropods also include crustaceans (krus•TAY•shuhnz) and arachnids (uh•RAK•nidz). Spiders are different from insects. One difference is that insects have six legs and spiders have eight legs.

PROBLEM SOLVING Using the table, tell about how many more types of insects there are than arachnids.

•DATA LINK•

ARTHROPODS		
Crustaceans: 40,000 types	**Insects: 750,000 types**	**Arachnids: 70,000 types**
Crabs	Grasshoppers	Spiders
Lobsters	Termites	Scorpions
Shrimps	Flies	Daddylonglegs
Crayfish	Butterflies	Mites
	Beetles	Ticks

Garden spider

Use this page to help you review and remember
important skills needed for Chapter 3.

✅ VOCABULARY

Choose the best term from the box.

1. In 24 + 31 = 55, the number 55 is the ? .

2. In 74 − 31 = 43, the number 43 is the ? .

3. An example of the ? is 3 + 5 = 5 + 3.

> sum
> difference
> Order Property
>   of Addition
> Grouping Property
>   of Addition

✅ TWO-DIGIT ADDITION AND SUBTRACTION
(For Intervention, see p. H5.)

Find the sum or difference.

4. 13 +24	**5.** 58 −29	**6.** 36 −14	**7.** 52 +11	**8.** 78 −43
9. 73 −19	**10.** 65 +36	**11.** 42 +68	**12.** 90 −28	**13.** 81 +26

14. 23 + 37 + 42 = ■   **15.** 42 + 31 + 63 = ■

16. 87 − 59 = ■   **17.** 64 − 29 = ■

✅ THREE-DIGIT ADDITION AND SUBTRACTION (For Intervention, see p. H5.)

Find the sum or difference.

18. 480 +520	**19.** 851 −216	**20.** 657 −198	**21.** 971 +210	**22.** 517 +286

23. 870 − 357 = ■   **24.** 900 − 237 = ■   **25.** 794 + 518 = ■

26. 623 + 547 = ■   **27.** 354 − 199 = ■   **28.** 582 + 346 = ■

✅ COLUMN ADDITION (For Intervention, see p. H6.)

Group the addends. Then find the sum.

29. 13 27 28 +12	**30.** 33 17 46 +24	**31.** 45 34 +71	**32.** 66 44 66 +44	**33.** 27 42 58 +78

Estimate Sums and Differences

▶ **Learn**

PET POWER   The table shows the results of a survey by the North Adams Animal Shelter. About how many people have cats, dogs, or cats and dogs?

Estimate 34,221 + 38,899 + 6,520.

Round each number to the nearest ten thousand, and then add.

$$
\begin{array}{rcr}
34{,}221 & \rightarrow & 30{,}000 \\
38{,}899 & \rightarrow & 40{,}000 \\
+\ 6{,}520 & \rightarrow & +10{,}000 \\
\hline
 & & 80{,}000
\end{array}
$$

So, about 80,000 people have cats, dogs, or both.

You can also use rounding to estimate a difference.

About how many more people in North Adams have fish than have birds?

Estimate 4,872 − 1,036.

Round each number to the nearest thousand, and then subtract.

$$
\begin{array}{rcr}
4{,}872 & \rightarrow & 5{,}000 \\
-\ 1{,}036 & \rightarrow & -\ 1{,}000 \\
\hline
 & & 4{,}000
\end{array}
$$

So, about 4,000 more people have fish than have birds.

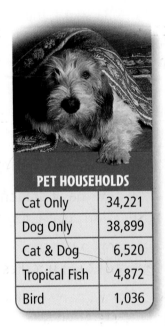

PET HOUSEHOLDS	
Cat Only	34,221
Dog Only	38,899
Cat & Dog	6,520
Tropical Fish	4,872
Bird	1,036

▶ **Check**

1. **Explain** how you would estimate the sum of 4,782 and 36,132.

Estimate the sum or difference.

2. 6,852
  − 4,257

3. 371,936
  + 427,483

4. $13,024
  + $58,417

5. $390,111
  − $ 23,187

▶ Practice and Problem Solving

Estimate the sum or difference.

6. 4,123
  + 2,381

7. $40,717
  + $74,910

8. 208,183
  + 642,275

9. 4,908
  + 8,600

10. 38,037
  − 4,047

11. 9,584
  − 6,102

12. 71,234
  + 12,736

13. $976,254
  − $105,224

Write the missing digit for the estimated sum or difference.

14. 2,146
  + ▇,211
  6,000

15. ▇04,125
  − 610,713
  200,000

16. $ 92,024
  − $ ▇8,121
  $ 40,000

17. $▇15,024
  + $ 498,103
  $ 700,000

USE DATA For 18–20, use the table.

18. About how many golden retrievers and cocker spaniels are registered in all?

19. About how many more German shepherds than poodles are there?

20. About how many beagles and poodles are registered in all?

21. REASONING Write two numbers that have an estimated sum of 1,000,000. Use the digits 1, 2, 3, 4, 5, and 6 in each number.

TOP U.S. DOG BREEDS

Breed	Number Registered
German Shepherd	79,076
Golden Retriever	68,993
Beagle	56,956
Poodle	56,803
Cocker Spaniel	45,305

Mixed Review and Test Prep

22. Order 26,812; 8,261; 28,612; and 28,599 from greatest to least. (p. 20)

23. Which number is greater, 827,156 or 826,156? (p. 18)

24. Write the missing numbers in the pattern.

  17, ▇, 37, 47, ▇, 67

25. What is 200,000 + 50,000 + 3,000 + 20 + 7 in standard form? (p. 6)

26. TEST PREP Which is two hundred eighty-six in expanded form? (p. 6)

  A 200 + 68
  B 280 + 6
  C 200 + 80 + 6
  D 2,000 + 800 + 6

EXTRA PRACTICE page H34, Set A

Use Mental Math Strategies

▶ **Learn**

THINK IT THROUGH Sometimes you don't need paper and pencil to compute. Use these strategies to help you compute mentally.

One Way *Break Apart* Strategy

You can break apart numbers to add the tens and ones separately.

Examples

Ⓐ Find the sum. 58 + 26   **Think:** 58 = 50 + 8
                                          26 = 20 + 6

  Add the tens.   50 + 20 = 70

  Add the ones.   8 + 6 = 14

  Add the sums.   70 + 14 = 84

  So, 58 + 26 = 84.

Ⓑ Find the difference. 46 − 25   **Think:** 46 = 40 + 6
                                          25 = 20 + 5

  Subtract the tens.   40 − 20 = 20

  Subtract the ones.   6 − 5 = 1

  Add the differences. 20 + 1 = 21

  So, 46 − 25 = 21.

More Examples

Ⓒ Find the sum. 39 + 48   Ⓓ Find the difference.
                                       97 − 52
  30 + 40 = 70
                                    90 − 50 = 40
   9 + 8 = 17
                                     7 − 2 = 5
  70 + 17 = 87
                                    40 + 5 = 45

50 + 20 = 70

8 + 6 = 14

70 + 14 = 84

• Explain how to use this strategy to find 62 + 29.

More Strategies

Another Way *Make a Ten* Strategy

You can change one number to a multiple of 10 and then adjust the other number.

Examples

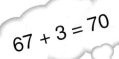
$67 + 3 = 70$

$24 - 3 = 21$

$70 + 21 = 91$

Ⓐ Find the sum. $67 + 24$

$67 + 3 = 70$ **Think:** 70 is a multiple of 10. Add 3 to 67 to get 70.

$24 - 3 = 21$ Subtract 3 from 24 to adjust the sum.

$70 + 21 = 91$ Add $70 + 21$.

Ⓑ Find the difference. $86 - 29$

$29 + 1 = 30$ **Think:** 30 is a multiple of 10. Add 1 to 29 to get 30.

$86 + 1 = 87$ Add 1 to 86 to adjust the difference.

$87 - 30 = 57$ Subtract $87 - 30$.

- What if you subtract 4 from 24 in Example A? How could you find $67 + 24$ mentally?

Subtraction is easier if the number you are subtracting is a multiple of 10. If you increase the number, you must add the same amount to adjust the answer.

More Examples

Ⓒ Find the difference. $75 - 38$

$38 + 2 = 40$

$$\begin{array}{r} 75 \\ -40 \\ \hline 35 \\ +\,2 \\ \hline 37 \end{array}$$

Ⓓ Find the sum. $84 + 76$

$84 + 6 = 90$

$$\begin{array}{r} 90 \\ +76 \\ \hline 166 \\ -\,6 \\ \hline 160 \end{array}$$

▶ Check

1. **Explain** how to find $83 - 37$ mentally.

2. **Write** an addition problem. Find the sum by breaking the numbers apart.

LESSON CONTINUES ▶

▶ Practice and Problem Solving

For 3–22, add or subtract mentally. Tell the strategy you used.

3. 85 − 17　　　　**4.** 72 + 28　　　**5.** 83 − 19　　　**6.** 95 + 28

7. 68 + 25　　　　**8.** 52 − 27　　　**9.** 74 + 32　　　**10.** 76 − 28

11. 78 − 15　　　**12.** 16 + 28　　　**13.** 26 − 12　　　**14.** 47 + 23

15. 27 + 48　　　**16.** 75 − 36　　　**17.** 38 + 85　　　**18.** 91 − 66

19. 62 − 29　　　**20.** 34 + 58　　　**21.** 44 − 17　　　**22.** 63 + 39

Find the sum or difference.

23. 168 − 59　　　**24.** 249 + 87　　　**25.** 216 + 79　　　**26.** 152 − 75

27. 261 + 88　　　**28.** 431 − 232　　　**29.** 441 + 263　　　**30.** 284 − 192

31. 758 − 453　　　**32.** 576 − 391　　　**33.** 713 + 428　　　**34.** 674 + 332

USE DATA For 35–37, use the table.

35. Which animal has a length of about 60 feet?

36. How many feet longer is the blue whale than 2 crocodiles placed end to end?

37. Which two animals have the greatest difference in length? the least difference?

38. John read that there were 672 species of animals in the Denver Zoo and 599 species in the Cleveland Zoo. How many more species are in the Denver Zoo?

LENGTH OF SEA ANIMALS (in feet)

Animal	Length
Blue Whale	110
Whale Shark	59
Asian Saltwater Crocodile	32
Atlantic Giant Squid	20
Japanese Spider Crab	9

39. On Friday, there were 2,999 people at the aquarium. On Saturday, there were 1,465 people. Use mental math to find how many people were at the aquarium during the two days. Explain your strategy.

40. **? What's the Question?** Use the table above. The answer is 90 feet.

41. Rex has 6 horses, 10 cows, 5 goats and 4 cars. He puts the cows, goats and cars in the barn. How many of his animals are in the barn?

42. Lisa's dog weighs 45 pounds. Reggie's dog weighs 27 pounds. How much more does Lisa's dog weigh than Reggie's dog?

Mixed Review and Test Prep

43. Write the numbers from greatest to least: 2,567; 2,763; 189,576; 187,487; 189,875. (p. 20)

44. Round 5,789,132 to the nearest hundred thousand. (p. 26)

Compare. Write <, >, or = for each ⬤.

45. 8,567 ⬤ 9,087 (p. 18)

46. 8,237,958 ⬤ 8,549,788 (p. 18)

47. TEST PREP Maria has 5 dimes, 1 nickel, and 3 pennies in her pocket. How much money does she have?

A 58¢   **B** 52¢   **C** 48¢   **D** 45¢

48. TEST PREP Which of the following is NOT true?

F $8 + 10 = 10 + 8$
G $12 + 3 = 11 + 4$
H $8 + 1 = 2 + 7$
J $7 + 9 = 6 + 8$

PROBLEM SOLVING Thinker's Corner

COMPATIBLE NUMBERS Two numbers that add up to sums like 10 or 100 are **compatible numbers**. Compatible numbers make mental math easy.

The Order Property of Addition says that numbers can be added in any order and the sum remains the same. Use the Order Property of Addition to add compatible numbers.

Find the sum. $68 + 230 + 32 + 170$

```
          100
      ┌──────────┐
      ↓          ↓
68 + 230 + 32 + 170
  ↑           ↑
      └──────────┘
         400
```

HINT: Look for compatible numbers.

$68 + 230 + 32 + 170 = 68 + 32 + 230 + 170$
$= 100 + 400 = 500$

So, the sum is 500.

> **HINT: These pairs of compatible numbers have a sum of 100.**
>
> **10 and 90**
> **20 and 80**
> **30 and 70**
> **40 and 60**
> **50 and 50**

Use compatible numbers to find each sum.

1. $175 + 25 + 61 + 39 = \blacksquare$

2. $82 + 18 + 60 + 40 = \blacksquare$

3. $78 + 250 + 122 + 48 = \blacksquare$

4. $302 + 168 + 32 + 175 = \blacksquare$

3 Add and Subtract 4-Digit Numbers

Quick Review

1. 460 + 218
2. 355 − 145
3. 175 + 250
4. 796 − 445
5. 804 + 257

▶ Learn

GREAT LAKES!   The area of Lake Erie is 9,910 square miles. The area of Lake Ontario is 7,340 square miles. What is the combined area of the two lakes?

Example 1

Find the sum. 9,910 + 7,340 = ■

Estimate. 10,000 + 7,000 = 17,000

STEP 1	STEP 2	STEP 3	STEP 4
Add the ones.	Add the tens.	Add the hundreds. Regroup 12 hundreds.	Add the thousands.
$\begin{array}{r} 9,910 \\ +7,340 \\ \hline 0 \end{array}$	$\begin{array}{r} 9,910 \\ +7,340 \\ \hline 50 \end{array}$	$\begin{array}{r} 1 \\ 9,910 \\ +7,340 \\ \hline 250 \end{array}$	$\begin{array}{r} 1 \\ 9,910 \\ +7,340 \\ \hline 17,250 \end{array}$

So, the combined area of the two lakes is 17,250 square miles. The answer is close to the estimate, so 17,250 is reasonable.

Example 2

Find the difference. 9,910 − 7,340 = ■

STEP 1	STEP 2	STEP 3	STEP 4
Subtract the ones.	Regroup 9 hundreds. Subtract the tens.	Subtract the hundreds.	Subtract the thousands.
$\begin{array}{r} 9,9\,1\,0 \\ -7,3\,4\,0 \\ \hline 0 \end{array}$	$\begin{array}{r} {}^{8\ 11} \\ 9,\,9\,1\,0 \\ -7,3\,4\,0 \\ \hline 7\,0 \end{array}$	$\begin{array}{r} {}^{8\ 11} \\ 9,\,9\,1\,0 \\ -7,3\,4\,0 \\ \hline 5\,7\,0 \end{array}$	$\begin{array}{r} {}^{8\ 11} \\ 9,\,9\,1\,0 \\ -7,3\,4\,0 \\ \hline 2,5\,7\,0 \end{array}$

So, the difference is 2,570.

1. **Explain** how you know when it is not necessary to regroup in subtraction.

Find the sum or difference. Estimate to check.

2. 6,899
   +2,267

3. 4,674
   −1,406

4. 8,902
   −5,730

5. 9,201
   + 1,321

► **Practice and Problem Solving**

Find the sum or difference. Estimate to check.

6. 6,798
   −4,127

7. $3,204
   −$2,413

8. 2,409
   5,762
   +4,005

9. 5,762
   5,243
   +1,111

10. 2,409
    +5,762

11. 5,320
    −1,375

12. 9,862
    −7,361

13. $3,228
    +$4,228

14. 2,409 + 1,952

15. $1,124 + $1,525 + $1,651 + $4,176

16. 6,230 − 4,651

17. 1,987 + 936

For 18–19, find the missing digit.

18. 2,90▪
    −1,894
    1,007

19. 3,486
    +▪,964
    9,450

USE DATA For 20–21, use the map.

20. The lengths of which two coastlines have a sum of about 4,000 miles? Find the actual sum.

21. About how many miles long are all the United States coastlines shown?

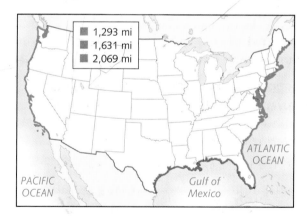

1,293 mi
1,631 mi
2,069 mi

PACIFIC OCEAN

ATLANTIC OCEAN

Gulf of Mexico

Mixed Review and Test Prep

For 22–23, round to the nearest hundred. (p. 26)

22. 563

23. 712

24. 562
    + 349

25. 736
    − 472

26. **TEST PREP** Find the missing number in the pattern. 135, 160, 185, ▪, 235

A 192

C 200

B 197

D 210

4 Subtract Across Zeros

Quick Review

1. 900 − 600

2. 1,000 − 500

3. 200 − 50

4. 500 − 250

5. 600 − 175

▶ Learn

ON THE MOVE Scientists track animals to find how far they migrate. A humpback whale migrated about 8,100 miles each year. A green sea turtle migrated 2,815 miles each year. About how much farther did the whale migrate than the turtle?

Technology Link

More Practice: Use E-Lab, *Subtracting Across Zeros.*

www.harcourtschool.com/ elab2002

Example

Find the difference. 8,100 − 2,815 = ■

Estimate. 8,000 − 3,000 = 5,000

STEP 1		**STEP 2**	
Subtract the ones. Regroup 1 hundred as 9 tens 10 ones.	$\begin{array}{r} {}^{9}_{0\ 10\ 10} \\ 8,1\,0\,0 \\ -\ 2,8\,1\,5 \\ \hline 5 \end{array}$	Subtract the tens.	$\begin{array}{r} {}^{9}_{0\ 10\ 10} \\ 8,1\,0\,0 \\ -\ 2,8\,1\,5 \\ \hline 8\,5 \end{array}$
STEP 3		**STEP 4**	
Subtract the hundreds. Regroup 8 thousands as 7 thousands 10 hundreds.	$\begin{array}{r} {}^{10\ \ 9}_{7\ \ 0\ 10\ 10} \\ 8,1\,0\,0 \\ -\ 2,8\,1\,5 \\ \hline 2\,8\,5 \end{array}$	Subtract the thousands.	$\begin{array}{r} {}^{10\ \ 9}_{7\ \ 0\ 10\ 10} \\ 8,1\,0\,0 \\ -\ 2,8\,1\,5 \\ \hline 5,2\,8\,5 \end{array}$

So, the humpback whale migrated about 5,285 miles farther than the green sea turtle. Since 5,285 is close to the estimate of 5,000, the answer is reasonable.

▶ Check

1. **Explain** how to regroup 40,000 to subtract 7,165.

Find the difference. Estimate to check.

2. 400
− 287

3. 3,700
− 1,692

4. $300
− $163

5. 2,100
− 594

▶ Practice and Problem Solving

Find the difference. Estimate to check.

6. 4,001
− 3,090

7. 6,008
− 4,009

8. $3,005
− $1,978

9. 5,004
− 859

10. 5,700
− 4,190

11. 7,001
− 3,090

12. 5,200
− 3,087

13. 8,600
− 6,123

14. 3,000
− 2,218

15. 8,000
− 2,319

16. 9,008
− 4,899

17. 1,000
− 919

Compare. Write <, >, or = for each ●.

18. 3,000 − 2,541 ● 4,200 − 3,756

19. 2,000 − 1,008 ● 2,100 − 1,097

20. One seal weighs 130 kilograms more than a second seal. If the second seal weighs 179 kilograms, how much does the first seal weigh?

21. An arctic tern flew 10,230 miles to Antarctica. A sea turtle swam 1,400 miles to South America. How much farther did the arctic tern migrate than the sea turtle?

22. A male walrus at the zoo weighs 1,390 kilograms. A female walrus weighs 1,122 kilograms. How much more does the male walrus weigh?

23. **? What's the Error?** Quan subtracted 9,910 from 23,000 and got 12,090. Describe and correct his error.

Mixed Review and Test Prep

24. 2,624 (p. 40)
+2,981

25. 6,422 (p. 40)
+8,600

26. Write four million, two hundred seventy-six thousand, one hundred three in standard form. (p. 8)

27. Order 100,430; 100,562; 99,650 from greatest to least. (p. 20)

28. **TEST PREP** The rectangular box for David's turtle is 90 cm long and 56 cm wide. How much shorter is the width than the length?

A 146 cm **C** 65 cm
B 94 cm **D** 34 cm

EXTRA PRACTICE page H34, Set D

Choose a Method

Quick Review

1. 263 − 44
2. 630 − 463
3. 32 + 28 + 12
4. 23 + 17 + 41
5. 854 − 392

▶ **Learn**

THINK BIG! In the United States, the two largest states are Alaska and Texas. What is the combined area in square miles of the two states?

Example

Estimate. Then choose a method of computation.

600,000 + 300,000 = 900,000

Use paper and pencil. Find the sum.

615,230 + 267,277 = ■

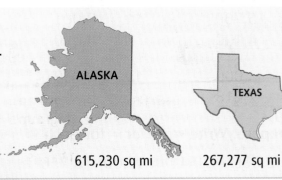

ALASKA

TEXAS

615,230 sq mi   267,277 sq mi

STEP 1

Add the ones and tens. Regroup 10 tens as 1 hundred 0 tens.

$$
\begin{array}{r}
\overset{1}{}615,230 \\
+267,277 \\
\hline
07
\end{array}
$$

STEP 2

Add the hundreds and thousands. Regroup 12 thousands as 1 ten thousand 2 thousands.

$$
\begin{array}{r}
\overset{1}{}\,\overset{1}{}615,230 \\
+267,277 \\
\hline
2,507
\end{array}
$$

STEP 3

Add the ten thousands.

$$
\begin{array}{r}
\overset{1}{}\,\overset{1}{}615,230 \\
+267,277 \\
\hline
82,507
\end{array}
$$

STEP 4

Add the hundred thousands.

$$
\begin{array}{r}
\overset{1}{}\,\overset{1}{}615,230 \\
+267,277 \\
\hline
882,507
\end{array}
$$

Use a calculator. Find the sum.

[6] [1] [5] [2] [3] [0] [+]

[2] [6] [7] [2] [7] [7] [Enter =]   `882507`

AK

CANADA

UNITED STATES

TX

MEXICO

So, Alaska and Texas have a combined area of 882,507 square miles. The answer is close to the estimate of 900,000. So, 882,507 is reasonable.

• Why is mental math not a good choice for finding the sum of these numbers?

Subtract Greater Numbers

Matterhorn Peak in Colorado is 12,264 feet high. Mt. McKinley in Alaska is 20,320 feet high. How much higher is Mt. McKinley than Matterhorn Peak?

▼ **Mt. McKinley, North America's highest mountain, is located in Denali National Park in Alaska.**

Example

Estimate. Then choose a method to compute.

20,000 − 10,000 = 10,000

USE PAPER AND PENCIL Find the difference.

20,320 − 12,264 = ■

STEP 1

Subtract the ones. Regroup 2 tens as 1 ten 10 ones.

```
      1 10
  2 0, 3 2 0
− 1 2, 2 6 4
            6
```

STEP 2

Subtract the tens and hundreds. Regroup 3 hundreds 1 ten as 2 hundreds 11 tens.

```
        11
      2 1 10
  2 0, 3 2 0
− 1 2, 2 6 4
        0 5 6
```

STEP 3

Subtract the thousands and ten thousands. Regroup 2 ten thousands 0 thousands as 1 ten thousand 10 thousands.

```
            11
  1 10 2 1 10
  2 0, 3 2 0
− 1 2, 2 6 4
    8, 0 5 6
```

USE A CALCULATOR Find the difference.

 (2) (0) (3) (2) (0) (−) (1) (2) (2) (6) (4) (Enter =)

```
  20320−12264
=         8056
```

So, Mt. McKinley is 8,056 feet higher than Matterhorn Peak. The answer is close to the estimate of 10,000. So, 8,056 is reasonable.

▶ Check

1. **Tell** how an estimate helps you decide if your answer is reasonable.

Find the sum or difference. Write the method you used to compute.

2. $98,634
   − $62,512

3.   437,702
   + 58,093

4.   500,600
   − 263,710

5.   261,710
   + 172,542

LESSON CONTINUES ▶

▶ Practice and Problem Solving

Find the sum or difference. Write the method you used to compute.

6. $983,368
    − $261,125

7. 51,237
    + 28,802

8. 200,000
    + 134,902

9. 265,710
    + 137,942

10. 412,533
    + 283,056

11. 584,201
    − 493,557

12. $784,200
    + $ 35,220

13. 80,702
    − 60,698

14. $761,357
    − $572,326

15. 546,071
    + 291,503

16. 63,200
    − 22,050

17. 268,412
    +   37,287

18. 33,333 + 214,142 + 527,998 = ■

19. 211,500 + 201,943 + 14,159 = ■

Compare. Write <, >, or = for each ●.

20. 637,124 − 215,275 ● 784,725 − 398,419

21. 323,125 + 125,362 ● 342,125 + 125,362

22. 211,345 + 467,311 ● 200,457 + 478,199

23. Find the difference of 723,503 and 500,497.

24. Find the sum of 864,235 and 252,214.

25. Find the difference of 974,210 and 473,689.

26. Find the sum of 670,000 and 220,925.

Find the missing digit.

27. 43■,257
    − 253,019
    186,238

28. 278,269
    + 921,57■
    1,199,843

29. 156,217
    + ■32,368
    788,585

30. In 1998, a theater spent $112,840 on equipment. In 1999, $97,560 was spent. What was the total amount spent? Was more money spent in 1998 or 1999?

31. The play *Clue* had 47,250 performances. The play *The Mousetrap* had 18,872 performances. How many more performances were given of the play *Clue*?

▲ Outdoor play in Palo Duro Canyon, Texas.

32. Lori wrote a number that is thirty-one thousand, five hundred twenty-three greater than 17,996. What number did she write?

33. ✎ **Write a problem** using the data. There are 479,743 people in Wyoming and 609,331 people in Alaska.

34. Kyle got on the elevator on the fourth floor. He went down 2 floors and then up 6 floors where he got off the elevator. On which floor did Kyle get off the elevator?

35. Mr. Randall sold 1,285 wooden cutouts at his crafts booth. He has 298 left. How many cutouts did Mr. Randall have to start with?

Mixed Review and Test Prep

USE DATA For 36–37, use the table. (p. 44)

FALL FOLIAGE FESTIVAL ATTENDANCE		
	1998	**1999**
Thursday	13,789	15,034
Friday	23,681	27,950
Saturday	34,625	41,393

36. What was the total attendance at the Festival in 1998 for all 3 days?

37. How many more people attended the Festival in 1999 than in 1998?

38. Which is less: 315,731 or 351,731? (p. 18)

39. How many people are between the fifth and twelfth in line for a movie?

40. **TEST PREP** What is five million, twenty-five thousand, ten in standard form? (p. 8)

   A 5,025,100   **C** 525,210
   B 5,025,010   **D** 525,100

41. **TEST PREP** What is the difference? (p. 44)

        489,211 − 16,179

   F 451,389   **H** 497,230
   G 473,032   **J** 505,390

PROBLEM SOLVING ThiNker's CorNer

CLUSTERING

Baseball stadiums seat tens of thousands of fans. Look at the data in the table. Notice that the numbers cluster, or gather, around 50,000. You can use this fact to help you estimate the total number of seats in the stadiums.

BASEBALL STADIUM SEATING CAPACITIES	
Atlanta Braves	50,062
Baltimore Orioles	48,876
Cincinnati Reds	52,953
Milwaukee Brewers	53,192
Minnesota Twins	48,678
Montreal Expos	46,500
Pittsburgh Pirates	47,687
St. Louis Cardinals	49,738

1. Estimate the number of seats in the 3 stadiums with the greatest number of seats.

2. Estimate the number of seats in the 3 stadiums that have the least number of seats.

3. Compare your totals. Explain why there is no difference between your totals.

EXTRA PRACTICE page H34, Set E

6

Problem Solving Skill
Estimate or Find Exact Answers

UNDERSTAND ▷ PLAN ▷ SOLVE ▷ CHECK

FANS IN THE STANDS   The table shows five games ranked by attendance.

BASEBALL GAME ATTENDANCE		
Rank	Attendance	Team
1	80,227	Montreal vs. Colorado
2	78,672	San Francisco vs. Los Angeles
3	74,420	Detroit vs. Cleveland
4	73,163	St. Louis vs. Cleveland
5	72,470	Philadelphia vs. Colorado

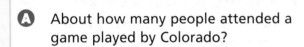

MATH IDEA   Whether you need an exact answer or an estimate depends on the situation.

Examples

A About how many people attended a game played by Colorado?

Estimate to find the number of people.

$$\begin{array}{r} 80{,}227 \\ +72{,}470 \end{array} \rightarrow \begin{array}{r} 80{,}000 \\ +70{,}000 \\ \hline 150{,}000 \end{array}$$

So, about 150,000 people attended.

B How many more people attended the Detroit game than the Philadelphia game?

Subtract to find an exact answer.

$$\begin{array}{r} {}^{3}\overset{13}{\cancel{4}}, \overset{3}{\cancel{4}}\,\overset{12}{2}\,0 \\ -72{,}470 \\ \hline 1{,}950 \end{array}$$

So, 1,950 more people attended.

Talk About It

• Tell whether you need an estimate or an exact answer to find how many more people attended the Detroit game than the Philadelphia game.

Problem Solving Practice

Brianne has $20.00 to spend at the ballpark.

Tell whether an estimate or an exact answer is needed. Solve.

1. Brianne bought a baseball hat and a pennant. How much change will she get from $20.00?

2. About how much money does Brianne need to buy a baseball and a bat at the ballpark?

3. Domingo wants to buy a CD for $12.98, a poster for $3.98, and a book for $4.95. He has $20.00. Does he have enough money for all three items? Explain how you decided.

4. The Photo Gallery has a total of 6,232 photographs. It has 5,760 photographs in photo albums. How many photographs are not in albums?

Mixed Applications

At a college, the students can choose to play a sport. This year, **7,825** students signed up for an indoor sport, and **16,984** different students signed up for an outdoor sport.

5. Which would you use to find the number of students who signed up for a sport?
   A 16,984 + 7,825
   B 17,000 + 8,000
   C 16,984 − 7,825
   D 17,000 − 8,000

6. If 48,676 students attend the college, how many students did NOT sign up for a sport?
   F 40,851 students
   G 30,000 students
   H 23,867 students
   J 70,000 students

7. At the college 29,470 students live in dorms on the campus. If there are 31,652 students, how many students do not live on campus?

8. In the football stadium, there are 6,322 seats. There are 1,974 students seated watching a practice game. How many seats are empty?

9. The School of Science bought a new telescope for $8,376. Write the number in word form.

10. **Write a problem** in which you need to find an estimated sum or difference.

Review/Test

✔ CHECK VOCABULARY AND CONCEPTS

For 1, choose the best term from the box.

> exact number
> estimate

1. When you round the numbers before you add or subtract, the sum or difference is an _?_. (pp. 34–35)

2. Show the three steps to find 47 + 32 by breaking the numbers apart. (pp. 36–39)

✔ CHECK SKILLS

Estimate the sum or difference. (pp. 34–35)

3. 32,895 + 27,982

4. $34,698 − $11,683

5. 53,625 − 24,219

6. 93,863 + 68,410

7. 538,195 + 273,982

8. $4,669 − $1,028

9. 7,516 − 2,495

10. 853,467 − 387,410

Add or subtract mentally. Tell the strategy you used. (pp. 36–39)

11. 43 − 27 **12.** 54 + 37 **13.** 89 + 17 **14.** 92 − 43 **15.** 65 + 49

Find the sum or difference. Write the method you used to compute. (pp. 40–47)

16. 3,258 + 1,784

17. 5,925 − 2,378

18. 8,586 − 4,455

19. 7,448 − 1,737

20. 5,000 + 3,718

21. 35,276 − 22,865

22. 784,109 − 426,545

23. 30,019 − 12,645

✔ CHECK PROBLEM SOLVING

Write whether an estimate or an exact answer is needed. **Solve.** (pp. 48–49)

24. The school cafeteria served 115 cartons of whole milk, 78 cartons of chocolate milk, and 35 cartons of orange juice. About how many drinks were served?

25. Ellie will win a prize if she sells 400 boxes of cookies this year. If she sold 113 in January, 107 in February, and 68 in March, how many more boxes does she need to sell?

Standardized Test Prep

 Understand the problem.
See item **9**.

Since you do not need an exact answer, read all the answer choices carefully. Think about which choice could be an estimate.

Also see problem **1**, p. H62.

For 1–9, choose the best answer.

1. Which is **not** true?
   A 6,411 < 6,141
   B 49,312 < 49,401
   C 3,702 = 3,702
   D 7,942 > 4,709

2. How is seventy-six thousand, five hundred twenty written in standard form?
   F 76,052   H 76,502
   G 706,520   J 76,520

3. Which is the best estimate for 729,312 − 284,306?
   A 400,000   C 500,000
   B 1,000,000   D 450,000

4. Which is the best estimate for 25,908 + 43,149?
   F 70,000   H 60,000
   G 80,000   J 50,000

5. Fernando bought groceries that cost $17. He paid with a $50 bill. Which would be a good estimate of the change Fernando should receive?
   A $10   C $30
   B $20   D $40

6. 8,014
   − 6,438
   F 1,676   H 1,776
   G 2,676   J NOT HERE

7. Which group lists the numbers in order from *least* to *greatest*?
   A 23,119; 29,131; 23,094
   B 23,094; 23,119; 29,131
   C 23,094; 29,131; 23,119
   D 29,131; 23,119; 23,094

8. 512,372
   + 327,819
   F 839,201   H 840,191
   G 840,201   J NOT HERE

9. For which of the following is an exact answer **not** needed?
   A giving change from a ten-dollar bill
   B paying for a purchase
   C deciding how much money to bring to the grocery store
   D deciding how many tickets to buy for your family to see a movie

Write What You Know

10. Describe a situation in which an exact answer is needed and one in which an estimate will do. Explain your thinking.

11. Explain how to find 38 + 92 mentally.

CHAPTER 4

Algebra: Use Addition and Subtraction

On April 2, 1792, Congress created the United States Mint. The United States Mint's primary mission is to produce enough coins for people to use. It produces six coins of different values, sizes, and masses.

PROBLEM SOLVING If you had one of each coin in your hand, what would be the total mass of all six coins?

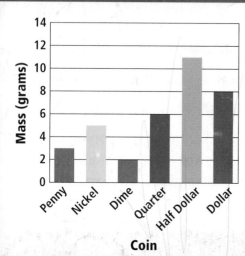

MASSES OF UNITED STATES COINS

Mass (grams)

Coin: Penny, Nickel, Dime, Quarter, Half Dollar, Dollar

DATA LINK

CHECK WHAT YOU KNOW

Use this page to help you review and remember
important skills needed for Chapter 4.

✓ VOCABULARY

Choose the best term from the box.

1. You can __?__ to find the difference.

2. You can __?__ to find the sum.

> add
> subtract
> difference

✓ MISSING ADDENDS (For Intervention, see p. H6.)

Find the missing number.

3. ■ + 4 = 12

4. 6 + ■ = 14

5. ■ + 5 = 12

6. 9 + ■ = 15

7. 7 + ■ = 16

8. 7 + ■ = 10

9. 8 + ■ = 16

10. ■ + 9 = 19

11. 9 + ■ = 17

12. 4 + ■ = 11

13. 7 + ■ = 19

14. ■ + 4 = 11

✓ FACT FAMILIES (For Intervention, see p. H7.)

Copy and complete each number sentence.

15. 5 + 2 = ■
    7 − 2 = ■

16. 6 + 3 = ■
    9 − 3 = ■

17. 8 + 1 = ■
    9 − 1 = ■

18. 7 + 6 = ■
    13 − 7 = ■

19. 9 + 5 = ■
    14 − 9 = ■

20. 7 + 8 = ■
    15 − 8 = ■

21. 8 + 8 = ■
    16 − 8 = ■

22. 4 + 7 = ■
    11 − 4 = ■

23. 17 + 3 = ■
    20 − 3 = ■

24. 14 + 5 = ■
    19 − 14 = ■

25. 12 + 6 = ■
    18 − 6 = ■

26. 7 + 4 = ■
    11 − 7 = ■

✓ NUMBER PATTERNS (For Intervention, see p. H7.)

Write the next three numbers in the pattern.

27. 10, 20, 30, 40, ■, ■, ■

28. 25, 30, 35, 40, ■, ■, ■

29. 14, 18, 22, 26, ■, ■, ■

30. 12, 18, 24, 30, ■, ■, ■

31. 33, 44, 55, 66, ■, ■, ■

32. 27, 30, 33, 36, ■, ■, ■

Expressions

Quick Review

Add 15 to each number.

1. 460   2. 523

3. 196   4. 837

5. 728

VOCABULARY

expression

▶ **Learn**

FANCY FLOWERS   Sue had 12 flowers. Lily gave her 4 more flowers. Sue gave her teacher 3 of her flowers. How many does she have left?

Using parentheses, you can write the expression $(12 + 4) - 3$, Think: 12 plus 4 more minus 3

An **expression** has numbers and operation signs. It does not have an equal sign.

Find the value of $(12 + 4) - 3$.

$(12 + 4) - 3$ Add 12 and 4.
↓
$16 - 3$ Subtract 3 from 16.
↓
13

Remember

Parentheses ( ) tell which operation to do first.

So, $(12 + 4) - 3$ is 13. Sue has 13 flowers left.

Examples

A   Find the value of $(5 - 3) + 6$.

$(5 - 3) + 6$ Subtract 3 from 5.
↓
$2 + 6$ Add 2 and 6.
↓
8

So, $(5 - 3) + 6$ is 8.

B   Find the value of $12 - (4 + 2)$.

$12 - (4 + 2)$ Add 4 and 2.
↓
$12 - 6$ Subtract 6 from 12.
↓
6

So, $12 - (4 + 2)$ is 6.

• How is Example A different from Example B?

1. **Tell** how parentheses help you find the value of an expression.

Tell what you would do first.

2. $10 - (3 + 4)$
3. $9 + (3 - 1)$
4. $6 + (4 - 1)$
5. $(16 - 2) + 41$

Find the value of each expression.

6. $3 + (6 - 2)$
7. $5 + (9 - 1)$
8. $(15 - 9) + 6$
9. $(8 + 13) - 8$

▶ **Practice and Problem Solving**

Tell what you would do first.

10. $12 + (8 - 4)$
11. $17 - (7 + 2)$
12. $20 + (13 - 5)$
13. $(12 + 6) - 8$

14. $24 + (9 - 3)$
15. $10 - (6 + 2)$
16. $7 - (6 + 3)$
17. $14 + (7 - 5)$

Find the value of each expression.

18. $7 + (19 - 6)$
19. $(8 + 7) - 5$
20. $6 + (8 - 1)$
21. $13 + (26 - 12)$

22. $(25 - 10) + 8$
23. $24 - (6 + 5)$
24. $(62 - 40) + 31$
25. $40 + (16 - 8)$

26. $53 - (26 + 8)$
27. $(21 + 62) - 30$
28. $36 - (2 + 9)$
29. $52 - (18 + 2)$

30. $(72 - 6) + 8 + 10$
31. $56 - (4 + 9 + 6)$
32. $(120 + 51 + 6) - 34$

33. $8,620 - (739 + 231)$
34. $2,934 + (695 - 356)$
35. $5,046 - (289 + 867)$

36. **USE DATA** Use the table at the right. How many more plants did the greenhouse sell in May than in March?

37. At the vegetable exhibit, there were 115 carrots and 76 cucumbers. A clerk sold 49 cucumbers. How many vegetables were left?

38. ✎ **Write About It** Explain how to find the value of $34 - (27 - 15)$.

GREENHOUSE SALES

Month	Plants Sold
March	1,903
April	1,567
May	3,000
June	2,540

Mixed Review and Test Prep

39. $\begin{array}{r} 7,984 \\ -6,825 \\ \hline \end{array}$ (p. 40)

40. $\begin{array}{r} 27,337 \\ +19,443 \\ \hline \end{array}$ (p. 44)

41. Write 367,490 in expanded form. (p. 6)

42. Round 67,832 to the nearest thousand. (p. 26)

43. **TEST PREP** Which shows the correct value of the digit 7 in 17,961,099? (p. 8)

   A 7,000   **C** 700,000
   B 70,000   **D** 7,000,000

Use Parentheses

Quick Review

Compare. Write $<$, $>$, or $=$ for each ●.

1. 46 ● 70   2. 936 ● 521

3. 623 ● 632 4. 846 ● 846

5. 758 ● 785

▶ Learn

SUPERSONIC STICKERS Jarrod now has 12 stickers. At first he had 10 stickers. He threw away 2 torn stickers. Then he found 4 more stickers.

Which expression shows that Jarrod has 12 stickers?

$$(10 - 2) + 4 \qquad 10 - (2 + 4)$$

Example 1 Find the value of each expression to find the one which equals 12.

$(10 - 2) + 4$ Subtract 2 from 10.	$10 - (2 + 4)$ Add 2 and 4.
↓	↓
$8 + 4$ Add 8 and 4.	$10 - 6$ Subtract 6 from 10.
↓	↓
12	4

So, $(10 - 2) + 4$ shows that Jarrod has 12 stickers.

Example 2 Which expression has the value of 7?

$(15 - 6) + 2$	$15 - (6 + 2)$
↓	↓
$9 + 2$	$15 - 8$
↓	↓
11 So, $(15 - 6) + 2$ does NOT equal 7.	7 So, $15 - (6 + 2)$ equals 7.

MATH IDEA The position of the parentheses can change the value of the expression.

▶ Check

1. **Explain** why the position of the parentheses is important.

Choose the expression that shows the given value. Write _a_ or _b_.

2. 19
   a. $(26 - 9) + 2$
   b. $26 - (9 + 2)$

3. 7
   a. $5 - (1 + 3)$
   b. $(5 - 1) + 3$

4. 13
   a. $(24 - 15) + 4$
   b. $24 - (15 + 4)$

Practice and Problem Solving

Choose the expression that shows the given value. Write *a* or *b*.

5. 5
   a. $(6 - 3) + 2$
   b. $6 - (3 + 2)$

6. 17
   a. $15 - (2 + 4)$
   b. $(15 - 2) + 4$

7. 52
   a. $(33 - 4) + 23$
   b. $33 - (4 + 23)$

8. 22
   a. $56 - (43 + 9)$
   b. $(56 - 43) + 9$

9. 67
   a. $(53 - 11) + 25$
   b. $53 - (11 + 25)$

10. 12
   a. $65 - (34 - 19)$
   b. $(65 - 34) - 19$

Show where the parentheses should be placed to make the expression equal to the given value.

11. $22 - 6 + 2$; 18

12. $54 - 4 + 19$; 31

13. $60 - 31 + 11$; 40

14. $63 - 14 + 8$; 57

15. $79 - 52 + 33$; 60

16. $143 - 16 + 36$; 91

Find the number that gives the expression a value of 25.

17. $(8 + 6) + \blacksquare$

18. $49 - (\blacksquare + 6)$

19. $(19 + \blacksquare) - 8$

20. Find the value of the expression.

$(160 - 25) + (14 - 3)$

21. What is 1,479 less than 5,000?

22. A post office sold 2,567 stamps on Monday and 1,869 stamps on Tuesday. How many more stamps were sold on Monday?

23. **? What's the Error?** George said the value of $98 - (25 + 16)$ is 89. Describe his error. Write the correct value.

Mixed Review and Test Prep

24. 1,463 (p. 40)
   742
   +2,896

25. 6,000 (p. 42)
   −4,756

26. Each week, Mr. Rogers delivers 1,000 newspapers. He delivers 785 newspapers on weekdays. How many newspapers does he deliver on the weekend? (p. 42)

27. Ines had 24 magazines. She gave 6 magazines to Ana. Then she bought 3 more magazines. Write an expression to show how many magazines Ines has now. (p. 56)

28. **TEST PREP** Which is greater than 35,921? (p. 18)
   A 35,192   **C** 35,902
   B 35,502   **D** 35,952

EXTRA PRACTICE page H35, Set B

3 Match Words and Expressions

▶ Learn

COOL COLLECTIONS Chris had $20. She spent $8 on a new frog figure and $5 on a book about frogs. How much money does Chris have left?

Example

Think: $20 minus $8 for frog figure and $5 for book

Which expression matches the meaning of the words?

($20 − $8) + $5   or $20 − ($8 + $5)

($20 − $8) + $5 ← Means first subtract the cost of the frog figure, and then add the cost of the book.

$20 − ($8 + $5) ← Means first add the cost of the frog figure and the book, and then subtract from $20.

The expression ($20 − $8) + $5 does NOT make sense for the words given.

The expression $20 − ($8 + $5) matches the meaning of the words.

Find the value. $20 − ($8 + $5)   Add $8 and $5.

↓

$20 − $13   Subtract $13 from $20.

↓

$7

So, Chris has $7 left.

 MATH IDEA Use the meaning of the words in a problem to help you place the parentheses in an expression when there is more than one operation.

▶ Check

1. **Write** the expression. Jake had $24. He bought a game for $12 and a book for $6. How much money did he have left?

Frogs of the World

Choose the expression that matches the words. Write a or b.

2. Rick had $8. He spent $5. He then earned $20.
  a. $8 − ($5 + $20)
  b. ($8 − $5) + $20

3. Teri had 18 stickers. She gave 4 to Susie and 3 to Jamie.
  a. 18 − (4 + 3)
  b. (18 − 4) + 3

▶ Practice and Problem Solving

Choose the expression that matches the words. Write a or b.

4. A room had 25 desks. Five desks were added and then 3 desks were removed.
  a. (25 + 5) − 3
  b. (25 − 5) + 3

5. Julie had $14. She spent $3 and then was given $4.
  a. $14 − ($3 + $4)
  b. ($14 − $3) + $4

6. April had 40¢. She found a dime and then bought a soda for 35¢.
  a. (40¢ + 10¢) − 35¢
  b. (40¢ − 10¢) + 35¢

7. Susan had 10 tickets. She gave 5 to her friends and then bought 2 more.
  a. 10 − (5 + 2)
  b. (10 − 5) + 2

8. Roy had 15 toy cars. He put 4 cars in one box and 5 cars in another box. How many cars are not in a box?

9. I am a number that is 19 tens and 9 hundreds greater than 389. What number am I?

10. Tommy was given 2 boxes with 10 trading cards in each box. He then gave his brother 4 trading cards. Write an expression for the words and then find the value.

11. 📖 Write a problem for the expression (3 + 6) − 2.

Mixed Review and Test Prep

12. 24,672 (p. 44)
  +19,823

13. 4,000 (p. 42)
  −1,534

14. A company has 6,000 employees. There are 1,275 employees in the Cleveland office. How many employees are not in Cleveland? (p. 42)

15. Order 451,980; 452,765; and 45,991 from least to greatest. (p. 20)

16. **TEST PREP** What is six hundred forty-seven more than 8,579? (p. 4)
  A 9,222   **C** 9,452
  B 9,226   **D** 9,623

EXTRA PRACTICE page H35, Set C

Use Variables

▶ **Learn**

FOOD FAIR Rhonda put 9 muffins on a tray. She put some more muffins on the tray. What expression can you write to show this?

When you do not know what number to use in an expression, you can use a symbol or a letter to represent the unknown number.

9 muffins + some more muffins
 ↓   ↓
 9   + b

The letter b in the expression $9 + b$ is called a variable. A **variable** can stand for a number.

VOCABULARY

variable

equation

Examples   Write an expression.

A   Jason had some pretzels. He gave away 5 pretzels.

some pretzels − 5 pretzels
     ↓   ↓
     p −   5

B   Tami puts 3 apple tarts in the basket. Mike puts some apple tarts in the basket.

3 apple tarts + some apple tarts
     ↓   ↓
     3   + a

MATH IDEA   You can use a variable to stand for an unknown number.

REASONING **What if** in Example B Tami added 4 apple tarts to the basket? How would you write the expression?

Equations with Variables

Remember that Rhonda had 9 muffins on a tray and put some more muffins on the tray. She counted her muffins. Rhonda now has a total of 14 muffins.

Write an equation to model this. An **equation** is a number sentence stating that two amounts are equal.

9 muffins plus some muffins equals 14 muffins.
$\downarrow$ $\downarrow$ $\downarrow$ $\downarrow$ $\downarrow$
9 + b = 14 ·

So, the equation is $9 + b = 14$.

Examples Write an equation. Choose a variable for the unknown.

Ⓐ 5 tickets plus some extra tickets are 45 tickets. Let t stand for the number of extra tickets.

5 + the number of extra tickets = 45
$\downarrow$ $\downarrow$ $\downarrow$
5 + t = 45

The equation is $5 + t = 45$.

Ⓑ There are some students. Eight students leave. Now there are 24 students. Let n stand for the beginning number of students.

the number of students − 8 = 24
$\downarrow$ $\downarrow$ $\downarrow$
n − 8 = 24

The equation is $n − 8 = 24$.

▶ Check

1. **Explain** the difference between an expression and an equation.

2. **Tell** whether you could use ▇ as the variable in Example A. Explain.

Write an expression. Choose a variable for the unknown.

3. 20 boxes of raisins plus some boxes

4. some crackers minus three crackers

Write an equation for each. Choose a variable for the unknown.

5. There are some boxes of cereal on a shelf. Sam takes 3 boxes. Now there are 12 boxes on the shelf.

6. There are 15 apples on the table. Five are red and the others are green.

LESSON CONTINUES

Practice and Problem Solving

Write an expression. Choose a variable for the unknown.

7. There are some birds in the tree. Seven birds flew away.

8. There are 17 students and some more students.

9. The bakery has some cookies on a tray. There are 5 more cookies in a box.

10. There are 5 dancers on stage. Some more dancers come on stage.

11. 21 CDs plus some CDs

12. some books minus 3 books

For 13–16, write an equation for each. Choose a variable for the unknown.

13. There are 25 sodas in a machine. Mr. Lee adds some sodas. Now there are 42 sodas.

14. Randy had 10 pieces of string. He gave some away. Now he has 3 pieces of string.

15. There are 35 crayons in a box. Four are red. The rest are other colors.

16. Mira added 12 pennies to a jar of pennies. Now there are 74 pennies in the jar.

17. Joe had 115 CDs. Kate gave him some more. Use the variable *n* to write an expression that shows how many CDs Joe has.

18. Javier had 420 points. He earned some more. Use the variable *p* to write an expression that shows how many points Javier has.

USE DATA **For 19–20, use the list.**

19. Jamie took some boxes of raisins. Write an expression to show how many boxes of raisins are left.

20. There were 5 apples in a basket. Some more apples were added. Write an equation to show how many apples are now in the basket.

21. ✎ **Write About It** Explain how to write an equation for the following. Bill has 20 cards. Seven of them are blue. The rest are other colors.

Fruit Salad List
12 apples
8 bananas
6 boxes of raisins
4 oranges

22. A teacher donated 35 folders, 17 books, and 45 pencils to the school carnival. How many items did the teacher donate?

23. Ms. Gray had $20.00. She bought socks for $5.46 and a shirt. Her change was $8.46. How much did the shirt cost?

Mixed Review and Test Prep

For 24–25, use the graph.

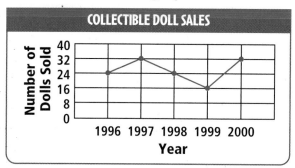

24. How many more dolls were sold in 1997 than in 1999?

25. What was the total number of dolls sold in the years 1996 to 2000?

26. Natalie read three books. One book had 368 pages, another had 331 pages, and the last book had 373 pages. Order the books from greatest number of pages to least number of pages. (p. 20)

For 27–28, use the numbers in the box.

65,743	56,437	64,347
73,645	57,463	75,364

27. Which numbers are greater than 65,574? (p. 18)

28. Which numbers are less than 64,357? (p. 18)

29. **TEST PREP** If 256 people attended the 5:00 P.M. showing of a movie and 324 people attended the 7:30 P.M. showing of the movie, how many people attended the movie? (p. 40)

A 680   **C** 580
B 600   **D** 500

PROBLEM SOLVING LiNKUP... to Health

The Food Guide Pyramid is a chart showing the food groups and the recommended number of servings from each group.

USE DATA For 1–2, use the Food Guide Pyramid.

1. Sherise ate some fruit today. She needs 1 more serving of fruit to have the highest recommended number of servings. What equation could you write to show the number of servings of fruit Sherise ate?

2. What expression could you write for the highest recommended number of servings from the vegetable group minus some number of servings of broccoli?

use sparingly

2–3 servings

2–3 servings

3–5 servings

2–4 servings

6–11 servings

5 Find a Rule

▶ **Learn**

NUMBER CRUNCHER When Mr. Wiley puts the number 12 in a number machine, the number 15 comes out. When he puts in 18, out comes 21, and when he puts in 24, out comes 27. What number comes out when he puts 30 into the machine?

Quick Review

Find the value of each expression.

1. $8 + (9 - 5)$

2. $(16 + 7) - 10$

3. $20 - (3 + 8)$

4. $(18 - 6) + 7$

5. $15 + (6 - 5)$

INPUT	OUTPUT
6	9
12	15
18	21
24	27
30	■

HINT: Look for a pattern to help you find a rule.

Pattern: Each output is 3 more than the input.

Rule: Add 3.

Input: 30   *Output*: 33

So, when Mr. Wiley puts in 30, the number 33 is the output.

You can write an equation to show the rule. Use variables to show the input and output.

input   output
↓   ↓
$x + 3 = y$ Think of the equation as a rule. To find the value of y, add 3 to x.

Examples   Find a rule. Write the rule as an equation.

Ⓐ

INPUT	OUTPUT
n	t
2	6
5	9
8	12
11	15

Test your rule on each pair of numbers in the table.

Ⓑ

INPUT	OUTPUT
p	r
37	31
24	18
18	12
12	6

Rule: Add 4. $n + 4 = t$

Rule: Subtract 6. $p - 6 = r$

1. **Explain** why it is important to test the rule with all the numbers in the table.

Find a rule. Write the rule as an equation.

2.

INPUT	x	6	15	18	23
OUTPUT	y	13	22	25	30

3.

INPUT	s	43	32	21	11
OUTPUT	t	37	26	15	5

► **Practice and Problem Solving**

Find a rule. Write the rule as an equation.

4.

INPUT	b	68	56	45	34
OUTPUT	c	57	45	34	23

5.

INPUT	d	39	47	55	60
OUTPUT	f	26	34	42	47

6.

INPUT	w	28	37	49	54
OUTPUT	z	48	57	69	74

7.

INPUT	r	20	37	28	43
OUTPUT	s	35	52	43	58

Use the rule and equation to make an input/output table.

8. Add 8.
$x + 8 = y$

9. Subtract 6.
$h - 6 = j$

10. Add 12.
$t + 12 = w$

11. Subtract 9.
$f - 9 = g$

12. Subtract 10.
$r - 10 = s$

13. Add 21.
$c + 21 = d$

14. Subtract 14.
$k - 14 = m$

15. Add 24.
$a + 24 = b$

16. Ling had some baseball cards. He gave 28 cards to a friend. He now has 89 cards. Write an equation to find the number of baseball cards Ling had.

17. **?** **What's the Question?** The answer is $n = 6$.

Mixed Review and Test Prep

18. 8,503 (p. 40)
   $-5,927$

19. 7,768 (p. 40)
   $+4,309$

Write the value of the blue digit. (p. 6)

20. 345,687

21. 982,521

22. **TEST PREP** Which is true? (p. 18)
   A $5,341 > 5,431$
   B $6,236 = 6,226$
   C $4,357 < 4,135$
   D $3,964 > 3,694$

6 Equations

▶ Learn

IT'S ALL THE SAME   Amounts on both sides of an equal sign have the same value. What happens when you add the same amount to both sides of an equation?

Marisha and Tony each have 1 dime. Marisha gets 1 nickel from Mrs. Maria, and Tony gets 5 pennies. Do they have the same amount? Use coins to model the problem.

Quick Review

Compare. Write <, >, or = for each ●.

1. 347 ● 374
2. 623 ● 632
3. 598 ● 589
4. 846 ● 846
5. 752 ● 725

HANDS ON

Activity

MATERIALS: coins, workmat

- Divide your paper into two parts. Think of each side of the paper as one side of an equation.
- Place 1 dime on each side. Then add 1 nickel to the left side. Place 5 pennies on the right side.

Marisha's money	Tony's money

- Compare the total value of the coins on each side. Are the values equal?

So, Tony and Marisha have the same amount of money.

Remember

In an equation, the left side and the right side are equal. Both sides have the same value.

Left Side   Right Side
↓   ↓
6 + 3   =   9

- Now add a nickel to each side. What are the values? Are both sides equal? Explain.
- **What if** you added a nickel to one side and 4 pennies to the other side? Are both sides equal?

MATH IDEA   If you add the same amount to both sides of a true equation, the values on both sides are still equal and the equation is still true.

Add Equal Amounts

This is a true equation because the values on both sides are equal.

$$5 + 4 = 9$$
$$\downarrow \qquad \downarrow$$
$$9 = 9$$

What if you add 3 to the left side of the equation? Does adding 3 to the right side of the equation keep the equation true?

$$5 + 4 = 9$$

Add 3 to the left side. $\rightarrow \; 5 + 4 + 3 = 9 + 3 \;\leftarrow$ Add 3 to the right side.

$$12 = 12 \;\leftarrow \text{Compare both sides.}$$

So, adding 3 to both sides keeps this equation true.

Examples

A

$$9 + 8 = 17$$

Add 7 to the left side. $\rightarrow \; 9 + 8 + 7 = 17 + 7 \;\leftarrow$ Add 7 to the right side.

$$24 = 24 \;\leftarrow \text{Compare both sides.}$$

B

$$46 = 15 + 31$$

Add 14 to the left side. $\rightarrow \; 46 + 14 = 15 + 31 + 14 \;\leftarrow$ Add 14 to the right side.

$$60 = 60 \;\leftarrow \text{Compare both sides.}$$

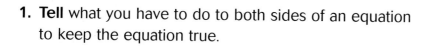

▶ Check

1. Tell what you have to do to both sides of an equation to keep the equation true.

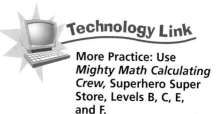

Technology Link

More Practice: Use *Mighty Math Calculating Crew,* Superhero Super Store, Levels B, C, E, and F.

Tell whether the values on both sides of the equation are equal. Write *yes* or *no*. Explain.

2. 2 nickels + 3 nickels = 5 dimes

3. 1 dime − 1 nickel = 5 pennies

LESSON CONTINUES ▶

Tell whether the values on both sides of the equation are equal. Write *yes* or *no*. Explain.

4. 1 dime = 5 pennies

5. 1 quarter = 2 dimes + 1 nickel

6. 2 nickels + 4 nickels = 5 dimes

7. 2 dimes − 3 nickels = 5 pennies

8. 2 dimes + 2 nickels = 1 dime + 4 nickels

9. 1 quarter + 1 dime = 1 dime + 1 nickel

10. 2 dimes + 4 nickels = 1 quarter + 5 pennies

11. 1 dime + 10 pennies = 1 dime + 2 nickels

Complete to make the equation true.

12. 2 nickels + 1 dime + 3 pennies = 4 nickels + _?_

13. 2 dimes + 3 nickels = 1 quarter + _?_

14. 1 nickel + 1 dime + 4 pennies = 3 nickels + _?_

15. 2 dimes + 2 nickels = 1 quarter + _?_

16. 5 nickels + 3 pennies = 2 dimes + _?_

17. 1 dime + 4 nickels = 1 quarter + _?_

18. 1 nickel + 1 dime + 6 pennies = 4 nickels + _?_

19. 3 dimes + 3 nickels = 1 quarter + _?_

20. 15 + 3 = 15 + ■

21. 8 + 1 = ■ + 8

22. 8 + 2 + ■ = 10 + 4

23. 10 + 9 + ■ = 19 + 8

24. 20 + ■ = 15 + 5 + 1

25. 7 + 10 = ■ + 5 + 2

26. 15 + 8 = 14 + ■

27. 7 + 9 = ■ + 8

28. 10 + 2 + ■ = 6 + 8

29. 11 + 6 + ■ = 19 + 5

30. **? What's the Error?** Stacey made this model. Describe her error. What must Stacey do to make the values equal?

31. In a box, Bob has 23 colored pencils, 37 markers, and 25 notepads. How many things are in Bob's box?

32. Carin had $14.00. Her mother gave her $1.75. She spent $2.45 on lunch. How much money did she have then?

33. Phan and Li each had 55¢. Mr. Lee gave Phan 4 coins and Li 2 coins. Could they still have the same amount of money? Explain.

34. Karen had 3 times as many coins as Jon. Could Jon have more money than Karen? Explain.

Mixed Review and Test Prep

For 35–36, use the graph.

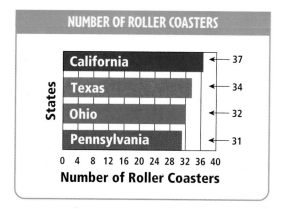

NUMBER OF ROLLER COASTERS

States
- California ← 37
- Texas ← 34
- Ohio ← 32
- Pennsylvania ← 31

0 4 8 12 16 20 24 28 32 36 40
Number of Roller Coasters

35. Complete: __?__ has 6 fewer roller coasters than __?__ has.

36. There are 427 roller coasters in the United States. How many roller coasters are there in the states not listed on the graph?

For 37–39, use the table. Each row and column has the same sum. Find the missing numbers. (p. 40)

112	157	164
213	A	159
108	215	B

37. What is the sum of each row and column?

38. What does **A** equal?

39. What does **B** equal?

40. **TEST PREP** What is five hundred five thousand, fifty-two in standard form? (p. 6)

   A 5,052   **C** 505,052

   B 55,052   **D** 5,005,052

PROBLEM SOLVING THINKER'S CORNER

BREAK THE CODE Look at the symbols at the right. Use the symbols to make equations.

If
▲ + ▲ = ■,
then
■ + ■ = ▲ + ▲ + ▲ + ▲

If
● + ▲ = ◆,
then
◆ + ▲ = ● + ▲ + ▲.

Code

▲ = ● + ●
◆ = ▲ + ●
■ = ▲ + ▲

ALGEBRA For 1, use the code.

1. Find the missing symbol(s).

If ▲ = ● + ●, then ▲ − ◆ = ● − __?__

2. Make up your own code. Write a problem that can be solved using your code. Give the problem to a partner to solve.

EXTRA PRACTICE page H35, Set E

Problem Solving Strategy
Make a Model

PROBLEM   Sally and Sam are top chess players. Sally won 8 games. Sam won 2 games and then won 6 more. The next day Sally won 3 more games. How many more wins does Sam need to be tied with Sally?

UNDERSTAND

- What are you asked to find?
- What information will you use?
- Is there any information you will not use? If so what?

PLAN

- What strategy can you use?
  Make a model to show how many wins Sam needs to tie with Sally. Use counters to model the number of games Sally won and Sam won.

SOLVE

- How can you use the strategy to solve the problem?
  Use a workmat and counters to model the number of games each person won.

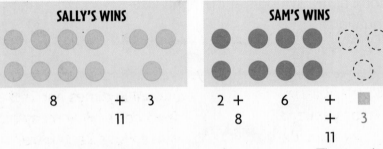

SALLY'S WINS	SAM'S WINS

$$8 \qquad + \quad 3$$
$$11$$

$$2 + \qquad 6 \qquad + \quad \blacksquare$$
$$8 \qquad\qquad + \quad 3$$
$$11$$

Make a model of the number of games won. Then write an equation to show the model:

$8 + 3 = 2 + 6 + 3$

So, Sam needs to win 3 more games.

CHECK

- How can you decide if your answer is correct?

▶ Problem Solving Practice

PROBLEM SOLVING STRATEGIES

Draw a Diagram or Picture
Make a Model or Act It Out
Make an Organized List
Find a Pattern
Make a Table or Graph
Predict and Test
Work Backward
Solve a Simpler Problem
Write an Equation
Use Logical Reasoning

Make a model and solve.

1. **What if** Sam wins 5 more chess games after he ties with Sally? If Sam and Sally end up winning the same number of games, how many more chess games will Sally win after the tie?

2. Bart had 6 checkers and found 7 more. Jenny has 8 checkers. How many more checkers will she need so that she and Bart have the same number of checkers?

3. Stacey gave 8 pens to Dan. Now she has 15 pens. What equation best describes the problem?

   A $n - 8 = 15$ **C** $n + 8 = 15$
   B $15 - 8 = n$ **D** $n + 15 = 2$

4. Lisa had 7 tapes. Mike gave her some more. Now she has 12 tapes. What equation best describes the problem?

   F $7 - t = 12$ **H** $12 - t = 19$
   G $7 + t = 12$ **J** $t - 7 = 12$

5. **? What's the Error?** John has captured 9 chess pieces. Carol has captured 4 chess pieces and has taken 3 more. Carol makes a model and decides she needs to capture 5 more pieces to have the same amount as John. Describe her error. Write the correct answer.

Mixed Strategy Practice

6. Ben is cooking hamburgers at a picnic. He bought 7 packages of hamburger buns. If each package holds 8 buns, how many buns did he get?

7. Kris has 3 quarters. Zach has an equal amount in dimes and nickels. Zach has 9 coins. How many dimes and nickels does Zach have?

8. A game has 24 squares. Each square is red or black. The number of red squares is 2 times the number of black squares. How many squares of each color are there?

9. Heather has 6 pencils and borrows 4 more. John has 10 pencils. If John gives away 2 pencils, what must Heather do to have the same amount?

Problem Solving Strategy

Review/Test

✔ CHECK VOCABULARY AND CONCEPTS

Choose the best term from the box.

equation
expression
variable

1. A letter or symbol that stands for an unknown number is a ? . (p. 60)

2. An ? combines numbers and operations but does not have an equal sign. (p. 54)

✔ CHECK SKILLS

Find the value of each expression. (pp. 54–55)

3. $135 - (20 + 12)$

4. $(251 - 123) + 289$

5. $44,876 - (317 + 1,288)$

6. $(240 + 52) - 119$

7. $485 - (142 + 93)$

8. $308 + (169 - 47)$

Show where the parentheses should be placed to make the expression equal to the given value. (pp. 56–57)

9. $50 - 3 + 12; 35$

10. $38 - 17 + 9; 30$

11. $212 - 117 + 14; 81$

12. $67 - 15 + 8; 60$

13. $40 + 48 - 17; 71$

14. $115 + 209 - 135; 189$

15. Joaquin had 50 sports banners. He traded 32 of his banners for 15 new banners. Write an expression. How many banners does he have left? (pp. 58–59)

16. 36 flowers minus some flowers equals 24 flowers. Write an equation. Choose a variable for the unknown. (pp. 60–61)

17. Complete to make the equation true. (pp. 66–69)

    5 dimes + 2 nickels =
    3 dimes + ? + 5 pennies

18. Find a rule. Write the rule as an equation. (pp. 64–65)

INPUT	c	47	62	84	96
OUTPUT	k	56	71	93	105

✔ CHECK PROBLEM SOLVING

For 19–20, make a model and solve. (pp. 70–71)

19. Karl and Erik worked on a project. Karl worked 8 hours. Together they worked 14 hours. How many hours did Erik work?

20. There were 23 members in Troop 102. Some new members joined, and now there are 30 members. How many new members joined?

Standardized Test Prep

 Get the information you need.
See item **9**.

Look for a relationship of *p* to *r* that is the same for each pair of numbers in the table. Find the equation that matches this relationship.

Also see problem **3**, H63.

For 1–10, choose the best answer.

1. What is 645,468 rounded to the nearest ten thousand?
   A 600,000   **C** 645,000
   B 640,000   **D** 650,000

2. Which symbol makes the following true?
   3,724,219 ● 3,723,722
   F <   **G** >   **H** =   **J** ÷

3. Which is the value of 12 − (7 + 2)?
   A 3   **B** 5   **C** 7   **D** 9

4. 462,307 + 255,294
   F 717,501   **H** 717,601
   G 717,591   **J** NOT HERE

5. Which number makes this equation true?
   6 + 5 + ■ = 19 − 5
   A 1   **B** 2   **C** 3   **D** 4

6. There are 10 boxes. Four of them are red, and the others are not red. Which equation describes this?
   F 4 + n = 10   **H** 10 + 4 = n
   G n + 10 = 4   **J** n + 10 = 14

7. How is three hundred four thousand, seventeen written in standard form?
   A 304,007   **C** 340,170
   B 304,017   **D** 342,017

8. Which set of coins is equivalent to 3 nickels 2 dimes?
   F 1 nickel 1 quarter
   G 1 dime 1 quarter
   H 3 nickels 1 quarter
   J 2 dimes 1 quarter

9. Which equation describes the rule in this table?

INPUT p	43	37	31	27
OUTPUT r	25	19	13	9

   A r − 18 = p   **C** r − p = 18
   B p + 18 = r   **D** p − 18 = r

10. Which expression has a value of 7?
    F 20 − (9 − 4)   **H** 20 − (9 + 4)
    G 20 + (9 − 4)   **J** 20 + (9 + 4)

Write What You Know

11. Chris had $12. He spent $1 on a drink and $3 on food. Use parentheses to write two different expressions to find how much money he had left. Then explain how to find the value of each expression.

12. Find a rule. Explain how you found the rule. Then use the rule to find the output when the input is 20.

INPUT x	3	12	18	27
OUTPUT y	28	37	43	52

PROBLEM SOLVING
MATH DETECTIVE

Mixed-Up Digits

7		5		6		1
	2		9		4	
	8		3			

REASONING The digits above were used to form two whole numbers until they got mixed up. Use the clues to find the two whole numbers. One digit will not be used. All other digits are used once and only once. Put on your thinking cap and use your reasoning skills. Good luck!

Case 1

Clue 1: The digit in the thousands place is 3 times the digit in the tens place.

Clue 2: The number rounds to 7,000 when rounded to the nearest thousand.

Clue 3: The digit in the hundreds place is 5 more than the digit in the ones place.

Clue 4: The number is even. What is the number?

Case 2

Clue 1: The number is greater than 5,000.

Clue 2: The digit in the hundreds place is 2 more than the digit in the tens place.

Clue 3: The sum of the digits is 21.

Clue 4: The number is odd. What is the number?

Think It Over!

- Write About It Explain how you solved each case.

- **STRETCH YOUR THINKING** Change one clue in Case 2 so that a different digit is left unused. Which number is left unused?

Challenge

Front-End Estimation

There are many ways to estimate. The front-end strategy is useful for addition. First, you find the sum of the front digits of each addend. Only digits of equal place value are added. Then, adjust the sum to find an estimate closer to the exact answer.

The Andersons drove 252 miles and then rode a train for 345 miles. About how far did they travel? Use front-end estimation with an adjustment to solve.

STEP 1

Add the front-end, or lead, digits.

$$252$$
$$+345$$
$$500 \leftarrow \text{front-end sum}$$

STEP 2

Adjust the front-end sum to give a more exact answer.

$$252$$
$$+345 \quad \text{Think: } 52 + 45$$
$$100 \quad \text{is about 100.}$$

STEP 3

Add the front-end sum and the adjustment to give a closer estimate.

$$500 \leftarrow \text{front-end sum}$$
$$+100 \leftarrow \text{adjustment}$$
$$600$$

So, the Andersons traveled about 600 miles.

Examples

A Estimate the sum.

$$432$$
$$79$$
$$+289$$

Front-end sum: 600
Adjustment: 32 + 79 + 89 is about 200.
Estimate: 600 + 200 = 800

B Estimate the sum.

$$4,328$$
$$2,416$$
$$+1,296$$

Front-end sum: 7,000
Adjustment: 328 + 416 + 296 is about 1,000.
Estimate: 7,000 + 1,000 = 8,000

 **MATH IDEA** Front-end addition with an adjustment gives an estimate close to the exact answer.

Try It

Solve by using front-end estimation with an adjustment.

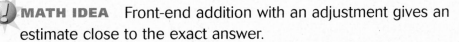

1. 595
$+293$

2. $\$3.98$
$+\$2.12$

3. $4,595$
$+3,293$

4. $1,405$
$3,127$
$+\ \ 326$

Study Guide and Review

VOCABULARY

Choose the best term from the box.

benchmark
equation
expression
variable

1. A known number of things that helps you understand the size or amount of a different number of things is a ? . (p. 2)

2. A symbol or letter that stands for an unknown number is a ? . (p. 60)

3. A part of a number sentence that combines numbers and operations but does not have an equal sign is an ? . (p. 54)

STUDY AND SOLVE

Chapter 1

Find the value of a digit.

Find the value of the blue digit.

4,359,687

The digit 5 is the ten thousands position.

So, the digit 5 has a value of 50,000.

Find the value of the blue digit.
(pp. 6–11)

4. 62,422   5. 382,755

6. 7,405,699   7. 1,469,302

8. 841,977,302 9. 354,821,638

Chapter 2

Compare and order numbers.

Order 6,824; 8,643; and 6,834 from *least* to *greatest*.

6,824	6,824	6,824
↓ 8 > 6	↓ 8 = 8	↓ 3 > 2
8,643	6,834	6,834
↓ 8 > 6		
6,834		

| 8,643 is the greatest number. | There are the same number of hundreds. | 6,834 > 6,824 |

So, 6,824 < 6,834 < 8,643.

Write < , >, or = for each ●.
(pp. 18–19)

10. 6,260 ● 6,620

11. 4,920 ● 4,290

12. 68,750 ● 68,750

13. 14,211 ● 4,321

14. 1,800,234 ● 1,801,254

15. Order the numbers 78,994; 87,497; and 78,499 from *least* to *greatest*. (pp. 20–23)

Chapter 3

Estimate and find sums and differences.

Find the difference.

$$\begin{array}{r} {\scriptstyle 9\ \ 9} \\ {\scriptstyle 4\ 10\ 10\ 10} \\ 5{,}0\ 0\ 0 \\ -3{,}5\ 6\ 7 \\ \hline 1{,}4\ 3\ 3 \end{array}$$

- Regroup when needed.
- Subtract the ones.
- Subtract the tens.
- Subtract the hundreds.
- Subtract the thousands.

Find the sum or difference. Estimate to check. (pp. 34–35, 40–47)

16. $\begin{array}{r} 99{,}452 \\ -\ 6{,}803 \end{array}$

17. $\begin{array}{r} 132{,}741 \\ +104{,}458 \end{array}$

18. $\begin{array}{r} 13{,}000 \\ -10{,}687 \end{array}$

19. $\begin{array}{r} 400{,}561 \\ +592{,}625 \end{array}$

Chapter 4

Evaluate expressions with parentheses.

Find the value. $(60 - 18) + 54$

$(60 - 18) + 54$
$\downarrow$
$42 + 54 = 96$

- Do the operation in the parentheses first.
- Subtract 18 from 60.
- Add 42 and 54.

Find the value of each expression. (pp. 54–57)

20. $91 - (53 + 35)$

21. $(92 - 67) + 6$

22. $(673 + 101) - 164$

Write expressions and equations that contain variables.

20 cups minus some cups plus 3 cups

$$(20 - c) + 3$$

Parentheses are used for $20 - c$ since some cups, c, were removed before 3 cups were added.

Write an expression or equation for each. Choose a variable for the unknown. (pp. 58–63)

23. Sandy had 5 cards. She gave some away.

24. Shanti had some stamps. She gave 23 to Penny. Shanti has 71 stamps left.

PROBLEM SOLVING PRACTICE (pp. 48, 70)

25. Mr. Miller has written 157 pages of a book. He wants the book to have about 550 pages. About how many more pages does he need to write? Tell whether an exact answer or estimate is needed, and solve.

26. Jack and Mary are playing a game. Jack has 7 cards and picks up 1 card from the unused stack. Mary has 5 cards. How many more cards does Mary need to have the same number as Jack?

PERFORMANCE ASSESSMENT

TASK A • POPULATION STUDY

The students in Maria's class are writing reports on Central American countries. Maria reads in an almanac that the population of El Salvador is 5,752,067; the population of Honduras is 5,861,955; the population of Costa Rica is 3,604,642; and the population of Nicaragua is 4,583,379.

a. Make a table to organize the population data for the four countries.

b. Describe the order in which you listed the populations. Where is Nicaragua on your list?

c. Maria read that the population of Panama is about 2,700,000. She thought that this number was probably rounded to the nearest hundred thousand. Describe the numbers that could represent the actual population of Panama. Where would you list Panama in your table?

TASK B • BUYING A COMPUTER

The Stockwells saved $1,900 to buy a computer, monitor, and printer. The table shows the cost of items they might choose.

Computer		Monitor		Printer	
Brand A	$1,029	15-inch	$239	ink jet	$289
Brand B	$1,179	17-inch	$379	laser jet	$429

a. Estimate to find one possible choice of computer, monitor, and printer the Stockwells could buy with the money they have. Then find the actual total cost of these items.

b. How much of the $1,900 will be left if they purchase the items you suggested?

c. Suppose the Stockwells decide to start a new savings account to buy software. They put $100 in the account. They take out $45 to buy a new game and then put in $20. Write an expression to find the amount of money remaining in the account. Then find the value of the expression.

Technology Linkup

E-Lab • Subtracting Across Zeros

There are 800 seats in the school auditorium. There are 684 people sitting in the auditorium. How many empty seats are there?

You can use E-Lab to subtract across zeros.

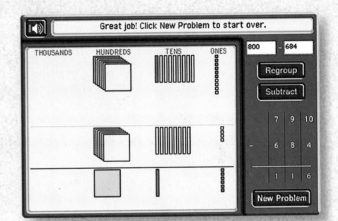

• Click on *Subtracting Across Zeros*.

• Click *New Problem* to begin.

• Type 800. Press *Enter*.

• Type 684. Press *Enter*.

• Click *Regroup* to regroup 1 hundred as 10 tens.

• Click *Regroup* to regroup 1 ten as 10 ones.

• Click *Subtract*.

• Record the difference.

So, there are 116 empty seats.

How many times would you need to *regroup* to subtract 705 − 271? What is the difference?

Practice and Problem Solving

Use E-Lab to find the difference.

1. 650 − 238   **2.** 970 − 468   **3.** 406 − 225   **4.** 803 − 275

5. 650 − 379   **6.** 307 − 138   **7.** 200 − 137   **8.** 700 − 411

9. Joanne has 300 tickets to sell. Will sells 138 of them for her. How many tickets does Joanne have left?

10. ✎ **Write a problem** that involves subtracting across zeros. Use E-Lab to solve the problem.

Multimedia Math Glossary www.harcourtschool.com/mathglossary

11. **Vocabulary** Look up *expression*, *variable*, and *equation*, in the Multimedia Math Glossary. Write an example of an expression and an equation using the variable *n*.

▲ The Mississippi River begins as a
brook, 20 feet wide, at the outlet
of Lake Itasca, Minnesota.

PROBLEM SOLVING ON LOCATION in
Minnesota

LAKES

Minnesota is called "the land of 10,000 lakes," but it
actually has more than 15,000 lakes.

USE DATA For 1–5, use the table.

1. **REASONING** Which two lakes are
   closest in size? What is the difference
   in their sizes?

2. What is the difference in size between
   the largest and smallest lake?

3. The sum of the sizes of which two
   lakes is between 54,000 and 55,000
   acres and is 3,983 acres less than
   Lake Winnibigoshish? What is the sum?

4. When you add the sizes of two lakes
   together, which two have a sum that
   is less than 30,000 acres? What is
   the sum?

5. The sizes of which two lakes have a sum between 72,000
   and 73,000 acres and a difference between 44,000 and
   45,000 acres? What are the sum and difference?

SIZES OF MINNESOTA LAKES	
Lake	**Size (in acres)**
Cass Lake	15,596
Lake Mille Lacs	132,516
Lake Minnetonka	14,004
Lake Vermilion	40,557
Lake Winnibigoshish	58,544
Leech Lake	111,527
Otter Tail Lake	13,725
Upper Red Lake	107,832

PARKS

The 2,552-mile long Mississippi River begins in Minnesota, where it flows from Lake Itasca. You can visit the headwaters of the Mississippi River in Itasca State Park. The Mississippi River travels past 10 states on its journey south to the Gulf of Mexico.

USE DATA For 1–5, use the table.

1. Put the states in order from *least* to *greatest* population. Which state has the third greatest population?

2. If you round the population of each state to the nearest thousand, for which states will the populations increase by 1 in the thousands place?

3. Write the population of Minnesota in expanded form.

4. **What if** some people, *p,* move to Minnesota? What expression can you write to show the number of people who would live there then?

5. **STRETCH YOUR THINKING** How many more people live in Minnesota than in Iowa?

POPULATIONS OF STATES THAT BORDER THE MISSISSIPPI RIVER	
State	**Population**
Arkansas	2,522,819
Illinois	11,895,849
Iowa	2,852,423
Kentucky	3,908,124
Louisiana	4,351,769
Minnesota	4,685,549
Mississippi	2,730,501
Missouri	5,402,058
Tennessee	5,368,198
Wisconsin	5,169,677

Lake Vermilion is the fifth largest lake in Minnesota. ▶

Collect and Organize Data

More than 250,000 students have participated in Seacamp in Florida. Students at Seacamp snorkel and scuba dive on reefs. While there, they count the fish they see. The tally table below shows the numbers of the different fish seen by one student on a dive. Make a bar graph of the information in the table.

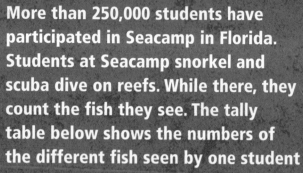

DATA LINK

TALLY TABLE

Type of Fish	Number Seen
Angelfish	卌 卌 IIII
Barracuda	III
Butterfly fish	IIII
Parrot fish	卌 卌
Triggerfish	卌 III

CHECK WHAT YOU KNOW

Use this page to help you review and remember
important skills needed for Chapter 5.

✔ VOCABULARY

Choose the best term from the box.

1. A _?_ uses numbers to show how often something happens.

2. Information collected about people or
   things is called _?_.

> data
> frequency table
> tally table

✔ READ PICTOGRAPHS (For Intervention, see p. H8.)

For 3–6, use the pictograph.

3. How many members of the Running Club
   are in fourth grade?

4. If 5 more third graders joined the
   Running Club, how many symbols would
   there be for third grade?

5. What is the total number of members
   in the Running Club?

6. How many fifth-grade members are
   in the Running Club?

RUNNING CLUB MEMBERS

Grade Levels	
Third	🏃🏃🏃
Fourth	🏃🏃🏃🏃🏃🏃
Fifth	🏃🏃🏃🏃🏃

Key: Each 🏃 stands for 2 members.

✔ TALLIES TO FREQUENCY TABLES (For Intervention, see p. H8.)

For 7–12, use the tables.

7. Copy and complete the frequency table.

8. How many pieces of fruit were sold in
   Weeks 1 and 2?

9. How many more pieces of fruit were sold
   in Weeks 1 and 2 than in Weeks 3 and 4?

10. How many pieces of fruit were sold during
    the four weeks?

11. If 7 more pieces of fruit were sold in
    Week 3, how many tally marks would be
    shown for that week?

12. In which week were 6 fewer pieces of fruit
    sold than in Week 1?

SCHOOL FRUIT STAND

Week	Pieces of Fruit Sold
1	ℍℍ ℍℍ lll
2	ℍℍ
3	ℍℍ lll ℍℍ
4	ℍℍ ll

SCHOOL FRUIT STAND

Week	Number
1	▨
2	5
3	▨
4	7

Collect and Organize Data

Quick Review

1. $315 + 70$

2. $817 - 209$

3. $257 + 43$ 4. $1,000 - 430$

5. $15 + 25 + 16 + 24$

VOCABULARY

survey   frequency

cumulative frequency

▶ Learn

TAKE YOUR PICK You are taking a **survey** when you ask different people the same questions and record their answers. Follow these rules to get the information you want:

• Make the questions clear and simple.

• Ask each person the questions only once.

• Use tally marks to record each person's answer or response.

Jason and Susie each wrote a question to find the class's favorite color for School Spirit Day decorations. Compare the results of their surveys.

Remember

In a tally table, tally marks are used to record data. The tally marks ⊮Ⅰ stand for 6.

What is your favorite color?

JASON'S SURVEY DATA

Color	Votes
Yellow	|||
Green	⊮⊮
Blue	⊮	
Red	⊮	
Orange	||||

Is your favorite color red, blue, or yellow?

SUSIE'S SURVEY DATA	
Color	Votes
Red	⊮ ⊮
Blue	⊮ ⊮	
Yellow	⊮ |||

Both surveys ask about favorite colors, but Jason's allows more color choices. His survey allows any colors, such as orange or green. Susie's question allows only 3 color choices.

Frequency Tables

A frequency table helps you organize the data from a tally table. The **frequency** is the number of times a response occurs. The **cumulative frequency** is a running total of the frequencies.

Use the table below to find the favorite meal and the number of students surveyed.

FAVORITE MEAL

Meal	Tally
Meat loaf	IIII
Pizza	⦀ ⦀
Taco	⦀ III
Spaghetti	⦀ I

FAVORITE MEAL

Meal	Frequency (Number of Students)	Cumulative Frequency
Meat loaf	4	4
Pizza	10	14
Taco	8	22
Spaghetti	6	28

← 4 + 10 = 14
← 14 + 8 = 22
← 22 + 6 = 28 ← Total number of students surveyed

Pizza has the greatest frequency. So, pizza is the favorite meal of the 28 students surveyed.

MATH IDEA Tables can be used to collect, organize, and display data.

Check

1. **Write** a survey question, with choices, to find the favorite type of pizza.

Use Jason's and Susie's survey results on page 82. Tell whether each statement is *true* or *false*. Explain.

2. Susie's data show that more students prefer blue than red.

3. Jason's data show that red is the students' favorite color.

For 4–6, use this frequency table.

4. How many slices were sold in Hour 2?

5. By the end of Hour 3, how many slices had been sold?

6. How many more slices were sold in Hour 1 than Hour 4?

Pizza Slices Sold

Hour	Frequency	Cumulative Frequency
1	16	16
2	20	36
3	12	48
4	9	57

LESSON CONTINUES ▶

Gina asked her friends, "What is your favorite kind of party?" She put the data in a table.

For 7–10, use the table to tell whether each statement is *true* or *false*. Explain.

FAVORITE KIND OF PARTY	
Party	**Votes**
Bowling	7
Skating	8
Movie	4
Pool	5

7. Gina's data show that more friends prefer a bowling party than a pool party.

8. Gina's data show that a pool party is the least favorite choice.

9. Gina's data show that a skating party is the favorite party of the greatest number of her friends.

10. Copy the party table and add a cumulative frequency column. How many friends did Gina survey?

11. Write a survey question to find the favorite fruit of your class. Survey your classmates. Make a tally table and a frequency table for your data.

12. 📓 **Write About It** Write a problem using this data: Rosa buys four items priced at $27, $12, $15, and $32.

For 13–15, use the table.

13. Copy and complete the table. How many tickets were sold on Wednesday?

14. How many tickets were sold during the 4 days?

15. On which two days were the most tickets sold?

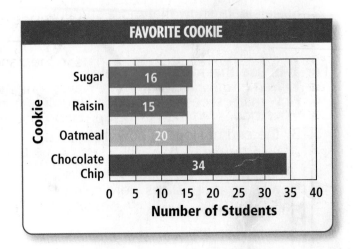

SKATING TICKETS SOLD		
Day	**Tickets Sold**	**Cumulative Frequency**
Tuesday	5	▨
Wednesday	▨	23
Thursday	10	▨
Friday	▨	45

16. Nancy planted a 4-year-old tree in the park in 1988. How old will the tree be in 2010?

17. Mrs. Barker was 34 years old in 1998. How old was she in 1979?

For 18–20, use the bar graph.

18. Which cookie is the students' favorite?

19. How many more students chose chocolate chip cookies than raisin cookies?

20. **REASONING** How many students were surveyed in all?

FAVORITE COOKIE

Cookie
- Sugar: 16
- Raisin: 15
- Oatmeal: 20
- Chocolate Chip: 34

Number of Students
0 5 10 15 20 25 30 35 40

84

For 21–23, use the table.

21. How many bottles of juice were sold Monday and Tuesday?

22. How many more bottles of juice were sold on Thursday than Tuesday?

23. How many bottles of juice were sold in all?

JOE'S JUICE STAND		
Day	Frequency	Cumulative Frequency
Monday	140	140
Tuesday	197	337
Wednesday	259	596
Thursday	238	834

Mixed Review and Test Prep

24. Round 56,341 to the nearest ten thousand. (p. 26)

25. Which is greater, 1,364,217 or 1,436,217? (p. 18)

26. Subtract 40,135 from 56,091. (p. 44)

27. Find the missing number in the pattern: 56,410; 45,410; ■; 23,410; 12,410 (p. 44)

28. Find the sum of 25,984 and the number 10,000 greater than 31,092. (p. 44)

29. Find the value of the expression $(56 - 39) + 48$. (p. 54)

30. 80,000 (p. 42)
    − 4,705

31. 67,813 (p. 44)
    + 14,589

32. **TEST PREP** $10 \times ■ = 8 \times 5$
    A 3   C 5
    B 4   D 6

33. **TEST PREP** $2,890 + 3,678 + 4,722$ (p. 40)
    F 9,180   H 11,180
    G 10,290   J 11,290

PROBLEM SOLVING THINKER'S CORNER

Taking a survey is a good way to predict how people will vote in an election. A random survey is one that asks a question of a group of people with different opinions. A survey that is not random may ask the same question of a group of people with similar interests.

Angela and Paul are running for class president. Angela has promised more computer time if elected, and Paul has a lot of friends in the band.

Use the survey results to answer each question.

1. Which group of people was probably chosen at random?

2. Explain why you think Kim's results were different from Seth's.

3. Who do you think will win the election? Tell why you think this.

KIM
Computer Club Members
Who would you vote for?
Angela 9 votes
Paul 1 vote

TOSHIO
Band Members
Who would you vote for?
Angela 2 votes
Paul 8 votes

SETH
Students at Lunch
Who would you vote for?
Angela 14 votes
Paul 20 votes

EXTRA PRACTICE page H36, Set A

HANDS ON

Find Median and Mode

▶ **Explore**

For a science experiment, Mr. Alber's class recorded the high temperatures for the first 11 days of October. Find the mode and median of this data.

VOCABULARY

mode   median

OCTOBER HIGH TEMPERATURES (in degrees Fahrenheit)										
Date	1	2	3	4	5	6	7	8	9	10	11
Temp.	65	62	62	62	65	60	59	59	57	58	60

Activity

MATERIALS: index cards

STEP 1

Find the mode.

Write the 11 temperatures on index cards. Sort the cards by numbers. The **mode** is the number that occurs most often. There may be more than one mode, or there may be no mode.

57   58   59   59   60

STEP 2

Find the median.

Order the index cards from least to greatest. Turn one card over on each end. Keep doing this, moving toward the middle, until only one number is showing. That number is the **median**, or middle number.

?

• What is the median temperature? What is the mode?

Try It

Include 57°F for October 12 and 59°F for October 13 in the data above.

   a. Find the mode.

   b. Find the median temperature.

We have added the temperature for October 12. Where do we place the temperature for October 13?

▶ Connect

You can find the median and mode without using cards. Use the Science Fair table to list all the ages from least to greatest to find the median and the mode.

9, 9, 9, 9, 9, 9, 10, 10, 10, 11, 11, 12, 12

↓ ↓

Mode = 9   Median = 10

The mode is most frequently used when dealing with groups of objects which are not numbers. Look at the Favorite Marine Animal table. Dolphin received the most responses, so dolphin represents the mode.

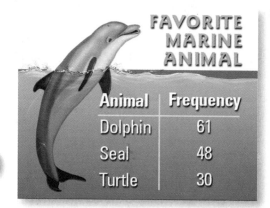

FAVORITE MARINE ANIMAL

Animal	Frequency
Dolphin	61
Seal	48
Turtle	30

▶ Practice and Problem Solving

For 1–2, find the median and mode.

1.

SCIENCE TEST SCORES					
Test	1	2	3	4	5
Score	88	95	67	86	95

2.

SHELLS COLLECTED							
Day	Sun	Mon	Tue	Wed	Thu	Fri	Sat
Shells	31	17	18	15	19	23	31

3. **? What's the Error?** Jim says the median of the data below is 220 and the mode is 170. Describe his error and write the median and mode.

PLAY TICKETS SOLD					
Week	1	2	3	4	5
Tickets	150	170	220	160	220

4. **REASONING** Carol scored 89, 88, 93, 88, 85, and 93 on six tests. After she took the seventh test, the mode was 93. Find the median of the seven scores.

5. Record the high temperatures for your city for a one-week period and put the data in a table. Find the median and the mode.

Mixed Review and Test Prep

For 6–7, write < or > for each ●.

6. 7,500 − 1,000 ● 7,400 (p. 40)

7. 8,395 ● 7,059 + 1,300 (p. 40)

8. Mr. Key is taking a trip of 832 miles. On Saturday he drives 453 miles. How much farther will he have to drive to finish the trip?

9. 287 + 109 + 91 (p. 36)

10. **TEST PREP** Find the value of (68 − 39) + 56. (p. 54)

A 51
B 75
C 85
D 87

3 Line Plot

▶ **Learn**

"X" MARKS THE SPOT A graph that shows the frequency of data along a number line is called a **line plot**. This line plot shows the number of runs Corey batted in during one baseball season.

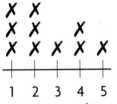

Each X on the line plot represents a tally mark.

Runs Batted In by Corey

You can use a line plot to find the range. The **range** is the difference between the greatest and the least values in a set of data.

$$5 - 1 = 4$$ The range is 4.

Look at the line plot of the number of brothers and sisters. Most of the data form a cluster, or group, from 0 to 3. The value 7 is called an outlier. An **outlier** is a value separated from the rest of the data.

▶ **Check**

1. **Explain** how a line plot and a tally table are alike.

For 2–3, use the library line plot.

2. What is the range of the data?

3. What value would be considered an outlier? Explain.

Quick Review

1. 480 + 120
2. 61 + 59   3. 100 − 43
4. 1,000 − 350 **5.** 165 + 135

VOCABULARY

line plot

range

outlier

RUNS BATTED IN BY COREY	
Runs	**Games**
1				
2				
3		
4			
5		

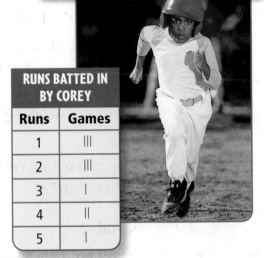

Number of Brothers and Sisters

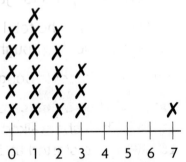

Library Books Checked Out

88

Practice and Problem Solving

For 4–6, use the first line plot.

4. How many teams won three races? How many races is that in all?

5. How many teams are shown? How do you know?

6. How many races were won in all?

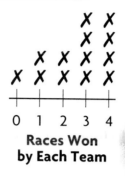

Races Won by Each Team

For 7–9, use the second line plot.

7. Each X in this line plot stands for one player. What do the numbers on the line plot stand for?

8. What value would be considered an outlier? Explain.

9. What is the range of the data on this line plot? What is the median?

Free-Throw Baskets Made

For 10–12, use this table.

10. Make a line plot of the data. Find the median, mode, and range of the data.

11. **REASONING** Find the total number of juice packs the students bought.

12. How many students bought juice packs?

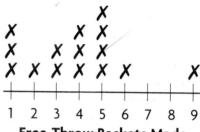

JUICE PACKS BOUGHT	
Juice Packs	Students
0	I
1	II
2	II
3	IIII

13. Jackie worked in the garden for 25 min. Then she walked the dog for 15 min. If she started at 11:30 A.M., what time did she finish?

14. **What's the Question?** Molly has some nickels. She gives 14 to her brother. The answer is $n - 14$.

Mixed Review and Test Prep

Round to the nearest thousand. (p. 26)

15. 75,391

16. 148,569

17. Write a number greater than 23,487 and less than 23,847. (p. 20)

18. $378 + (813 - 72)$ (p. 54)

19. **TEST PREP** What is 1,276 subtracted from 1,493? (p. 40)

A 216   C 223

B 217   D 769

EXTRA PRACTICE page H36, Set B

4 Stem-and-Leaf Plot

> ## Learn

BRANCH OUT A **stem-and-leaf plot** shows groups of data arranged by place value. Use the data for sit-ups completed by Mr. Clark's class to make a stem-and-leaf plot.

VOCABULARY
stem-and-leaf plot
stem   **leaf**

Example

STEP 1

Group the data by the tens digits.
13 17 11
21 26 29
34 34 30
41 42

SIT-UPS (in 1 minute)					
13	21	26	17	34	41
42	29	34	30	11	

STEP 2

Order the tens digits in a column from least to greatest to form the stems. Draw a line to the right of the stems.

1
2   Each tens digit is
3   called a **stem**.
4

STEP 3

Write each ones digit to the right of its tens digit.

1	3 7 1
2	1 6 9	Each ones
3	4 4 0	digit is called
4	1 2	a **leaf**.

STEP 4

Order the leaves from least to greatest. Include a title, labels, and a key to show what each stem and leaf stands for.

Sit-Ups (in 1 minute)

Stem	Leaves
1	1 3 7
2	1 6 9
3	0 4 4
4	1 2

2|6 = 26 sit-ups.

• What are the mode and median of the data? What is the range of the data?

> ## Check

For 1–2, use the stem-and-leaf plot above.

1. **Describe** how you found the median of the data in the stem-and-leaf plot.

2. A new student completed 50 sit-ups. How would you include her sit-ups in the plot?

90

For 3–4, use the stem-and-leaf plot of Lee's golf scores.

Lee's Golf Scores

Stem	Leaves
7	6 7 9 9 9
8	0 2 3 5 6 8 9
9	0 0 1 2 3

3. What are Lee's lowest and highest golf scores? What is the range?

4. What is the mode? the median?

For 5–7, use the stem-and-leaf plot of miniature golf scores.

Miniature Golf Scores

Stem	Leaves
2	8
3	0 5 5 7
4	1 2 2 4 8
5	0 0 1 3 3 3 7

5. Which digits are stems?

6. What is the mode? the median?

7. What are the lowest and highest miniature golf scores? What is the range?

REASONING For 8–9, use the stem-and-leaf plot of Kay's bowling scores.

Kay's Bowling Scores

Stem	Leaves
10	0 3 8
11	0 4 5 6
12	3 7

10|0 = 100.

8. Write all of the scores shown on this stem-and-leaf plot.

9. What is the median of Kay's bowling scores?

10. The heights of Mr. Jin's students are 53, 48, 55, 49, 49, 51, 55, 57, 57, 59, 54, 55, and 48 inches. Make a stem-and-leaf plot of these data. Find the median and the mode of the heights.

11. Josie is making a pictograph with a key in which each symbol stands for 5 students. How many symbols stand for 20 students? 25 students?

12. ✎ **Write About It** Explain the difference between stems and leaves on a stem-and-leaf plot.

Mixed Review and Test Prep

13. 15,982 (p. 44)
    +16,775

14. 3,901 (p. 40)
    +2,881

15. Write an expression using the variable n for a number of cards minus 8 cards. (p. 58)

16. Order 86,962; 86,923; 85,816 from least to greatest. (p. 20)

17. **TEST PREP** Choose the number that does *not* round to 8,600. (p. 26)

A 8,596   C 8,547

B 8,623   D 8,647

EXTRA PRACTICE page H36, Set C

Compare Graphs

Quick Review

1. 349 + 690
2. 921 − 487
3. 14 + ■ + 6 = 38
4. 3 × 4   5. 5 × 8

▶ **Learn**

RAISE THE BAR   Graphs A and B show the same data. However, the graphs look different.

The **scale** of a graph is a series of numbers placed at fixed distances. Both graphs have a scale of 0–50. The top value of the scale should be greater than the greatest value of the data.

The **interval** of a graph is the difference between two numbers on the scale. Graph A's scale has an interval of 5. Graph B's scale has an interval of 10. It is easier to compare the lengths of the bars in Graph A.

VOCABULARY

scale

interval

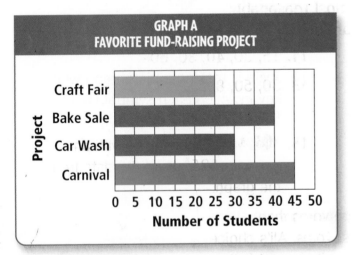

GRAPH A
FAVORITE FUND-RAISING PROJECT

Project: Craft Fair, Bake Sale, Car Wash, Carnival

0 5 10 15 20 25 30 35 40 45 50
Number of Students

GRAPH B
FAVORITE FUND-RAISING PROJECT

Project: Craft Fair, Bake Sale, Car Wash, Carnival

0 10 20 30 40 50
Number of Students

▶ **Check**

1. **Compare** the lengths of the bars in Graph A to the lengths of the bars in Graph B. Why do you think they are different?

For 2–3, explain how the length of the bars would change in Graph B above.

2. if the interval were 20

3. if the interval were 2

► Practice and Problem Solving

FAVORITE BOOKS

Type of Book: Fiction, Nonfiction, Poetry, Science Fiction
Number of Students: 0 10 20 30 40 50

For 4–7, explain how the length of the bars would change in the graph above.

4. if the interval were 1

5. if the interval were 2

6. if the interval were 20

7. if the interval were 5

8. What is the mode of the data in the graph above?

9. What is the scale of the graph above?

For 10–13, choose 5, 10, or 100 as the most reasonable interval for each set of data. Explain your choice.

10. 5, 15, 20, 10, 15

11. 15, 30, 40, 30, 60

12. 100, 200, 200, 450, 500, 300

13. 30, 50, 80, 60, 100

For 14–16, use the graph above.

14. Make a new graph with an interval of 5. Explain how the length of the bars changed.

15. Write a problem with an answer of 25, using the data from the graph.

16. REASONING Josef's choice had 15 more votes than nonfiction. Zoe's choice had the fewest votes. Ali's choice was different from Josef's and Zoe's, but it was not fiction. What did each student choose?

Mixed Review and Test Prep

17. Write eight hundred seventy-two thousand, one hundred six in standard form. (p. 6)

18. $\begin{array}{r} 6{,}000 \\ -1{,}753 \\ \hline \end{array}$ (p. 42)

19. $\begin{array}{r} 4{,}000 \\ -2{,}995 \\ \hline \end{array}$ (p. 42)

20. Find the value of $(36 - 12) + 14$. (p. 54)

21. TEST PREP Choose the equation that best fits these words: Nine eggs plus some more eggs are 12 eggs. (p. 66)

A $9 - n = 12$

C $9 + n = 12$

B $12 + 9 = n$

D $12 + n = 9$

EXTRA PRACTICE page H36, Set D

Problem Solving Strategy
Make a Graph

PROBLEM The table shows data for professional basketball players. A record for the most points scored by an individual in a playoff game, 63, was set in 1986. Between which two years did the number of points scored decrease the most?

UNDERSTAND

- What are you asked to find?
- What information will you use?
- Is there any information you will not use? If so, what?

PLAN

- What strategy can you use to solve the problem?

  You can *make a graph* to help you see the information easily.

SOLVE

- What graph or plot can you make to help you solve the problem?

  Make a bar graph. Then compare the lengths of the bars.

  Look at the graph from bottom to top. The greatest decrease in the length of two side-by-side bars is between 1997 and 1998.

  So, the number of points decreased the most between 1997 and 1998.

CHECK

- How can you check to see whether your graph is correct?

MOST POINTS SCORED IN A PLAYOFF GAME

Year	Points
1993	55
1994	56
1995	48
1996	46
1997	55
1998	45
1999	37

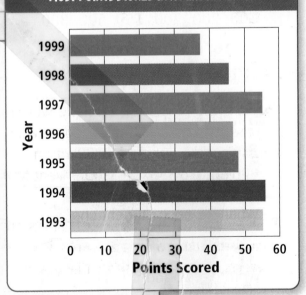

MOST POINTS SCORED IN A PLAYOFF GAME

Problem Solving Practice

Make a graph or a plot of the data. Then solve.

1. The ages of the 12 children in Mrs. Kling's fitness class are 10, 11, 10, 14, 12, 11, 12, 13, 12, 10, 11, and 12. Find the mode.

2. The numbers of pets owned by Mr. Hier's students are 0, 1, 2, 2, 0, 0, 0, 1, 3, 5, 3, 1, 1, 1, 2, 0, 1, 0, and 2. Find the median.

3. The number of minutes Sierra exercised for 11 days are 22, 23, 21, 19, 25, 21, 20, 24, 18, 21, and 22. Find the mode.

For 4–5, use the table.

4. How many trees were planted by Tate School?

   A 20   **C** 24
   B 23   **D** 25

5. How many more trees did Tate School plant in 2000 than in 1998?

   F 2   **H** 4
   G 3   **J** 5

PROBLEM SOLVING STRATEGIES

Draw a Diagram or Picture
Make a Model or Act It Out
Make an Organized List
Find a Pattern
▶ Make a Table or Graph
Predict and Test
Work Backward
Solve a Simpler Problem
Write an Equation
Use Logical Reasoning

TREES PLANTED BY TATE SCHOOL

Year	Number of Trees
1998	ⵘⵘ |||
1999	ⵘⵘ	
2000	ⵘⵘ ⵘⵘ

Mixed Strategy Practice

6. Henri found that 986 people went to a movie on Friday, 1,453 people went on Saturday, and 1,622 went on Sunday. How many people went to the movie in all?

7. Emma surveyed 60 students. She found that 28 like soccer, 17 like basketball, and 15 like hockey. Make a bar graph. Find the mode.

8. There are 57 shirts. Twenty shirts are blue and the others are yellow. Write an equation using the variable y to show this.

9. Ben has three times as many marbles as Jon. Together they have between 18 and 25 marbles. How many marbles might Jon have?

10. Carmen bought a shirt for $14.99. She bought a sweater for $25.39. Estimate to the nearest dollar the amount that Carmen spent.

11. The product of two numbers is 36. Their sum is 13. What are the numbers? What is their difference?

Problem Solving Strategy

Review/Test

✓ CHECK VOCABULARY AND CONCEPTS

Choose the best term from the box.

> stem-and-leaf plot
> line plot
> survey

1. You are taking a __?__ when you ask several people the same questions and record their answers. (p. 82)

2. A __?__ shows groups of data arranged by place value. (p. 90)

✓ CHECK SKILLS

For 3–4, use the frequency table. (pp. 82–85)

3. Copy and complete the cumulative frequency table at the right. How many cans were collected in all?

4. How many more cans were collected in Weeks 1 and 2 than in Week 4?

CANS OF FOOD COLLECTED		
Week	Frequency	Cumulative Frequency
1	10	▨
2	12	▨
3	14	▨
4	18	▨

For 5–7, use the table. (pp. 86–87, 88–89)

5. Make a line plot of Laura's test scores.

6. What is the median of Laura's test scores? What is the mode?

7. Find the range of Laura's test scores.

LAURA'S TEST SCORES							
Test	1	2	3	4	5	6	7
Score	86	90	95	84	94	86	92

✓ CHECK PROBLEM SOLVING

For 8–10, use the table. (pp. 92–93, 94–95)

8. The table shows the number of games won by a hockey team from 1996–1999. Use the data in the table to make a bar graph. Decide on a scale and interval to use, and label your graph.

9. Between which two years did the number of games won increase the most?

10. How many games did the Sharks win in all from 1996 to 1999?

SHARKS' WIN RECORD	
Year	Games Won
1996	35
1997	40
1998	30
1999	44

Standardized Test Prep

Eliminate choices.
See item **4**.

Think about the meaning of *mode*. Each list of scores includes the number 85, but only one has a mode of 85.

Also see problem **5**, p. H64.

For 1–6, choose the best answer.

1. Which expression has a value of 18?
   A $22 - (6 + 2)$
   B $22 + (6 - 2)$
   C $22 - (6 - 2)$
   D $22 + (6 + 2)$

2. $719,776$
   $- 482,143$

   F 237,019   **H** 273,919
   G 237,633   **J** NOT HERE

3. Jim wrote the ages of all his cousins.

   Ages: 2, 3, 11, 11, 12, 13, 13, 25, 26

   What is the median age of his cousins?
   A 11   **C** 13
   B 12   **D** 25

4. Sarah made a list of her last 7 test scores. The mode is 85. Which could be Sarah's test scores?
   F 100, 100, 90, 85, 70, 70, 65
   G 100, 95, 90, 85, 80, 75, 70
   H 90, 90, 90, 85, 80, 80, 80
   J 100, 100, 90, 90, 85, 85, 85

5. How many cans were collected by the end of week 2?

	CANS OF FOOD COLLECTED BY JUSTIN'S CLASS	
Week	Number of Cans	Cumulative Frequency
1	56	56
2	23	79
3	72	151

   A 56   **B** 43   **C** 72   **D** 79

   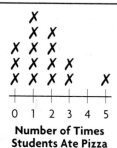

   Number of Times Students Ate Pizza

6. How many students ate pizza more than two times?
   F 8   **G** 5   **H** 3   **J** 1

Write What You Know

7. Use the table. In which week was the 100th can collected? Explain how you can tell.

8. Use the line plot. How many times did students eat pizza in all? Tell how you know.

Analyze and Graph Data

Steam locomotive

The first United States railroad company, the Baltimore & Ohio Railroad, was founded in 1827. Since then, railroads have been built across the United States. The line graph below shows how the number of miles of railroad line in the U.S. has changed over time.

PROBLEM SOLVING
During which 20-year period was the increase in miles of railroad line the greatest?

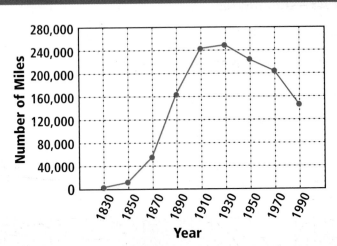

RAILROAD LINE IN THE UNITED STATES

CHECK WHAT YOU KNOW

Use this page to help you review and remember
important skills needed for Chapter 6.

✔ **PARTS OF A GRAPH** (For Intervention, *see* p. H9.)

For 1–3, use the first bar graph.

1. What is the title of this graph?

2. What label would you place at the bottom?

3. What label would you place at the left side?

✔ **READ BAR GRAPHS** (For Intervention, *see* p. H9.)

For 4–7, use the second bar graph.

4. Which way of going to school is used by the most students?

5. How many students ride to school in a car?

6. How many students were surveyed for this bar graph?

7. How many students go to school by car or by bicycle?

✔ **IDENTIFY POINTS ON A GRID**

(For Intervention, *see* p. H10.)

For 8–15, use the map. Write the ordered pair for each animal.

8. camel   9. tiger   10. monkey

11. bear   12. elephant   13. horse

Name the animal.

14. Start at the monkey. Go right 5. Then go down 2.

15. Start at the camel. Go left 1. Then go up 2.

FAVORITE PLACES TO VISIT

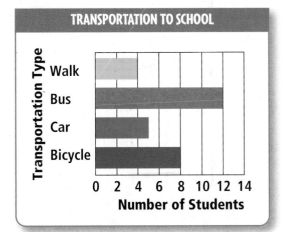

TRANSPORTATION TO SCHOOL

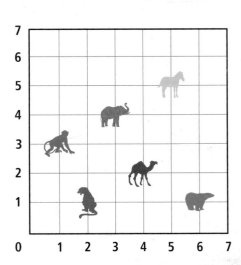

Double-Bar Graphs

Quick Review

Find the value of the expression for $n = 17$.

1. $21 - n$ **2.** $72 - n$

3. $n - 9$ **4.** $n + 15$

5. $n + 27$

VOCABULARY

double-bar graph

▶ Explore

Double-bar graphs are used to compare similar kinds of data. Make a double-bar graph that shows the data from the table.

Activity

MATERIALS: bar-graph pattern, two different-colored crayons

STEP 1

Decide on a title, labels, and a scale for the graph. For these data, use a scale of 0–16 with an interval of 4.

MONTHLY SNOWFALL (in inches)			
City	**Jan**	**Feb**	**Mar**
Chicago	11	8	7
Cleveland	12	12	10

STEP 2

Make the graph. Use one color for Chicago and another color for Cleveland. Make a key to show which color stands for each city.

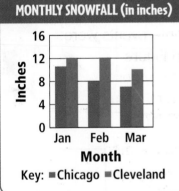

• What does the graph show about the monthly snowfall in Chicago and Cleveland?

What scale and interval should I use for my Winter Activity graph?

Try It

• Use the table to make a double-bar graph comparing the data for boys and girls.

FAVORITE WINTER ACTIVITY		
Activity	**Boys**	**Girls**
Sledding	77	60
Ice-skating	35	78
Skiing	75	70

▶ Connect

The data from the "Favorite Winter Activity" table are shown in two bar graphs. The same labels, intervals, and scales are used for both graphs. A key is not needed when the data are graphed separately.

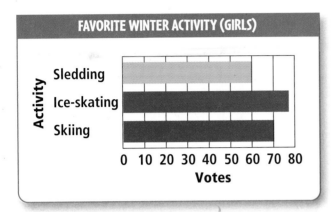

FAVORITE WINTER ACTIVITY (GIRLS)

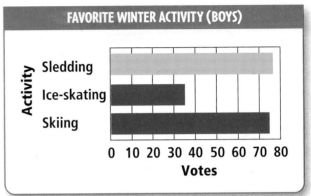

FAVORITE WINTER ACTIVITY (BOYS)

▶ Practice and Problem Solving

For 1–4, use the table.

1. Make a double-bar graph to compare the data for the two classes.

2. What scale and interval did you use? Explain your choices.

3. **REASONING** The fifth-grade class's favorite events were: Bobsledding–12, Ski Jumping–16, and Figure Skating–11. Make a triple-bar graph using these data and the table above. What is the total number of third, fourth, and fifth graders surveyed?

4. Find the mode for the third graders and the mode for the fourth graders.

FAVORITE WINTER OLYMPIC EVENT		
Event	Third Graders	Fourth Graders
Bobsledding	17	15
Ski Jumping	14	16
Figure Skating	9	10

Technology Link

More Practice: Use E-Lab, *Exploring Double-Bar Graphs.*

www.harcourtschool.com/elab2002

Mixed Review and Test Prep

5. Which is greater, 314,689 or 341,869? (p. 18)

6. Round 415,906 to the nearest ten thousand. (p. 26)

7. What is the value of 6 in the number 48,602,751? (p. 8)

8. What is 13,847 subtracted from 23,005? (p. 44)

9. **TEST PREP** What is the sum of 415,903 and 58,769? (p. 44)
   - **A** 463,662
   - **B** 464,572
   - **C** 474,672
   - **D** 474,772

Read Line Graphs

Quick Review

1. 17 + (15 − 12)

2. 45 − (18 + 12)

3. (25 + 25) − 10

4. (23 − 7) + 16

5. 76 − (48 + 17)

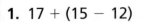

▶ **Learn**

READ BETWEEN THE LINES You can show how data change over a period of time by using a **line graph**.

Example Look at the graph. The line connecting the points shows the changes in the normal temperatures. What is the normal temperature for March?

STEP 1

Find the line labeled March. Follow that line up to the point (•).

STEP 2

Move left to the scale to locate the temperature for March.

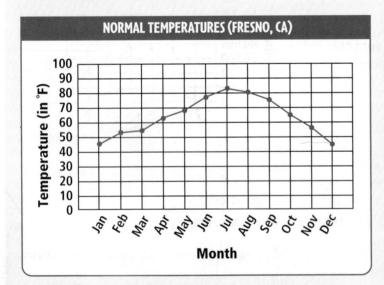

NORMAL TEMPERATURES (FRESNO, CA)

Temperature (in °F) vs. Month

The point for March is about in the middle of 50°F and 60°F. So, the March temperature is about 55°F.

REASONING **What if** the normal temperature for July were 93°F? Explain how to plot the point for this temperature in the line graph.

▶ Check

For 1–4, use the line graph at the right.

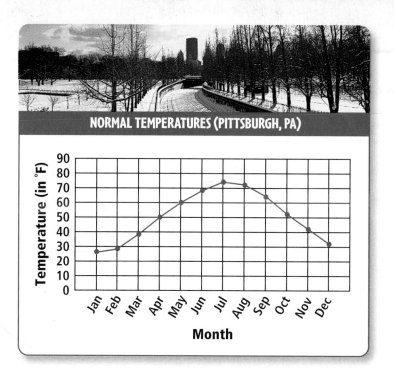

NORMAL TEMPERATURES (PITTSBURGH, PA)

1. **Name** the two months on the line graph in which the normal temperature increases from 39°F to 50°F.

2. What is the lowest normal temperature for Pittsburgh?

3. What month has the highest normal temperature?

4. Use the graph on page 102 to compare the January and February temperatures for Fresno and Pittsburgh.

▶ Practice and Problem Solving

For 5–7, use the line graph at the right.

5. How many people visited the museum in the year with the least attendance?

6. The museum opened in 1993 with 15,000 visitors. About how many people have visited the museum since 1993?

7. Between which two years did the attendance stay about the same?

8. ✎ **Write a problem** using the temperature data for Pittsburgh, PA.

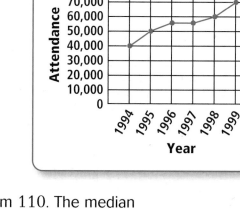

MUSEUM VISITORS

9. **REASONING** A set of 3 numbers has the sum 110. The median of the numbers is 40. The range is 30. What are the numbers?

Mixed Review and Test Prep

10. 35,842 − 18,764 = ■ (p. 44)

11. Which is greater, 46,731 or 46,781? (p. 18)

12. Find the mode of 2, 2, 3, 4, 6, 6, 7, 7, 9, 11, 11, 11. (p. 86)

13. Find the median of 2, 4, 7, 6, 9, 8, 10, 11, 5, 4, 6, 9, 9. (p. 86)

14. **TEST PREP** Find the sum 37,870 + 29,602. (p. 44)

   A 56,472   C 67,472

   B 57,472   D 67,572

EXTRA PRACTICE page H37, Set A

LESSON 3

Make Line Graphs

Quick Review
1. 200 − 34   2. 100 − 53
3. 400 − 47   4. 500 − 68
5. 300 − 81

VOCABULARY
trends

▶ Explore

Use the table for The Railroad Museum of Pennsylvania to make a line graph.

Activity
MATERIALS: line graph pattern

STEP 1

Choose a scale and an interval. Write the scale numbers along the left side of the graph. Write the labels and title on the graph.

STEP 2

Plot a point to show each year's attendance. Connect the points from left to right using a line.

THE RAILROAD MUSEUM OF PENNSYLVANIA			
Year	Attendance	Year	Attendance
1995	142,000	1997	134,000
1996	150,000	1998	135,000

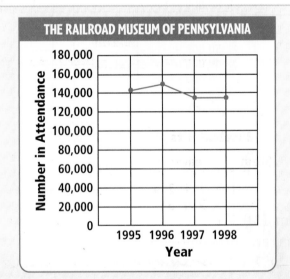

I'm starting my line graph of train passengers. What scale and interval should I use?

Try It

• Make a line graph for the train data. Be sure to title and label your graph. Choose an appropriate interval and scale.

TRAIN PASSENGERS					
Day	Mon	Tues	Wed	Thu	Fri
Number	69	46	52	85	120

▶ Connect

In the Model Railroad Club Membership line graph, you can see **trends**, or areas where the data increase, decrease, or stay the same over time. An increase in membership is seen between 1999 and 2000.

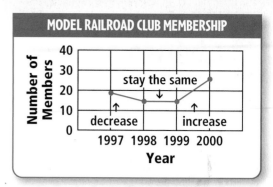

▶ Practice and Problem Solving

For 1–2, make a line graph.

1.

TIME SPENT ON HOMEWORK					
Day	Mon	Tue	Wed	Thu	Fri
Minutes	25	55	50	30	30

2.

PLAYERS IN LEAGUE				
Year	1997	1998	1999	2000
Players	150	225	300	375

For 3–5, use the line graph.

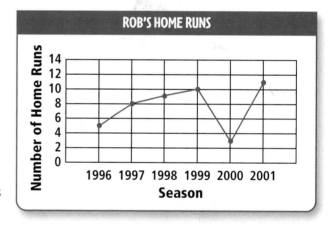

3. How many more home runs did Rob hit during 1999–2001 than during 1996–1998? How many home runs did Rob hit during the six seasons?

4. During which years did Rob hit more than 8 home runs?

5. ✎ **Write About It** Describe any trends in Rob's home run data. Between which two years did the number of home runs hit by Rob increase the most?

6. Find the mode of the data in the stem-and-leaf plot. How many players had fewer than 23 hits in the season?

Baseball Hits

Stem	Leaves
1	1 1 1 2 4
2	1 2 4 5
3	1

Mixed Review and Test Prep

7. Which is greater, 1,238 or 1,543 − 308? (p. 40)

8. Round 744,478 to the nearest thousand. (p. 26)

9. 4,951 − 3,729 (p. 44)

10. Find 100,000 more than 2,843,715. (p. 8)

11. **TEST PREP** Which expression shows 5 less than a number, *x*? (p. 54)

   A 5 − *x*   **C** *x* + 5

   B *x* − 5   **D** *x* − 5 − *x*

Chapter 6 **105**

Choose an Appropriate Graph

▶ **Learn**

NATURE NAP Some animals sleep most of the day, and others hardly sleep at all. Look at how Cindy, Joe, and Elayna showed the data in the table. Which graph or plot works best for the data?

Quick Review

Find the missing number.

1. $7 + 10 = \blacksquare + 5$

2. $13 + 27 = 50 - \blacksquare$

3. $8 + 9 = \blacksquare - 4$

4. $\blacksquare - 9 = 11 + 7$

5. $23 - \blacksquare = 16$

HOURS OF SLEEP	
Animal	**Hours of Sleep in a Day**
Giraffe	4
Raccoon	13
Squirrel	14
Bear	8

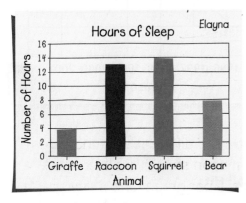

A bar graph is used to compare data about different groups.

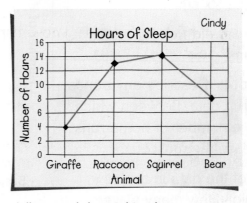

A line graph is used to show changes over time.

Joe

Hours of Sleep

Stem	Leaves
0	4 8
1	3 4

A stem-and-leaf plot arranges data by place value.

The data show the number of hours different groups of animals sleep each day. The data do not show changes over time, and some important information is not shown by the stem-and-leaf plot. So, Elayna's bar graph is best for displaying the data.

- Which graph is best for displaying the number of hours you sleep each night for a week? Explain.

Line Plot or Line Graph

The graph you choose sometimes depends on the type of information that you want to show.

Elsa recorded her science test scores each week for nine weeks. She displayed the data in two different ways.

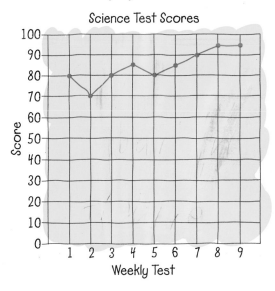

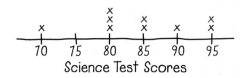

Elsa's line plot shows all the scores, but the line graph shows the week each test was taken.

- **REASONING** Which display is best for finding how many times Elsa got an 80 on a science test? Explain.

Check

1. **Choose** which display Elsa should use to show how her science scores have improved.

2. **Describe** the differences in the types of information shown in a stem-and-leaf plot, a bar graph, and a line graph.

For 3–6, write the kind of graph or plot you would choose.

3. to compare the speeds of five different animals

4. to show the total rainfall each day for a week

5. to show the weekly math grades of your classmates

6. to show your height each year since birth

7. Lynnette found that there were 36 cars, 4 vans, 6 bikes, 1 bus, and 3 trucks on a street. Would a bar graph or a line graph be a better choice for Lynnette's data?

8. The temperature was recorded each hour from 6 A.M. to 6 P.M. Would a bar graph or a line graph be a better choice to show the data?

9. **Write About It** Explain which display would be easier to use to find the median and mode of Elsa's scores.

Practice and Problem Solving

For 10–17, write the kind of graph or plot you would choose.

10. to compare the favorite subjects of students in two fourth-grade classes

11. to show the number of soccer goals scored by all team members

12. to keep a record of plant growth

13. to show monthly temperatures

14. to show the number of pets owned by each student in fourth grade

15. to compare the favorite ice cream flavors of students in two classes

16.

CLASS TEST SCORES				
79	83	91	88	94
96	85	77	81	92

17.

NUMBER OF NEW STUDENTS	Sep	Oct	Nov	Dec
1999	31	11	10	6
2000	18	22	2	10

For 18–21, use the graph and the plot.

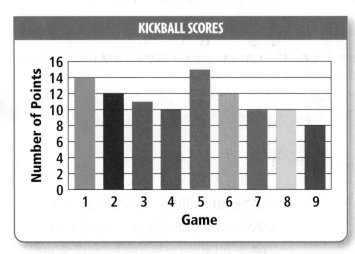

18. Which display shows how many points the team scored in the third game?

19. In which game did the team score more points than in the first game? How many more points were scored?

20. What is the median of the team's scores? What is the mode?

21. In how many games did the team score more than 10 points?

22. **?** **What's the Error?** To look for trends, Frank made a line plot of the number of hours he slept each night for a week. Describe his error. Write the correct answer.

23. **Write About It** Write a question and survey 20 students. Choose and make an appropriate display for the data. Tell why you chose that kind of display.

24. A graph showed increases in the cost of postage stamps over the past 30 years. What kind of graph do you think it was? Explain.

26. NUMBER SENSE The least two-digit number that rounds to 100 is 50. What is the least three-digit number that rounds to 1,000? What is the greatest four-digit number that rounds to 1,000?

25. A graph compares the languages spoken by students in two fourth-grade classes. What kind of graph do you think it was? Explain.

27. A total of 2,615 concert tickets were sold Monday–Wednesday. If 543 tickets were sold Monday and 876 tickets were sold Tuesday, how many more tickets were sold on Wednesday than Tuesday?

Mixed Review and Test Prep

28. 16,083 (p. 44)
   + 8,564

29. 60,004 (p. 42)
   −46,937

30. 34,985 − 12,607 (p. 44)

31. Write an expression using the variable y to show 25 beads minus some beads. (p. 60)

32. TEST PREP Which is greater than 11,463 and less than 11,600? (p. 20)

   A 11,375   **C** 11,552
   B 11,459   **D** 11,673

33. TEST PREP Cheri's class has 5 cages with 3 hamsters in each cage. How many hamsters does her class have?

   F 8   **G** 9   **H** 12   **J** 15

PROBLEM SOLVING

Thinker's Corner

A scientist recorded notes about the length of animal tails. Help the scientist make inferences and predictions.

When you make an **inference,** you draw conclusions based on information you have.

When you make a **prediction,** you guess what might happen based on information you have.

1. Use the notes to make inferences and to complete the table.

2. Predict what the length of a baby African elephant's tail will be when the elephant is full-grown.

3. Tell why you made the prediction you did.

Notes

A. The tails of the red kangaroo, giraffe, and African buffalo are the same length.

B. The Asian elephant's tail is eight inches longer than the African elephant's tail.

C. The leopard's tail is four inches shorter than the Asian elephant's tail.

LENGTH OF MAMMAL TAILS	
Mammal	**Length (in inches)**
?	51
?	43
?	59
?	43
?	43
?	55

EXTRA PRACTICE page H37, Set B

Problem Solving Skill
Draw Conclusions

UNDERSTAND ⟩ PLAN ⟩ SOLVE ⟩ CHECK ⟩

RAINY SEASON Bangladesh is located in Asia on the Indian Ocean. Sometimes the weather is very dry, and sometimes there are heavy rains. During the rainy—or monsoon—season, people often use waterways instead of roads for transportation.

You can use a graph to compare data. Use the data and what you know to answer questions and draw conclusions.

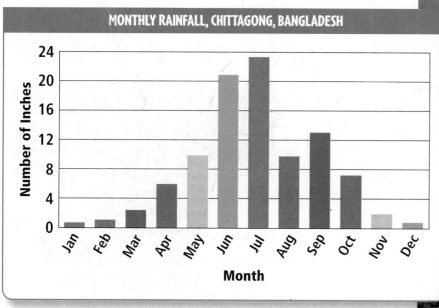

MONTHLY RAINFALL, CHITTAGONG, BANGLADESH

Number of Inches / Month

Which months of the year do you think make up the monsoon season?

Monsoon season is a time of very heavy rains. From April to October, Bangladesh usually receives over 5 inches of rain each month. During June and July, monthly rainfall can be greater than 20 inches.

So, June and July probably make up the monsoon season in Bangladesh.

• Estimate the yearly rainfall in Chittagong.

For 1–4, use the graph on page 110.

1. In which two months does Bangladesh receive the most rainfall?

2. In which two months does Bangladesh receive the least rainfall?

3. During which months are the roads in Bangladesh likely to be flooded?

4. Make a double-bar graph to compare the monthly rainfall in Chittagong with the monthly rainfall in your town.

For 5–6, use the graph.

Jennifer made a line graph to show the snowfall from November to February.

5. Which conclusion can you NOT make about the data?

   A More snow fell in January than in February.

   B The snowfall decreased from November to February.

   C Six more inches of snow fell in January than in December.

   D Eighteen inches of snow fell in December.

6. How many total inches of snow fell?

   F 64 inches

   G 65 inches

   H 66 inches

   J 69 inches

MONTHLY SNOWFALL

Problem Solving Skill

Mixed Applications

7. Mrs. Porter wants to buy one sticker for each of her 32 students. If stickers come in packs of 10, how many packs of stickers will she need to buy? Explain.

8. **REASONING** Write an expression with a value of 893, using only addition and subtraction, but do not use 3, 8, or 9 as digits.

9. Mr. Milo is taking a trip of 731 miles. He travels 458 miles on Saturday and completes the trip on Sunday. How many fewer miles did he travel on Sunday?

10. **? What's the Question?** Look at the Monthly Snowfall line graph. The answer is 18.

Review/Test

✓ CHECK VOCABULARY AND CONCEPTS

Choose the best term from the box.

1. A graph that uses a line to show how something changes over a period of time is a _?_. (p. 102)

2. A graph used to compare similar kinds of data is called a _?_. (p. 100)

> double-bar graph
> line graph
> stem-and-leaf plot

✓ CHECK SKILLS

For 3–4, use the line graph. (pp. 102–105)

3. The line graph shows how many push-ups Anna did each day for 6 days. On which days did Anna do fewer than 20 push-ups?

4. What is the total number of push-ups Anna did during the six days?

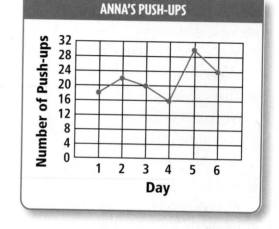

For 5–8, write the kind of graph or plot you would choose. (pp. 106–109)

5. to compare the favorite movies of two classes

6. to record the height of a tree during six months

7. to compare the favorite ice cream flavors of your classmates

8. to record students' scores on a recent history test

✓ CHECK PROBLEM SOLVING

For 9–10, use the Favorite Animal frequency table. (pp. 100–101, 110–111)

9. Make a double-bar graph to compare the data of the first-grade and third-grade classes.

10. Can you conclude that more first-grade students like dogs and horses than third-grade students? Explain.

FAVORITE ANIMAL		
Animal	First-Grade Students	Third-Grade Students
Cat	3	5
Dog	7	9
Fish	2	7
Horse	9	6

Standardized Test Prep

TIP! **Understand the problem.**
See item **3.**

Think about the meaning of *median*.
You will have to do something to the
data set before you can find the median.

Also see problem **1,** p. H62.

For 1–7, choose the best answer.

1. Which is true?
   A 483,207 > 439,718
   B 416,063 > 418,603
   C 425,596 < 420,096
   D 438,706 < 437,806

2. $\begin{array}{r} 32,456 \\ -26,812 \\ \hline \end{array}$

   F 6,644   **H** 5,444
   G 5,644   **J** NOT HERE

This data set shows the number of books
some of Kyle's friends have.

   16, 13, 18, 14, 19, 18, 14, 18, 13

3. What is the median number of books?
   A 7   **B** 16   **C** 18   **D** 19

4. What kind of graph would be best to
   show the population growth of a city
   over several years?
   F double-bar graph
   G stem-and-leaf plot
   H line graph
   J bar graph

For 5–6, use the line graph.

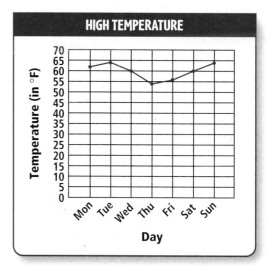

5. Which of the following is true?
   A Monday was cooler than Saturday.
   B Wednesday was cooler than
   Sunday.
   C Friday was warmer than Tuesday.
   D Sunday was cooler than Thursday.

6. What is the interval of the graph?
   F 5   **H** 66
   G 8   **J** 64

7. What is 3,278,458 rounded to the
   nearest ten thousand?
   A 3,300,000
   B 3,280,000
   C 3,278,000
   D 3,270,000

Write What You Know

8. Use the data for Problem 3. Find the
   mode. Then add two numbers to the
   data and tell what the new mode is.
   Explain.

9. Use the line graph. Find the range
   of high temperatures for the week.
   Then choose a day and tell what
   temperature for that day would make
   a greater range. Explain.

CHAPTER 7

Understand Time

The clock tower on the Country Club Plaza in Kansas City is an example of Spanish architecture in the city. In fact, the Country Club Plaza was modeled after marketplaces in Spain. The Plaza is the oldest shopping center in the United States. About 10 million tourists visit the Country Club Plaza every year.

PROBLEM SOLVING
The graph shows flight times from various cities to Kansas City. If a plane leaves Chicago at 9:55 A.M., at what time will it arrive in Kansas City?

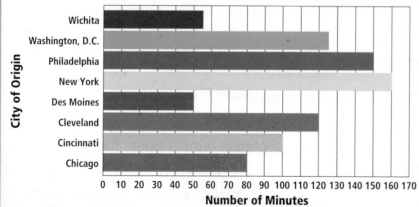

FLIGHT TIMES TO KANSAS CITY

City of Origin: Wichita, Washington, D.C., Philadelphia, New York, Des Moines, Cleveland, Cincinnati, Chicago

Number of Minutes: 0 10 20 30 40 50 60 70 80 90 100 110 120 130 140 150 160 170

Use this page to help you review and remember important skills needed for Chapter 7.

✓ TIME TO THE HALF AND QUARTER HOUR (For Intervention, see p. H10.)

Write the time.

1.

2.

3.

4.

Write the time as it would look on a digital clock.

5. three thirty

6. quarter after one

7. 15 minutes before ten

✓ TIME TO THE MINUTE (For Intervention, see p. H11.)

Write the time.

8.

9.

10.

11.

Write the time as it would look on a digital clock.

12. 23 minutes after ten

13. two minutes before one

14. 17 minutes before eight

✓ USE A CALENDAR (For Intervention, see p. H11.)

For 15–18, use the calendar.

15. What day of the week is November 14?

16. What is the date of the third Tuesday in November?

NOVEMBER						
Sun	Mon	Tue	Wed	Thu	Fri	Sat
					1	2
3	4	5	6	7	8	9
10	11	12	13	14	15	16
17	18	19	20	21	22	23
24	25	26	27	28	29	30

17. If you circled all of the dates for Saturdays and Sundays in November, how many dates would be circled?

18. If soccer practices are on Mondays and Wednesdays, how many soccer practices will there be in November?

Before and After the Hour

▶ **Learn**

ON TIME? This morning, Aaron's family will take a train to see his aunt. The train leaves at eleven o'clock. Aaron's family is in the station waiting room. The time right now is shown on the clock. Has Aaron's family missed the train?

The time on the clock is 10:48, or 12 minutes before 11:00. So, Aaron's family has not missed the train.

Examples

Ⓐ

Read or write this time as:

• 9:50
• 50 minutes after nine
• 10 minutes before ten

Ⓑ

Read or write this time as:

• 2:24 and 16 seconds
• 24 minutes 16 seconds after two
• 44 seconds before 2:25

Use what you know about units of time to estimate how long an activity will last.

• What activities can you do in about 1 second? in about one minute?

UNITS OF TIME
60 **seconds (sec)** = 1 minute (min)
60 minutes = 1 hour (hr)
24 hours = 1 day

▶ **Check**

1. Find the number of seconds in 2 minutes.

Write the time as shown on a digital clock.

2. 38 minutes after six   **3.** 16 minutes before two   **4.** 19 minutes after four

Write the time as shown on a digital clock.

5. 17 minutes after four   **6.** 10 minutes before five   **7.** 13 minutes before seven

Write the time shown on the clock in 2 different ways.

8.

9.

10.

11.

12.

Write the letter of the unit used to measure the time.
Use each answer only once.

> **a.** days   **b.** hours   **c.** minutes   **d.** seconds

13. to blink your eyes

14. to get dressed for school

15. from sunrise to sunset

16. go to summer camp

17. REASONING Mark spends 3 hours doing chores and eating meals. He wants to visit friends for 4 hours, shop for 2 hours, read for 3 hours, and sleep for 10 hours. Will Mark be able to do everything in one day? Explain.

18. Normally, a person blinks 1 time about every 6 seconds. About how many times does a person blink in 30 seconds? About how many times does a person blink in 1 minute?

Mixed Review and Test Prep

19. Write three hundred sixty-five thousand, eight hundred forty-two in standard form. (p. 6)

Find the missing number. (p. 66)

20. $16 + \blacksquare = (11 + 5)$

21. $14 + \blacksquare = 7 + 7 + 6$

22. Lowell School has 513 students. There are 284 girls. How many fewer boys than girls are there?

23. TEST PREP Wendy's bank has 7 quarters, 11 dimes, 16 nickels, and 37 pennies in it. How much is in her bank?
A $3.57 **B** $3.77 **C** $4.02 **D** $4.27

A.M. and P.M.

Quick Review

Name the time that is 15 minutes after each time.

1. 12:45 **2.** 4:30 **3.** 1:00

4. 11:15 **5.** 10:45

VOCABULARY

A.M.   P.M.

▶ Learn

AROUND THE CLOCK   Pete is competing in a school Geography Bee at 7:00 P.M. How do you know if the competition is in the morning or evening?

A.M. and P.M. name the part of day. **A.M.** means "before noon." **P.M.** means "after noon." The hour hand on an analog clock goes around once for the A.M. hours and once for the P.M. hours.

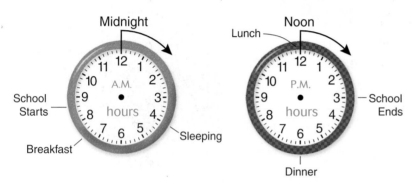

Think: The hours between midnight and noon are A.M. hours. The hours between noon and midnight are P.M. hours.

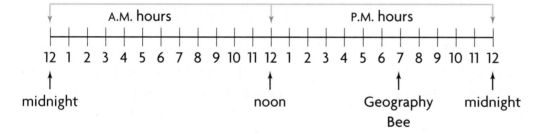

So, the Geography Bee begins at 7:00 in the evening.

▶ Check

1. **Explain** how you can remember if the time is A.M. or P.M. on a given day.

Write the time, using A.M. or P.M.

2. when you eat dinner   3. when school starts   4. when the sun rises

Write the time, using A.M. or P.M.

5. when you wake up

6. when you go to bed

7. when the sun sets

8. when you get home from school

9. when you eat breakfast

10. when the library opens

Write A.M. or P.M.

11. Kyle has soccer practice from 4:00 _?_ to 6:15 _?_.

12. Angela eats dinner at 6:15 _?_.

13. John went to work at 12:20 in the afternoon.

14. Heather went for a jog at 5:30 in the morning.

USE DATA For 15–18, use the information below.

On Saturday morning, Bill bought Penny's gift at the mall. At 12:15 P.M. he bought flowers for his mom. Bill took the gift to Penny's house at 2:30 P.M. The mall opened at 10:00 A.M.

15. Did the mall open before noon or after noon?

16. Did Bill buy Penny's gift before going to the flower shop? Explain.

17. Did Bill buy the flowers after going to Penny's house? Explain.

18. If Bill ate breakfast at 9:30 A.M., could he have eaten breakfast at the mall? Why or why not?

19. Jessica has 8 quarters, 3 dimes, 4 nickels, and 2 pennies. Round the amount of money to the nearest dollar.

20. 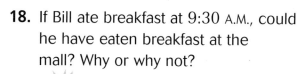 **What's the Error?** Glenn is given a note telling him he has an after-school meeting at 3:00 A.M. Describe the error. Write the correct answer.

Mixed Review and Test Prep

Order the numbers from *greatest* to *least*. (p. 20)

21. 56,397,456; 5,907,654; 56,937,465

22. 8,492,056; 4,823,904; 8,942,341

23. Round seven hundred forty-five thousand, three hundred to the nearest ten thousand. (p. 26)

24. What is four hundred thirteen million, eight hundred thousand, forty-nine written in standard form? (p. 8)

25. **TEST PREP** Which number is less than 13,432? (p. 18)

A 13,423   **C** 13,623

B 13,501   **D** 13,732

▶ Learn

START TO FINISH   Trail Elementary School is having a Cultural Fair. Use the schedule to find how long the fair will last.

Elapsed time is the time that passes from the start to the end of an activity. You can use a clock to count forward from the starting time to the ending time.

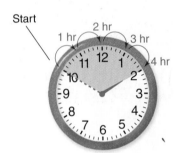

Think: From 10:00 A.M. to 2:00 P.M., 4 full hours have passed.

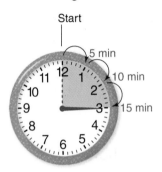

Think: From 2:00 P.M. to 2:15 P.M., 15 minutes have passed.

So, the fair will last 4 hr 15 min.

Quick Review

Name the time that is ten minutes later.

1. 2:45 P.M.   **2.** 4:20 P.M.

3. 11:30 A.M.   **4.** 7:05 A.M.

5. 8:55 A.M.

VOCABULARY
elapsed time

CULTURAL FAIR

Parade
10:00 A.M.– 10:45 A.M.

Slide Show
10:45 A.M.– 11:30 A.M.

Food Tasting
11:30 A.M.– 12:30 P.M.

Show and Tell
12:30 P.M. – 1:45 P.M.

Dance Show
1:45 P.M. – 2:15 P.M.

Example 1

A karate presentation has been added at 2:30 P.M. If it lasts 35 minutes, at what time will it end?

One Way

Use addition.

$$
\begin{array}{r}
2 \text{ hr } 30 \text{ min} \\
+ \quad\quad 35 \text{ min} \\
\hline
2 \text{ hr } 65 \text{ min} \\
3 \text{ hr } \ 5 \text{ min}
\end{array}
$$

Think: 60 min = 1 hr
2 hr + 1 hr + 5 min

Another Way

Count forward on a clock.

Both ways show that the karate presentation will end at 3:05 P.M.

Example 2

The principal wants to move the dance show to 10:15 A.M. The parade must end by 10:10 A.M. The parade will last 45 minutes. Find the start time of the parade.

One Way

Use subtraction.

Think: 1 hr = 60 min.
Rename 10 hr 10 min
as 9 hr 70 min.
60 min + 10 min = 70 min

$$
\begin{array}{r}
9 \quad 70 \\
\cancel{10}\ \text{hr}\ \cancel{10}\ \text{min} \\
-\qquad 45\ \text{min} \\
\hline
9\ \text{hr}\ 25\ \text{min}
\end{array}
$$

Another Way

Count backward on a clock.

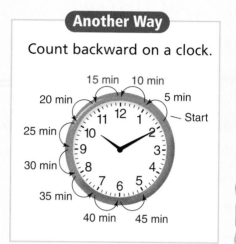

The parade should start at 9:25 A.M.

- Rana's mom can attend only the slide show and the food tasting. Will she be at school for more or less than 2 hours? Explain.

Technology Link

More Practice: Use E-Lab, *Elapsed Time on a Clock*

www.harcourtschool.com/ elab2002

▶ Check

1. **Find** the elapsed time from the beginning of the slide show to the end of show and tell at the Cultural Fair. Use the data on page 120.

Find the elapsed time.

2. **start:** 9:30 A.M.
   end: 4:50 P.M.

3. **start:** 11:10 A.M.
   end: 2:45 P.M.

4. **start:** 1:20 P.M.
   end: 9:05 P.M.

USE DATA For 5–6, use the movie schedule.

5. At about what time does each movie end?

6. Can you see all the movies in one day? Why or why not?

7. The movie *In a Flash* ends at 9:10 P.M. What time does it begin?

Each movie is 90 min.

MOVIE SCHEDULE

Storm at Sea	1:15 P.M.
Eagle's Flight	2:35 P.M.
Mountain Journey	4:30 P.M.
Race to the Finish	6:05 P.M.

REEL #1

Find the elapsed time.

8. start: 7:35 A.M.
   end: 2:15 P.M.

9. start: 9:10 A.M.
   end: 11:53 A.M.

10. start: 7:45 P.M.
    end: 1:10 A.M.

11. start: 11:35 A.M.
    end: 3:10 P.M.

12. start: 6:40 A.M.
    end: 10:17 A.M.

13. start: 4:25 P.M.
    end: 12:38 A.M.

Copy and complete the tables.

	START TIME	END TIME	ELAPSED TIME
14.	10:50 A.M.	■	3 hr 5 min
16.	■	2:30 P.M.	2 hr 40 min
18.	10:45 A.M.	2:22 P.M.	■

	START TIME	END TIME	ELAPSED TIME
15.	■	9:42 P.M.	4 hr 30 min
17.	8:12 A.M.	■	6 hr 33 min
19.	4:50 P.M.	8:05 P.M.	■

20. ✷ **? What's the Error?** Jim says the elapsed time from 7:35 A.M. to 9:45 P.M. is 2 hr 10 min. Describe his error. Write the correct answer.

21. **REASONING** Horace and his family left on Sunday at 9:10 A.M. to visit friends. They returned on Monday at 2:00 P.M. How long were they gone in hours and minutes?

USE DATA For 22–23, use the flight schedule.

22. **ESTIMATION** Use the table to find about how long each flight lasts. For about how long is Kelly in Atlanta?

23. How many minutes longer is the flight from Atlanta to Boston than the flight from Miami to Atlanta?

KELLY'S FLIGHT SCHEDULE	
Miami, FL, to Atlanta, GA	**Atlanta, GA, to Boston, MA**
LEAVES: 10:30 A.M.	LEAVES: 1:40 P.M.
ARRIVES: 12:18 P.M.	ARRIVES: 4:18 P.M.

24. Look at the table below. The 1896 Olympic marathon distance was 40,000 meters. Since 1924, the distance has been 2,195 meters longer. What is the total distance that Carlos and Josia ran? How much faster were their times than Spiridon's?

OLYMPIC MARATHONS		
Year	**Name**	**Time**
1896	Spiridon Loues	2 hr 58 min 50 sec
1984	Carlos Lopes	2 hr 9 min 21 sec
1996	Josia Thugwane	2 hr 12 min 36 sec

USE DATA For 25–27, use the schedules.

25. At about what time does each flight arrive in New York City?

26. If Ms. Lane needs to be in New York City for a 3:30 P.M. meeting, on which airline can she fly?

27. Mr. Wright arrived in New York at about 4:15 P.M. At what time did he leave Miami?

FLIGHTS FROM MIAMI, FL, TO NEW YORK CITY, NY	
Each flight lasts about 2 hours and 45 minutes.	
Airline	**Departure Time**
Airline A	9:05 A.M.
Airline B	11:10 A.M.
Airline C	1:30 P.M.
Airline D	2:45 P.M.

Mixed Review and Test Prep

Write the value of the blue digit. (pp. 4–11)

28. 45,623

29. 83,238

30. 172,908

31. 2,498,762

32. **TEST PREP** 27 + 35 + 118 (p. 36)

A 160   **B** 170   **C** 179   **D** 180

33. **TEST PREP** Ryan had 32 baseball cards. Ken gave Ryan 18 cards. Ryan gave 5 of those cards to Tran. Choose an expression for the number of cards Ryan has now. (p. 58)

F 32 − (18 − 5)   **H** 32 + (18 + 5)

G 32 + (18 − 5)   **J** 32 − (18 + 5)

PROBLEM SOLVING LiNKUP ... to Reading

STRATEGY • USE GRAPHIC AIDS Graphic aids are tables, time lines, maps, and diagrams that organize information so it is easy to read. Use the weather map to answer the questions.

1. Name the types of information shown on the map.

2. Use the compass rose to find the cities with rain that are east of Chicago.

3. Today is November 21. Eight days ago, Seattle and Minneapolis were 9 degrees warmer than shown. Name the date eight days ago, and give the temperatures in both cities.

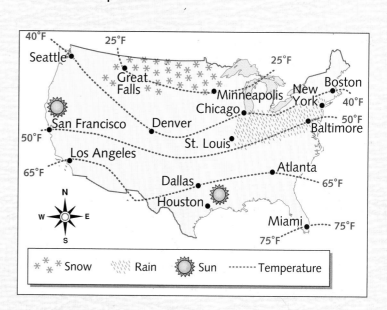

EXTRA PRACTICE page H38, Set C

Problem Solving Skill
Sequence Information

UNDERSTAND ⟩ PLAN ⟩ SOLVE ⟩ CHECK

PLAN YOUR DAY   Sanjay plans to write a report about his visit to a wild animal park. He *must* see both movies and the alligator show. He will be at the park from 8:30 A.M. to 2:30 P.M. and will have lunch from 11:30 to noon. How can Sanjay schedule his day to complete these activities?

One way to sequence information is to arrange the data in order of importance.

To help you use time wisely, things that must be done should be at the top of the list. Things you could do another day should go closer to the bottom.

Things I must do:
- movie about endangered animals
- movie about rainforests
- alligator show
- lunch 11:30 - noon

Things I want to do:
- bird show
- dolphin feeding
- animal behavior

Sanjay completed his schedule as shown.

- In what other way could Sanjay plan his activities? Explain.

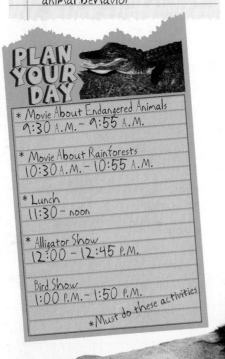

PLAN YOUR DAY

* Movie About Endangered Animals
  9:30 A.M. - 9:55 A.M.

* Movie About Rainforests
  10:30 A.M. - 10:55 A.M.

* Lunch
  11:30 - noon

* Alligator Show
  12:00 - 12:45 P.M.

Bird Show
1:00 P.M. - 1:50 P.M.

*Must do these activities

ANIMAL PARK schedule

Dolphin Feeding (15 min)
9:00 A.M.; 1:30 P.M.

Movie About Endangered
Animals (25 min)
9:30 A.M.; 11:00 A.M.; 1:00 P.M.

Animal Behavior (45 min)
10:00 A.M.; 11:30 A.M.; 2:30 P.M.

Bird Show (50 min)
11:00 A.M.; 1:00 A.M.; 3:30 P.M.

Movie About Rain Forests
(25 min)
10:30 A.M.; 11:00 A.M.; 12:00 noon

Alligator Show (45 min)
10:00 A.M.; 12:00 noon; 2:00 P.M.

For 1–4, use the schedule and the information below.

Evelyn has movie theater gift certificates. She will be at the theater from noon to 5:30 P.M., and really wants to see one new movie.

MOVIE SCHEDULE

Movie	Playing Time
Safari	12:20–2:30; 2:30–4:40
Gentle Journey	12:40–3:10; 2:30–5:00
Lions New!	12:40–3:00; 2:40–5:00
Wild Adventures	11:30–2:30; 2:40–5:40
Raging River New!	12:20–2:40; 2:50–5:10
Whale Wonders	2:50–4:55; 3:20–5:25

1. How would you schedule Evelyn's afternoon?

2. Which movie is longer, *Lions* or *Safari*?

3. Which movies would Evelyn be able to watch?
   A *Raging River, Wild Adventures*
   B *Safari, Wild Adventures*
   C *Safari, Raging River*
   D *Lions, Gentle Journey*

4. Which two movies are the same length?
   F *Safari, Whale Wonders*
   G *Gentle Journey, Lions*
   H *Wild Adventures, Gentle Journey*
   J *Lions, Raging River*

Mixed Applications

USE DATA For 5–6, use the table and the information below.

Amy is planning a surprise party for her friend. She wants the party to last four hours. The table shows the activities she would like to include. The stars indicate activities that must be done.

SURPRISE PARTY

Activity	Time
*Eat lunch and cake	30 min
*Open presents	45 min
*Go swimming	1 hr 30 min
Play charades	1 hr 30 min
Play volleyball	1 hr 15 min

5. Can the party include all of the starred activities? Why or why not?

6. Can the party include all of the activities? How would you plan the surprise party? Make a schedule.

7. Raul bought a tennis racket for $40.50 and 3 cans of tennis balls for $3.00 each. He had $7.37 left. How much money did he have before this purchase?

8. One hundred twelve students try out for the Youth Orchestra. If 42 boys and 34 girls are chosen, how many students are not chosen?

9. 📖 **Write a problem** about elapsed time, using the movie schedule above.

Elapsed Time on a Calendar

Quick Review

What day of the week is 9 days

1. after Monday?

2. before Sunday?

3. after Thursday?

4. before Friday?

5. before Tuesday?

▶ **Learn**

TIME AFTER TIME The school play, *Time Trek,* will be presented on March 7. Rehearsals will begin on February 25 and last for 8 days. There will be no rehearsals on Saturday or Sunday. Find the day and date of the last rehearsal.

To find the elapsed time, count the number of days. Start counting with February 25.

VOCABULARY

century   **decade**

FEBRUARY						
Sun	Mon	Tue	Wed	Thu	Fri	Sat
					1	2
3	4	5	6	7	8	9
10	11	12	13	14	15	16
17	18	19	20	21	22	23
24	25	26	27	28		

MARCH						
Sun	Mon	Tue	Wed	Thu	Fri	Sat
					1	2
3	4	5	6	7	8	9
10	11	12	13	14	15	16
17	18	19	20	21	22	23
24/31	25	26	27	28	29	30

Remember

Units of Time
1 week = 7 days
1 year = 12 months
1 year = about 52 weeks
1 year = 365 days

So, Wednesday, March 6, is the last rehearsal.

• Will the rehearsals last for more than or less than one week? by how many days?

Tito practices his voice exercises every day for 4 weeks. Using the fact 1 week = 7 days, Tito practices his voice exercises for 4 × 7 days, or 28 days.

• **What if** Tito continues to practice his voice exercises for 9 more weeks? How many more days does he practice?

Century or Decade

Ford's Theatre in Washington, D.C. opened in 1863. The theater is 1 century 4 decades 1 year old in 2004. How old, in years, is the theater in 2004?

> 1 **century** = 100 years
>
> 1 **decade** = 10 years

▲ Ford's Theatre, Washington, D.C.

Example

You can multiply to change from a larger unit of time to a smaller unit of time

1 decade is 10 years, so 4 decades is 4 × 10, or 40, years.

$$
\begin{array}{rr}
1 \text{ century} = & 100 \text{ years} \\
4 \text{ decades} = + & 40 \text{ years} \\
\hline
& 140 \text{ years}
\end{array}
$$

1 century 4 decades 1 year is 140 years + 1 year.

So, the theater is 141 years old in 2004.

Technology Link

To learn more about units of time, watch the Harcourt Math Newsroom Video *Leap Day.*

 Check

1. Find the number of decades in a century.

USE DATA For 2–3, use the calendars here and on page 126.

2. Report cards were mailed on March 15. They will be mailed again in 9 weeks. On what date will report cards be mailed?

3. In a leap year, February has 29 days instead of 28. How long is it from February 24 to March 4 in a regular year? in a leap year?

Find the missing number.

4. 5 weeks = ■ days   **5.** 800 years = ■ centuries   **6.** 7 decades = ■ years

7. 2 years = ■ months   **8.** 3 decades = ■ years   **9.** 7 weeks = ■ days

APRIL						
Sun	Mon	Tue	Wed	Thu	Fri	Sat
	1	2	3	4	5	6
7	8	9	10	11	12	13
14	15	16	17	18	19	20
21	22	23	24	25	26	27
28	29	30				

MAY						
Sun	Mon	Tue	Wed	Thu	Fri	Sat
			1	2	3	4
5	6	7	8	9	10	11
12	13	14	15	16	17	18
19	20	21	22	23	24	25
26	27	28	29	30	31	

LESSON CONTINUES

USE DATA For 10–13, use the calendars.

SEPTEMBER						
Sun	Mon	Tue	Wed	Thu	Fri	Sat
1	2	3	4	5	6	7
8	9	10	11	12	13	14
15	16	17	18	19	20	21
22	23	24	25	26	27	28
29	30					

OCTOBER						
Sun	Mon	Tue	Wed	Thu	Fri	Sat
		1	2	3	4	5
6	7	8	9	10	11	12
13	14	15	16	17	18	19
20	21	22	23	24	25	26
27	28	29	30	31		

Come and see
The Old Time Theatre Exhibit
September 2 to October 9

10. The theater exhibit is from September 2 through October 9. About how many weeks will the exhibit last?

11. The exhibit began advertising 2 months before it came to town. In what month did it begin advertising?

12. Cliff bought a ticket 5 weeks before the last day of the exhibit. On what day and date did Cliff buy his ticket?

13. The next exhibit will begin 2 weeks 2 days after October 9. What is the opening day and date of the new exhibit?

Find the missing number.

14. 2 years 3 weeks = about ■ weeks

15. 5 centuries 3 decades = ■ decades

16. 3 centuries 4 decades = ■ years

17. 6 decades 7 years = ■ years

18. About how many years is 4 decades 52 weeks?

19. What year was 1 century 2 decades before 1913?

20. What year was 2 centuries 3 decades after 1728?

21. What year was 1 century 4 decades 3 years after 1789?

22. Before his birthday, Carl had some CDs. He was given 5 more on his birthday. Then he had 27 CDs. How many CDs did he have before his birthday?

23. Joel has read 134 pages in his whale book. He has 153 more pages to read. Write a number sentence using *n* to show how many pages are in the book.

24. 📓 **Write About It** Explain how to find two dates. The first is 10 days after October 6. The second is 2 weeks after the first. What are the two dates?

25. Liz left on a trip May 23 and returned on June 7. Bridget left on a trip June 4 and returned on June 22. Whose trip was longer? How much longer?

26. If today is September 12, what date and day of the week is 11 days from now?

27. **PATTERNS** Look at the dates in any of the columns of one of the calendars. Describe the pattern.

Mixed Review and Test Prep

28. Round 853,902 to the nearest hundred thousand. (p. 26)

29. Add 64,217 and 36,371. (p. 44)

30. Find the sum of 37,954 and the number 10,000 greater than 82,149. (p. 44)

31. Find the value of the expression $92 + (35 + 6)$. (p. 54)

32.   34,000 (p. 42)
   − 9,651

33.   81,643 (p. 44)
   + 27,285

34. **TEST PREP** $3 \times \blacksquare = 6 \times 5$
  A 1   **C** 5
  B 3   **D** 10

35. **TEST PREP** $943 + \blacksquare = 8,352$ (p. 40)
  F 7,409   **H** 8,409
  G 8,295   **J** 9,295

PROBLEM SOLVING LiNKUP ... to Science

A time line shows the order of events. Events are listed by date, from the earliest to the latest. This time line shows some of the events in the history of the United States Space Program.

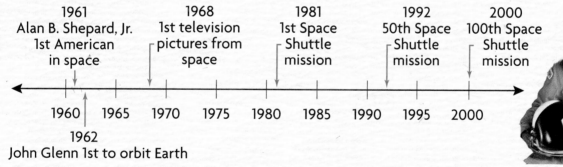

1961 Alan B. Shepard, Jr. 1st American in space

1968 1st television pictures from space

1981 1st Space Shuttle mission

1992 50th Space Shuttle mission

2000 100th Space Shuttle mission

1960 1965 1970 1975 1980 1985 1990 1995 2000

1962 John Glenn 1st to orbit Earth

Dr. Mae Jemison

For 1–4, use the time line.

1. How many decades passed between the first American in space and the first space shuttle mission?

2. How many years before 2000 were the first television pictures taken in space?

3. Dr. Mae Jemison, born in 1956, was a mission specialist on the fiftieth space shuttle mission. How old was Dr. Jemison the year of this flight?

4. **REASONING** John Glenn returned to space in 1998, when he was 77 years old. How old was he when he orbited the Earth?

EXTRA PRACTICE page H38, Set D

Review/Test

✓ CHECK VOCABULARY AND CONCEPTS

Choose the best term from the box.

A.M.
P.M.
decade
century

1. ___?___ means "before noon." (p. 118)

2. A ___?___ is 100 years. A ___?___ is 10 years. (p. 127)

✓ CHECK SKILLS

**Write the time shown on the clock in two different ways.
Include seconds when given.** (pp. 116–117)

3.

4.

5.

Write the time, using A.M. or P.M. (pp. 118–119)

6. when the mall opens

7. when you eat lunch at school

8. when you go to sleep

Find the elapsed time. (pp. 120–123)

9. **start:** 8:22 A.M.
   end: 5:09 P.M.

10. **start:** 3:18 P.M.
    end: 11:59 P.M.

11. **start:** 8:43 P.M.
    end: 3:33 A.M.

12. **start:** 7:26 A.M.
    end: 9:12 A.M.

✓ CHECK PROBLEM SOLVING

13. Use the table to make a schedule for Sharon between 10:30 A.M. and 3:00 P.M. (pp. 124–125)

THINGS TO DO	
Activity	**Time**
Violin practice	1 hr
* Haircut	1:30–2:00 P.M.
* Homework	45 min
Movie	12:30–2:20 P.M.
Lunch	30 min

*Starred activities must be done.

14. The county fair will be open for two weeks. If the fair ends on September 27, on what date will it open? (pp. 124–125)

September						
Sun	Mon	Tue	Wed	Thu	Fri	Sat
1	2	3	4	5	6	7
8	9	10	11	12	13	14
15	16	17	18	19	20	21
22	23	24	25	26	27	28
29	30					

Standardized Test Prep

 Get the information you need.
See item **5.**

A *wait between flights* is the amount of time you are in one city. Find how long it is from the arrival time to departure time.

Also see problem **3**, p. H63.

For 1–5, choose the best answer.

1. 319,426
    +477,309

    A 796,725   **C** 797,735
    B 796,735   **D** NOT HERE

2. What time does the clock show?

   F 10 minutes after 9
   G 10 minutes before 9
   H quarter of ten
   J 10 minutes before 8

3. Which would you most likely be doing at 8:00 A.M.?
   A eating dinner
   B going to a movie
   C going home from school
   D eating breakfast

For 4–5, use the table.

Suzanne's Flight Schedule

Baltimore to Jacksonville
Leaves 9:00 A.M.   Arrives 11:38 A.M.

Jacksonville to Houston
Leaves 1:50 P.M.   Arrives 3:46 P.M.

4. According to her schedule, how long will it take Suzanne to get from Baltimore to Jacksonville?
   F 2 hr 12 min   **H** 2 hr 28 min
   G 2 hr 22 min   **J** 2 hr 38 min

5. According to her schedule, how long will Suzanne wait between flights?
   A 2 hr 12 min   **C** 1 hr 56 min
   B 3 hr 12 min   **D** 2 hr 56 min

Write What You Know

6. There are 30 days in April and June, and there are 31 days in May. How many days are there from April 14 to the last day of June? Explain your reasoning.

7. Tom left his house at 8:40 A.M. and walked 25 minutes to a friend's house. He left the friend's house at 10:45 A.M. How long did Tom spend at the friend's house? Explain how you found your answer.

PROBLEM SOLVING
MATH DETECTIVE

Picture This

Four students drew graphs to represent their trips home from school. Read the descriptions given, and think about what each graph is showing. Match each description to the graph that communicates the same information.

Keisha:

I was in a hurry to get home after school to watch my favorite television program. I ran all the way home to get there quickly.

Tom:

I was walking home at a steady pace but decided to stop at the store for a soda. After resting for a while I continued my walk home.

Juan:

I walked home with my friend Pedro. We were so busy talking that it took us a long time to get to my house.

Maria:

I walked quickly with my friend to her house. We stayed there a while and then went to our friend's house for a quick visit. I realized I was late and I ran home.

Think It Over!

• ✎ **Write About It** Explain how you determined which graph matched each story.

• Draw a graph that tells the following story: Jackie planted bean seeds. The seeds grew steadily for a while. Jackie forgot to water the plant. This caused the growth to slow down.

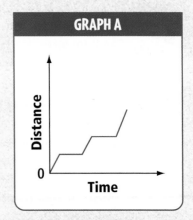

GRAPH A

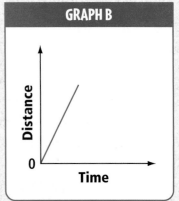

GRAPH B

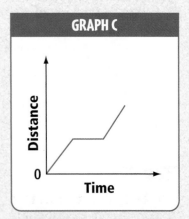

GRAPH C

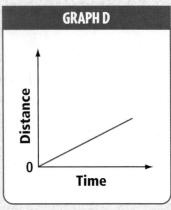

GRAPH D

Challenge

Time Zones

The continental United States is divided into four time zones. Look at the map. When moving from one time zone to the next, there is a difference of 1 hour.

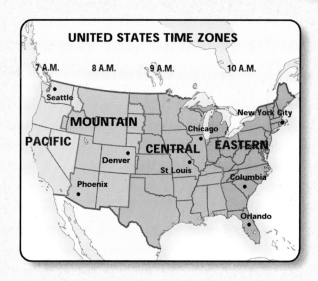

UNITED STATES TIME ZONES

7 A.M. 8 A.M. 9 A.M. 10 A.M.

Seattle

MOUNTAIN

New York City

PACIFIC

CENTRAL EASTERN

Denver

Chicago

St Louis

Columbia

Phoenix

Orlando

Examples

A What time is it in Columbia, South Carolina, when it is 8:00 A.M. in Phoenix, Arizona?

Columbia, South Carolina, is in the eastern time zone. Phoenix, Arizona, is in the mountain time zone. The difference between the mountain time zone and the eastern time zone is 2 hours.

So, it is 10:00 A.M. in Columbia, South Carolina.

B What time is it in the Pacific time zone when it is 8:00 P.M. in the central time zone?

CENTRAL	5 P.M.	6 P.M.	7 P.M.	8 P.M.
PACIFIC	3 P.M.	4 P.M.	5 P.M.	

From the central time zone to the Pacific time zone the pattern is subtract 2 hours.

So, it is 6 P.M. in the Pacific time zone when it is 8 P.M. in the central time zone.

Try It

For 1–4, use the map or a pattern to solve.

1. What time is it in the central time zone when it is 8:00 P.M. in the Pacific time zone?

2. What is the time difference between New York City and Denver?

3. If it is 2 P.M. in Orlando, what time is it in Chicago?

4. If it is 8:30 P.M. in St. Louis, what time is it in Seattle?

Find the next time for each pattern.

5. 12:00 P.M., 2:00 P.M., 4:00 P.M., ▓

6. 3:15 P.M., 6:15 P.M., 9:15 P.M., ▓

Study Guide and Review

VOCABULARY

Choose the best term from the box.

1. In a set of data that is ordered from least to greatest, the number in the middle is the ? . (pp. 86-87)

2. A ? is 100 years. A ? is 10 years. (pp. 126-129)

cumulative
frequency
median
mode
century
decade

STUDY AND SOLVE

Chapter 5

Use cumulative frequency tables.

How many pies were sold at the festival on Day 1 and Day 2?

PIES SOLD AT ART FESTIVAL		
Day	**Frequency**	**Cumulative Frequency**
1	16	16
2	23	39
3	14	53

So, 39 pies were sold on Day 1 and Day 2.

Interpret data from graphs using a variety of scales.

How would the heights of the bars on the graph change if the interval were 5?

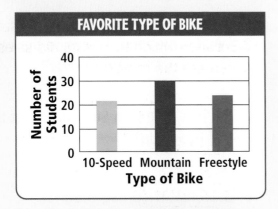

FAVORITE TYPE OF BIKE

The bars would be taller if the interval were 5.

For 3–5, use the cumulative frequency table. (pp. 82-85)

3. How many pies were sold on the first day?

4. How many more pies were sold on Day 2 than Day 1?

5. What was the total number of pies sold at the end of 3 days?

For 6–9, choose 5, 10, or 100 as the most reasonable interval for each set of data. (pp. 92-93)

6. 0, 50, 100, 100

7. 25, 30, 50, 75, 80

8. 5, 15, 20, 30, 35

9. 100, 150, 200, 200, 350

10. Use the bar graph at the left. How would the heights of the bars on the graph change if the interval were 20?

Chapter 6

Choose an appropriate graph.

Choose a graph to compare the number of red folders sold by two stores in one week.

| A line graph shows changes over time. | A bar graph compares different groups of data. |

A bar graph would be the better display since two groups of data are being compared.

Write the kind of graph or plot you would choose. (pp. 106–109)

11. to show the test scores of everyone in a class

12. to show a jogger's time at each mile on a 3-mile run

13. to compare the favorite sports of fourth-grade boys and girls

Chapter 7

Read and write time. Write the elapsed time.

Find the elapsed time.

Start Time   End Time

- Count ahead one hour at a time from 10:45 A.M. to 2:45 P.M., or 4 hours.

- Count ahead one minute at a time from 2:45 P.M. until you get to 2:52 P.M., or 7 minutes.

So, the elapsed time is 4 hours 7 minutes.

14. Write the time shown on the clock two ways. Include seconds.
(pp. 116–117)

Find the elapsed time. (pp. 120–123)

15. Start: 8:15 A.M.
    End: 6:48 P.M.

16. If a movie started at 11:50 A.M. and was 2 hours 18 minutes long, at what time did the movie end?

PROBLEM SOLVING PRACTICE

Solve. (pp. 94–95, 110–111, 124–125)

17. Use the table to make a schedule for Beth from 1:00–4:00 P.M. The schedule must include lunch.

ACTIVITY	TIME
Watch video	1 hour 45 minutes
Lunch	45 minutes
Ride bike	45 minutes
Play board game	30 minutes

18. Make a graph of the data in the table. Between which two years did sales increase the most?

YEAR	SALES
1997	$250
1998	$180
1999	$220
2000	$270

PERFORMANCE ASSESSMENT

TASK A • FAVORITE ACTIVITIES

Caroline took a survey of the favorite after-school activities of some of her classmates. She asked the students this question, "Would you rather go bike riding, play soccer, or play a video game?"

FAVORITE AFTER-SCHOOL ACTIVITY		
Activity	**Boys**	**Girls**
Bike Riding	⊩⊩ IIII	⊩⊩⊩⊩ I
Playing Soccer	⊩⊩⊩ II	⊩⊩⊩ III
Playing Video Game	⊩⊩⊩ I	⊩ III

a. Did Caroline ask a good survey question? Why or why not?

b. Caroline recorded the results of her survey in the table. Make a plot or graph Caroline could use to display the data.

c. Write a question your classmates could answer by looking at the graph or plot you drew.

TASK B • TIME FOR FUN

Daniel and his family went to the airport to meet his cousin, Michael, who was coming for a visit. Michael's plane was scheduled to arrive at 9:35 A.M. It arrived 10 minutes late.

a. The family stopped for a quick snack and then drove for 1 hour and 45 minutes to get back to Daniel's house. Draw a clock face to show the time you think they arrived at Daniel's house.

b. The boys want to plan an afternoon of activities to last 4 hours. The table shows the activities they would like to include. There is a star next to the activities they definitely want to include. What other activities would they be able to add?

c. Make a schedule for the boys' afternoon. Give the starting time, ending time, and elapsed time of each activity.

Afternoon Activities	
Activity	Time
Hiking	1 hr 15 min
*Lunch	30 min
Swimming	1 hr
*Car Show	1 hr 15 min
Fishing	45 min

Technology Linkup

E-Lab • Finding the Median and Mode

The students were divided into 9 teams for field day. The teams won 12, 13, 12, 13, 14, 13, 9, 10, and 15 ribbons. What was the median number of ribbons won?

You can use E-Lab to find median and mode.

- Click on *Finding the Median and Mode*.
- Click *New Problem* to begin.
- Type 12. Press *Enter*.
- Type 13. Press *Enter*.
- Keep doing this for the other numbers of ribbons won.
- Click *Sort*.
- Record the median.

So, the median number of ribbons won was 13.

What is the mode of these data? How does using E-Lab help you find the mode?

Practice and Problem Solving

Use E-Lab to find the median and mode for the set of data. Draw each line plot.

1. 6, 7, 5, 3, 12, 9, 10, 1, 9

2. 9, 4, 5, 10, 13, 11, 14, 3

3. 6, 4, 2, 3, 2, 2, 1, 2, 6

4. 13, 12, 14, 10, 15, 10, 15, 13, 10

5. The table shows the number of hits Chad got on his new web site. What is the median number of hits? What is the mode?

HITS ON CHAD'S WEB SITE						
Day	1	2	3	4	5	6	7
Hits	5	7	6	5	9	5	8

6. What if Chad got only 4 hits on Day 7? Then what would the median and mode be?

Multimedia Math Glossary www.harcourtschool.com/mathglossary

7. Vocabulary Look up *mode* in the Multimedia Math Glossary. Write a problem that can be answered using the example shown in the glossary.

PROBLEM SOLVING ON LOCATION
in Oregon

▲ The lowest temperature ever recorded in Oregon, ⁻54°F, happened in Prineville on August 10, 1898.

TEMPERATURE

Oregon has a variety of climates. The areas along the coast are more mild, and the temperatures in the mountains are more extreme.

The table shows the normal temperature for 6 months in Portland, Oregon.

USE DATA For 1–6, use the table.

PORTLAND NORMAL TEMPERATURES	
Month	**Temperature (in °F)**
January	40
February	44
March	47
April	51
May	57
June	64

1. Make a graph to compare the normal temperatures for these months.

2. What scale and interval did you use in your graph? Explain your choice.

3. What is the range of temperatures? Explain how you find a range.

4. Between which two consecutive months is there the greatest difference in normal temperature? What is the difference?

5. Which month has a normal temperature that is 17° below the normal temperature in June?

6. For which months do the normal temperatures differ by exactly 20 degrees?

7. **STRETCH YOUR THINKING** Find out the normal temperatures of some cities in your area. Make a double-bar graph to display your data.

PRECIPITATION

Most people agree that Oregon has a rainy climate. Actually, the eastern side of the Cascades Mountain range has a much drier climate than the western side of the range.

The line plot shows the number of days it rained or snowed during each month of one year in Klamath Falls, Oregon.

Crater Lake, the deepest lake in the U.S., is less than a day's drive from Klamath Falls. ▼

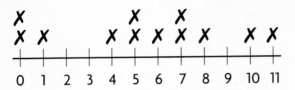

Days of Rain or Snow in Klamath Falls

USE DATA For 1–6, use the line plot.

1. What is the mode for the data? Explain how you know.

2. What is the range of the data? How would the range change if there had been 1 month with 12 rainy or snowy days rather than 11?

3. Can you tell by looking at the line plot which months had 5 rainy or snowy days? Explain.

4. Explain why there are no marks on the line plot above 2, 3, and 9 days of rain or snow.

5. What was the total number of days it rained or snowed in Klamath Falls in this year?

6. **Write About It** Write a question that can be answered using the line plot. Answer your question.

Practice Multiplication and Division Facts

Marching bands have instruments ranging from piccolos to drums. To form the letter M, the musicians arrange themselves in lines of 4 and 10.

PROBLEM SOLVING Look at the table. How many musicians will it take to form two letter Ms of the same size?

LETTER M TABLE

Line	Musicians in Each Row
1	10
2	10
3	4
4	4
5	4
6	10
7	10

CHECK WHAT YOU KNOW ✓

Use this page to help you review and remember
important skills needed for Chapter 8.

✓ VOCABULARY

Choose the best term from the box.

1. When you multiply, the answer is called the _?_ .

2. In $24 \div 6 = 4$, 4 is the _?_ .

> product
> quotient
> multiply

✓ MULTIPLICATION AND DIVISION FACTS (For Intervention, see p. H12.)

Find the product or quotient.

3. 6×10	**4.** 4×12	**5.** 5×6	**6.** 3×9
7. 8×2	**8.** 7×8	**9.** 9×9	**10.** 5×11
11. $49 \div 7$	**12.** $63 \div 7$	**13.** $45 \div 5$	**14.** $54 \div 6$

✓ MEANING OF MULTIPLICATION (For Intervention, see p. H12.)

Use the array to find the value of each expression.

15.

3×6

16.

3×5

17.

3×4

18.

2×4

19.

2×7

20.

2×6

✓ MEANING OF DIVISION (For Intervention, see p. H12.)

Answer the questions for each picture.

21.

How many counters in all?

How many groups?

How many in each group?

22.

How many counters in all?

How many groups?

How many in each group?

Relate Multiplication and Division

▶ Learn

HALFTIME NOTES The band played 6 songs during the halftime of the football game. Each song was 3 minutes long. How long did the band play?

$6 \times 3 = n$
$\downarrow$
$6 \times 3 = 18$

$n = 18$

So, the band played for 18 minutes.

The band played for 18 minutes at another football game. Each song was 3 minutes long. How many songs did the band play?

$18 \div 3 = n$

$\begin{array}{cccccc} 6 & \times & 3 & = & 18, & \text{so} & 18 & \div & 3 & = & 6 \\ \downarrow & & \downarrow & & \downarrow & & \downarrow & & \downarrow & & \downarrow \\ \text{factor} & & \text{factor} & & \text{product} & & \text{dividend} & \text{divisor} & & \text{quotient} \end{array}$

$n = 6$

So, the band played 6 songs.

MATH IDEA Multiplication and division are opposite, or **inverse operations**. One operation undoes the other.

A set of related multiplication and division equations using the same numbers is a **fact family**.

$6 \times 3 = 18 \qquad 18 \div 3 = 6$

$3 \times 6 = 18 \qquad 18 \div 6 = 3$ ← fact family for 3, 6, 18

▶ Check

1. **Write** a multiplication equation that you can use to find $24 \div 4$.

Find the value of the variable. Write a related equation.

2. $2 \times 4 = n$ **3.** $12 \div 3 = x$ **4.** $28 \div 4 = y$ **5.** $5 \times 3 = z$

▶ Practice and Problem Solving

Find the value of the variable. Write a related equation.

6. $16 \div 2 = n$ **7.** $20 \div 4 = b$ **8.** $3 \times 4 = y$ **9.** $5 \times 4 = c$

10. $6 \times 3 = p$ **11.** $6 \times 5 = n$ **12.** $36 \div 4 = a$ **13.** $27 \div 3 = y$

14. $b \div 4 = 8$ **15.** $18 \div n = 2$ **16.** $y \times 4 = 40$ **17.** $8 \times n = 24$

Write the fact family for each set of numbers.

18. 2, 3, 6   **19.** 4, 7, 28   **20.** 3, 7, 21

21. Name 2 fact families that have only two equations. Explain.

USE DATA For 22–24, use the table.

22. Michael collected baseball cards for 3 months. He collected 44 cards in the first month and 29 cards in the second month. How many cards did Michael collect in the third month?

23. Michael collected the same number of soccer cards in each of 4 weeks. How many soccer cards did he collect each week?

24. What is the total number of cards Michael collected?

25. ✎ **Write About It** Explain how you can use multiplication to solve a division problem. Give an example.

MICHAEL'S CARD COLLECTION	
Sport	**Number Collected**
Football	64
Soccer	28
Baseball	110
Basketball	35

Multiply and Divide Facts Through 5

▶ **Learn**

PUT IT IN REVERSE!   Mrs. Frazier asked her students to use models to show that division is the inverse of multiplication. This is how her students showed that division is the inverse of multiplication.

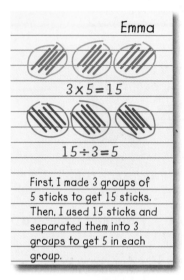

Emma

$3 \times 5 = 15$

$15 \div 3 = 5$

First, I made 3 groups of 5 sticks to get 15 sticks. Then, I used 15 sticks and separated them into 3 groups to get 5 in each group.

Blake

0 1 2 3 4 5 6 7 8 9
$3 \times 3 = 9$

0 1 2 3 4 5 6 7 8 9
$9 \div 3 = 3$

I started at 0 and made jumps of 3 on a number line to land at 9. Then, starting at 9 I took jumps of 3 back to 0.

Carlos

$5 \times 4 = 20$

$20 \div 4 = 5$

I made 5 rows of 4 blocks to make 20 blocks. Then I divided the 20 blocks into 4 columns to get 5 in each column.

Latoya

→ To multiply, I looked across row 6 and down column 4 to find the product 24.

→ To divide, I found 24 by looking down column 4. Then I looked left to find the quotient.

$6 \times 4 = 24$,
so $24 \div 4 = 6$.

✕	0	1	2	3	4	5	6	7	8	9
0	0	0	0	0	0	0	0	0	0	0
1	0	1	2	3	4	5	6	7	8	9
2	0	2	4	6	8	10	12	14	16	18
3	0	3	6	9	12	15	18	21	24	27
4	0	4	8	12	16	20	24	28	32	36
5	0	5	10	15	20	25	30	35	40	45
6	0	6	12	18	24	30	36	42	48	54

▶ **Check**

1. **Draw** a model that shows the inverse of 5×6.
   Then write the related equation.

Find a related multiplication or division equation.

2. $10 \div 2 = 5$

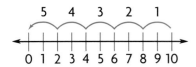

3. $4 \times 4 = 16$

4. $4 \times 3 = 12$

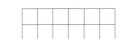

Practice and Problem Solving

Find a related multiplication or division equation.

5. $6 \times 2 = 12$

6. $2 \times 3 = 6$

7. $4 \times 2 = 8$

Find the product or quotient.

8. $14 \div 2$

9. 5×5

10. 5×7

11. $32 \div 4$

12. $28 \div 4$

13. 8×4

14. $36 \div 4$

15. 3×0

16. $8 \div 2$

17. 2×5

 ALGEBRA Find the value of the variable.

18. $4 \times 2 = 8$, so $4 \times 2 \times 2 = n$.

19. $5 \times 5 = 25$, so $(5 \times 5) + 10 = a$.

20. $45 \div 5 = 9$, so $(45 \div 5) \times 2 = b$.

21. $3 \times 4 = 12$, so $3 \times 4 \times 2 = n$.

22. There are 16 friends bowling. The same number of friends are bowling in each of 4 lanes. How many friends are bowling in each lane?

23. Felipe's radio-controlled car can go 2 blocks and return in about one minute. If Felipe has his car do this 2 times, how many blocks will the car travel?

24. ✎ **Write About It** Write a problem using division that you can solve by using the multiplication equation $7 \times 5 = 35$.

Mixed Review and Test Prep

25. $354 + 1{,}234$ (p. 40)

26. $586 + 4{,}821$ (p. 40)

27. Round 8,754 to the nearest hundred.

28. Round 14,842 to the nearest thousand. (p. 26)

29. **TEST PREP** At the store, Martha bought corn for $1.19, an onion for $0.35, and lettuce for $0.89. How much change did she receive from $5.00? (p. 40)

A $2.43

B $2.57

C $2.87

D $7.43

EXTRA PRACTICE page H39, Set B

Multiply and Divide Facts Through 10

▶ Learn

DIVIDE AND CONQUER You can break apart numbers to make them easier to multiply. Find the product of 6 and 8. What is 6×8?

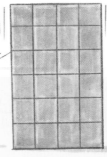

Activity
MATERIALS: centimeter grid paper

What is 6×8?

STEP 1	STEP 2	STEP 3
Outline a rectangle that is 6 units high and 8 units wide.	Cut apart the rectangle to make two arrays for products you know.	Find the sum of the products of the two smaller rectangles.

 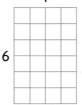 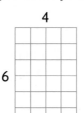

STEP 3:
$6 \times 4 = 24$
$6 \times 4 = 24$
$24 + 24 = 48$

STEP 2: The factor 8 is now 4 plus 4.

So, $6 \times 8 = 48$.

• What two small rectangles can you make if you cut equal parts of the 6×8 rectangle horizontally?

Use grid paper and break apart one factor to find each product.

a. 7×6 **b.** 8×10 **c.** 6×9

Technology Link

To learn more about *Multiply and Divide Facts Through 10,* watch the Harcourt Math Newsroom Video *Area Code Solutions.*

 MATH IDEA If you forget a multiplication fact, you can break apart one of the factors so that you can use multiplication facts that you know.

More Strategies

Strategies can help you learn the multiplication and division facts that you do not know.

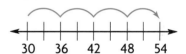

Think of the inverse.

What is $36 \div 9$?

Think: $4 \times 9 = 36$

So, $36 \div 9 = 4$.

Use the Order Property.

What is 8×5?

Think: $5 \times 8 = 40$

So, $8 \times 5 = 5 \times 8 = 40$.

$\uparrow$
Facts to memorize.

Use a pattern.

What is 6×9?

Think: $6 \times 5 = 30$, so I can count on from 30 by 6 for the remaining 4 times.

Count: 30 . . . 36, 42, 48, 54.

So, $6 \times 9 = 54$.

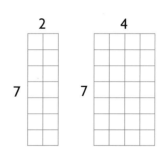

Use the *break apart* strategy.

What is 7×6?

Think: $6 = 2 + 4$

$7 \times 2 = 14$ and $7 \times 4 = 28$,
$7 \times 6 = 14 + 28$

So, $7 \times 6 = 42$.

▶ **Check**

1. **Explain** how breaking apart numbers can help you find the product of greater numbers.

Find the product or quotient. Show the strategy you used.

2. 9×6 **3.** 3×10 **4.** 6×6 **5.** $42 \div 6$ **6.** $72 \div 9$

LESSON CONTINUES

Show how the arrays can be used to find the product.

7. What is 9×8?

$9 \times 4 = \blacksquare$

$9 \times 4 = \blacksquare$

$9 \times 8 = \blacksquare + \blacksquare$

So, $9 \times 8 = \blacksquare$.

8. What is 7×9?

$7 \times 3 = \blacksquare$

$7 \times 3 = \blacksquare$

$7 \times 3 = \blacksquare$

$7 \times 9 = \blacksquare + \blacksquare + \blacksquare$

So, $7 \times 9 = \blacksquare$.

Find the product or quotient. Show the strategy you used.

9. 5×9	**10.** 4×9	**11.** 4×6
12. 9×9	**13.** 8×8	**14.** 10×7
15. 9×7	**16.** 7×8	**17.** 8×9
18. $48 \div 6$	**19.** $72 \div 8$	**20.** $48 \div 8$
21. $70 \div 10$	**22.** $54 \div 6$	**23.** $63 \div 9$
24. $60 \div 6$	**25.** $49 \div 7$	**26.** $28 \div 7$

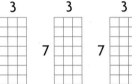

Technology Link

More Practice: Use *Mighty Math Number Heroes, Quizzo* Level L.

27. If you know that $7 \times 10 = 70$, how can you find the product 7×9?

For 28–31, look for patterns in the Facts of Nine table.

28. How does the pattern of the tens digits in the products relate to the pattern of the first factors?

29. How does each product relate to the second factor?

30. Explain how you can use the patterns you found to find 9×9 without the table.

31. **❓ What's the Error?** Gwyn used the *break apart* strategy to solve 8×9. Describe her error. Show a correct way to use the strategy.

$$8 \times 3 = 24 \quad \text{and} \quad 8 \times 3 = 24$$

$$24 + 24 = 48$$

Facts of Nine
$1 \times 9 = 9$
$2 \times 9 = 18$
$3 \times 9 = 27$
$4 \times 9 = 36$
$5 \times 9 = 45$
$6 \times 9 = 54$
$7 \times 9 = 63$
$8 \times 9 = 72$
$9 \times 9 = 81$

32. Sandra shared some jacks equally with Maria and Tom. There were 2 jacks left. Maria got 11 jacks. How many jacks did Sandra have to start?

33. Lola used 42 cups of sugar to make grape jelly. If each batch used 7 cups of sugar, how many batches did she make?

Mixed Review and Test Prep

34. Write six hundred seventy thousand, fifty-four in standard form. (p. 6)

35. Write 5,241,211 in expanded form. (p. 8)

36. Order 111, 456, and 237 from greatest to least. (p. 20)

37. Order 35,314; 34,919; and 35,335 from least to greatest. (p. 20)

38. 47,345 (p. 44)
   $+24,399$

39. 90,260 (p. 44)
   $-18,435$

40. **TEST PREP** In batting practice, Ellie hit 10 balls in a row before getting a strike. Then she hit 12 balls, then 16 balls. How many balls did she hit altogether? (p. 36)

   A 10   **C** 26
   B 22   **D** 38

41. **TEST PREP** At the art fair, May bought 2 rings for $3.50 each. She also bought a picture for $7.50 and a drink for $1.25. How much money did May spend at the art fair? (p. 40)

   F $8.75   **H** $15.75
   G $12.25   **J** $17.75

PROBLEM SOLVING LINKUP ... to Art

The Romans often covered floors with small colored tiles made of material such as marble or stone. The picture shows part of a Roman floor. Works of art done in this style are known as *mosaics*.

You can use what you know about multiplying to find the total number of tiles you would need to make a design to cover part of your desk.

1. Draw a section of the design shown or make your own design with pattern blocks.

2. Record the number of each of the different shapes you used in your design.

3. Write a multiplication sentence to find the total number of each shape of tile you would need if you made 8 designs like the one you drew.

Multiplication Table Through 12

HANDS ON

Quick Review

1. 9×6 2. 8×5
3. $63 \div 9$ 4. 7×7
5. $54 \div 6$

MATERIALS blank multiplication table through 12

▶ **Explore**

Use strategies and patterns to learn new facts.

Make your own multiplication table for the facts from 0 through 12.

• What pattern do you see in the facts for 11?

×	0	1	2	3	4	5	6	7	8	9	10	11	12
0	0	0	0	0	0	0	0	0	0	0	0		
1	0	1	2	3	4	5	6	7	8	9	10		
2	0	2	4	6	8	10	12	14	16	18	20		
3	0	3	6	9	12	15	18	21	24	27	30		
4	0	4	8	12	16	20	24	28	32	36	40		
5	0	5	10	15	20	25	30	35	40	45	50		
6	0	6	12	18	24	30	36	42	48	54	60		
7	0	7	14	21	28	35	42	49	56	63	70		
8	0	8	16	24	32	40	48	56	64	72	80		
9	0	9	18	27	36	45	54	63	72	81	90		
10	0	10	20	30	40	50	60	70	80	90	100		

Activity

THE ELEVENS

Complete the column for 11 to 10×11. Use break-apart numbers to find 11×11 and 12×11.

Think:

$10 \times 11 = 110$ $10 \times 11 = 110$
$1 \times 11 = 11$ $2 \times 11 = 22$
So, $11 \times 11 = 121$. So, $12 \times 11 = 132$.

Complete the row for 11.

THE TWELVES

Complete the column for 12 to 10×12. Use break-apart numbers to find 12×12.

Think:

$10 \times 12 = 120$ $2 \times 12 = 24$
$12 \times 12 = 120 + 24$
So, $12 \times 12 = 144$.

Complete the row for 12.

I found the row for 8 and the column for 11. What's the product?

Try It

Use your multiplication table to find the product.

a. 8×11 **b.** 3×11 **c.** 4×10

▶ Connect

Since division facts are related to multiplication facts, you can use your multiplication table to help you find the quotient in a division problem.

$9 \times 12 = 108$, so $108 \div 12 = 9$.

Technology Link

More Practice: Use E-Lab, *Modeling Multiplication Facts.*

www.harcourtschool.com/ elab2002

Talk About It

• Explain how you can use the multiplication table to find $99 \div 11$.

▶ Practice and Problem Solving

Use the multiplication table to find the product or quotient.

1. 1×12 **2.** 2×10 **3.** 3×11 **4.** 4×12 **5.** 5×9

6. 6×10 **7.** 7×11 **8.** 8×12 **9.** $120 \div 12$ **10.** $88 \div 11$

11. $90 \div 10$ **12.** $144 \div 12$ **13.** $110 \div 11$ **14.** $121 \div 11$ **15.** $100 \div 10$

 **ALGEBRA** **Find the value of the variable.**

16. $r \times 12 = 120$ **17.** $11 \times n = 121$ **18.** $100 \div s = 10$ **19.** $p \div 11 = 7$

20. REASONING Describe a way to help you remember the facts $88 \div 11$ and $77 \div 11$ and the facts $132 \div 11$ and $121 \div 11$.

21. Josh wants to pick 108 apples. So far, he has counted 5 dozen. How many more dozen apples does he need to meet his goal?

22. Use a pattern or strategy to find the product 14×12. Explain your answer.

23. **? What's the Question?** The answer is that one factor is 11 and the product is 132.

Mixed Review and Test Prep

For 24–25, find the value of the expression. (p. 54)

24. $(15 - 9) + 3$

25. $(27 + 2) - 12$

26. What is three hundred seventy-two thousand, twenty-five in standard form? (p. 6)

27. How much time elapses between 11:30 A.M. and 1:05 P.M.? (p. 120)

28. TEST PREP Marty had $10.00. He spent $5.49. How much does Marty have left? (p. 42)

  A $4.32   **C** $4.51

  B $4.49   **D** $4.61

Multiply 3 Factors

Quick Review

1. 4×3 2. 9×9

3. 10×8 4. 7×5

5. 12×4

▶ **Learn**

MUSICAL NUMBERS   For the spring concert, the music teacher, Mr. Griffiths, chose 4 pieces by each of 2 composers. Each piece takes 5 minutes to play. How many minutes will the music at the concert last?

$$4 \times 2 \times 5 = m$$

You can use the Grouping Property of Multiplication to help you find the product.

VOCABULARY

Grouping Property of Multiplication

GROUPING PROPERTY OF MULTIPLICATION

When the grouping of factors is changed, the product remains the same.

Remember

You perform the operations in the parentheses ( ) first.

Example

Use parentheses ( ) to group the factors you multiply first.

$(4 \times 2) \times 5 = m$	Multiply 4 by 2.
$\downarrow$	
$8 \quad \times 5 = 40$	Then multiply that product by 5.
$4 \times (2 \times 5) = m$	Multiply 2 by 5.
$\downarrow$	
$4 \times \quad 10 \quad = 40$	Then multiply that product by 4.

So, the music will last 40 minutes.

▲ Duke Ellington, a famous composer of jazz, played over 20,000 performances worldwide in his lifetime.

▶ **Check**

1. **Name** two ways you can group $2 \times 3 \times 4$ to find the product. Are the products the same? Explain.

Find the product.

2. $(6 \times 2) \times 5$ 3. $(2 \times 5) \times 2$ 4. $(6 \times 2) \times 6$ 5. $(5 \times 2) \times 5$

Show two ways to group by using parentheses ( ).
Find the product.

6. $3 \times 4 \times 2$ **7.** $3 \times 3 \times 3$ **8.** $4 \times 2 \times 3$ **9.** $5 \times 2 \times 4$

▶ Practice and Problem Solving

Find the product.

10. $(6 \times 2) \times 1$ **11.** $4 \times (3 \times 3)$ **12.** $9 \times (2 \times 4)$ **13.** $7 \times (4 \times 3)$

14. $11 \times (5 \times 2)$ **15.** $(2 \times 4) \times 2$ **16.** $6 \times (4 \times 3)$ **17.** $(2 \times 5) \times 10$

18. $(3 \times 2) \times 7$ **19.** $(2 \times 2) \times 2$ **20.** $(8 \times 1) \times 9$ **21.** $9 \times (2 \times 5)$

Show two ways to group by using parentheses.
Find the product.

22. $5 \times 2 \times 3$ **23.** $9 \times 1 \times 5$ **24.** $3 \times 2 \times 6$ **25.** $9 \times 0 \times 12$

26. $2 \times 2 \times 3$ **27.** $5 \times 2 \times 5$ **28.** $2 \times 2 \times 6$ **29.** $4 \times 3 \times 4$

30. $3 \times 4 \times 2$ **31.** $6 \times 1 \times 5$ **32.** $4 \times 2 \times 3$ **33.** $5 \times 2 \times 6$

Write <, >, or = for each ●.

34. $1 \times 2 \times 3 \ ● \ 3 \times 1 \times 2$

35. $2 \times (2 \times 5) \ ● \ (2 \times 2) \times 5$

36. $9 \times (2 \times 3) \ ● \ 3 \times (6 \times 2)$

37. $10 \times (3 \times 1) \ ● \ 3 \times (3 \times 2)$

38. Aaron walked his neighbor's 3 dogs for 3 days. He earned $2 per dog each day. How much money did Aaron earn?

39. For $(9 \times 2) \times 3$, explain how using the Grouping Property of Multiplication makes it easier to find the product.

40. Emanuel practices the guitar 3 hours each week. Lois practices the piano 5 hours each week. How many more hours does Lois practice in 4 weeks than Emanuel?

Mixed Review and Test Prep

41. $\begin{array}{r} 105 \\ 28 \\ + \ 82 \\ \hline \end{array}$

42. $\begin{array}{r} 4,569 \ {\scriptstyle(p.\ 40)} \\ 2,382 \\ +4,375 \\ \hline \end{array}$

43. $\begin{array}{r} 43,259 \ {\scriptstyle(p.\ 44)} \\ -18,513 \\ \hline \end{array}$

44. $\begin{array}{r} 1,003 \ {\scriptstyle(p.\ 42)} \\ - \ \ 426 \\ \hline \end{array}$

45. TEST PREP Rosa collected 3 dozen eggs and Peter collected 7 dozen. How many eggs did they collect together? (p. 148)

A 10 **B** 110 **C** 120 **D** 130

EXTRA PRACTICE page H39, Set D

Problem Solving Skill
Choose the Operation

UNDERSTAND 〉 PLAN 〉 SOLVE 〉 CHECK 〉

OPERATION RAINFALL Study the problems. Use the chart to help you choose the operation needed to solve each problem.

Add	Join groups of different sizes
Subtract	Take away or compare groups
Multiply	Join equal-size groups
Divide	Separate into equal-size groups or find how many in each group

A. What if Orange County gets the same amount of rainfall for the next 5 months as in September? What would be the total rainfall?

B. About how much rain fell each week in April?

C. How much more rain fell in August and September than in May through July?

D. What is the total amount of rainfall for Orange County from April to September?

• Solve Problems A–D.

• What two different operations could you use to solve Problem A? to solve Problem B?

Talk About It

• What words in the box at the top of the page help you decide which operation to use for each of Problems A–D?

• What operation or operations did you use to solve Problems C–D?

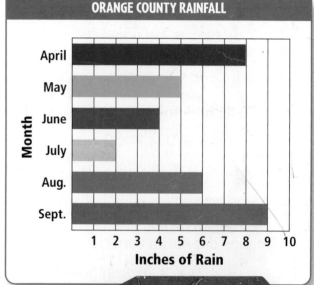

ORANGE COUNTY RAINFALL

Problem Solving Practice

Solve. Name the operation or operations you used.

1. During the past 9 weeks, the school chorus practiced a total of 36 hours. If they practiced the same number of hours each week, how many hours did they practice each week?

2. The cafeteria served 124 school lunches. There were a total of 11 pizzas cut into 12 slices. If each student received 1 slice, how many slices were left?

3. Mrs. Ling ordered 12 pizzas cut into 8 slices each. How many slices did she order?

4. Mr. Davis cut a sheet cake into 6 rows of 8 pieces. How many pieces of cake are there in all?

Choose the letter of the correct answer.

Before the concert, Michele sold 12 umbrellas. Each umbrella was shared by 4 people. How many people used the umbrellas?

5. Which expression could you use to solve the problem?

   A $12 \div 4$ **C** 12×4
   B $12 - 4$ **D** $12 + 4$

6. How many people used the umbrellas?

   F 4   **G** 8   **H** 16   **J** 48

Mixed Applications

USE DATA For 7–10, use the table.

7. Ben buys 2 board games and 4 books. How much change does he get from $5.00?

8. Tyler bought 3 tapes and 2 books. How much did he spend at the sale?

9. ✎ Write a problem using the information in the table.

10. Caro bought 2 tapes, a book, and a board game. Chad bought 4 tapes. How much more did Chad spend then Caro?

11. A train left Oakville at 7:45 A.M. It arrived in Bay City at 1:10 P.M. How long was the train trip?

Item	Price
Board Games	$0.50
Books	$0.25
Tapes	$0.75

Review/Test

✅ CHECK VOCABULARY AND CONCEPTS

Choose the best term from the box.

1. The _?_ states that when the grouping of factors is changed, the product remains the same. (p. 150)

2. Multiplication and division are opposite, or _?_ operations. One operation undoes the other. (p. 140)

3. A set of related multiplication and division sentences using the same numbers is a _?_. (p. 140)

> inverse
> fact family
> Grouping Property
>   of Multiplication
> quotient

✅ CHECK SKILLS

Find the value of the variable. Write a related equation. (pp. 140–141)

4. $18 \div 3 = n$ **5.** $6 \times 7 = a$ **6.** $y \div 4 = 5$ **7.** $9 \times c = 63$

Draw a model that shows the inverse of the model shown. Then write the related equation. (pp. 142–143)

8. $3 \times 5 = 15$ **9.** $3 \times 3 = 9$ **10.** $4 \times 2 = 8$

      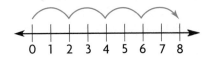

Find the product or quotient. (pp. 144–151)

11. 4×7 **12.** $18 \div 3$ **13.** $56 \div 8$ **14.** 9×8

15. $144 \div 12$ **16.** 8×11 **17.** $8 \times (2 \times 3)$ **18.** $5 \times (6 \times 2)$

✅ CHECK PROBLEM SOLVING

Solve. Write the operation or operations you used. (pp. 152–153)

19. For a class field trip, 42 students went to a museum. If the museum takes groups of 7 students, how many groups were there?

20. Jack wants 6 packs of cards. If a store sells 1 pack of cards for $3 and a set of 6 packs of cards for $15, how much money will he save if he buys the set?

⭐ Standardized Test Prep

 Choose the answer.
See item **1**.

If your answer doesn't match one of the choices, check your computation. If your computation is correct, mark the letter that shows NOT HERE.

Also see problem **6**, p. H64.

For 1–11, choose the best answer.

1. $30,000 - 14,582$

    A 15,417   **C** 15,528

    B 15,418   **D** NOT HERE

2. An assembly started at 12:55 and ended at 2:47. How long did it last?

    F 2 hr 52 min   **H** 1 hr 8 min

    G 2 hr 8 min   **J** 1 hr 52 min

3. $3 \times 2 \times 7$

    A 28   **B** 35   **C** 42   **D** 49

4. $8 \times 2 \times 5$

    F 56   **G** 72   **H** 80   **J** 92

5. Which expression is **not** equivalent to 36?

    A 5×8 **C** 6×6

    B 9×4 **D** 3×12

6. What is the value of n if $7 \times n = 63$?

    F 7   **G** 8   **H** 9   **J** 12

7. What interval is used for this bar graph?

FOOD DRIVE

Week / Number of Cans

    A 1   **B** 2   **C** 5   **D** 45

8. What is the value of n if $48 \div n = 8$?

    F 6   **G** 7   **H** 8   **J** 12

9. Which expression is **not** equivalent to 4?

    A $16 \div 4$ **C** $28 \div 7$

    B $24 \div 6$ **D** $36 \div 4$

10. Which of these facts is **not** related to the others?

    F $20 \div 4 = 5$ **H** $5 \times 4 = 20$

    G $32 \div 8 = 4$ **J** $20 \div 5 = 4$

11. Jane divided each of two cakes into eight pieces. Which expression can be used to find the total number of pieces of cake?

    A $(8 \div 2) \times 2$ **C** $(8 \times 2) \div 2$

    B $8 \div 2$ **D** 8×2

Write What You Know

12. How can you use the Grouping Property of Multiplication to find $(7 \times 4) \times 2$ mentally?

13. Katalina earned $32. She worked 3 hours on Saturday and 1 hour on Monday. How much did Katalina earn for each hour she worked? Explain.

Algebra: Use Multiplication and Division Facts

In 1902 President Theodore "Teddy" Roosevelt was shown in a cartoon with a bear cub. The cartoon inspired a shopkeeper to create the stuffed animal we now call the teddy bear.

PROBLEM SOLVING The bar graph shows the number of teddy bears five different collectors own. If each bear cost about $10, about how much did the largest collection cost? the smallest collection?

DATA LINK

TEDDY BEAR COLLECTIONS

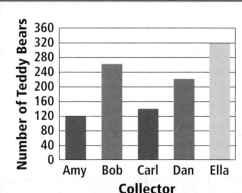

Number of Teddy Bears

360
320
280
240
200
160
120
80
40
0

Amy   Bob   Carl   Dan   Ella

Collector

CHECK WHAT YOU KNOW ✓

Use this page to help you review and remember
important skills needed for Chapter 9.

✓ VOCABULARY

Choose the best term from the box.

1. When you __?__, the answer is called the product.

2. In $36 \div 4 = 9$, 9 is the __?__.

multiply
quotient
product

✓ MISSING FACTORS (For Intervention, see p. H13.)

Find the missing factor.

3. ■ $\times 4 = 12$

4. $6 \times$ ■ $= 48$

5. ■ $\times 5 = 35$

6. $28 = 4 \times$ ■

7. $9 \times$ ■ $= 54$

8. $63 = 7 \times$ ■

9. $7 \times$ ■ $= 21$

10. $7 \times$ ■ $= 84$

11. $64 = 8 \times$ ■

12. ■ $\times 9 = 90$

13. $72 = 9 \times$ ■

14. ■ $\times 4 = 44$

✓ FACT FAMILIES (For Intervention, see p. H16.)

Copy and complete each number sentence.

15. $5 \times 2 =$ ■
    $10 \div$ ■ $= 2$

16. $6 \times 3 =$ ■
    $18 \div$ ■ $= 3$

17. $8 \times 5 =$ ■
    $40 \div$ ■ $= 5$

18. $7 \times 6 =$ ■
    $42 \div$ ■ $= 6$

19. $12 \times 5 =$ ■
    $60 \div$ ■ $= 5$

20. $8 \times 9 =$ ■
    $72 \div$ ■ $= 8$

21. $6 \times 6 =$ ■
    $36 \div$ ■ $= 6$

22. $8 \times 11 =$ ■
    $88 \div$ ■ $= 8$

23. $4 \times 8 =$ ■
    ■ $\div 8 = 4$

24. $9 \times 7 =$ ■
    ■ $\div 7 = 9$

25. $6 \times 8 =$ ■
    ■ $\div 8 = 6$

26. $7 \times 8 =$ ■
    ■ $\div 8 = 7$

✓ NUMBER PATTERNS (For Intervention, see p. H7.)

Write the next three numbers in the pattern.

27. 100, 90, 80, 70, ■, ■, ■

28. 75, 70, 65, 60, ■, ■, ■

29. 27, 30, 33, 36, ■, ■, ■

30. 8, 16, 24, 32, ■, ■, ■

31. 18, 27, 36, 45, ■, ■, ■

32. 96, 84, 72, 60, ■, ■, ■

33. 145, 260, 375, 490, ■, ■, ■

Expressions with Parentheses

▶ **Learn**

BUNCHES OF BEARS At 9:00 A.M., the Teddy Bear Workshop had 12 stuffed animals on each of 3 racks. By 11:00 A.M., 8 animals had been sold. How many stuffed animals were left to sell?

Think: 12 times 3 minus 8

Notice that the expression $(12 \times 3) - 8$ has two different operations.

Find the value of $(12 \times 3) - 8$.

$(12 \times 3) - 8$ Do what is inside the
  ↓   parentheses first.
$36 - 8$ Multiply. Then subtract.
  ↓
  28

So, $(12 \times 3) - 8$ is 28. The Teddy Bear Workshop has 28 stuffed animals left to sell.

Remember

An *expression* is a part of a number sentence that has numbers and operation signs. It does not have an equal sign.

Examples Find the value of the expression.

Ⓐ $(5 \times 3) + 6$ Multiply 5 and 3.
    ↓
  $15 + 6$ Add 15 and 6.
    ↓
    21

So, $(5 \times 3) + 6$ is 21.

Ⓑ $(49 \div 7) - (4 - 2)$ Divide 49 by 7.
    ↓
  $7 - (4 - 2)$ Subtract 2 from 4.
         ↓
    $7 - 2$ Subtract 2 from 7.
       ↓
       5

So, $(49 \div 7) - (4 - 2)$ is 5.

Choose an Expression

The Teddy Bear Workshop had 55 bears. They sold 10 bears and packed the rest in 5 boxes with an equal number in each box.

Which expression shows that the Teddy Bear Workshop packed 9 bears in each box?

$$(55 - 10) \div 5 \quad \text{or} \quad 55 - (10 \div 5)$$

Find the value of each expression. Which one equals 9?

$(55 - 10) \div 5$	$55 - (10 \div 5)$
↓	↓
$45 \div 5$	$55 - 2$
↓	↓
9	53

So, $(55 - 10) \div 5$ shows that the Teddy Bear Workshop packed 9 bears in 5 boxes.

MATH IDEA   The position of the parentheses can change the value of an expression.

More Examples   Which placement of the parentheses in the expression gives the value of 6?

A $(10 - 2) \times 2$
↓
8×2
↓
16

B $10 - (2 \times 2)$
↓
$10 - 4$
↓
6

So, $10 - (2 \times 2)$ is 6.

▶ Check

1. Show where you would place the parentheses so that $8 \times 6 - 3$ is 24.

Find the value of the expression.

2. $4 \times (6 - 2)$ **3.** $24 \div (3 \times 2)$ **4.** $(45 \div 9) - 1$ **5.** $(8 + 9) - (6 - 3)$

LESSON CONTINUES ▶

Find the value of the expression.

6. $(9 \times 2) + 4$ **7.** $(15 \div 3) + 46$ **8.** $99 - (12 \times 8)$ **9.** $7 \times (54 \div 6)$

10. $7 + (9 \times 2)$ **11.** $(35 \div 5) \times 9$ **12.** $8 + (16 \div 4)$ **13.** $3 \times (12 - 2)$

14. $(8 + 7) \div 3$ **15.** $24 \div (6 + 2)$ **16.** $(26 - 6) \div 4$ **17.** $12 + (5 \times 3)$

18. $8 \times (7 + 2)$ **19.** $(54 \div 6) - 2$ **20.** $16 - (2 \times 5)$ **21.** $43 - (3 \times 8)$

Choose the expression that shows the given value.

22. 43
   a. $1 + (6 \times 7)$
   b. $(1 + 6) \times 7$

23. 25
   a. $5 \times (6 - 5)$
   b. $(5 \times 6) - 5$

24. 17
   a. $14 + (6 \div 2)$
   b. $(14 + 6) \div 2$

25. 34
   a. $(4 + 6) \times 5$
   b. $4 + (6 \times 5)$

26. 18
   a. $16 + (4 \div 2)$
   b. $(16 + 4) \div 2$

27. 21
   a. $3 \times (5 + 2)$
   b. $(3 \times 5) + 2$

28. 27
   a. $(4 + 5) \times 3$
   b. $4 + (5 \times 3)$

29. 36
   a. $(6 \times 8) - 2$
   b. $6 \times (8 - 2)$

30. 63
   a. $8 + (1 \times 7)$
   b. $(8 + 1) \times 7$

Find the value of each expression.

31. $(36 \div 4) + (15 - 5)$ **32.** $(30 - 6) \div (17 - 14)$

33. $(3 \times 8) - (12 - 8)$ **34.** $(12 \div 3) \times (12 - 4)$

USE DATA For 35–36, use the table.

35. MENTAL MATH How much do 3 pencils, 1 pen, and 1 note pad cost?

36. REASONING Jackie bought 5 items. She spent $2.80. What school supplies did she buy?

37. ✎ Write a problem that involves finding the value of an expression.

School Store	
Pen	$0.55
Pencil	$0.20
Notepad	$0.75

38. Mr. Lily worked 8 hours each day for 5 days. He also worked 4 hours on Saturday. How many hours did he work?

39. Trevor bought a lamp for $28.45. He got $11.55 in change. How much money did he give the clerk?

40. After he bought 8 erasers for $0.05 each, Shawn spent $2.35 on paper. Use the expression $(8 \times \$0.05) + \$2.35$ to find how much he spent.

41. In class, Jessica sits 2 rows in front of Arthur and 3 rows behind Stephanie. How many rows are between Arthur and Stephanie?

Mixed Review and Test Prep

Write <, >, or = for each ●. (p. 18)

42. 1,945 ● 1,899

43. 34,785 ● 34,885

Find the product. (p. 148)

44. 5×12 **45.** 11×12

Write the time as shown on a digital clock. (p. 116)

46. 12 minutes past two

47. 15 minutes before four

48. TEST PREP Which is a related multiplication equation for $24 \div 8 = 3$? (p. 140)
A $24 \div 6 = 4$ **C** $24 \div 4 = 6$
B $8 \times 4 = 32$ **D** $3 \times 8 = 24$

49. TEST PREP Brandon put 8 ounces of cheese on each of 9 large pizzas. If he bought 96 ounces of cheese, how many ounces of cheese were left over? (p. 152)
F 24 ounces   **H** 88 ounces
G 72 ounces   **J** 168 ounces

PROBLEM SOLVING Thinker's Corner

VALUABLE PLATES Write an expression using the numbers in each license plate. Copy the numbers in order from left to right. Place parentheses and operation signs (+, −, ×, or ÷) to make the expression equal the given value.

1.
— THE FIRST STATE —
729 411
— DELAWARE —
Value: 52

2.
NEW YORK
819 364
Value: 81

3.
Illinois   Land of Lincoln
125 104
Value: 20

4.
— Massachusetts —
674 183
— The Spirit of America —
Value: 69

EXTRA PRACTICE page H40, Set A

Match Words and Expressions

▶ **Learn**

WHAT'S LEFT?   Tyna worked 3 hours and earned $6 each hour. She spent $4 of the money she earned on postcards. Find how much money she had left.

Think: 3 hours times $6 each hour minus $4

Which expression matches the meaning of the words?

$(3 \times 6) - 4$ ← First find the money earned, (3×6), and then subtract the money spent from the money earned.

$3 \times (6 - 4)$ ← First subtract the money spent from the hourly rate, $(6 - 4)$, and then multiply by the hours worked.

$(3 \times 6) - 4$ Matches the meaning.

$3 \times (6 - 4)$ Does not match the meaning.

To find how much money Tyna had left, find the value of the expression $(3 \times 6) - 4$.

$(3 \times 6) - 4$ Multiply 3 and 6.
↓
$18 - 4$ Subtract 4 from 18.
↓
14

So, Tyna had $14 left.

MATH IDEA   The meaning of the words in the problem tells you where to place the parentheses if there is more than one operation.

Example   Choose the expression that matches the words.

Sue bought 6 tickets at $4 each and food for $3. $(6 \times 4) + 3$ or $6 \times (4 + 3)$

Match an expression to the words.
6 tickets at $4 each and food for $3
↓   ↓   ↓
(6×4) +   3

So, the expression $(6 \times 4) + 3$ matches the words.

Choose the expression that matches the words.

1. Rick had $8 and then worked 4 hours for $5 each hour.

   a. $(8 + 4) \times 5$ **b.** $8 + (4 \times 5)$

2. Teri had 4 pages with 7 stickers on each and then gave away 5 stickers.

   a. $(4 \times 7) - 5$ **b.** $4 \times (7 - 5)$

► **Practice and Problem Solving**

Choose the expression that matches the words.

3. April had 40¢. She gave away 2 nickels.

   a. $40 - (2 \times 5)$ **b.** $(40 - 2) \times 5$

4. Julie had $14. She spent $3 and then was given $4.

   a. $(14 - 3) + 4$ **b.** $14 - (3 + 4)$

5. There are 2 desks not in rows and 6 rows with 5 desks in each row.

   a. $(2 + 6) \times 5$ **b.** $2 + (6 \times 5)$

6. Sue bought 4 tickets for $7 each. She paid $2 in sales tax on the total.

   a. $(4 \times 7) + 2$ **b.** $4 \times (7 + 2)$

For 7–8, choose an expression from the box to match the words. Then find the value of the expression.

7. There are 4 desks not in rows and 3 rows with 5 desks in each row. How many desks are there in all?

$$(4 + 3) \times 5 \qquad 4 + (3 \times 5)$$
$$(15 - 8) + 2 \qquad 15 - (8 + 2)$$

8. Robbie had 15 postcards. He gave away 8 of these, and then he bought 2 more. How many cards did he have then?

9. Albert gave 122 of his trading cards to James. Albert now has 385 trading cards. How many trading cards did Albert have before he gave some to James?

10. **? What's the Error?** John claims that $5 \times (4 + 3) = 23$. Describe his error. Give the correct answer.

Mixed Review and Test Prep

11. 473 (p. 40)
    $\underline{+592}$

12. $1,007$ (p. 40)
    $\underline{-\ 829}$

13. $8 \times (3 \times 2)$ (p. 150)

14. $25 - (4 + 12)$ (p. 56)

15. **TEST PREP** A new movie began advertising 3 months before opening. The movie opened on July 4. In what month did advertising begin? (p. 126)

   A March   **C** May
   B April   **D** June

3 Equations

Quick Review

1. 3×2 2. 5×4

3. 3×4 4. 6×3

5. 3×7

▶ **Learn**

IS IT EQUAL? In this activity, you will multiply both sides of an equation by the same number to test if both sides stay equal to each other.

 Activity

MATERIALS: coins, pieces of paper

Make a workmat.

$1 \times \$0.10$ $2 \times \$0.05$

Each side of the workmat represents one side of an equation.

Place 1 dime on the left side and 2 nickels on the right side.

- Compare the two sides. Are the values of the coins equal? Explain.

- Multiply the value of the coins on each side by 4. Compare the two sides. Are the values still equal? Explain.

$4 \times \$0.10$ $8 \times \$0.05$

- **What if** you multiply the value of the coins on one side by 4 and the value of the coins on the other side by 2? Will the values of the two sides still be equal? Explain.

> **Remember**
>
> An **equation** is a number sentence. It uses an equal sign to show that two amounts are equal.

Multiply Both Sides

Use the workmat. Decide if both sides stay equal.

a. Left side: 5 pennies; right side: 1 nickel;
multiply both sides by 3.

b. Left side: 3 nickels; right side: 2 nickels
5 pennies; multiply the left side by 2 and
the right side by 3.

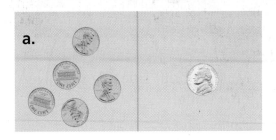

a.

b.

MATH IDEA   When you multiply both sides of an
equation by the same number, the two sides
stay equal.

Examples   Multiply both sides of the equation by the
given number. Find the values.

Ⓐ   6 = 6; multiply by 2.

$$6 = 6$$
$$6 \times 2 = 6 \times 2 \quad \leftarrow \text{Multiply by 2.}$$
$$\downarrow \qquad \downarrow$$
$$12 \quad = \quad 12$$

Ⓑ   12 = 2 × 6; multiply by 4.

$$12 = 2 \times 6$$
$$12 \times 4 = (2 \times 6) \times 4 \quad \leftarrow \text{Multiply by 4.}$$
$$\downarrow \qquad \qquad \downarrow$$
$$48 \quad = \quad 48$$

Ⓒ   (3 + 5) = (1 + 7); multiply by 3.

$$(3 + 5) = (1 + 7)$$
$$(3 + 5) \times 3 = (1 + 7) \times 3 \quad \leftarrow \text{Multiply by 3.}$$
$$\downarrow \qquad \qquad \downarrow$$
$$8 \times 3 \quad = \quad 8 \times 3$$
$$\downarrow \qquad \qquad \downarrow$$
$$24 \quad = \quad 24$$

Ⓓ   (3 × 3) = (27 ÷ 3); multiply by 5.

$$(3 \times 3) = (27 \div 3)$$
$$(3 \times 3) \times 5 = (27 \div 3) \times 5 \quad \leftarrow \text{Multiply by 5.}$$
$$\downarrow \qquad \qquad \downarrow$$
$$9 \times 5 \quad = \quad 9 \times 5$$
$$\downarrow \qquad \qquad \downarrow$$
$$45 \quad = \quad 45$$

▶ Check

1. **Show** that when you multiply both sides of the
equation (4 + 5) = (6 + 3) by the number 3,
the values of the sides stay equal.

**Multiply both sides by the given number. Find the
new value.**

2. 7 = 7; multiply both sides by 5.

3. 6 = 2 × 3; multiply both sides by 8.

LESSON CONTINUES ▶

Practice and Problem Solving

Multiply both sides by the given number. Find the new value.

4. 3 nickels = 3 nickels; multiply both sides by 4.

5. 1 dime 2 pennies = 2 nickels 2 pennies; multiply both sides by 5.

6. 2 nickels = 1 dime; multiply both sides by 7.

7. 5 pennies = 1 nickel; multiply both sides by 6.

8. $9 = 3 \times 3$; multiply both sides by 9.

9. $8 = 2 \times 4$; multiply both sides by 8.

10. $4 \times 3 = 12$; multiply both sides by 5.

11. $12 = 3 \times 4$; multiply both sides by 7.

12. $(4 + 2) = (2 \times 3)$; multiply both sides by 2.

13. $(2 + 6) = (2 \times 4)$; multiply both sides by 4.

14. $(3 + 7) = (2 \times 5)$; multiply both sides by 3.

15. $(8 + 4) = (6 \times 2)$; multiply both sides by 6.

USE DATA For 16–19, use the graph.

16. How many more apple pies are baked than coconut pies?

17. The bakery has an order for 40 apple pies. Are there enough pies baked to fill the order? Explain.

18. Mrs. Holiday wants 24 peach pies for a picnic. How many more peach pies does the bakery need to make?

19. The museum bought all the coconut pies for a party. How many pies did the museum buy?

20. **What if** you multiply both sides of the equation $(4 + 3) = (49 \div 7)$ by 3? Are the values of the sides of the equation still equal?

21. **? What's the Error?** Jan claims that if you multiply the left side of $5 = 5$ by 3 and the right side by 2, the equation will stay equal. What is her error?

NUMBER OF PIES BAKED

Type of Pie

Peach

Blueberry

Coconut

Apple

Each 🥧 = 6 pies.

22. Ian invited 10 friends bowling. Each friend bowled 3 games. Ian bowled 2 games. Write an expression to show how many games were bowled.

23. There were 12 rows of plants. Each row had 4 flowers. Mr. Williams picked 3 flowers. Write an expression to show how many flowers are in the garden now.

24. There are 18 boxes. There are twice as many large boxes as small boxes. How many large boxes are there? How many small boxes are there?

Mixed Review and Test Prep

25. Write $800,000 + 40,000 + 300 + 9$ in standard form. (p. 6)

26. $123 + 656$ (p. 36)

27. $2,389 - 921$ (p. 40)

28. Denise has 3 cases of pens. Each case has 6 boxes. Each box has 2 pens. How many pens does she have? (p. 150)

29. $\begin{array}{r} 2,173 \text{ (p. 40)} \\ +3,584 \\ \hline \end{array}$

30. $\begin{array}{r} 9,314 \text{ (p. 40)} \\ -6,763 \\ \hline \end{array}$

31. **TEST PREP** A local station is showing a movie that begins at 7:00 P.M. It is 90 minutes long. At what time will it end? (p. 120)

A 5:00 P.M.   **C** 8:00 P.M.

B 6:30 P.M.   **D** 8:30 P.M.

PROBLEM SOLVING LiNKUP... to Reading

STRATEGY • CAUSE AND EFFECT Sometimes one detail in a problem has an effect on another detail. The **cause** is the reason something happens. The **effect** is the result.

Study Karen's party plans on the clipboard.

What if four more people came to Karen's party? Does she have enough chairs? If not, how many more chairs does she need?

Copy and complete the table.

1. List the causes in the **Cause** column. List the effects in the **Effect** column.

Karen's Party

29 people
8 tables
4 chairs at each table

CAUSE	EFFECT
4 more people came to the party.	There are more people at the party than she planned for.
Think: How many people were at the party?	**Think:** What will happen?

2. Solve the problem.

3. Felix bought 7 packs of bottled water with 6 bottles in each. He drank two bottles of water. How many bottles does he have left? List the cause and effect.

Expressions with Variables

Quick Review

Solve for *r*.

1. $6 + r = 16$ **2.** $7 + 5 = r$

3. $r + 8 = 17$ **4.** $r - 5 = 9$

5. $23 - r = 21$

▶ **Learn**

***B* IS FOR BASKET**   Tara made some baskets from behind the 3-point line. What expression can you write to tell the total number of points she scored?

You can use a variable to represent the number of 3-point baskets that Tara made.

<u>number of 3-point baskets</u> × <u>3 points</u>
                ↓   ↓
               b   ×   3

The variable *b* stands for the number of 3-point baskets that were made.

Suppose Tara made four 3-point baskets. To find the value of the expression, replace *b* with the number of baskets made.

$b \times 3$

$4 \times 3 = 12$ ← Replace *b* with 4 since she made 4 baskets.

So, Tara scored 12 points.

Remember

A ***variable*** can stand for any number.

Examples   During a basketball game, Fred made eight 3-point baskets and Juan made six 3-point baskets.

A Use the expression $b \times 3$ to find how many points Fred scored.	**B** Use the expression $b \times 3$ to find how many points Juan scored.
$b \times 3$ Replace *b* with 8.	$b \times 3$ Replace *b* with 6.
↓	↓
8×3 Evaluate the expression.	6×3 Evaluate the expression.
↓	↓
24	18
So, Fred scored 24 points.	So, Juan scored 18 points.

1. **Write** an expression you could use to find the number of points scored for any number of 2-point baskets.

Find the value of the expression.

2. $4 \times y$ if $y = 7$ **3.** $4 \times y$ if $y = 8$ **4.** $18 \div x$ if $x = 3$ **5.** $54 \div x$ if $x = 6$

▶ **Practice and Problem Solving**

Find the value of the expression.

6. $9 \times y$ if $y = 7$ **7.** $9 \times y$ if $y = 9$ **8.** $96 \div x$ if $x = 12$ **9.** $56 \div y$ if $y = 8$

10. $a \times 9$ if $a = 12$ **11.** $b \times 4$ if $b = 7$ **12.** $x \div 5$ if $x = 35$ **13.** $x \div 5$ if $x = 40$

Choose the expression that matches the words.

14. 4 times a number of cards, c, in a pack
  a. $4 \times c$ **b.** $4 + c$

15. $24 divided by a number of people, p
  a. $p \div 24$ **b.** $24 \div p$

Write an expression that matches the words.

16. 7 times a number of pages, p, in a book

17. 18 players divided into a number of teams, t

18. 36 boys divided by a number of baseballs, b

19. a number of cartons, c, times 12 cans

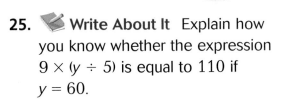

Find the value of the expression if $y = 4$, if $y = 8$, and if $y = 10$.

20. $8 \times y$

21. $12 \times y$

22. $40 \div y$

23. $y \div 2$

24. Erin has 5 quarters in each of 4 rows. Jocelyn has 8 quarters in each of 3 rows. How much more money does Jocelyn have than Erin?

25. ✍ **Write About It** Explain how you know whether the expression $9 \times (y \div 5)$ is equal to 110 if $y = 60$.

Mixed Review and Test Prep

For 26–27, write a related division equation. (p. 140)

26. $6 \times 9 = 54$ **27.** $9 \times 11 = 99$

28. $3,894 - 2,085$ (p. 40)

29. $185 - 165$ (p. 36)

30. **TEST PREP** There are 6 boxes with 12 computer disks in each box. How many disks are there? (p. 148)
  A 64 disks   **C** 78 disks
  B 72 disks   **D** 86 disks

Quick Review
1. 6×6 2. 6×4
3. 4×7 4. 9×2
5. 3×5

▶ **Learn**

THE MISSING PIECE   A variable can be used in an equation to represent missing or unknown information. Any letter can be used as a variable.

Notice how the equations change when the missing information changes.

A 6 marbles in each of 8 bags is a total number of marbles.
↓   ↓   ↓
6×8 = t ← t is the total number of marbles.

B 6 marbles in each of a number of bags is a total of 48 marbles.
↓   ↓   ↓
$6 \times b$ =   48   ← b is the number of bags.

C A number of marbles in each of 8 bags is a total of 48 marbles.
↓   ↓   ↓
$m \times 8$ =   48   ← m is the number of marbles in each bag.

▶ **Check**

1. **Choose** the equation that shows that a number of dimes divided equally by 4 people is 10 dimes each.
   a. $4 \times d = 10$ **b.** $d \div 4 = 10$ **c.** $10 \div d = 4$

Write an equation for each. Choose a variable for the unknown. Tell what the variable represents.

2. 3 shelves with the same number of prizes on each shelf is 18 prizes.

3. 18 tickets divided equally for 3 people is the number of tickets for each person.

4. Some cans of soda for each of 4 people is a total of 8 cans of soda.

Write an equation for each. Choose a variable for the
unknown. Tell what the variable represents.

5. 10 pencils in each of 4 boxes is the
total number of pencils.

6. Some pencils in each of 4 boxes
is a total of 40 pencils.

7. 40 pencils in each of a number of
boxes is 10 pencils in each box.

8. A number of cars divided equally
among 5 rows is 9 cars in each row.

9. The number of ice cubes in each of
5 glasses is a total of 30 ice cubes.

10. 6 students in each of 9 groups is the
total number of students.

11. A number of crayons divided equally
among 4 boxes is 8 crayons in
each box.

12. 50 grapes divided equally among
a number of bags is 10 grapes in
each bag.

USE DATA For 13–15, use the table.

13. At the school carnival, Carlos and 4 of
his friends ate some hot dogs. They
spent $20.00. Write an equation
to show the number of hot dogs
they ate.

Burgers & Bites	
Burger	$3
Hot Dog	$2
Fries	$2
Soda	$1

14. ✎ **Write a problem** about the
school carnival that can be solved
by writing an equation with one
variable. The equation can involve
multiplication or division.

15. In the first hour, the concession stand
collected a total of $36 just for
burgers. Write an equation to show
the total number of burgers sold.

Mixed Review and Test Prep

16. 46 (p. 36)
    27
    +41

17. 788 (p. 36)
    −321

18. 32,989 (p. 44)
    −13,294

19. 22,509 (p. 44)
    +49,486

20. ⭐ **TEST PREP** In the parking lot, there
are 8 rows with 12 spaces in each.
Thirteen spaces are taken. How many
more cars can park in the parking lot?
(p. 148)

A 66   C 83

B 72   D 96

▶ **Learn**

INPUT/OUTPUT The Math Factory sorts numbers into boxes. Use the input/output table to find what number comes out when 45 is put into the machine.

INPUT	OUTPUT
25	5
30	6
35	7
40	8
45	■

HINT: Look for a pattern to help you find a rule.

Pattern: Each output is the input divided by 5.

Rule: Divide by 5.

← *Input:* 45   *Output:* 45 ÷ 5 = 9

So, when you put in 45, the Math Factory puts out 9.

You can write an equation to show the rule. Use variables to show the input and output.

input   output
↓   ↓
x ÷ 5 = y Think of the equation as a rule. To find the value of y, divide x by 5.

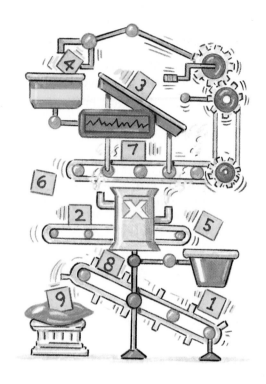

Examples
Find a rule. Write the rule as an equation.

A
INPUT	n	3	4	5	6
OUTPUT	t	18	24	30	36

Rule: Multiply by 6.
    $n \times 6 = t$

Test your rule on each pair of numbers in the table.

B
INPUT	p	72	63	54	45
OUTPUT	r	8	7	6	5

Rule: Divide by 9.
    $p ÷ 9 = r$

1. **Explain** why it is important to test a rule with all the number pairs in the table.

Find a rule. Write the rule as an equation.

2.
INPUT	x	21	28	35	42
OUTPUT	y	3	4	5	6

3.
INPUT	s	8	9	10	11
OUTPUT	t	40	45	50	55

► **Practice and Problem Solving**

Find a rule. Write the rule as an equation.

4.
INPUT	b	12	11	9	7
OUTPUT	c	24	22	18	14

5.
INPUT	d	16	32	48	64
OUTPUT	f	2	4	6	8

6.
INPUT	w	6	8	10	12
OUTPUT	z	72	96	120	144

7.
INPUT	r	18	24	30	36
OUTPUT	s	6	8	10	12

Use the rule and the equation to make an input/output table.

8. Divide by 11.
$x \div 11 = y$

9. Multiply by 4.
$h \times 4 = j$

10. Divide by 10.
$t \div 10 = w$

11. Multiply by 8.
$f \times 8 = g$

12. Multiply by 7.
$r \times 7 = s$

13. Divide by 6.
$c \div 6 = d$

14. Multiply by 3.
$k \times 3 = m$

15. Divide by 12.
$a \div 12 = b$

16. Eva made 48 note cards. She divided the cards equally among 4 friends. Find the number of note cards Eva gave each friend.

17. A magnet store had 1,000 magnets. In the first week, five hundred twenty-six magnets were sold. How many magnets are left?

Mixed Review and Test Prep

18. $34,578 + 27,916$ (p. 44)

19. $98,521 - 85,769$ (p. 44)

20. Over the weekend, the Garden Shop sold 9,754 plants in all. It sold 4,675 plants on Friday and 1,987 plants on Saturday. How many plants did the Garden Shop sell on Sunday? (p. 40)

21. What is 12,784 written in expanded form? (p. 4)

22. **TEST PREP** Which is **not** true? (p. 18)
   A $6,431 > 6,341$
   B $6,236 = 6,236$
   C $6,357 < 6,135$
   D $6,964 > 6,694$

Problem Solving Strategy
Work Backward

Quick Review

1. $3 \times n = 27$

2. $n \div 3 = 5$

3. $36 \div n = 9$

4. $5 \times n = 30$

5. $5 + n = 9 \times 8$

PROBLEM Tony had some dimes in his bank. He added 7 nickels and then had a total of 75¢. How much money did Tony have in dimes?

UNDERSTAND

- What are you asked to find?

   What information will you use?

   Is there information you will not use? Explain.

PLAN

- What strategy can you use to solve the problem?

   You can write an equation with a variable. Then solve the equation by *working backward*.

SOLVE

- How can you solve the problem?

   Write an equation. Let *d* represent the amount of money in dimes.

   $$d + (7 \times 5) = 75 \leftarrow \text{75¢ in all}$$
   $\llcorner$ amount in dimes

   To find the value of *d*, work backward.

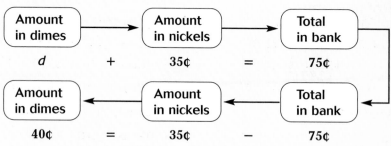

Amount in dimes		Amount in nickels		Total in bank
d	+	35¢	=	75¢

Amount in dimes		Amount in nickels		Total in bank
40¢	=	35¢	−	75¢

   So, the variable, *d*, has a value of 40. Tony had 40¢ in dimes.

CHECK

- How can you use the expression $d + (7 \times 5)$ to check the solution?

▶ Problem Solving Practice

Write an equation and work backward to solve.

1. **What if** Tony had a total of 85¢ after adding 7 nickels? How much would Tony have in dimes?

2. Sid had 5 trading cards. Then his mother gave him some packages with 8 cards in each package. Sid now has 37 cards. How many packages did his mother give him?

Joey and Nicole are playing a board game. In the first three turns, Joey moves 6 spaces forward, 3 back, and 4 forward. Nicole moves 5 spaces forward, 1 back, and 5 forward.

3. Who is ahead in the game?
   - **A** Joey
   - **B** Nicole
   - **C** Lucy
   - **D** They are on the same space.

4. How many spaces are between Joey and Nicole?
   - **F** 1 space
   - **G** 3 spaces
   - **H** 4 spaces
   - **J** 5 spaces

PROBLEM SOLVING STRATEGIES

Draw a Diagram or Picture
Make a Model or Act It Out
Make an Organized List
Find a Pattern
Make a Table or Graph
Predict and Test
▶ **Work Backward**
Solve a Simpler Problem
Write an Equation
Use Logical Reasoning

Problem Solving Strategy

LOSE 1 TURN   GO AHEAD 2   TAKE EXTRA TURN

Mixed Strategy Practice

5. The Snack Bar at the local skating rink uses 8 lemons for every 2 quarts of lemonade. How many lemons are used to make 8 quarts of lemonade?

6. Ty had some money in his coin bank. He put 3 dimes and 7 pennies in the bank and now has $1.17. How much money was in the bank to begin with?

7. The Snack Bar sold 341 drinks on Saturday. On Sunday, 85 drinks were sold in the morning and 163 in the afternoon. How many more drinks were sold on Saturday than Sunday?

8. Taylor had 127 baseball cards. He gave 18 cards to Felisha. Felisha then gave Taylor some cards. Taylor then had a total of 114 cards. How many cards did Felisha give Taylor?

9. Sally's dance lesson begins at 4:25 P.M. and lasts for 1 hr 15 min. What time does her lesson end?

10. ✏️ **Write a problem** involving coins that can be solved using the strategy *work backward*. Then solve the problem.

Review/Test

✓ CHECK CONCEPTS

1. Tell which operation to do first to find the value of $2 + (3 \times 4)$. How do you know? (pp. 158–161)

2. Write an expression you could use to find the number of quarters that have the same value as any number of dollars. (pp. 168–169)

✓ CHECK SKILLS

Find the value of the expression. (pp. 158–161)

3. $(18 \div 3) + 2$

4. $(6 \times 7) - 5$

5. $(63 \div 7) - 9$

6. $25 + (4 \times 6)$

7. $(32 - 8) \div 8$

8. $7 - (72 \div 12)$

9. $45 - (7 \times 6)$

10. $(8 \times 9) + 21$

Write an expression that matches the words. (pp. 162–163)

11. a number of books, b, divided by 9 readers

12. $42 divided by a number of people, p

Multiply both sides by the given number. Find the new value. (pp. 164–167)

13. $6 = (2 \times 3)$; multiply both sides by 9.

14. $(6 \times 2) = (4 + 8)$; multiply both sides by 11.

Find a rule. Write the rule as an equation. (pp. 172–173)

15.

INPUT	OUTPUT
a	b
4	28
5	35
8	56

16.

INPUT	OUTPUT
x	y
72	9
56	7
40	5

17.

INPUT	OUTPUT
m	n
5	55
7	77
9	99

18.

INPUT	OUTPUT
q	r
144	12
132	11
120	10

✓ CHECK PROBLEM SOLVING

Write an equation and work backward to solve. (pp. 174–175)

19. Tommy had 3 nickels. Angela gave him some quarters. Now he has 90¢. How much money does Tommy have in quarters?

20. Jack had some dimes. He spent 4 dimes on candy and had 30¢ left over. How many dimes did Jack have before buying the candy?

Standardized Test Prep

TIP! Eliminate choices.
See item **2.**

Look at the operation signs in each expression to find the most likely choices. Find the values of only the most likely expressions.

Also see problem **5**, p. H64.

For 1–9, choose the best answer.

1. What is 3,743,279 rounded to the nearest million?

  A 3,000,000   **C** 3,800,000

  B 3,700,000   **D** 4,000,000

2. Which expression has a value of 0?

  F $24 - (19 - 5)$

  G $24 - (19 + 5)$

  H $24 + (19 - 5)$

  J $24 + (19 + 5)$

3. Which equation is true?

  A 5 dimes = 10 nickels

  B 10 dimes = 5 nickels

  C 20 dimes = 30 nickels

  D 5 dimes = 20 nickels

4. What is the value of $6 + (4 \times 5)$?

  F 15   **H** 50

  G 26   **J** 120

5. What is the value of $4 \times (6 - 2)$?

  A 2   **C** 22

  B 16   **D** 26

6. What is the value of $9 \times y$ for $y = 8$?

  F 56   **H** 72

  G 63   **J** 81

7.   32,184

    $- \underline{27,366}$

  A 14,818   **C** 4,728

  B 4,818   **D** NOT HERE

8. There is a total of 72 pencils in 6 boxes. Each box has the same number of pencils in it. Which equation can be used to find p, the number of pencils in each box?

  F $6 \times 72 = p$ **H** $p \times 72 = 6$

  G $p \div 6 = 72$ **J** $p \times 6 = 72$

9. Which expression can be used to find the cost of photocopying a number of pages at 15¢ per page?

  A $15 \times p$ **C** $15 + p$

  B $p \div 15$ **D** $p - 15$

Write What You Know

10. Find an equation that describes a rule for this table. Use the data in the table to check your equation.

INPUT	b	2	4	5	7
OUTPUT	c	14	28	35	49

11. Eight adults and 5 children went to a county fair. Adult admission cost $6 and child admission cost $4. Write an expression you can use to find the total cost of admission for this group. Find the cost.

Three Cheers for Multiplication

You can add two numbers to find their sum. You can also multiply the same two numbers to find their product.

Think about the numbers 2 and 3.

What is their sum? $2 + 3 = 5$

What is their product? $2 \times 3 = 6$

For each problem, use your reasoning skills to find the two numbers that give the sum and product shown. The first one is done for you.

Copy and complete the table.

	SUM	PRODUCT	FIRST NUMBER	SECOND NUMBER
1.	5	6	2	3
2.	7	12	▨	▨
3.	7	10	▨	▨
4.	7	6	▨	▨
5.	13	36	▨	▨
6.	10	24	▨	▨
7.	15	50	▨	▨
8.	13	30	▨	▨
9.	13	12	▨	▨
10.	20	100	▨	▨
11.	25	100	▨	▨
12.	12	0	▨	▨

Think It Over!

- **STRETCH YOUR THINKING** Give a sum and a product for which it is **not** possible to find two whole numbers that give that sum and product.

- Find three pairs of numbers that give a sum of 13 and a product not given in the table. List each pair of numbers and their product.

Challenge

Square Numbers

A number that can be modeled with a square array is called a square number.

A square number is a number that is the product of any number and itself.

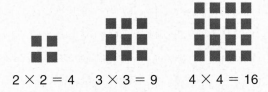

$2 \times 2 = 4$ $3 \times 3 = 9$ $4 \times 4 = 16$

To find some square numbers, start by multiplying 1 by 1.

$1 \times 1 = 1$, so 1 is a square number.

Continue multiplying to find other square numbers.

$2 \times 2 = 4$, so 4 is a square number.

$3 \times 3 = 9$, so 9 is a square number.

Another way to show 3×3 is 3^2, which is read "three squared."

5^2, or "5 squared" means 5×5, or 25.

The **square root** of a number is one of the 2 equal factors of that number.

$5 \times 5 = 25$, so 5 is the square root of 25.

$3 \times 3 = 9$, so 3 is the square root of 9.

×	1	2	3	4	5	6	7	8	9	10	11	12
1	1	2	3	4	5	6	7	8	9	10	11	12
2	2	4	6	8	10	12	14	16	18	20	22	24
3	3	6	9	12	15	18	21	24	27	30	33	36
4	4	8	12	16	20	24	28	32	36	40	44	48
5	5	10	15	20	25	30	35	40	45	50	55	60
6	6	12	18	24	30	36	42	48	54	60	66	72
7	7	14	21	28	35	42	49	56	63	70	77	84
8	8	16	24	32	40	48	56	64	72	80	88	96
9	9	18	27	36	45	54	63	72	81	90	99	108
10	10	20	30	40	50	60	70	80	90	100	110	120
11	11	22	33	44	55	66	77	88	99	110	121	132
12	12	24	36	48	60	72	84	96	108	120	132	144

▲ The square numbers form a diagonal line in the multiplication table.

Try It

Copy and complete the table.

	FACTORS	SQUARE NUMBER	SQUARE ROOT
1.	3×3	9	3
2.	4×4	▪	▪
3.	5×5	▪	▪
4.	6×6	▪	▪

	FACTORS	SQUARE NUMBER	SQUARE ROOT
5.	7×7	▪	▪
6.	8×8	▪	▪
7.	9×9	▪	▪
8.	10×10	▪	▪

Study Guide and Review

VOCABULARY

Choose the best term from the box.

1. Multiplication and division undo each other. They are _?_ operations. (p. 140)

2. The _?_ states that when the grouping of factors is changed, the product remains the same. (p. 150)

> inverse
> **Grouping Property of Multiplication**
> **fact family**

STUDY AND SOLVE

Chapter 8

Write related multiplication and division facts.

> Find the value of n. Write a related multiplication or division equation.
>
> $28 \div 4 = n$
>
> Multiplication: $4 \quad \times \quad 7 \quad = \quad 28$
> $\qquad\qquad\downarrow \qquad\quad \downarrow \qquad\quad \downarrow$
> $\qquad$ factor $\quad$ factor $\quad$ product
>
> Division: $\quad 28 \quad \div \quad 4 \quad = \quad 7$
> $\qquad\qquad\downarrow \qquad\quad \downarrow \qquad\quad \downarrow$
> $\qquad$ dividend $\quad$ divisor $\quad$ quotient
>
> $4 \times 7 = 28$ is related to $28 \div 4 = 7$.
>
> So, the value of n is 7.

Find the value of the variable. Write a related multiplication or division equation. (pp. 140–141, 142–149)

3. $88 \div 11 = y$ **4.** $9 \times 6 = h$

5. $3 \times 5 = s$ **6.** $18 \div 6 = n$

7. $90 \div 10 = p$ **8.** $6 \times 8 = a$

9. $8 \times 10 = x$ **10.** $144 \div 12 = b$

11. $72 \div 8 = t$ **12.** $6 \times 7 = k$

Find the value of expressions with three factors.

> Find the value.
>
> $2 \times (5 \times 4)$
> $\qquad\quad \downarrow$
> $2 \times \quad 20$
> $\qquad 40$
>
> • Find the product in parentheses first.
>
> • Multiply.
>
> So, $2 \times (5 \times 4) = 40$.

Find the product. (pp. 150–151)

13. $(3 \times 2) \times 9$ **14.** $5 \times (3 \times 4)$

15. $6 \times (3 \times 4)$ **16.** $(6 \times 2) \times 7$

17. $(3 \times 3) \times 3$ **18.** $6 \times (2 \times 3)$

19. $3 \times (2 \times 4)$ **20.** $(5 \times 2) \times 8$

21. $(2 \times 4) \times 5$ **22.** $4 \times (3 \times 4)$

Chapter 9

Write an expression for the words, and then find the value.

Erika had 12 pencils. Then she bought 3 more packs with 5 pencils in each pack.

$12 + (3 \times 5)$
↓
$12 + \quad 15$
↓
27

- Write an expression for the number of pencils.
- Multiply the number of pencils that Erika bought.
- Find the value of the expression.

So, Erika now has 27 pencils.

Write an expression for the words. Use a variable.

8 times the number of people, *m*

8 times the number of people
↓ ↓   ↓
8   ×   *m*

- Choose an operation.
- Write the expression.

Write an expression. Then find the value. (pp. 158–161, 162–163)

23. Maryann bought 4 hair ribbons for $3 each and then spent $6 on lunch. How much did she spend in all?

24. Robin had 50¢. She gave away 2 dimes. How much money does she have now?

25. Carole had 6 pennies and 3 nickels. How much money did she have?

Write an expression that matches the words. (pp. 168–169)

26. 7 times the number of people, *p*, in a room

27. $60 divided by the number of equally-sized bank accounts, *b*

28. 24 kittens in several same-size litters, *l*

PROBLEM SOLVING PRACTICE

Solve. (pp. 152, 174)

29. Toni earns $2.00 baby-sitting and $4.00 mowing lawns each week. If she works for 8 weeks, how much money will she have earned? Name the operation or operations you used.

30. Charlie had 12¢. His father gave him some nickels, and now he has 47¢. How many nickels did Charlie's father give him?

31. The Garden Club plants flowers around the park. They planted an equal number of plants in 6 gardens. If the club had 72 trays of plants, how many trays were planted in each garden? Name the operation or operations you used.

PERFORMANCE ASSESSMENT

TASK A • OUTDOOR GAMES

After school 36 students stay to play outdoor games. The students are divided into teams. There are at least two players on a team, and all the teams have the same number of players.

a. Draw arrays to show two different ways 36 students can be divided into teams of equal size.

b. Write a multiplication sentence and a division sentence for each array you drew. Explain how the sentences you wrote show that division is the inverse of multiplication.

c. Suppose 4 more students want to play the outdoor games. Will the teams be able to stay the same size as before, or will different-sized teams need to be formed?

TASK B • WORK IT OUT

After school and on weekends, Nicole and her sister Amy earn spending money. Nicole takes some of her neighbors' dogs for walks and Amy mows lawns.

a. Nicole earns $2 for each dog she walks. Write an expression to find the number of dollars she will earn for walking any number of dogs. (Use d to represent the number of dogs.) Then use the expression you wrote to find out how much Nicole will earn if she walks 4 dogs.

b. Amy made this input/output table to help her know how much money (d) she will earn for mowing different numbers of lawns (m). How much does Amy earn for each lawn she mows? Write an equation that describes a rule for the table.

INPUT	OUTPUT
m	d
1	6
2	12
3	18
4	24

c. Nicole and Amy want to buy a CD that costs $16. What is one way they can earn $16 together? Tell how many dogs Nicole would have to walk and how many lawns Amy would have to mow.

Mighty Math Calculating Crew • Multiplication and Division Facts

Use *Nick Knack, SuperTrader* to practice multiplication and division facts. Click on Captain Nick Knack to go to Planet Havarti to help him take inventory.

Click  and choose Grow Slide Level E. Then click OK.

Captain Nick Knack will show you a model and ask for your help to solve multiplication problems. Look at the model, then type your answer using the key pad. Continue answering all the questions for Captain Nick Knack.

Practice and Problem Solving

1. Choose Grow Slide Level J. Answer at least 5 questions. Draw a model for each division problem and write the expression.

Multiply or divide.

2. 3×8	**3.** $63 \div 9$	**4.** 6×7	**5.** $60 \div 5$
6. $108 \div 12$	**7.** 9×5	**8.** $80 \div 10$	**9.** 12×4
10. 11×11	**11.** $144 \div 12$	**12.** $132 \div 11$	**13.** 9×11
14. $81 \div 9$	**15.** 8×7	**16.** 7×12	**17.** $48 \div 8$

18. **REASONING** Campbell multiplied $7 \times 4 \times 5$ by grouping $(7 \times 4) \times 5$. Erika multiplied by grouping $7 \times (4 \times 5)$. Whose way is easier? Why?

Multimedia Math Glossary www.harcourtschool.com/mathglossary

19. **Vocabulary** Find out more about the terms *factor* and *product* in the Multimedia Math Glossary. Write other multiplication or division expressions and label them using the examples shown in the glossary.

PROBLEM SOLVING ON LOCATION

at the Beach

TIDE POOLS

Tide pools are pools that form where the ocean meets the land. Tide pools provide homes for many small ocean plants and animals.

For 1–3, write an equation with a variable that matches each statement. Tell what the variable represents. Find the value of the variable.

1. 5 arms each on some starfish are 35 arms.

2. Some arms on each of 10 starfish are 60 arms.

3. 12 blue crabs divided equally among 3 tide pools is the number of blue crabs in each pool.

4. Suppose you see 4 hermit crabs in each of 4 tide pools and 5 hermit crabs in each of 2 tide pools. How many hermit crabs have you seen in all? Write an equation you can use to solve the problem.

5. The table shows the number of tentacles on 1 to 5 octopods. Find a rule. Write the rule as an equation, and complete the table.

INPUT o	OUTPUT t
1	8
2	16
3	24
4	32
5	■

▲ **Hilton Head Island, SC** is the second largest barrier island in the U.S. It is about 12 miles long and 3 miles wide.

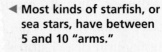

◄ Most kinds of starfish, or sea stars, have between 5 and 10 "arms."

BOTTLENOSE DOLPHINS

Bottlenose dolphins are the most common kind of dolphin you will find along the United States from Cape Cod through the Gulf of Mexico. You can often see bottlenose dolphins if you visit the beaches in South Carolina.

1. A baby calf nurses 4 times each hour for the first week of its life. How many times does it nurse in 8 hours during this time?

2. Bottlenose dolphins can swim about 20 miles per hour in short bursts. If a dolphin could keep up this speed, how long would it take the dolphin to swim the 60 miles from Hilton Head Island to Folly Beach?

3. Bottlenose dolphins usually dive from 3 to 46 feet deep. If a dolphin dives 3 feet in its first dive and 33 feet in its second dive, how many times as far did it dive the second time than the first time?

4. Even though dolphins usually dive to as little as 3 feet, they can dive deeper. A trained dolphin dove about 600 times as deep as 3 feet. About how deep did the trained dolphin dive?

5. Dolphins eat fish, squids, and crustaceans. A 300-pound dolphin eats around 12 pounds of food each day. About how much food does a 300-pound dolphin eat in 5 days? in a week?

6. There are about 67,000 bottlenose dolphins in the U.S. Gulf of Mexico and 11,700 in the U.S. waters of the western North Atlantic. How many dolphins is this in all? How many more are in the Gulf of Mexico?

A dolphin opens its blowhole ▶ and starts to breathe out under water. When it jumps out of the water, the dolphin breathes in and then closes the blowhole.

Multiply by 1-Digit Numbers

Lightning strikes occur somewhere on Earth over 100 times every second. The bar graph shows the number of lightning flashes per square mile during a storm.

PROBLEM SOLVING Find the number of flashes in a 5-square-mile area for each location.

DATA
LINK

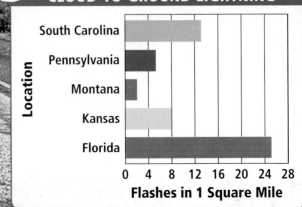

CLOUD-TO-GROUND LIGHTNING

Location	Flashes in 1 Square Mile
South Carolina	
Pennsylvania	
Montana	
Kansas	
Florida	

0 4 8 12 16 20 24 28
Flashes in 1 Square Mile

Use this page to help you review and remember important skills needed for Chapter 10.

✔ VOCABULARY

Choose the best term from the box.

1. If you say the product of 4 and 187 is about 800, your answer is an _?_ .

2. A number sentence with an equal sign to show that two amounts are equal, such as $3 + 7 = 10$, is an _?_ .

| round |
| equation |
| estimate |

✔ MODEL MULTIPLICATION (For Intervention, see p. H13.)

Write a multiplication sentence for the model.

3.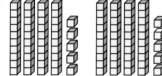

4.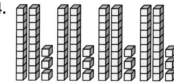

5.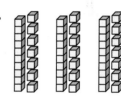

6.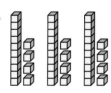

7.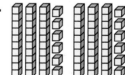

8.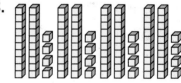

✔ MULTIPLICATION FACTS (For Intervention, see p. H12.)

Find each product.

9.   5
  × 6

10.   8
  × 3

11.   9
  × 5

12.   9
  × 8

13.   9
  × 7

14.   2
  × 6

15.   7
  × 4

16.   3
  × 5

17.   5
  × 8

18.   5
  × 7

19.   8
  × 8

20.   6
  × 4

Mental Math: Multiplication Patterns

▶ Learn

FOUND MONEY Basic facts and patterns can be used to help you find products mentally.

Suppose you want to find the total number of pennies in 7 rolls of pennies. You know that each roll has 50 pennies. How can you find the total?

You can multiply 7 and 50 to find the total.

$$\begin{array}{r} 50 \\ \times\ 7 \\ \hline 350 \end{array}$$

Think: $50 = 5 \times 10$, or 5 tens
7×5 tens $= 35$ tens
35 tens $= 35 \times 10$, or 350

So, there are 350 pennies.

You can use mental math to multiply greater numbers when you know basic facts and patterns.

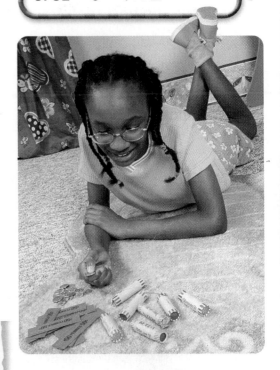

Examples

A
$7 \times 20 = 140$
$7 \times 200 = 1,400$
$7 \times 2,000 = 14,000$
$7 \times 20,000 = 140,000$

B
$5 \times 80 = 400$
$5 \times 800 = 4,000$
$5 \times 8,000 = 40,000$
$5 \times 80,000 = 400,000$

MATH IDEA As the number of zeros in a factor increases, the number of zeros in the product increases.

▶ Check

1. **Tell** what rule you can use to multiply with multiples of 10, 100, 1,000, and 10,000.

Use a basic fact and a pattern to write each product.

2. **a.** 3×20
   b. 3×200

3. **a.** 4×40
   b. 4×400

4. **a.** 2×800
   b. $2 \times 8,000$

5. **a.** $5 \times 4,000$
   b. $5 \times 40,000$

Quick Review

1. $6 \times 7 = \blacksquare + 39$

2. $\blacksquare \times 3 = 25 - 4$

3. $5 \times 2 = 60 \div \blacksquare$

4. $4 \times \blacksquare = 10 \times 2$

5. $32 - 8 = 4 \times \blacksquare$

▶ Practice and Problem Solving

Use a basic fact and a pattern to write each product.

6. a. 2×40
  b. 2×400

7. a. 4×30
  b. 4×300

8. a. 2×500
  b. $2 \times 5,000$

9. a. $2 \times 6,000$
  b. $2 \times 60,000$

10. a. 8×60
   b. 8×600

11. a. 5×40
   b. 5×400

12. a. 4×800
   b. $4 \times 8,000$

13. a. $3 \times 9,000$
   b. $3 \times 90,000$

Multiply mentally. Write the basic multiplication fact and the product.

14. 6×400

15. $7 \times 4,000$

16. 3×600

17. 5×600

18. $7 \times 60,000$

19. 6×90

20. $3 \times 7,000$

21. $3 \times 30,000$

 ALGEBRA Find the value of n.

22. $5 \times 40,000 = n$

23. $n = 2 \times 3,000$

24. $6 \times n = 4,200$

25. $8 \times 6,000 = n$

26. $n = 7 \times 5,000$

27. $3 \times n = 2,700$

USE DATA Copy and complete each table to show the number of coins in the coin rolls.

28. There are 40 nickels in 1 roll of nickels. NICKELS NICKELS

Number of Rolls	1	2	3	4	5	6
Number of Nickels	40	80	▪	▪	▪	▪

29. There are 50 dimes in 1 roll of dimes. DIMES DIMES

Number of Rolls	1	2	3	4	5	6
Number of Dimes	50	100	▪	▪	▪	▪

30. Carmen has three $20 bills and five $10 bills. How much money does Carmen have?

31. REASONING What is $80,000 \times 7,000$? Tell how you found the answer.

32. ? **What's the Error?** Renee has 20 rolls of pennies. Each roll has 50 pennies. Renee calculates that she has 100 pennies. Describe her error. Write the correct answer.

Mixed Review and Test Prep

33. 12×12 (p. 148)

34. $8 \times 2 \times 5$ (p. 150)

35. Wendy's karate class begins at 11:40 A.M. and lasts 1 hour and 50 minutes. When does Wendy's class end? (p. 120)

36. Evaluate $4 \times (8 - 2)$. (p. 158)

37. TEST PREP What is the value of the blue digit in $4\textbf{3}5,269$? (p. 6)

  A 300   **C** 13,000

  B 3,000   **D** 30,000

EXTRA PRACTICE page H41, Set A

2 Estimate Products

Quick Review

Round to the nearest hundred.

1. 572   **2.** 921

3. 1,726   **4.** 2,135

5. 5,834

▶ Learn

FLOUR POWER The neighborhood bakery sold 289 loaves of bread last month. If it takes 4 cups of flour to make each loaf of bread, about how many cups of flour were used?

Round the greater factor. Then use basic facts and patterns to estimate products.

Example Estimate. 4×289

STEP 1

Round 289 to the nearest hundred.

4×289
$\downarrow$
4×300

STEP 2

Use basic facts and patterns.

Think: $4 \times 3 = 12$
   $4 \times 300 = 1,200$
   ↑   ↑↑   ↑↑

0 zeros + 2 zeros = 2 zeros

So, the bakery used about 1,200 cups of flour.

Remember

To round a number:
- Find the place to which you want to round. Look at the digit to its right.
- If the digit is *less than 5*, the digit in the rounding place stays the same.
- If the digit is *5 or more,* the digit in the rounding place increases by 1.

More Examples   Estimate the products.

A $5,479 \rightarrow 5,000$
$\underline{\times \quad 7} \quad \underline{\times \quad 7}$
$\quad\quad\quad\quad 35,000$

B $\$5.82 \rightarrow \$6$
$\underline{\times \quad 6} \quad \underline{\times 6}$
$\quad\quad\quad\quad \$36$

▶ Check

1. Explain how you can tell if the estimated products in the examples are greater than or less than the exact products.

Round one factor. Estimate the product.

2. 6×23 **3.** 9×507 **4.** $3 \times 8,126$ **5.** $4 \times \$57.63$

Round one factor. Estimate the product.

6. 187
    × 4

7. 87
    × 6

8. 764
    × 5

9. 679
    × 4

10. $11.89
    × 6

11. 247
    × 8

12. 389
    × 7

13. $6.24
    × 7

14. 8 × 26

15. 5 × $65.13

16. 4 × 749

17. 9 × 1,789

18. 5 × 359

19. 9 × $0.63

20. 4 × 3,322

21. 5 × 4,789

22. 3 × $19.85

23. 4 × 341

24. 7 × $812.15

25. 2 × 6,789

Choose two factors from the box for each estimated product.

26. ■ × ▲ = 1,600

27. ■ × ▲ = 600

28. ■ × ▲ = 7,000

29. ■ × ▲ = 2,400

4	7	392
6	123	989

USE DATA For 30–32, use the data on the right.

30. Kai bought 4 loaves of French bread. About how much did she spend?

31. Jana has $5.00. Can she buy a strawberry cake and 2 sugar cookies? Explain.

32. Which costs more, a peach pie or 10 sugar cookies? Explain.

33. Ms. Williams wants to order one cupcake for each student. There are 6 classes with 24 students in each class. Will 120 cupcakes be enough? Explain.

34. ❓ **What's the Error?** Jacob estimated the product of 225 and 6 as 12,000. Describe and correct his error.

Mixed Review and Test Prep

35. Write <, >, or = for ●. (p. 18)
    107,216 ● 170,206

36. (7 × 5) − 5 (p. 158)

37. Round 2,678,213 to the nearest hundred thousand. (p. 26)

38. What are the missing numbers in the pattern? (p. 36)
    101, 131, 161, ■, ■, 251

39. **TEST PREP** Find the median for 23, 23, 15, 27, and 40. (p. 86)
    A 2   **B** 15   **C** 23   **D** 40

Multiply 2-Digit Numbers

MATERIALS base-ten blocks

▶ **Learn**

OUT OF STYLE Years ago, fruits and vegetables were sold in units called dry measures. These units are seldom used today.

DRY MEASURES	
1 quart	= 2 pints
1 peck	= 8 quarts or 16 pints
1 bushel	= 4 pecks or 32 quarts

Quick Review

Find the value of the expression.

1. $7 \times y$ if $y = 4$

2. $35 \div a$ if $a = 5$

3. $8 \times z$ if $z = 6$

4. $b \div 7$ if $b = 42$

5. $c \times 9$ if $b = 8$

MATERIALS base-ten blocks

There were 4 bushels of pears at a fruit stand. The pears were sold in quart boxes. How many quarts of pears are in 4 bushels?

Multiply. 4×32

Estimate. $4 \times 30 = 120$

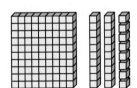

Activity
Make a model to find the product.

STEP 1	Model 4 groups of 32.	
STEP 2	Combine the ones. 4×2 ones = 8 ones	
STEP 3	Combine the tens. 4×3 tens = 12 tens Regroup 12 tens as 1 hundred 2 tens.	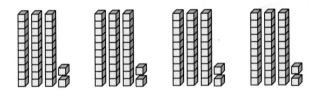
STEP 4	Record the product. 1 hundred 2 tens 8 ones	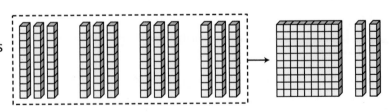

So, there are 128 quarts of pears in 4 bushels.

• In which place-value position was regrouping needed? Why?

Example Strawberries are sold in pint boxes. There are 16 pints in a peck. How many pints of strawberries are in 5 pecks?

Multiply. 5 × 16 =   or   16
                              × 5

STEP 1

Model 5 groups of 16. Multiply the ones.

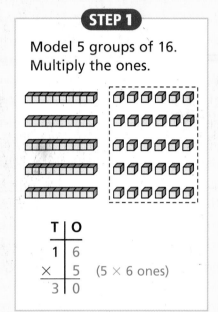

```
T | O
1 | 6
× |  5    (5 × 6 ones)
3 | 0
```

STEP 2

Multiply the tens.

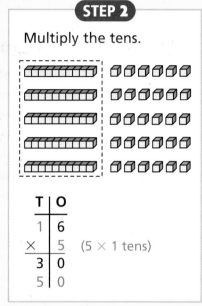

```
T | O
1 | 6
× |  5    (5 × 1 tens)
3 | 0
5 | 0
```

STEP 3

Add to find the product.

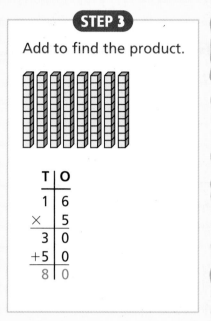

```
T | O
1 | 6
× |  5
3 | 0
+5 | 0
8 | 0
```

So, there are 80 pints of strawberries in 5 pecks.

More Examples

Ⓐ
```
    64
 ×   6
    24    (6 × 4 ones)
 + 360    (6 × 6 tens)
   384
```

Ⓑ
```
    47
 ×   6
    42    (6 × 7 ones)
 + 240    (6 × 4 tens)
   282
```

Ⓒ
```
    72
 ×   3
     6    (3 × 2 ones)
 + 210    (3 × 7 tens)
   216
```

▶ **Check**

1. **Model** 3 × 37 with base-ten blocks. Use paper and pencil to record what you did.

Multiply. Tell which place-value positions need to be regrouped.

2. 52	3. 24	4. 82	5. 36	6. 49
× 3	× 6	× 4	× 2	× 5

Find the product. Estimate to check.

7. 3 × 43   **8.** 6 × 51   **9.** 4 × 24   **10.** 5 × 38   **11.** 7 × 14

LESSON CONTINUES ▶

► Practice and Problem Solving

Multiply. Tell which place-value positions need to be regrouped.

12. 46
     × 2

13. 57
     × 4

14. 39
     × 7

15. 82
     × 4

16. 27
     × 3

Find the product. Estimate to check.

17. 62
     × 4

18. 48
     × 5

19. 59
     × 3

20. 36
     × 6

21. 74
     × 8

22. 93
     × 3

23. 35
     × 7

24. 67
     × 4

25. 81
     × 5

26. 49
     × 6

27. 32
     × 9

28. 85
     × 7

29. 78
     × 2

30. 17
     × 8

31. 83
     × 6

32. 8×26

33. 5×34

34. 7×56

35. 3×29

36. 4×92

37. 6×43

38. 9×62

39. 5×84

40. 7×49

41. 4×73

 **ALGEBRA**   **For 42–45, use base-ten blocks to find the missing factor.**

42. $5 \times \blacksquare = 230$ **43.** $4 \times \blacksquare = 304$ **44.** $\blacksquare \times 65 = 195$ **45.** $\blacksquare \times 37 = 222$

Compare. Write <, >, or = for each ●.

46. 3×45 ● 6×27 **47.** 5×62 ● 6×57 **48.** 4×81 ● 7×35

49. 8×37 ● 5×79 **50.** 6×27 ● 3×54 **51.** 9×58 ● 5×98

52. There are 4 pecks in a bushel. How many pecks are in 48 bushels?

53. REASONING Rachel says that 5×36 is the same as 90×2. Do you agree or disagree? Explain.

54. How many pecks are in 35 bushels and 3 pecks?

55. There are 16 pints in a peck. How many pints are in a bushel?

56. **?** **What's the Error?** Cory says that $7 \times 52 = 354$. Describe Cory's error, and find the correct product.

57. Write a problem that involves multiplying a 2-digit number by a 1-digit number with regrouping of both the ones and tens.

58. How many minutes are in 3 hours 25 minutes?

59. How many days are in 6 weeks less 5 days?

60. Mrs. Kuwana bought 12 packages with 8 buns in each package for a neighborhood party. How many packages of 10 hot dogs will she need?

61. Will you have more cookies if you buy the sandwich cookies or the sugar cookies shown in the picture?

Mixed Review and Test Prep

62. 23 + 48 (p. 36)   **63.** 39 − 16 (p. 36)

64. Round 3,270,516 to the nearest hundred thousand. (p. 26)

65. Find the value of the expression. 42 − (25 + 17) (p. 56)

Complete to make the equation true.

(p. 68)

66. 24 + 37 = ▧ + 31

67. ▧ − 45 = 14 + 28

68. **TEST PREP** Don wants to go to the movie that starts at 1:15 P.M. The movie lasts for 1 hour 52 minutes. When will the movie end? (p. 120)
A 1:52 P.M.   **C** 2:15 P.M.
B 2:07 P.M.   **D** 3:07 P.M.

69. **TEST PREP** Betty saved some quarters. Her mother gave her 8 nickels. Now she has 90¢. How many quarters had she saved? (p. 174)
F 1 quarter   **H** 3 quarters
G 2 quarters   **J** 4 quarters

PROBLEM SOLVING Thinker's Corner

HIDE-AND-SEEK   Many children play the game of hide-and-seek. In some places, the person who is "It" yells "A bushel of wheat, a basket of rye, who's not ready, holler I." Did you know that a bushel is a way to measure? In the multiplication problems below, there are some hidden numbers. Use what you know about multiplying to find the hidden numbers.

1.
```
   2▧
×   4
  92
```

2.
```
  35
× ▧
 105
```

3.
```
  ▧7
×  7
 329
```

4.
```
  6▧
×  8
 512
```

5.
```
  49
× ▧
 245
```

6.
```
  ▧4
×  6
 32▧
```

7.
```
  9▧
×  3
 ▧79
```

8.
```
  36
× ▧
 2▧6
```

9.
```
  ▧5
×  8
 36▧
```

10.
```
  ▧▧
×  5
 365
```

HANDS ON

Model Multiplication

Quick Review

Find the product.

1. 3×15 **2.** 4×22

3. 5×14 **4.** 2×32

5. 6×21

MATERIALS base-ten blocks

▶ **Explore**

You can make a model to find a product.

Activity 1

Multiply. 4×124

Estimate. $4 \times 100 = 400$

STEP 1
Model 4 groups of 124.
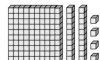 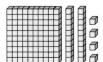 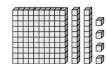

STEP 2
Combine the ones.
4×4 ones $= 16$ ones
Regroup 16 ones as 1 ten 6 ones.

STEP 3
Combine the tens.
4×2 tens $= 8$ tens
Add the regrouped ten.
8 tens + 1 ten = 9 tens

STEP 4
Combine the hundreds.
4×1 hundred $= 4$ hundreds
Record the product.
4 hundreds 9 tens 6 ones, or 496
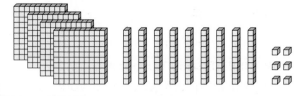

Write: $4 \times 124 = 496$
Read: 4 times 124 equals 496.

Try It

Use base-ten blocks to multiply. Record the product.

a. 3×125 **b.** 4×213

c. 2×105 **d.** 3×276

I am multiplying 3×125.
I have 3 groups of 125.
What is my next step?

▶ **Connect**

You may need to regroup both the ones and the tens.

Activity 2
Multiply. 3 × 138
Estimate. 3 × 100 = 300

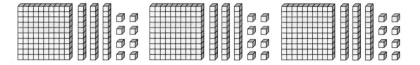

STEP 1
Model 3 groups
of 138.

STEP 2
Combine the ones.
3 × 8 ones = 24 ones
Regroup 24 ones as 2 tens 4 ones.

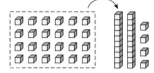

STEP 3
Combine the tens.
3 × 3 tens = 9 tens
Add the 2 regrouped tens.
Regroup 11 tens as 1 hundred 1 ten.

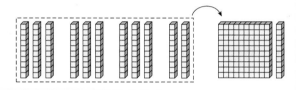

STEP 4
Combine the hundreds.
3 × 1 hundred = 3 hundreds
Add the 1 regrouped hundred.
Record the product.
4 hundreds 1 ten 4 ones, or 414

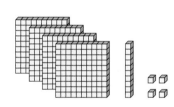

▶ **Practice and Problem Solving**

Use base-ten blocks to multiply. Record the product.

1. 4 × 104   **2.** 3 × 134   **3.** 3 × 123   **4.** 2 × 451

Multiply. You may wish to use base-ten blocks.

5. 3 × 144   **6.** 3 × 309   **7.** 5 × 116   **8.** 4 × 208

9. ❓ **What's the Error?** Zachary wrote 826 for the product of 4 times 234. Describe and correct his error.

Mixed Review and Test Prep

10. Sara's flute lesson is from 4:50 P.M. to 5:20 P.M. Her piano lesson is from 6:00 P.M. to 6:45 P.M. How many minutes does Sara spend in lessons? (p.120)

11. Write 56,730 in expanded form. (p. 4)

12. **TEST PREP** Which time is 12 minutes before 3:00? (p. 116)
   A 1:48   **B** 2:30   **C** 2:48   **D** 2:50

Quick Review
Regroup.

1. 5 tens 13 ones

2. 4 hundreds 12 tens

3. 2 hundreds 18 tens

4. 6 hundreds 16 tens

5. 4 hundreds 14 tens

▶ **Learn**

MIX IT UP! Tom is organizing his mother's recipes into 2 boxes. If 156 recipe cards will fit into each box, how many cards can Tom use?

Example
You can multiply 2 × 156 to find the number of recipe cards Tom can use.

Estimate. 2 × 200 = 400

One Way Use models to find the product.

STEP 1	STEP 2	STEP 3
Multiply the ones.	Multiply the tens.	Multiply the hundreds.
2 × 6 ones = 12 ones	2 × 5 tens = 10 tens	2 × 1 hundred = 2 hundreds
Regroup 12 ones as 1 ten 2 ones.	Add the regrouped ten. 10 tens + 1 ten = 11 tens Regroup 11 tens as 1 hundred 1 ten.	Add the regrouped hundred. 2 hundreds + 1 hundred = 3 hundreds

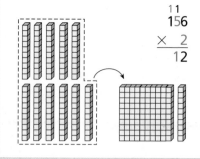

$$\begin{array}{r} \overset{1}{1}56 \\ \times\ 2 \\ \hline 2 \end{array}$$

$$\begin{array}{r} \overset{11}{1}56 \\ \times\ 2 \\ \hline 12 \end{array}$$

$$\begin{array}{r} \overset{11}{1}56 \\ \times\ 2 \\ \hline 312 \end{array}$$

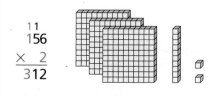

So, Tom can use 312 recipe cards.

• **Explain** why you estimate.

Another Way Use place value and expanded form to find the product.

Multiply 368 by 4. $4 \times 368 = \blacksquare$

Think: $368 = 300 + 60 + 8 \leftarrow$ expanded form

Multiply each addend by 4, and then find the sum.

Think: $\quad 4 \times 300 \qquad 4 \times 60 \qquad 4 \times 8$

$\qquad\qquad \downarrow \qquad\qquad\quad \downarrow \qquad\qquad \downarrow$

Write: $\quad 1,200 \quad + \quad 240 \quad + \quad 32 = 1,472$

- Look at the example above. How does using the expanded form of 368 help you find the product of 4×368?

 Technology Link

More Practice: Use Mighty Math Number Heroes, *Quizzo*, Levels N, O.

More Examples

A
$$\begin{array}{r} {}^{1} \\ 104 \\ \times\ \ 4 \\ \hline 416 \end{array}$$

B
$$\begin{array}{r} {}^{11} \\ 234 \\ \times\ \ 3 \\ \hline 702 \end{array}$$

C
$$\begin{array}{r} {}^{22} \\ 645 \\ \times\ \ 5 \\ \hline 3,225 \end{array}$$

D
$$\begin{array}{r} {}^{21} \\ 242 \\ \times\ \ 7 \\ \hline 1,694 \end{array}$$

 Check

1. **Explain** how you can use expanded form to find the product 6×274.

Multiply. Tell which place-value positions need to be regrouped.

2. $\begin{array}{r} 136 \\ \times\ \ 5 \\ \hline \end{array}$
3. $\begin{array}{r} 254 \\ \times\ \ 2 \\ \hline \end{array}$
4. $\begin{array}{r} 321 \\ \times\ \ 3 \\ \hline \end{array}$
5. $\begin{array}{r} 125 \\ \times\ \ 9 \\ \hline \end{array}$
6. $\begin{array}{r} 106 \\ \times\ \ 5 \\ \hline \end{array}$

7. $\begin{array}{r} 237 \\ \times\ \ 5 \\ \hline \end{array}$
8. $\begin{array}{r} 184 \\ \times\ \ 9 \\ \hline \end{array}$
9. $\begin{array}{r} 365 \\ \times\ \ 7 \\ \hline \end{array}$
10. $\begin{array}{r} 279 \\ \times\ \ 3 \\ \hline \end{array}$
11. $\begin{array}{r} 142 \\ \times\ \ 6 \\ \hline \end{array}$

Find the product. Estimate to check.

12. 3×437
13. 6×201
14. 7×173
15. 5×253
16. 8×771

LESSON CONTINUES ▶

Multiply. Tell which place-value positions need to be regrouped.

17. 253
× 2

18. 39
× 5

19. 38
× 3

20. 387
× 6

21. 618
× 4

Find the product. Estimate to check.

22. 241
× 2

23. 632
× 4

24. 318
× 3

25. 653
× 5

26. 204
× 8

27. 729
× 7

28. 214
× 9

29. 516
× 2

30. 806
× 5

31. 422
× 9

32. 2×825

33. 7×402

34. 4×973

35. 8×531

36. 2×925

37. 5×327

38. 3×432

39. 4×573

40. 2×875

41. 4×246

Multiply. Then add to find the product.

42. 3×472

$(3 \times 400) + (3 \times 70) + (3 \times 2)$
↓　　　　↓　　　　↓
■　+　■　+　■ = ■

43. 5×647

$(5 \times 600) + (5 \times 40) + (5 \times 7)$
↓　　　　↓　　　　↓
■　+　■　+　■ = ■

Compare. Write <, >, or = for each ●.

44. 4×326 ● 3×467

45. 8×199 ● 5×321

46. 2×375 ● 3×250

47. 5×272 ● 6×231

48. 7×408 ● 6×476

49. 4×742 ● 8×369

50. **REASONING** Using only the digits 4, 5, 7, and 8, find the greatest product and the least product possible where one factor is a 3-digit number.

51. **? What's the Question?** Jamal bought 3 rolls of film at $4.79 each. The answer is $14.37.

53. There are 52 weeks in 1 year. How many weeks are in 3 years 2 weeks?

52. How many hours are in one week 2 days?

54. ✎ **Write a problem** that involves multiplying a 3-digit number by a 1-digit number. Explain which place-value positions need to be regrouped.

55. ✎ **Write About It** Can two different multiplication problems have the same estimated product? Explain.

56. Ms. Lash bought 5 boxes of plastic forks for the school's spaghetti dinner. Each box had 175 forks. How many forks did she buy?

57. For the spaghetti dinner, 114 tables were set up in the gym. Each table had 4 chairs. How many chairs were there?

58. Each salad at the spaghetti dinner weighed 6 ounces. How many ounces of salad were needed for 425 salads?

Mixed Review and Test Prep

59.
$$6,899 \text{ (p. 40)}$$
$$+ 2,267$$

60.
$$8,902 \text{ (p. 40)}$$
$$- 5,730$$

61. Find the value of $2 \times (30 + 6)$. (p. 158)

62. $6 \times 5,000$ (p. 186)

63. Katrina went to the 11:30 A.M. movie. The movie ended at 2:10 P.M. How long did the movie last? (p. 120)

64. **TEST PREP** Jason has 16 stamps, Carlos has 6 stamps, and Nick has 12 stamps. How many stamps do they have in all?

A 30   **B** 34   **C** 42   **D** 46

65. **TEST PREP** There are 60 napkins in a package. Ms. Willis bought 8 packages. How many napkins did Ms. Willis buy? (p. 186)

F 240   **G** 360   **H** 420   **J** 480

PROBLEM SOLVING   Thinker's Corner

SLEEPY TIME   To solve the riddle below, find each product. Match the letter to its product in the blanks below. The letters will spell the answer to the riddle.

Why was the computer so tired when it got home?

It had a $\underset{634}{\underline{\quad?\quad}}$ $\underset{984}{\underline{\quad?\quad}}$ $\underset{1,064}{\underline{\quad?\quad}}$ $\underset{1,245}{\underline{\quad?\quad}}$ $\underset{1,557}{\underline{\quad?\quad}}$ $\underset{1,064}{\underline{\quad?\quad}}$ $\underset{1,569}{\underline{\quad?\quad}}$ $\underset{2,940}{\underline{\quad?\quad}}$ $\underset{1,592}{\underline{\quad?\quad}}$.

D	**A**	**V**	**H**
249	246	420	317
$\times$ 5	$\times$ 4	$\times$ 7	$\times$ 2

R	**D**	**E**	**I**
152	519	199	523
$\times$ 7	$\times$ 3	$\times$ 8	$\times$ 3

EXTRA PRACTICE page H41, Set D

Multiply 4-Digit Numbers

▶ **Learn**

PEN PALS   There are 3 classes of fourth graders at Vancouver Elementary School. Each class sent out 1,276 pen pal surveys to schools around the world. How many surveys were sent?

Example 1

Multiply. 3 × 1,276

Estimate. 3 × 1,300 = 3,900

STEP 1	STEP 2	STEP 3	STEP 4
Multiply the ones.	Multiply the tens.	Multiply the hundreds.	Multiply the thousands.
$\overset{1}{1,276}$	$\overset{21}{1,276}$	$\overset{21}{1,276}$	$\overset{21}{1,276}$
$\times\quad 3$	$\times\quad 3$	$\times\quad 3$	$\times\quad 3$
8	28	828	3,828

So, 3,828 surveys were sent. Since 3,828 is close to the estimate of 3,900, it is reasonable.

More Examples

A 8 × 1,775	**B** 3 × 2,814	**C** 7 × 3,964
$\overset{664}{1,775}$	$\overset{2\ 1}{2,814}$	$\overset{6\ 42}{3,964}$
$\times\quad 8$	$\times\quad 3$	$\times\quad 7$
14,200	8,442	27,748

• In Example C, how do you record the 28 ones when you multiply 4 by 7?

Multiply Money

Ms. Kreuz wants to buy 8 ink cartridges for her classroom printers to print the results of the survey. Each ink cartridge costs $19.95. How much will she have to pay?

You can multiply amounts of money in the same way you multiply whole numbers. Then write the product in dollars and cents.

Example 2

Multiply. 8 × $19.95

Estimate. 8 × $20 = $160

STEP 1	STEP 2	STEP 3
Write the problem, using whole numbers.	Multiply to find the product.	Write the product in dollars and cents.
$$\begin{array}{r} 1995 \\ \times\ \ \ \ 8 \\ \hline \end{array}$$	$$\begin{array}{r} {}^{7\,7\,4} \\ 1995 \\ \times\ \ \ \ 8 \\ \hline 15960 \end{array}$$	The product is $159.60.

So, Ms. Kreuz will pay $159.60 for the printer cartridges.

• Compare the estimate to the answer. Is your answer reasonable?

More Examples

A 6 × $11.03

$$\begin{array}{r} {}^{1} \\ \$11.03 \\ \times\ \ \ \ \ 6 \\ \hline \$66.18 \end{array}$$

B 4 × $34.16

$$\begin{array}{r} {}^{1\ 2} \\ \$34.16 \\ \times\ \ \ \ \ 4 \\ \hline \$136.64 \end{array}$$

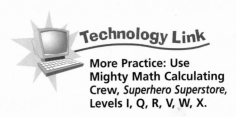

Technology Link

More Practice: Use Mighty Math Calculating Crew, *Superhero Superstore*, Levels I, Q, R, V, W, X.

• How is multiplying money like multiplying whole numbers?

▶ Check

1. **Explain** where to place the decimal point in the product 5 × $34.45.

Find the product. Estimate to check.

2. 5 × $24.16   **3.** 3 × 1,223   **4.** 2 × 4,381   **5.** 3 × $3.04

LESSON CONTINUES

Find the product. Estimate to check.

6. 6,025
× 5

7. $3.14
× 2

8. $46.19
× 7

9. 9,842
× 2

10. $4.02
× 5

11. 8,734
× 3

12. $38.27
× 8

13. 2,761
× 4

14. 4,714
× 6

15. 7,105
× 8

16. $14.79
× 2

17. $29.35
× 3

18. $2 \times 3,322$

19. $4 \times \$8.41$

20. $4 \times 2,068$

21. $5 \times 4,861$

22. $3 \times 2,507$

23. $8 \times \$46.19$

Find the product. Estimate to check.

24. $(5 \times \$7.64) \times 4$

25. $(5 \times 4,997) \times 2$

26. $(3 \times \$15.98) \times 2$

27. $(9 \times 3,004) \times 6$

28. $(4 \times \$36.12) \times 3$

29. $(7 \times 5,341) \times 6$

30. $3 \times 1,509$

$(3 \times 1,000) + (3 \times 500) + (3 \times 9)$
↓   ↓   ↓
▨   +   ▨   +   ▨ = ▨

31. $6 \times \$25.40$

$(6 \times \$20) + (6 \times \$5) + (6 \times \$0.40)$
↓   ↓   ↓
▨   +   ▨   +   ▨ = ▨

32. **ᵃ⁺ᵇ⁄c ALGEBRA** Find the missing numbers. Explain how to find the numbers.

Keisha

4,623
× ▨
36,9▨4

33. **? What's the Error?** Look at Jeff's work. Describe his error. Write the correct answer.

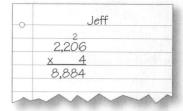

Jeff

2
2,206
× 4
8,884

USE DATA For 34–36, use the line graph.

34. How many months had more than 18 sunny days?

35. How many more sunny days were there in June than February?

36. What was the total number of sunny days in February and March?

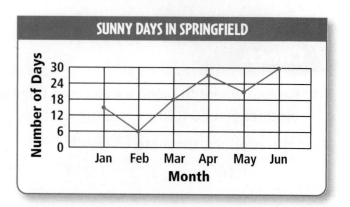

SUNNY DAYS IN SPRINGFIELD

USE DATA For 37–38, use the table.

37. **REASONING** Mr. Jenson needs to buy 3 mouse pads for his classroom. If he has $30, does he have enough money? Explain.

38. Isabel bought 3 packs of paper and an ink cartridge. What was the total cost of these items?

COMPUTER SUPPLIES	
Pack of Paper	$1.99
Mouse Pad	$9.50
Ink Cartridge	$18.25
Box of Disks	$7.39

Mixed Review and Test Prep

39.   3,896 (p. 40)   40.   4,568 (p. 40)
    +9,215   −3,675

41. Round 216,453 to the nearest hundred thousand. (p. 26)

42. Ben's quiz scores are 86, 77, 93, 88, and 84. What is the median score? (p. 86)

43. **TEST PREP** Find the value of $(56 − 18) + 13$. (p. 56)

   A 51   **B** 55   **C** 61   **D** 62

44. **TEST PREP** Find the sum.
   $7,856 + 2,309 + 9,427$ (p. 40)

   F 18,582   **H** 19,592
   G 19,582   **J** 19,692

PROBLEM SOLVING LiNKUP ... to Reading

STRATEGY • ANALYZE INFORMATION To solve a problem, you need to **analyze** the information by breaking it into different parts.

The members of the computer club have $100.00 to spend on a pizza party. They plan to order 5 pizzas, 3 orders of garlic bread, and 8 bottles of fruit punch. Do they have enough money?

Information	Analyze
The computer club has $100.00 to spend on a pizza party.	Tells how much money they have to spend.
5 pizzas for $15.25 each	Multiply to find the total cost of the pizza. 5 × $15.25
3 orders of garlic bread for $3.75 each	Multiply to find the total cost of the garlic bread. 3 × $3.75
8 bottles of fruit punch for $1.35 each	Multiply to find the total cost of the fruit punch. 8 × $1.35
Do they have enough money?	Add the amounts together and compare to $100.00.

1. Solve the problem. Explain.

Analyze the problem below. Solve.

2. The band members need to raise $300.00 for a project. They washed cars for $4.50 each. One group washed 15 cars, another group 32 cars, and the third group washed 19 cars. Did the members raise more money or less money than they needed? How much more or less?

Problem Solving Strategy
Write an Equation

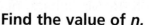

PROBLEM   Ray uses his computer to find information on the Internet. Each month of Internet service costs $15.95. What is the total cost for Ray to have Internet service for 6 months?

UNDERSTAND

- What are you asked to find?
- What information will you use?
- Is there any information you will not use?

PLAN

- What strategy can you use to find the answer?

  You can *write an equation* to find the total cost of the Internet service. An equation can show how the facts in the problem are related.

SOLVE

- How can you use the strategy to find the answer?

  You can *write an equation* that uses *n* for the product.

number of months		cost per month		total cost of Internet service
↓		↓		↓
6	×	$15.95	=	*n*
		$95.70	=	*n*

  $$\begin{array}{r} \$15.95 \\ \times \quad 6 \\ \hline \$95.70 \end{array}$$

  So, it costs $95.70 for 6 months of service.

CHECK

- Look back at the problem. Does the answer make sense for the problem? Explain.
- Is the product 6 × $15.95 the same as $15.95 × 6? Explain.

Problem Solving Practice

PROBLEM SOLVING STRATEGIES

Draw a Diagram or Picture
Make a Model or Act It Out
Make an Organized List
Find a Pattern
Make a Table or Graph
Predict and Test
Work Backward
Solve a Simpler Problem
▶ **Write an Equation**
Use Logical Reasoning

For 1–3, write an equation and solve.

1. **What if** Internet access cost $19.95 per month? How much would it cost Ray to have Internet access for 5 months?

2. Leon bought 4 books that cost $12.45 each. How much did he spend?

3. Ms. Davies allows her computer class 25 minutes each day to try new computer programs. How many minutes is this in 5 days?

Adam ran a 10-mile race at a pace of 7 minutes per mile. How many minutes did it take Adam to complete the race?

4. What equation can you use to help you answer the question?
   - **A** $10 \times 7 = n$
   - **B** $7 \times 5 = n$
   - **C** $10 = 7 \times n$
   - **D** $7 = 10 \times n$

5. What solution answers the question?
   - **F** 55 minutes
   - **G** 60 minutes
   - **H** 70 minutes
   - **J** 75 minutes

Mixed Strategy Practice

6. Four students are standing in line. Jane is standing 2 meters behind Kara and 1 meter in front of Chrissy. Chrissy is standing an equal distance between Kara and Paul. How far away is the first student standing from the last?

7. A computer lab has 8 rows of computers. Each row has 4 computers. There are also 4 computers grouped in the center of the room. How many computers are in the lab?

8. Jason paid $78.69 for a computer book, software, and speakers. Use the prices shown to find the cost of the speakers.

9. Karen left her house at the time shown on the clock. She arrived at Kaitlyn's house 15 minutes later. They spent 30 minutes eating lunch and then took a 15-minute walk. What time did they finish their walk?

Problem Solving Strategy

Review/Test

✓ CHECK VOCABULARY AND CONCEPTS

Choose the best term from the box.

1. To solve 5×16, the regrouped 3 tens is _?_ to the product 5×1. (p. 194)

2. To estimate a product, you can _?_ one of the factors. (p. 188)

3. When you have the expression 4×600, you can find the value by using _?_ and _?_. (p. 186)

> basic facts
> added
> patterns
> round

✓ CHECK SKILLS

Round one factor. Estimate the product. (pp. 188–189)

4. 5×294 5. 5×66 6. 3×834 7. $6 \times \$5.36$

Multiply. Tell which place-value positions need to be regrouped. (pp. 196–199)

8. $\begin{array}{r} 97 \\ \times\ 3 \\ \hline \end{array}$ 9. $\begin{array}{r} 402 \\ \times\ 7 \\ \hline \end{array}$ 10. $\begin{array}{r} 389 \\ \times\ 8 \\ \hline \end{array}$ 11. $\begin{array}{r} 612 \\ \times\ 5 \\ \hline \end{array}$

Find the product. Estimate to check. (pp. 190–193, 196–199, 200–203)

12. $\begin{array}{r} 45 \\ \times\ 4 \\ \hline \end{array}$ 13. $\begin{array}{r} 1,068 \\ \times\ 4 \\ \hline \end{array}$ 14. $\begin{array}{r} 42 \\ \times\ 4 \\ \hline \end{array}$ 15. $\begin{array}{r} 863 \\ \times\ 3 \\ \hline \end{array}$

16. $\begin{array}{r} 3,122 \\ \times\ 2 \\ \hline \end{array}$ 17. $\begin{array}{r} 56 \\ \times\ 3 \\ \hline \end{array}$ 18. $\begin{array}{r} 3,897 \\ \times\ 5 \\ \hline \end{array}$ 19. $\begin{array}{r} 3,045 \\ \times\ 9 \\ \hline \end{array}$

20. $\begin{array}{r} \$8.21 \\ \times\ 4 \\ \hline \end{array}$ 21. $\begin{array}{r} \$14.96 \\ \times\ 8 \\ \hline \end{array}$ 22. $\begin{array}{r} \$7.52 \\ \times\ 3 \\ \hline \end{array}$ 23. $\begin{array}{r} \$38.15 \\ \times\ 7 \\ \hline \end{array}$

✓ CHECK PROBLEM SOLVING

Solve. (pp. 204–205)

24. Carmen buys eight magazines. Each magazine sells for $3.28. How much does Carmen spend on magazines? Write a number sentence and solve.

25. There are 500 sheets of paper in each package. If 9 packages are ordered, how many sheets of paper is this?

⭐Standardized Test Prep

Decide on a plan.
See item **7.**

Find the information you need in the table. Think about the relationship of the numbers given and find the equation that shows it.

Also see problem **4,** p. H63.

For 1–8, choose the best answer.

1. Which expression is **not** equivalent to 16?

   A $12 + (9 - 5)$ **C** $20 - (7 - 3)$

   B $8 \times (5 - 3)$ **D** $27 - (7 - 4)$

2. John ordered a drink and 2 sandwiches. The cost of the drink was $1. Which expression shows the total cost if *s* is the cost of one sandwich?

   F $2 \times (s + \$1)$ **H** $(2 \times s) + \$1$

   G $(2 \times s) + (2 \times \$1)$ **J** $(s \div 2) + \$1$

3. Ed earns $27.50 each week baby-sitting. How much does Ed earn in 4 weeks?

   A $31.50   **C** $68.00

   B $55.00   **D** $110.00

4. Ted rented videos for $3.95 each. Which equation shows the total cost of renting 5 videos?

   F $5 \times n = \$3.95$ **H** $\$3.95 \div n = 5$

   G $5 \times \$3.95 = n$ **J** $\$3.95 + n = 5$

5. 8×436

   A 3,248   **C** 3,488

   B 3,448   **D** NOT HERE

For 6–7 use the table.

SPORTS STORE PRICE TABLE	
Item	**Cost**
1 can of tennis balls	$4.12
1 pack of golf tees	$2.87
1 pair of running shorts	$13.72
1 baseball	$1.69

6. Susie wants to buy 3 of the same item. She estimates her total will be about $12. Which item does she want to buy?

   F tennis balls   **H** running shorts

   G golf tees   **J** baseball

7. What is the cost of 2 pairs of running shorts and 3 baseballs?

   A $15.41   **C** $29.13

   B $17.10   **D** $32.51

8. Which is the best estimate for this product?

   $$\begin{array}{r} \$19.85 \\ \times \quad 4 \\ \hline \end{array}$$

   F $70   **H** $80

   G $100   **J** $120

Write What You Know

9. Use the table. Jeff is going to buy 5 of the same item. He estimates that he will have about $5 left if he pays with a $20 bill. Which item is he going to buy? Explain.

10. Explain how to use mental math to find the product $5 \times 4,000$.

Understand Multiplication

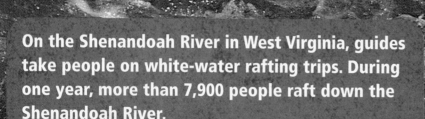

On the Shenandoah River in West Virginia, guides take people on white-water rafting trips. During one year, more than 7,900 people raft down the Shenandoah River.

PROBLEM SOLVING Look at the graph. It shows the number of white-water rafting trips four companies led during one year. If each rafting trip had 6 guests, how many guests went rafting?

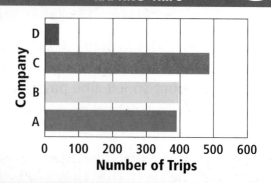

RAFTING TRIPS

Company (vertical axis): D, C, B, A

Number of Trips (horizontal axis): 0 100 200 300 400 500 600

Use this page to help you review and remember important skills needed for Chapter 11.

✓ VOCABULARY

Choose the best term from the box.

factors
multiples
product
quotient

1. The numbers 20, 30, 40, and 50 are _?_ of 10.

2. In $2 \times 8 = 16$, the number 16 is the _?_ .

3. In the equation $7 \times 8 = 56$, the numbers 7 and 8 are the _?_.

✓ MULTIPLY BY TENS, HUNDREDS, AND THOUSANDS (For Intervention, see p. H14.)

Multiply.

4. 4×10

5. 10×40

6. 10×400

7. 40×10

8. 5×10

9. 50×10

10. 50×100

11. $50 \times 1,000$

12. $\begin{array}{r} 100 \\ \times\ \ \ 9 \\ \hline \end{array}$

13. $\begin{array}{r} 100 \\ \times 900 \\ \hline \end{array}$

14. $\begin{array}{r} 90 \\ \times 10 \\ \hline \end{array}$

15. $\begin{array}{r} 1,000 \\ \times\ \ \ \ 90 \\ \hline \end{array}$

✓ MODEL MULTIPLICATION (For Intervention, see p. H13.)

Use the base-ten blocks to find the product.

16. $\begin{array}{r} 14 \\ \times\ 3 \\ \hline \end{array}$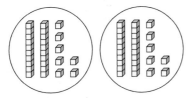

17. $\begin{array}{r} 25 \\ \times\ 2 \\ \hline \end{array}$

18. $\begin{array}{r} 27 \\ \times\ 2 \\ \hline \end{array}$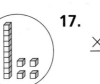

19. $\begin{array}{r} 17 \\ \times\ 3 \\ \hline \end{array}$

20. $\begin{array}{r} 12 \\ \times\ 5 \\ \hline \end{array}$

Mental Math: Patterns with Multiples

Quick Review

1. 5×10
2. 20×10
3. 50×10
4. 10×600
5. $3,000 \times 10$

▶ Learn

CROSS-COUNTRY Luis and his family drove from New York City to the Grand Canyon. They drove for a total of 50 hours at an average speed of 50 miles per hour. About how far did they drive?

Multiply. 50×50

Use a basic fact and a pattern.

Factors		Product	
5×5	$=$	25	**Think:** Use the basic fact $5 \times 5 = 25$.
5×50	$=$	250	Look for a pattern of zeros.
50×50	$=$	$2,500$	

↑ ↑   ↑
1 **zero**   1 **zero**   2 **zeros**

So, Luis and his family drove about 2,500 miles.

• What do you notice about the pattern of zeros in the factors and the products?

▲ The Grand Canyon is located in northern Arizona and surrounds 277 miles of the Colorado River.

Examples Use a basic fact and a pattern to find the product.

Ⓐ
$4 \times 2 = 8$
$4 \times 20 = 80$
$4 \times 200 = 800$
$4 \times 2,000 = 8,000$

Ⓑ
$7 \times 6 = 42$
$70 \times 60 = 4,200$
$70 \times 600 = 42,000$
$70 \times 6,000 = 420,000$

Ⓒ
$6 \times 5 = 30$
$60 \times 5 = 300$
$60 \times 50 = 3,000$
$600 \times 50 = 30,000$

 MATH IDEA Basic facts and patterns can be used to help you find products when you multiply by multiples of 10, 100, or 1,000.

▶ Check

1. **Explain** why the products in Example C have more zeros than the factors.

Use a basic fact and a pattern to find the products.

2. 4×70
4×700
$4 \times 7,000$

3. 6×60
60×60
600×60

4. 8×50
80×50
800×50

5. 30×9
300×90
$300 \times 9,000$

▶ Practice and Problem Solving

Use a basic fact and a pattern to find the product.

6. 80×600

7. 700×500

8. 60×30

9. 50×20

10. $80 \times 1,200$

11. $100 \times 1,000$

12. $300 \times 5,000$

13. $8 \times 3,000,000$

14. $\begin{array}{r} 90 \\ \times\ 6 \\ \hline \end{array}$

15. $\begin{array}{r} 8,000 \\ \times\quad 60 \\ \hline \end{array}$

16. $\begin{array}{r} 40 \\ \times 30 \\ \hline \end{array}$

17. $\begin{array}{r} 600 \\ \times\ 12 \\ \hline \end{array}$

ALGEBRA Find the value of *n*.

18. $n \times 800 = 7,200$

19. $80 \times n = 720,000$

20. $8,000 \times n = 40,000$

21. $7,000 \times 10 = n$

22. On Tori's vacation, her dad drove about 40 miles per hour for 6 hours each day. About how far did he drive in 10 days?

23. Kim exercised 6 minutes one week, 9 minutes the second week, 12 minutes the third week, and 15 minutes the fourth week. If the pattern continues, how long will she exercise in the sixth week?

24. **? What's the Error?** Describe Sheldon's error. Write the correct answer.

Sheldon

$100,000 \times 10 = 1,100,000$

Mixed Review and Test Prep

JULY						
Sun	Mon	Tue	Wed	Thu	Fri	Sat
				1	2	3
4	5	6	7	8	9	10
11	12	13	14	15	16	17
18	19	20	21	22	23	24
25	26	27	28	29	30	31

25. Dana went to camp for 3 weeks. She returned home July 31. When did she leave for camp? (p. 126)

26. $64.70 - \$35.87$

27. Compare. Write $<$, $>$, or $=$ for the ●.
(p. 44)
$15,970 - 10,820$ ● $80,700 - 75,812$

28. $3 \times \$25.62$ (p. 200)

29. **TEST PREP** What is the range for 2, 3, 2, 6, 5? (p. 88)

A 2   **B** 3   **C** 4   **D** 5

EXTRA PRACTICE page H42, Set A

Multiply by Multiples of 10

▶ **Learn**

ROUNDUP At the roundup, 30 ranchers each herded 52 cows. How many cows were herded in all?

Example Multiply. 30×52

STEP 1

Multiply by the ones.
Place a zero in the ones place.

$$\begin{array}{r} 52 \\ \times 30 \\ \hline 0 \end{array} \leftarrow \text{0 ones} \times 52$$

STEP 2

Multiply by the tens.

$$\begin{array}{r} 52 \\ \times 30 \\ \hline 1,560 \end{array} \leftarrow \text{3 tens} \times 52$$

So, the ranchers herded 1,560 cows.

MATH IDEA When you multiply a whole number by a multiple of 10, the digit in the ones place of the product is always a zero.

▶ **Check**

1. **Tell** how many zeros are in the product of 18 and 20. How many zeros are in the product of 12 and 50? Explain how you can tell before you multiply.

Find the product.

2. $\begin{array}{r} 12 \\ \times 10 \\ \hline \end{array}$ 3. $\begin{array}{r} 18 \\ \times 30 \\ \hline \end{array}$ 4. $\begin{array}{r} 28 \\ \times 20 \\ \hline \end{array}$ 5. $\begin{array}{r} 32 \\ \times 40 \\ \hline \end{array}$ 6. $\begin{array}{r} 47 \\ \times 30 \\ \hline \end{array}$

7. $\begin{array}{r} 46 \\ \times 40 \\ \hline \end{array}$ 8. $\begin{array}{r} 91 \\ \times 20 \\ \hline \end{array}$ 9. $\begin{array}{r} 55 \\ \times 60 \\ \hline \end{array}$ 10. $\begin{array}{r} 72 \\ \times 30 \\ \hline \end{array}$ 11. $\begin{array}{r} 33 \\ \times 30 \\ \hline \end{array}$

Find the product.

12. 16×10	**13.** 19×20	**14.** 34×40	**15.** 27×30	**16.** 48×50
17. 55×70	**18.** 78×80	**19.** 84×60	**20.** 34×50	**21.** 99×90
22. 45×50	**23.** 93×30	**24.** 56×20	**25.** 45×40	**26.** 87×30

27. 20×14 **28.** 30×24 **29.** 26×40 **30.** 36×50 **31.** 20×52

32. 70×39 **33.** 52×80 **34.** 50×54 **35.** 69×70 **36.** 90×18

Find the missing digits.

37. $\blacksquare0 \times 40 = 400$ **38.** $20 \times \blacksquare0 = 600$ **39.** $\blacksquare7 \times 40 = 680$

40. $53 \times \blacksquare0 = 3,180$ **41.** $4\blacksquare \times 50 = 2,250$ **42.** $77 \times 3\blacksquare = 2,\blacksquare10$

USE DATA For 43–45, use the bar graph.

43. How many shows does each person do in 20 weeks?

44. How many shows does Carrie do in 50 weeks?

45. In 30 weeks, how many more shows does Jake do than Floyd?

46. There are 5,000 seats at each show. At the noon show only 3,475 seats were filled. How many seats were not filled?

47. ✎ **Write a problem** that can be solved by multiplying by a multiple of 10 using data from the bar graph.

48. **ALGEBRA** Mr. Cano wrote the expression ▲ $\times 80 = 64,000$. Find the value of ▲.

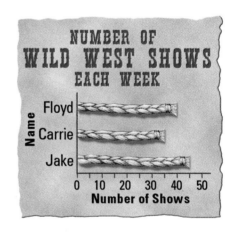

NUMBER OF
WILD WEST SHOWS
EACH WEEK

Name — Floyd, Carrie, Jake
0 10 20 30 40 50
Number of Shows

Mixed Review and Test Prep

49. $23,426$ (p. 44) $+ \ 2,213$

50. $6,124$ (p. 44) $- 4,750$

51. Find the median for the data: 40, 82, 27, 39, and 52. (p. 86)

52. In what place-value position is the 7 in 7,324,980? (p. 8)

53. **TEST PREP** $(3 \times 4) \times 2 = n$ (p. 150)

     A $n = 8$ **C** $n = 14$

     B $n = 12$ **D** $n = 24$

EXTRA PRACTICE page H42, Set B

3 Estimate Products

▶ **Learn**

JUICE BREAK   During the summer, the Roadside Market sold 176 gallons of lemonade. If 12 lemons were used to make each gallon, about how many lemons were used in all?

Example   Estimate to find about how many.

STEP 1

176×12 Round each factor.

$$\begin{array}{c} 176 \\ \times\ 12 \end{array} \rightarrow \begin{array}{c} 200 \\ \times\ 10 \end{array}$$

STEP 2

Multiply. $\begin{array}{r} 200 \\ \times\ 10 \\ \hline 2{,}000 \end{array}$

So, about 2,000 lemons were used in all.

Remember

To round a number:

• Find the place to which you want to round. Look at the digit to its right.

• If the digit is less than 5, the digit in the rounding place stays the same.

• If the digit is 5 or greater, the digit in the rounding place increases by 1.

More Examples

A $\begin{array}{r} 73 \\ \times 42 \end{array} \rightarrow \begin{array}{r} 70 \\ \times 40 \\ \hline 2{,}800 \end{array}$ **B** $\begin{array}{r} \$254 \\ \times\ 65 \end{array} \rightarrow \begin{array}{r} \$300 \\ \times\ 70 \\ \hline \$21{,}000 \end{array}$

Technology Link

To learn more about estimating products, watch the Harcourt Math Newsroom Video, *New Microscope.*

▶ **Check**

1. **Explain** how you can estimate the product 52×168.

Round each factor. Estimate the product.

2. $\begin{array}{r} 18 \\ \times 29 \end{array}$ 3. $\begin{array}{r} 389 \\ \times\ 64 \end{array}$ 4. $\begin{array}{r} \$45 \\ \times\ 12 \end{array}$ 5. $\begin{array}{r} \$259 \\ \times\ 41 \end{array}$

6. $\begin{array}{r} 52 \\ \times 27 \end{array}$ 7. $\begin{array}{r} 76 \\ \times 31 \end{array}$ 8. $\begin{array}{r} 410 \\ \times\ 78 \end{array}$ 9. $\begin{array}{r} 197 \\ \times\ 16 \end{array}$

▶ Practice and Problem Solving

Round each factor. Estimate the product.

10.	$19 × 12	**11.**	278 × 33	**12.**	$548 × 45	**13.**	38 ×27	**14.**	32 ×61
15.	419 × 72	**16.**	78 ×36	**17.**	64 ×67	**18.**	219 × 23	**19.**	634 × 55
20.	527 × 34	**21.**	56 ×39	**22.**	67 ×46	**23.**	915 × 32	**24.**	742 × 44

25. 13×85 **26.** 76×852 **27.** $\$49 \times 24$ **28.** 90×412 **29.** $18 \times \$319$

30. 27×32 **31.** 41×582 **32.** 72×775 **33.** $12 \times \$605$ **34.** 78×540

Use estimation to compare. Write <, >, or = for each ●.

35. 20×132 ● $3,000$ **36.** $13,000$ ● 26×645 **37.** 49×42 ● $1,800$

38. USE DATA The table shows the average number of apples in each size bag and the number of bags sold. Estimate the number of apples sold for each of the bag sizes.

39. Each month the average person in the United States uses about 46 pounds of paper. Estimate the amount of paper used by the average person in the United States in 1 year.

40. There are 60 packages of drawing paper. Each package has 200 sheets of paper. How many sheets of paper are there?

41. **Write About It** Explain how you would estimate the product 16×934.

Bags	Average Number of Apples	Number of Bags Sold
Small	11	27
Medium	18	21
Large	26	12

Mixed Review and Test Prep

42. 2,357 (p. 40)
    +4,987

43. 6,998 (p. 40)
    −4,736

44. Write the time that is 1 hour 50 minutes later. (p. 120)

45. $7 \times (2 \times 3) = n$ (p. 150)

46. TEST PREP There are 4 stacks of 20 baseball cards and 5 other baseball cards that are not in a stack. Which expression shows the total number of baseball cards? (p. 162)

A $(5 \times 20) + 4$ **C** $(20 \times 5) + 4$
B $(4 \times 20) + 5$ **D** $(4 \times 5) + 20$

HANDS ON

Model Multiplication

▶ Explore

Whitney and Dora used one-inch tiles to cover a rectangular tabletop. The tabletop is 24 tiles long and 18 tiles wide. How many tiles did they use?

MATERIALS grid paper, different-color markers

Activity 1
Multiply. 18×24

You can make a model to find the product.

STEP 1	**STEP 2**	**STEP 3**
Outline a rectangle that is 18 units wide and 24 units long. Think of the area as 18×24.	Break apart the model to make rectangles to show factors that are easy to multiply.	Multiply. Add the partial products to find how many squares are in the model.

STEP 2:
$$24 = 20 + 4$$
$$\times 18 = 10 + 8$$

STEP 3:
partial products
$8 \times 4 = 32$
$8 \times 20 = 160$
$10 \times 4 = 40$
$10 \times 20 = 200$

$32 + 160 + 40 + 200 = 432$

Write: $18 \times 24 = 432$
Read: 18 times 24 equals 432.

So, Whitney and Dora used 432 tiles.

Try It

Make a model to find the product. Use grid paper and markers.

  a. 14×16

  b. 15×21

  c. 34×17

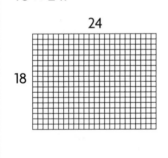

I am multiplying 14 × 16. I have found all of the partial products. What do I do next?

▶ Connect

Here is a way to record multiplication.

Activity 2 Multiply. 23 × 16

STEP 1

Show the model.

16

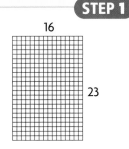

23

Record.

23
×16

STEP 2

Break apart the model to make rectangles to show factors that are easy to multiply.

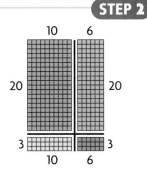

10   6

20   20

3   3

10   6

Separate the tens place and the ones place for each factor.

↓   ↓

23 = 20 + 3
×16 = 10 + 6

STEP 3

Multiply. Add the partial products to find how many squares are in the model.

23 = 20 + 3
×16 = 10 + 6

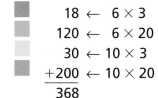

18 ← 6 × 3
120 ← 6 × 20
30 ← 10 × 3
+200 ← 10 × 20
368

▶ Practice and Problem Solving

Make a model, record, and solve.

1.	2.	3.	4.	5.
16	19	22	21	26
×15	×12	×18	×14	×21

Mixed Review and Test Prep

6. The rule is *add 4*. Use the rule and input 5 to find the output. (p. 64)

7. Corey bought 3 CDs that each cost $15.88. How much change did he receive from $50.00? (p. 200)

8. Order 23,465; 32,465; 23,645; and 23,456 from least to greatest. (p. 20)

9. **TEST PREP** (48 ÷ 8) − 2 (p. 158)

   A 8   **B** 6   **C** 4   **D** 2

Problem Solving Strategy
Solve a Simpler Problem

PROBLEM Ms. Alexander is buying art supplies for the 148 fourth-grade students at her school. She estimates that each student will use 50 sheets of drawing paper during the school year. How many sheets of drawing paper should she buy?

UNDERSTAND

- What are you asked to find?
- What information will you use?
- Is there any information you will not use? Explain.

PLAN

- What strategy can you use to solve the problem?
  You can find the product 50×148 by breaking apart 148 into numbers that are easier to multiply and then *solving the simpler problem*.

$$\begin{array}{r} 148 \\ \times\ 50 \\ \hline \end{array}$$

SOLVE

- How can you use the strategy to solve the problem?
  Rewrite 148 as $100 + 40 + 8$, multiply each addend by 50, and add the partial products.

$$
\begin{array}{rl}
148 = & 100 + 40 + 8 \\
\times\ 50 = & \\
\hline
& 50 \quad \leftarrow 50 \times (100 + 40 + 8) \\
& 400 \quad \leftarrow 50 \times 8 \\
& 2{,}000 \quad \leftarrow 50 \times 40 \\
+ & 5{,}000 \quad \leftarrow 50 \times 100 \\
\hline
& 7{,}400
\end{array}
$$

So, Ms. Alexander should buy 7,400 sheets of paper.

CHECK

- Look back at the problem. Does the answer make sense for the problem? Explain.

▶ Problem Solving Practice

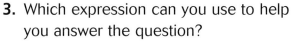

PROBLEM SOLVING STRATEGIES

Draw a Diagram or Picture
Make a Model or Act It Out
Make an Organized List
Find a Pattern
Make a Table or Graph
Predict and Test
Work Backward
▶ **Solve a Simpler Problem**
Write an Equation
Use Logical Reasoning

Break the problem into simpler parts and solve.

1. **What if** there were 62 fourth-grade students in an art class? About how many sheets of paper would be needed for the year?

2. There were 37 students in a summer art class. Each student made 50 drawings. How many drawings did they make?

The art teacher bought 3,300 crayons. She gave 20 crayons to each of the 157 students in her classes. How many crayons did she give to the students?

3. Which expression can you use to help you answer the question?
   A $3,300 \times (100 + 50 + 7)$
   B $20 \times (3,000 + 300)$
   C $20 \times (100 + 50 + 7)$
   D $157 + 20$

4. Which shows the total number of crayons the teacher gave her students?
   F 3,240 crayons
   G 3,140 crayons
   H 3,014 crayons
   J 2,904 crayons

Mixed Strategy Practice

USE DATA For 5 and 6, use the bar graph.

5. How many books did Mr. Matthew's class, Mr. Stevens's class, and Miss Kelsie's class read altogether?

6. **? What's the Question?** The difference in the number of books read is 75 books.

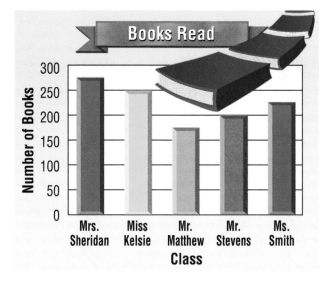

7. Julia takes ballet after gymnastics. Gymnastics begins at 3:30 P.M. and lasts 1 hour 15 minutes. It takes 10 minutes to walk to ballet class. At what time does Julia get to ballet?

8. The ninth-grade English teacher asked her students to write a poem 125 words long. There are 80 students in her classes. About how many words will the teacher read?

9. Mr. James bought 30 pieces of chalk for each classroom. There are 15 classrooms on the second floor and 22 on the first floor. How many pieces of chalk did Mr. James buy?

Problem Solving Strategy

Review/Test

✓ CHECK CONCEPTS

1. Explain how a basic fact and a pattern can help you find the product 60×400. (pp. 210–211)

2. Make a model to show how to break apart the numbers 32 and 14 to find the product. (pp. 216–217)

✓ CHECK SKILLS

Use a basic fact and a pattern to find the product. (pp. 210–211)

3. 20×900	4. 500×600	5. $\begin{array}{r} 9{,}000 \\ \times\ \ \ \ 60 \end{array}$	6. $\begin{array}{r} 300 \\ \times 700 \end{array}$

Find the product. (pp. 212–213)

7. $\begin{array}{r} 66 \\ \times 30 \end{array}$	8. $\begin{array}{r} 44 \\ \times 60 \end{array}$	9. $\begin{array}{r} 22 \\ \times 90 \end{array}$	10. $\begin{array}{r} 75 \\ \times 80 \end{array}$
11. $\begin{array}{r} 73 \\ \times 40 \end{array}$	12. $\begin{array}{r} 36 \\ \times 30 \end{array}$	13. $\begin{array}{r} 52 \\ \times 20 \end{array}$	14. $\begin{array}{r} 64 \\ \times 50 \end{array}$

15. 24×50 16. 86×20 17. 19×40 18. 93×70

Round each factor. Estimate the product. (pp. 214–215)

19. $\begin{array}{r} 222 \\ \times\ \ 48 \end{array}$	20. $\begin{array}{r} 252 \\ \times\ \ 14 \end{array}$	21. $\begin{array}{r} 931 \\ \times\ \ 56 \end{array}$	22. $\begin{array}{r} 79 \\ \times 68 \end{array}$

✓ CHECK PROBLEM SOLVING

For 23–25, break the problem into simpler parts and solve.
(pp. 218–219)

23. A choir has 98 members and needs to buy 20 sheets of music for each member. How many sheets do they need to buy?

24. The Toy Warehouse has 248 shelves. Each shelf has 16 cases of toys. How many cases are on the shelves?

25. Kevin is reading a series of books. Each book has 155 pages, and there are 25 books in the series. How many total pages will Kevin have read when he completes the series?

Standardized Test Prep

Look for important words.
See item **7.**
The important word is *estimate.*
Estimate each product before you choose the greatest.

Also see problem **2,** p. H62.

For 1–10, choose the best answer.

1. What is the product 12×27?
   A 324   **C** 296
   B 312   **D** 282

2. 7×326
   F 2,142   **H** 2,252
   G 2,182   **J** 2,282

3. What is the value of *n* for
   $70 \times 800 = n$?
   A 5,600   **C** 560,000
   B 56,000   **D** 5,600,000

4. 45×40
   F 180   **H** 1,800
   G 1,600   **J** 18,000

5. Which expression is equivalent
   to 21?
   A $24 - (8 - 5)$
   B $24 + (8 - 5)$
   C $24 - (8 + 5)$
   D $(24 + 8) - 5$

6. Which number makes this equation
   true?
   $\blacksquare 7 \times 60 = 2,220$
   F 1   **H** 3
   G 2   **J** 4

7. Estimate each product. Which is
   the greatest?
   A 67×33 **C** 58×32
   B 78×22 **D** 61×39

8.   98
      $\times\ 20$
   F 1,960   **H** 1,860
   G 1,640   **J** NOT HERE

9. During one week the librarian
   recorded that these numbers of
   books were checked out: 30, 30,
   31, 38, 39, 51, 57. What is the
   median number of books that were
   checked out?
   A 11   **C** 38
   B 30   **D** 41

10. Which equation describes the rule in
   this table?

INPUT	*n*	6	7	9	11
OUTPUT	*p*	18	21	27	33

   F $p = n + 12$ **H** $p = n + 14$
   G $p = n \times 3$ **J** $p = 2 \times n + 6$

Write What You Know

11. Explain how you could find the
product $80 \times 5,000$ mentally. Then
find it.

12. Describe how you can make a model
using grid paper to find the product
18×15. Tell what the product is.

Multiply by 2-Digit Numbers

DATA LINK

GIANT PUMPKIN WEIGHOFF

Weight (in pounds)

900
800
700
600
500
400
300
200
100
0

Adams Calai Gibson Kubiac Moss Pastor
Pumpkin Growers

What is the biggest pumpkin you have ever seen? Every year, growers of giant pumpkins compete in a "Giant Pumpkin Weighoff." The graph shows the weight of some of the pumpkins grown in Ohio in one year.

PROBLEM SOLVING Estimate the weight of the heaviest pumpkin and then find that estimated weight in ounces.

Use this page to help you review and remember
important skills needed for Chapter 12.

✓ VOCABULARY

Choose the best term from the box.

1. In $4 \times 15 = 60$, the number 60 is the ___?___.

2. In $35 + 12 = 47$, the number 12 is an ___?___.

3. In $3 \times 123 = 369$, the number 123 is a ___?___.

4. The symbol used to separate dollars from cents in money is called a ___?___.

5. Multiplying the ones, tens, and hundreds separately and then adding the products is called the ___?___ method.

addend
factor
product
partial products
decimal point

✓ MULTIPLY BY 1-DIGIT NUMBERS (For Intervention, see p. H15.)

Find the product. Estimate to check.

6.	12	7.	43	8.	39	9.	61
	$\times 6$		$\times 4$		$\times 3$		$\times 5$

10.	35	11.	25	12.	125	13.	163
	$\times 7$		$\times 5$		$\times 2$		$\times 3$

14.	350	15.	100	16.	49	17.	101
	$\times 3$		$\times 5$		$\times 7$		$\times 9$

✓ ESTIMATE PRODUCTS (For Intervention, see p. H15.)

Round the number in blue. Estimate the product.

18.	12	19.	14	20.	21	21.	18
	$\times 9$		$\times 6$		$\times 8$		$\times 5$

22.	272	23.	350	24.	649	25.	212
	$\times 3$		$\times 7$		$\times 4$		$\times 7$

26.	322	27.	817	28.	444	29.	9,145
	$\times 9$		$\times 2$		$\times 3$		$\times 2$

30. $4 \times 2,781$ 31. 5×550 32. $6 \times 7,317$ 33. 4×385

Multiply by 2-Digit Numbers

Quick Review

1. 3×12 2. 2×25
3. 2×31 4. 2×40
5. 2×32

▶ Learn

TOTALLY TOMATOES   Mr. Henson grows tomatoes. There are 35 plants on each tray. Mr. Henson has 88 trays. How many tomato plants does he have in all?

Example 1

Multiply. 35×88
                    ↓   ↓
Estimate. $40 \times 90 = 3,600$

You can use place value and regrouping to find 35×88.

STEP 1	STEP 2	STEP 3
Think of 35 as 3 tens 5 ones. Multiply by 5 ones.	Multiply by 3 tens, or 30.	Add the products.
$\begin{array}{r} \overset{4}{88} \\ \times 35 \\ \hline 440 \end{array}$ ← 5×88	$\begin{array}{r} \overset{2}{\overset{4}{88}} \\ \times 35 \\ \hline 440 \\ 2640 \end{array}$ ← 30×88	$\begin{array}{r} \overset{2}{\overset{4}{88}} \\ \times 35 \\ \hline 440 \\ +2\,640 \\ \hline 3,080 \end{array}$

So, there are 3,080 tomato plants. Since 3,080 is close to the estimate of 3,600, the answer is reasonable.

More Examples

A
$$\begin{array}{r} \overset{2}{\overset{7}{29}} \\ \times 38 \\ \hline 232 \\ +870 \\ \hline 1,102 \end{array}$$
232 ← 8×29
+870 ← 30×29

B
$$\begin{array}{r} \overset{3}{\overset{1}{\$55}} \\ \times 62 \\ \hline 110 \\ +3\,300 \\ \hline \$3,410 \end{array}$$
110 ← 2×55
+3 300 ← 60×55

• When you multiply by 2-digit numbers, explain how you know in which place to begin.

Find Products

The greenhouse has 45 bags of potting soil.
Each bag has enough soil to pot 29 plants.
How many plants can be potted?

Example 2

Multiply.   45 × 29
             ↓   ↓
Estimate.   50 × 30 = 1,500

One Way	Another Way
Colleen used regrouping to find the product.	Brad used partial products.

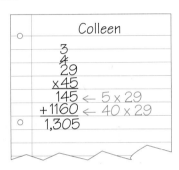

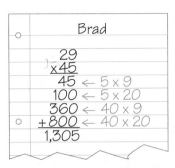

So, there is enough soil to pot 1,305 plants. Since 1,305 is
close to the estimate of 1,500, the answer is reasonable.

MATH IDEA   You can find products using place
value and regrouping or you can use the partial
products method.

- Explain how the partial products method to find
  products is different from using regrouping.

Technology Link
More Practice: Use E-Lab,
Modeling Multiplication.
www.harcourtschool.com/
elab2002

▶ Check

1. **Tell** which method of recording multiplication
   you like better. Explain your choice.

Choose either method to find the product. Estimate to check.

2. 37 ×22	3. 54 ×31	4. 42 ×26	5. 78 ×41	6. $23 × 34
7. 67 ×14	8. 93 ×76	9. 82 ×47	10. 51 ×79	11. 38 ×64

LESSON CONTINUES ▶

Practice and Problem Solving

Choose either method to find the product. Estimate to check.

12. 44 ×35

13. 67 ×14

14. $63 × 42

15. 81 ×22

16. 72 ×59

17. $38 × 29

18. 76 ×45

19. 68 ×79

20. 97 ×65

21. 82 ×35

22. 52 ×17

23. 69 ×42

24. $53 × 74

25. 22 ×85

26. $71 × 61

27. 51 × 28

28. 38 × 17

29. 23 × 14

30. 61 × 51

31. 41 × $16

32. 61 × $87

33. 74 × 11

34. 76 × $68

35. 14 × 65

36. 38 × 55

37. 46 × 34

38. 63 × 59

MENTAL MATH Write the missing product.

39. 20 × 16 = 320, so 20 × 17 = ▥

40. 45 × 28 = 1,260, so 45 × 29 = ▥

41. 13 × 50 = 650, so 13 × 49 = ▥

42. 45 × 17 = 765, so 45 × 16 = ▥

Copy and complete.

43.
```
      45
     ×72
      10  ←▥ × ▥
      80  ←▥ × ▥
     350  ←▥ × ▥
  +2,800  ←▥ × ▥
   3,240
```

44.
```
      23
     ×98
      24  ←▥ × ▥
     160  ←▥ × ▥
     270  ←▥ × ▥
  +1,800  ←▥ × ▥
   2,254
```

45. ✳ **ALGEBRA** Find the missing numbers. Explain.
```
      35
     ×6▥
     1▥5
  +2 100
   2,275
```

46. ❓ **What's the Error?** Describe Emilia's error. Write the correct answer.

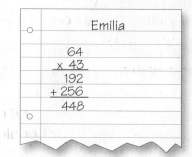

Emilia
```
    64
  x 43
   192
 + 256
   448
```

47. REASONING When you multiply a two-digit number by a two-digit number, what is the greatest number of digits you can have in the product? Explain.

USE DATA For 48–49, use the graph.

48. Grace and her father belong to a gardening club. How many more members were there in 2000 than in 1997?

49. Each member pays $32 in dues each year. How much money was paid in dues in 1999?

50. Emily rides her bike 22 miles each week for exercise. What is the total number of miles Emily rides in a year?

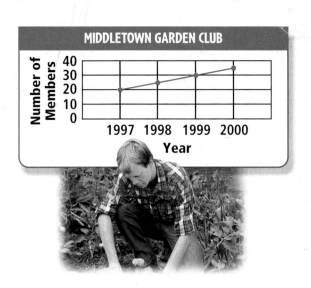

MIDDLETOWN GARDEN CLUB

Mixed Review and Test Prep

51. 9,213 (p. 200)   52.   942 (p. 196)
    × 6   × 5

53. One load of laundry in a washing machine uses 49 gallons of water. Mr. Porter washed 17 loads of laundry last month. How much water did he use? (p. 224)

54. What is the median? 18, 19, 19, 26, 28, 35, 35, 40, 42 (p. 86)

55. **TEST PREP** Al has 14 rows of stamps. Each row has 9 stamps. Al's brother has 56 stamps. How many stamps does Al have?
    A 94 **B** 126 **C** 182 **D** 443

56. **TEST PREP** Mr. Konrad has 56 tulip bulbs. He plants them in 8 rows with the same number in each row. How many bulbs are in each row? (p. 144)
    F 10 **G** 9 **H** 8 **J** 7

PROBLEM SOLVING LINKUP ... to Science

NETWORKS

A **network** is a system of parts that are connected. For example, computers in an office that are connected are in a network.

This diagram shows a network of paths. Each path has a value.

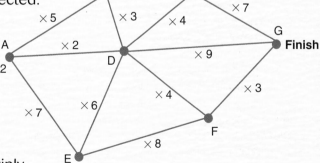

a. Find a path from **Start** to **Finish.**
b. Begin at Start with the number 2.
c. As you move from letter to letter, multiply your results by the number along the path.

1. Name a path that has a product greater than 500.

2. Name a path that has a product less than 100.

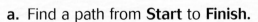

EXTRA PRACTICE page H43, Set A

More About Multiplying by 2-Digit Numbers

▶ Learn

FIELDS OF FRUIT   Kevin's family has 36 rows of trees in an orange grove. Each row has 128 orange trees. How many trees does Kevin's family have?

Example

Multiply. 36×128
              ↓   ↓
Estimate. $40 \times 100 = 4,000$

STEP 1	**STEP 2**	**STEP 3**
Multiply by the ones.	Multiply by the tens.	Add the products.
$\begin{array}{r} {\scriptstyle 14} \\ 128 \\ \times\ 36 \\ \hline 768 \end{array}$ ← 6 × 128	$\begin{array}{r} {\scriptstyle 2} \\ {\scriptstyle 14} \\ 128 \\ \times\ 36 \\ \hline 768 \\ 3840 \end{array}$ ← 30 × 128	$\begin{array}{r} {\scriptstyle 2} \\ {\scriptstyle 14} \\ 128 \\ \times\ 36 \\ \hline 768 \\ +3\ 840 \\ \hline 4,608 \end{array}$

So, Kevin's family has 4,608 orange trees. Since 4,608 is close to the estimate of 4,000, the answer is reasonable.

More Examples

A	**B**	**C**
$\begin{array}{r} {\scriptstyle 1} \\ 204 \\ \times\ 41 \\ \hline 204 \\ +8\ 160 \\ \hline 8,364 \end{array}$ $\begin{array}{l} \leftarrow 1 \times 204 \\ \leftarrow 40 \times 204 \end{array}$	$\begin{array}{r} {\scriptstyle 5} \\ 109 \\ \times\ 60 \\ \hline 000 \\ +6\ 540 \\ \hline 6,540 \end{array}$ These zeros can ← be omitted. ← 60 × 109	$\begin{array}{r} {\scriptstyle 1} \\ {\scriptstyle 31} \\ \$562 \\ \times\ 35 \\ \hline 2\ 810 \\ +16\ 860 \\ \hline \$19,670 \end{array}$ $\begin{array}{l} \leftarrow 5 \times 562 \\ \leftarrow 30 \times 562 \end{array}$

▶ Check

1. **Explain** what happened to the regrouped digit, 5, in Example B when the 0 was multiplied by 60.

Find the product. Estimate to check.

| 2. 237 × 21 | 3. $103 × 29 | 4. 187 × 35 | 5. 417 × 72 | 6. 532 × 20 |

Find the product. Estimate to check.

| 7. 888 × 22 | 8. $794 × 25 | 9. 204 × 41 | 10. 437 × 70 | 11. $837 × 21 |

| 12. 357 × 41 | 13. $627 × 30 | 14. 904 × 86 | 15. $790 × 32 | 16. 252 × 53 |

17. 23×256 18. 52×236 19. $85 \times \$299$ 20. $80 \times \$567$ 21. 50×108

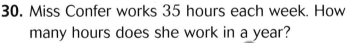 **ALGEBRA**   Find the value for n that makes the equation true.

22. $20 \times 543 = n$ 23. $30 \times 147 = n$ 24. $80 \times 209 = n$

25. $n \times 276 = 2{,}760$ 26. $n \times 900 = 54{,}000$ 27. $n \times 500 = 40{,}000$

USE DATA For 28–29, use the graph.

28. Medium fruit baskets sold for $16.75. What was the total sales in December for medium baskets?

29. Small fruit baskets sold for $7.49. How much more did Fruit Galore make on sales of small fruit baskets in December than in November?

30. Miss Confer works 35 hours each week. How many hours does she work in a year?

31. **REASONING** Find the missing digit in the equation $908 \times 3\blacksquare = 31{,}780$. Explain how you found your answer.

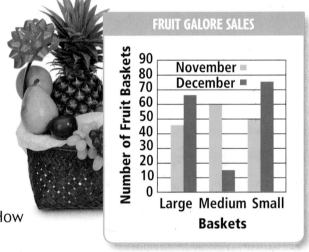

FRUIT GALORE SALES

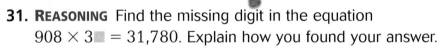

—**Mixed Review and Test Prep**—

32. 3,447 (p. 40) +1,725

33. 4,007 (p. 42) −1,341

34. $60 + (8{,}420 - 1{,}650)$ (p. 56)

35. Compare. Write $<$, $>$, or $=$ for ●. (p. 18)
2,076,355 ● 2,085,325

36. **TEST PREP** Peter is 8 years older than Mark. If Mark is 29 years old, how old is Peter? (p. 58)

A 21 years old
B 27 years old
C 31 years old
D 37 years old

3 Choose a Method

▶ Learn

FOREIGN EXCHANGE   In the United States, dollars are used for money. In Italy lire are used. When Angela visited her grandparents in Italy last year, one dollar was equal to 1,699 lire. If Angela spent 42 dollars on gifts, how many lire did she spend?

Example

Estimate. Then choose a method of computation that will be useful for the numbers given.

$40 \times 2,000 = 80,000$

Use Paper and Pencil Find the product. $42 \times 1,699 = $ ▪

STEP 1	STEP 2	STEP 3
Multiply by the ones.	Multiply by the tens.	Add the products.
$\begin{array}{r} {\scriptstyle 1\,1\,1} \\ 1,699 \\ \times\quad 42 \\ \hline 3\ 398 \end{array}$ ← 2 × 1,699	$\begin{array}{r} {\scriptstyle 2\ 3\ 3} \\ {\scriptstyle 1\ 1\ 1} \\ 1,699 \\ \times\quad 42 \\ \hline 3\ 398 \\ 67\ 960 \end{array}$ ← 40 × 1,699	$\begin{array}{r} {\scriptstyle 2\ 3\ 3} \\ {\scriptstyle 1\ 1\ 1} \\ 1,699 \\ \times\quad 42 \\ \hline 3\ 398 \\ +67\ 960 \\ \hline 71,358 \end{array}$

So, Angela spent 71,358 lire. Since 71,358 is close to the estimate of 80,000, the answer is reasonable.

More Examples

Ⓐ Use Mental Math
$40 \times 5,025 = $ ▪

Think: $40 \times 5,000 = 200,000$
and $40 \times 25 = 1,000$

So, $40 \times 5,025 = 201,000$.

Ⓑ Use a Calculator
$24 \times \$3,701 = $ ▪

So, $24 \times \$3,701 = \$88,824$.

💡 **MATH IDEA**   You can find a product by using paper and pencil, a calculator, or mental math. Look at the numbers in the problem before you choose a method.

1. **Explain** when you might choose mental math instead of paper and pencil.

Find the product. Write the method you used to compute.

2. 152
    × 11

3. $31
    × 22

4. 1,700
    ×   5

5. $317
    × 72

6. 5,502
    ×   24

► **Practice and Problem Solving**

Find the product. Estimate to check.

7. 434
    × 28

8. $287
    ×   7

9. 56
    × 60

10. $504
    × 31

11. 7,200
    ×   8

? **What's the Error?** Exercises 12–15 show 4 different common errors. Describe each error and correct it.

12.   1,274
       ×   67
       8,918
     + 7,644
      16,562

13.   5,782
         ×   88
        40,646
     + 406,460
       446,006

14.   2,500
       ×   32
         500
     + 7,500
       8,000

15.   4,306
         ×   39
        38,754
     + 129,180
       157,834

16. A magazine company has 9,822 customers. Each customer is sent 2 issues each month. How many magazines does the company send each year?

17. Angela bought some stamps that cost $0.79 and some postcards that cost $0.55. She spent $5.12. How many of each did she buy?

18. The average person should drink 8 cups of water a day. How many cups would that be in a week? in 30 days?

19. **REASONING** Find the product. Explain your method.   684
    × 306

Mixed Review and Test Prep

Find the product. (p. 212)

20. 42
    ×20

21. 57
    ×30

22. The river cruise began at 9:15 A.M. and ended at 2:50 P.M. How long was the cruise? (p. 120)

23. 19 × 27 = *n* (p. 224)

24. **TEST PREP** What year is 2 centuries 1 decade after 1927? (p. 126)

    A 1937   C 2037

    B 2027   D 2137

Practice Multiplication

▶ Learn

BIKE BONANZA Bill owns a bike shop. Last month at his bike sale, Bill sold 24 bikes for $75.99 each. How much did he take in on the sale of these bikes?

Remember

Multiply money the way you multiply whole numbers. Then you can write the product in dollars and cents.

Example

Multiply. $24 \times \$75.99$

Estimate. $20 \times \$80 = \$1,600$

STEP 1	STEP 2	STEP 3	STEP 4
Multiply the ones.	Multiply the tens.	Add the products.	Write the answer in dollars and cents.
$\begin{array}{r} ^{2\,3\,3} \\ 7,599 \\ \times\quad 24 \\ \hline 30\ 396 \leftarrow 4 \times \\ 7,599 \end{array}$	$\begin{array}{r} ^{1\ 11}_{2\,3\,3} \\ 7,599 \\ \times\quad 24 \\ \hline 30\ 396 \\ 151\ 980 \leftarrow 20 \times \\ 7,599 \end{array}$	$\begin{array}{r} ^{1\ 11}_{2\,3\,3} \\ 7,599 \\ \times\quad 24 \\ \hline 30\ 396 \\ +151\ 980 \\ \hline 182\ 376 \end{array}$	The product is $1,823.76.

So, Bill took in $1,823.76. Since $1,823.76 is close to the estimate of $1,600, the answer is reasonable.

More Examples

A
$\begin{array}{r} ^{1}_{4} \\ 4,006 \\ \times\quad 27 \\ \hline 28\ 042 \leftarrow 7 \times 4,006 \\ +80\ 120 \leftarrow 20 \times 4,006 \\ \hline 108,162 \end{array}$

B
$\begin{array}{r} ^{3}_{2} \\ \$5.09 \\ \times\quad 43 \\ \hline 1\ 527 \leftarrow 3 \times 509 \\ +20\ 360 \leftarrow 40 \times 509 \\ \hline \$218.87 \leftarrow \text{Add decimal point} \\ \text{and dollar sign.} \end{array}$

▶ Check

1. **Explain** how you know where to put the decimal point in a money problem.

Find the product. Estimate to check.

2. 892
   × 37

3. 5,637
   ×   51

4. $25.68
   ×   21

5. 2,015
   ×   24

6. $48.03
   ×   39

▶ Practice and Problem Solving

Find the product. Estimate to check.

7. 1,735
   ×   43

8. 97
   ×68

9. $37.45
   ×   28

10. $2.56
   ×   81

11. 1,505
   ×   79

12. $3 \times 2,954$ **13.** 56×8221 **14.** $20 \times 1,076$ **15.** $31 \times \$2.25$ **16.** 48×126

17. $29 \times 4,151$ **18.** $72 \times 1,407$ **19.** $68 \times 7,521$ **20.** $99 \times 9,998$ **21.** $55 \times \$30.63$

ALGEBRA   Find the number for n that makes the equation true.

22. $45 \times 236 = n$ **23.** $36 \times 2,087 = n$ **24.** $93 \times \$47 = n$ **25.** $68 \times 795 = n$

26. $24 \times n = 24,000$ **27.** $80 \times n = 72,000$ **28.** $12 \times n = 1,440$

USE DATA   For 29–32, use the sign.

29. How much will Train Museum tickets cost for 25 students?

30. How much more are two adult tickets than two student tickets?

31. Bradley has $15. Can he buy a student ticket and a book that costs $4.50? Explain.

32. ✎ Write a problem about buying tickets for more than 1 adult and more than 1 student who visit the museum. Show the solution.

TRAIN MUSEUM	
children (under 6)	$5.25
students	$10.75
adults	$13.50

Mixed Review and Test Prep

33.   7,805 (p. 40)
   +2,678

34.   2,900 (p. 42)
   −   407

35. $50,225 + 23,186$ (p. 44)

36. $25,030 − 21,089$ (p. 44)

37. $84,582 + 37,329$ (p. 44)

38. **TEST PREP** Rachel scored 15 points and then some more points. By the end of the basketball game, she had scored 32 points. Which equation describes the situation? (p. 60)

A $15 − p = 32$ **C** $32 + 15 = p$
B $15 + p = 32$ **D** $p − 15 = 32$

EXTRA PRACTICE page H43, Set D

Problem Solving Skill
Multistep Problems

UNDERSTAND > PLAN > SOLVE > CHECK

VOCABULARY

multistep problem

REACH YOUR GOAL The soccer players sold bottles of water to earn money for equipment. They charged $2.25 for each bottle of water. They sold 52 bottles on Saturday and 45 bottles on Sunday. How much money did they collect?

MATH IDEA Sometimes it takes more than one step to solve a problem. To solve **multistep problems**, decide *what* the steps are and *in what order* you should do them.

Example

To find how much money the soccer players collected, multiply the total number of bottles sold by $2.25.

STEP 1	STEP 2
Add to find the total number of bottles sold.	Multiply to find the amount of money collected.
$52 \leftarrow$ bottles sold on Saturday $+45 \leftarrow$ bottles sold on Sunday 97	$\$2.25 \leftarrow$ price for each bottle $\times \quad 97 \leftarrow$ total number of bottles $\$218.25$
The players sold 97 bottles of water.	So, the players collected $218.25.

Talk About It

• Could you use the equation $(52 + 45) \times \$2.25 = n$ to solve the problem? Explain.

• Would you get the same answer if you first multiplied $52 \times \$2.25$ and $45 \times \$2.25$ and then added the two products? Explain.

1. **What if** the soccer players had charged $3.25 for each bottle of water they sold? How much more money would they have collected?

2. The soccer coach bought 12 new uniforms for $17.25 each and 6 soccer balls for $8.25 each. How much did he spend in all?

USE DATA For 3–4, use the table.

During soccer season, Darren's Snack Bar sells lunches. Darren kept this record of the number of items sold at last Saturday's game.

3. Which equation can you use to find the amount of money Darren took in from the sale of fruit salads and turkey sandwiches?
   A $2.59 + $5.87 = n
   B ($2.59 × 33) + ($5.87 × 29) = n
   C ($2.59 × 33) − ($5.87 × 29) = n
   D ($2.59 + $5.87) × (33 + 29) = n

4. How much more money did Darren take in from the sale of fruit salads than from the sale of veggie sandwiches?
   F $4.37 more   H $6.63 more
   G $4.43 more   J $17.43 more

DARREN'S SNACK BAR - Saturday Specials		
Item	Price	Number Sold
Fruit Salad	$2.59	33
Turkey Sandwich	$5.87	29
Veggie Sandwich	$4.38	18
Chicken Soup	$1.39	15
Chocolate Pie	$1.21	12

Mixed Applications

5. Mrs. Ling had a bucket of crayons. She gave 14 crayons to each of 21 students. If 131 crayons were left in the bucket, how many crayons were there to start?

6. Norman spent $15.00 on a pizza and 2 salads. The pizza cost twice as much as the 2 salads. What was the price of each item?

7. Tommy and Helen were playing a game. First Helen picked a number and added 4. Then she multiplied by 6. Last she subtracted 3. The result was 51. What number did Helen pick?

8. Harry, Eli, Macy, and Sandy are standing in line at the movies. Harry is just behind Eli. Macy is between Harry and Sandy. Who is first in line?

9. Band members practice from 3:00 P.M. to 5:00 P.M. twice a week. For how many hours do they practice in 4 weeks?

10. Write a problem that requires more than two steps to solve. Show the solution.

Problem Solving Skill

Review/Test

✓ CHECK VOCABULARY AND CONCEPTS

Choose the best term from the box.

> multistep problem
> estimate
> partial products
> ones

1. A _?_ requires two or more steps to find the solution.
   (p. 234)

2. To check whether your answer is reasonable, compare the answer to the _?_. (p. 224)

3. When multiplying by two-digit numbers, you add the _?_ to get the final product. (p. 224)

✓ CHECK SKILLS

Find the product. Estimate to check. (pp. 224–233)

4. 39 ×16	5. 54 ×33	6. 143 × 62	7. 472 × 73
8. 92 ×58	9. 61 ×29	10. 2,303 × 67	11. 8,845 × 53
12. 67 ×82	13. $297 × 34	14. 9,709 × 36	15. $62.11 × 85
16. 604 × 55	17. 253 × 31	18. $43.26 × 41	19. $7,132 × 18
20. 5,853 × 24	21. 77 ×46	22. 935 × 25	23. 720 × 99

✓ CHECK PROBLEM SOLVING

Solve. (pp. 234–235)

24. Josh plays on a bowling team. There are 13 players. If 5 players scored an average of 145 each and 8 players scored an average of 134 each, what is the total score of the team?

25. A box can hold 36 markers. How many markers can 25 boxes hold?

Standardized Test Prep

Check your work.
See item **4.**

If your answer doesn't match the choices given, use a different method to compute. Try using the partial products method.

Also see problem **7,** p. H65.

For 1–8, choose the best answer.

1. Which is the best estimate for this sum?
4,712 + 2,934
- **A** 7,000
- **B** 8,000
- **C** 9,000
- **D** 10,000

2. Which is the best estimate for this product?

216
$\times$ 38

- **F** 6,000
- **G** 80,000
- **H** 8,000
- **J** 12,000

3. 81 × 457
- **A** 37,017
- **B** 32,517
- **C** 36,917
- **D** NOT HERE

4. 76 × 44
- **F** 608
- **G** 304
- **H** 3,704
- **J** NOT HERE

5. Which time does the clock show?

- **A** 12:07
- **B** 12:52
- **C** 1:07
- **D** 1:52

6. Jack wants to mentally calculate how much it would cost for 24 students to see a play that costs $7 per ticket. Which is a way to calculate 24 × $7?
- **F** (20 × $7) + 4
- **G** 20 × 4 × $7
- **H** (20 × $7) + (4 × $7)
- **J** 20 + 4 + $7

7. What is the value of *n* if
14 × 5,682 = *n*?
- **A** 76,248
- **B** 79,648
- **C** 77,248
- **D** 79,548

8. Jean had $15. She paid $5 to get into the fair and $4 for food. Which expression describes this?
- **F** ($15 − $5) + $4
- **G** $15 − ($5 − $4)
- **H** $15 − ($5 + $4)
- **J** $15 + $5 − $4

Write What You Know

9. There are 365 days in most years. Tell how old you are, and estimate the number of days that have passed since you were born.

10. Meg earns $12 each hour. Last week she worked 8 hours on Monday and Wednesday and 7 hours on Tuesday, Thursday, and Friday. Did Meg earn $300 last week? Explain your answer.

Squeeze Play

29	214	42	361	9	62

	84	102	8	19	

Use estimation skills and your knowledge of multiplication to solve each problem. Determine which factors from the factor boxes above will give a product for each range. Once a number is used in one problem, it may not be used in another. Put on your thinking cap. Good luck!

1. Product is between 700 and 800.

2. Product is between 700 and 800.

3. Product is between 800 and 900.

4. Product is between 6,000 and 7,000.

Think It Over!

• Which factors were left unused in the factor box?

• Find the greatest product possible for the missing digits at the right. Is the product nearly 1,000, nearly 10,000, nearly 100,000, or nearly 1,000,000?

• **STRETCH YOUR THINKING** **What if** the product for Problem 4 was between 13,000 and 14,000? Which factors would be left unused in the factor box?

Challenge

Mental Multiplication

There is an easy way to find products using mental math.

Printers are on sale at Computers Plus for $99. How much will it cost to buy 7 printers for the computer lab?

Multiply. 7×99

Think: 99 is 1 less than 100.

$7 \times 1 = 7$

$7 \times 99 = (7 \times 100) - 7$

$\quad\quad\quad = 700 - 7$

$\quad\quad\quad = 693$

So, it will cost $693 to buy 7 printers.

Examples

A 6×79
Think: 79 is 1 less than 80.
$\quad\quad 6 \times 1 = 6$
$6 \times 79 = (6 \times 80) - 6$
$\quad\quad\quad = 480 - 6$
$\quad\quad\quad = 474$

B $8 \times \$298$
Think: $298 is 2 less than $300.
$\quad\quad 8 \times 2 = \$16$
$8 \times \$298 = (8 \times \$300) - \$16$
$\quad\quad\quad\quad = \$2,400 - \$16$
$\quad\quad\quad\quad = \$2,384$

C 3×147
Think: 147 is 3 less than 150.
$\quad\quad 3 \times 3 = 9$
$3 \times 147 = (3 \times 150) - 9$
$\quad\quad\quad\quad = 450 - 9$
$\quad\quad\quad\quad = 441$

D $5 \times \$245$
Think: $245 is 5 less than $250.
$\quad\quad 5 \times 5 = \$25$
$5 \times \$245 = (5 \times \$250) - \$25$
$\quad\quad\quad\quad = \$1,250 - \$25$
$\quad\quad\quad\quad = \$1,225$

Try It
Use mental math to find each product.

1. 5×49 2. 2×85 3. $7 \times \$25$ 4. 78×3 5. 8×99

6. 8×17 7. 58×6 8. $2 \times \$39$ 9. 29×9 10. $\$65 \times 3$

11. 197×8 12. $7 \times \$49$ 13. 598×9 14. $\$59 \times 4$ 15. $4 \times \$146$

Study Guide and Review

VOCABULARY

Choose the best term from the box.

1. To find products mentally, you can use _?_ and _?_
   (pp. 186, 210)

2. To help you decide where to break apart a factor, you can use _?_. (p. 216)

| basic facts |
| multiple |
| place value |
| patterns |

STUDY AND SOLVE

Chapter 10

Write products of multidigit numbers multiplied by one-digit numbers.

$$\begin{array}{r} \overset{2\ 3}{259} \\ \times\quad 4 \\ \hline 1,036 \end{array}$$

- Multiply the ones. $9 \times 4 = 36$
- Multiply the tens. $5 \times 4 = 20$
- Multiply the hundreds. $2 \times 4 = 8$
- Regroup as needed.

Multiply. (pp. 190–203)

3. 61×2 4. 398×5

5. $2,608 \times 4$ 6. $5,312 \times 3$

7. $\begin{array}{r} 456 \\ \times\quad 9 \\ \hline \end{array}$ 8. $\begin{array}{r} 2,862 \\ \times\qquad 6 \\ \hline \end{array}$

Chapter 11

Write products using models.

17×21

- Outline a 17×21 rectangle on grid paper.
- Break apart the model into rectangles whose factors are easy to multiply. Multiply.
- Add the partial products.

$200 + 10 + 140 + 7 = 357$

Make a model to find the product. (pp. 216–217)

9. 15×14 10. 26×23

11. 12×31 12. 19×16

13. $\begin{array}{r} 19 \\ \times 36 \\ \hline \end{array}$ 14. $\begin{array}{r} 24 \\ \times 31 \\ \hline \end{array}$

15. $\begin{array}{r} 18 \\ \times 25 \\ \hline \end{array}$ 16. $\begin{array}{r} 22 \\ \times 34 \\ \hline \end{array}$

Chapter 12

Multiply two-digit numbers.

$$\begin{array}{r} \overset{1}{72} \\ \times 46 \\ \hline 432 \\ +2,880 \\ \hline 3,312 \end{array}$$

← 6 × 72   • Multiply the ones.
← 40 × 72   • Multiply the tens.
• Add the products.

Find the product. (pp. 224–229)

17. 84
×68

18. 93
×48

19. 705
× 27

20. 259
× 65

21. 51 × 24

22. 895 × 39

Multiply greater numbers.

$$\begin{array}{r} \overset{2}{} \overset{4}{} \\ 5,407 \\ \times 61 \\ \hline 5,407 \\ +324,420 \\ \hline 329,827 \end{array}$$

← 1 × 5,407   •• Multiply the ones.
← 60 × 5,407   • Multiply the tens.
• Add the products.

Find the product. (pp. 230–233)

23. 5,904
× 64

24. 6,007
× 81

25. 8,753
× 26

26. 1,098
× 79

27. 6,509
× 75

28. 7,216
× 52

PROBLEM SOLVING PRACTICE

Solve. (pp. 204–205, 218–219, 234–235)

29. There were 246 people at the arcade and each person won 20 tickets. How many total tickets were won? Use a simpler problem to solve.

30. Martina walks 15 minutes in the morning, 15 minutes in the afternoon, and 45 minutes in the evening. If she walks everyday for 30 days, how many minutes does she walk?

31. Thao borrowed 6 history videos to watch for homework. Each video was 75 minutes long. Write an equation to show how long it would take to watch all the videos. Solve.

32. Mr. Uri bought shirts for his daughter's basketball team. There were 14 girls on the team. Each shirt cost $12.48. How much did Mr. Uri spend on the shirts?

PERFORMANCE ASSESSMENT

TASK A • FISH STORY

Andy received an aquarium for his birthday. He read these directions for setting up the aquarium.

Directions:

• Add between 7 and 10 pounds of gravel.
• Add 4 or 5 plants.
• Add your choice of fish.

Andy has saved $35.00 to buy gravel, plants, and fish for the aquarium. He wants to have at least two of each kind of fish he chooses.

SUPPLY	PRICE
Gravel	$0.65 per pound
Plants	$3.25 each

FISH	PRICE PER FISH
Tetra	$1.79
Platy	$2.19
Molly	$2.99

a. Estimate the total cost of the gravel and plants Andy needs. Then use estimation to make a list of number and kinds of fish that Andy could buy so that supplies and fish together cost about $35.00.

b. What is the actual total cost of the items you chose?

c. Explain how you can use multiplication to help you solve this problem.

TASK B • SPEED WRITING

MATERIALS: stopwatch or clock with second hand

This task is a contest to see who can write the fastest.

a. With a partner timing you, write the word *math* as many times as you can in one minute. Then count and record the number of times you wrote *math*.

b. If you continued to write at this rate, how many times could you write the word *math* in 30 minutes? How many times could you write the word *math* in 2 hours?

c. Use what you found out in part **b** to explain how you could predict how many times you could write the word *multiply* in 45 minutes.

Technology Linkup

Calculator • Number Patterns

Steve makes a number pattern. His rule is: add 17 and then multiply by 3. The first number in the pattern is 1. What are the next three numbers in the pattern?

You can use a calculator to extend number patterns.

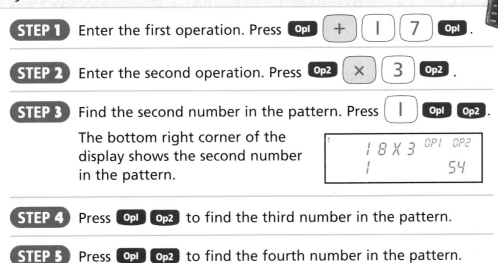

STEP 1 Enter the first operation. Press **Op1** **+** **1** **7** **Op1** .

STEP 2 Enter the second operation. Press **Op2** **×** **3** **Op2** .

STEP 3 Find the second number in the pattern. Press **1** **Op1** **Op2** .

The bottom right corner of the display shows the second number in the pattern.

```
 1 8 X 3   OP1  OP2
 1              54
```

STEP 4 Press **Op1** **Op2** to find the third number in the pattern.

STEP 5 Press **Op1** **Op2** to find the fourth number in the pattern.

So, the next three numbers in the pattern are 54, 213, and 690.

Practice and Problem Solving

Use the TI-15 Keys to find the next three numbers in the pattern.

1. **rule:** add 7 and then multiply by 2
   first number: 2

2. **rule:** subtract 5 and then multiply by 3
   first number: 19

3. **rule:** add 6 and then divide by 2
   first number: 40

4. **rule:** subtract 30 and then divide by 4
   first number: 1,000

5. **Write a problem** Make up a rule for a pattern. The rule should include two operations. List the first 5 numbers in your pattern.

Multimedia Math Glossary www.harcourtschool.com/mathglossary

6. **Vocabulary** Look up *multistep problem* in the Multimedia Math Glossary. Write a problem that can be answered by using the example shown in the glossary.

PROBLEM SOLVING ON LOCATION

in New Jersey

HIKING AND BIKING

New Jersey is the fifth-smallest state in the United States. The scenery in New Jersey includes mountains, valleys, rivers, and beaches. There are outdoor activities for people of all ages.

1. Leslie, Joanne, and Will plan to hike for 3 hours in Allamuchy State Park. Each friend brings 8 ounces of water for each hour they plan to hike. Write an equation you can use to show how much water they bring altogether. How much water do they bring altogether?

▲ Preserved sand dunes near the New Jersey shore

2. José rides his bike on the Old Mine Road in the Delaware Water Gap. He bikes at an average speed of 19 miles per hour. Will it take him more or less than 2 hours to bike the 37-mile round trip trail? Explain how you know.

3. The members of the Outing Club are hiking in High Point State Park. They average a speed of 3 miles per hour. They would like to hike a total of 16 hours over two days. Will they be able to complete 50 miles of hiking? Explain how you know.

4. Wawayanda State Park is one of the best places to go mountain biking in New Jersey. Suppose mountain bikes can be rented for $5.25 per hour or $20 per day. If you want to ride a mountain bike for 3 hours, would it be less expensive to rent by the hour or by the day? Explain how you know.

CANOEING AND ROCK CLIMBING

Canoeing and rock climbing are other popular outdoor activities in New Jersey.

1. Visitors like to canoe on the Delaware River in the Delaware Water Gap. What if 2-hour canoe rentals cost $35 and a group of visitors rents 8 canoes for 2 hours? Write an equation you can use to find the total cost. What is the total cost?

2. Cindy leads a group on a trip along Cedar Swamp Creek. There are 3 people in each of 19 canoes. Cindy is alone in her canoe. Write an equation you can use to find the total number of people, including Cindy. How many people are there, including Cindy?

USE DATA For 3–5, use the table.

3. Two good climbs in the Delaware Water Gap are called Heights of Madness and Morning Sickness. If the rental shop rents 54 pairs of rock-climbing shoes and 45 backpacks, how much will it collect in rental fees?

RENTAL EQUIPMENT	
Item	**Price**
rock-climbing shoes	$6
backpack	$8
sleeping bag	$10

4. How much will it cost a family of four if each member rents rock-climbing shoes, a backpack, and a sleeping bag?

5. Suppose the shop clerk collects $320 from a group in rental fees for rock-climbing shoes, backpacks, and sleeping bags. What is one possible group of items that might have been rented?

▼ **The Delaware River makes an S-shaped curve between New Jersey and Pennsylvania. This area is known as the Delaware Water Gap.**

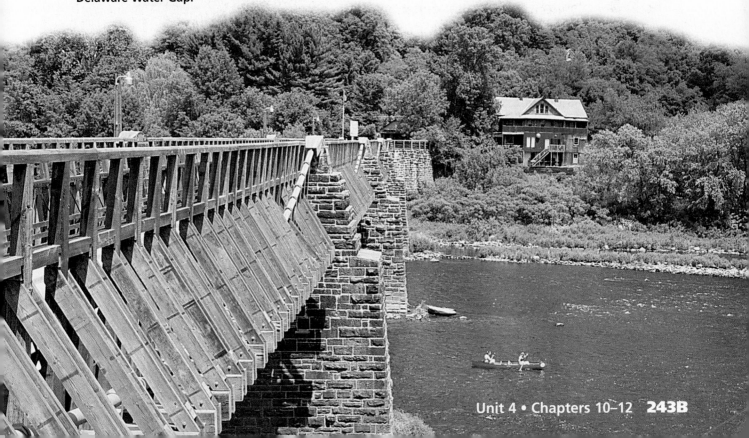

ANIMAL MIGRATION

Animal	Estimated Migration Distance (km)
Green turtle	2,200
Arctic tern	20,000
Pacific gray whale	20,000

Each year, thousands of Atlantic green turtles travel thousands of kilometers from the coast of Brazil to Ascension Island in the Atlantic Ocean.

PROBLEM SOLVING   If one green turtle travels about 100 km each day, about how long will it take to migrate 2,200 kilometers?

Atlantic green turtle

CHECK WHAT YOU KNOW ✓

Use this page to help you review and remember
important skills needed for Chapter 13.

✓ VOCABULARY

Choose the term from the box that names a number from
the example.

divisor
dividend
quotient
divide

1. The __?__ is 8.

2. The __?__ is 7.

EXAMPLE:

$$7\overline{)56}$$ with quotient 8

✓ DIVISION FACTS (For Intervention, see p. H12.)

Write the division fact that each picture represents.

3.

4.

5.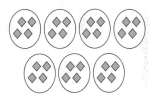

Find the quotient.

6. $50 \div 5$ **7.** $45 \div 9$ **8.** $66 \div 6$ **9.** $54 \div 6$

10. $100 \div 10$ **11.** $64 \div 8$ **12.** $60 \div 5$ **13.** $144 \div 12$

14. $72 \div 8$ **15.** $49 \div 7$ **16.** $81 \div 9$ **17.** $36 \div 6$

✓ FACT FAMILIES (For Intervention, see p. H16.)

Find the missing numbers.

18. $5 \times 8 = 40$
$8 \times 5 = \blacksquare$
$40 \div \blacksquare = 5$
$40 \div \blacksquare = 8$

19. $3 \times 9 = 27$
$9 \times 3 = \blacksquare$
$27 \div \blacksquare = 3$
$27 \div \blacksquare = 9$

20. $8 \times 7 = 56$
$7 \times 8 = \blacksquare$
$56 \div \blacksquare = 7$
$56 \div \blacksquare = 8$

21. $4 \times 9 = 36$
$9 \times 4 = \blacksquare$
$36 \div \blacksquare = 4$
$36 \div \blacksquare = 9$

22. $6 \times 4 = 24$
$4 \times 6 = \blacksquare$
$24 \div \blacksquare = 6$
$24 \div \blacksquare = 4$

23. $7 \times 5 = 35$
$5 \times 7 = \blacksquare$
$\blacksquare \div 7 = 5$
$35 \div \blacksquare = 7$

24. $9 \times 7 = 63$
$7 \times 9 = \blacksquare$
$63 \div \blacksquare = 7$
$\blacksquare \div 7 = 9$

25. $6 \times 8 = 48$
$8 \times 6 = \blacksquare$
$48 \div \blacksquare = 8$
$48 \div 8 = \blacksquare$

Divide with Remainders

▶ Learn

MODEL IT! Rico has 19 model airplanes to put on some shelves. What is the greatest number of airplanes that he can put on each of 3 shelves? How many airplanes will be left over?

Sometimes a number cannot be divided evenly. The amount left over is called the **remainder**.

VOCABULARY

remainder

HANDS ON Activity

Make a model to divide 19 by 3. Write $19 \div 3$ or $3\overline{)19}$.

MATERIALS: counters

STEP 1

Use 19 counters.   Write: $3\overline{)19}$

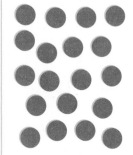

STEP 2

Draw three circles. Divide the 19 counters into 3 equal groups.

Record:

$$\begin{array}{r} 6\ r1 \\ 3\overline{)19} \\ -18 \\ \hline 1 \end{array}$$

↑
remainder

The quotient is 6. The remainder is 1.
So, $19 \div 3 = 6\ r1$.

So, Rico can put 6 airplanes on each shelf. There will be 1 airplane left over.

MATH IDEA The remainder is a number less than the divisor.

▶ Check

1. **Explain** why the answer could not be 5 r4 for the model in the activity.

Make a model, record, and solve.

2. $15 \div 2$ **3.** $20 \div 3$ **4.** $22 \div 4$ **5.** $37 \div 6$

▶ Practice and Problem Solving

Make a model, record, and solve.

6. $29 \div 3$ **7.** $35 \div 4$ **8.** $57 \div 8$ **9.** $45 \div 6$

Divide. You may wish to use counters.

10. $13 \div 4$ **11.** $65 \div 7$ **12.** $30 \div 4$ **13.** $39 \div 5$

14. $3\overline{)28}$ **15.** $4\overline{)37}$ **16.** $5\overline{)42}$ **17.** $5\overline{)49}$

18. $23 \div 4$ **19.** $75 \div 8$ **20.** $29 \div 4$ **21.** $53 \div 7$

22. $2\overline{)17}$ **23.** $5\overline{)22}$ **24.** $6\overline{)47}$ **25.** $4\overline{)26}$

26. $8\overline{)44}$ **27.** $4\overline{)19}$ **28.** $3\overline{)28}$ **29.** $7\overline{)38}$

30. Mrs. Dawson has 39 balloons to decorate her classroom. What is the greatest number of balloons that she can put into each of the 4 corners? How many will be left over?

31. Eric took 28 photos at sporting events for the *Herald*. He took the same number of photos at 4 different events. How many photos did he take at each event?

32. I am an even number that has more ones than tens. My thousands digit is the quotient $9 \div 3$ and my tens digit is the sum of 3 and 3. I have no hundreds. What number am I?

33. **?** **What's the Error?** Bryan made this model for $4\overline{)23}$. Describe his error. Draw a correct model.

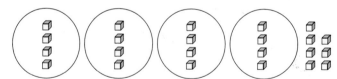

Mixed Review and Test Prep

Compare. Use $<$ or $>$ for each ●. (p. 144)

34. 9×3 ● 6×4

35. $42 \div 7$ ● $56 \div 8$

36. $\begin{array}{r} 328 \\ \times \quad 4 \end{array}$ (p. 196)   **37.** $\begin{array}{r} 295 \\ \times \quad 3 \end{array}$ (p. 196)

38. **TEST PREP** The model castle has 2 floors with 14 doors on each floor, 3 floors with 12 doors on each floor, and 1 floor with 10 doors. How many doors are in the castle? (p. 196)

A 36   **C** 74

B 41   **D** 89

EXTRA PRACTICE page H44, Set A

HANDS ON

Model Division

▶ **Explore**

TAKE A LOOK!   The Berkshire Museum has pictures of 48 extinct animals in 3 different rooms. Each room has the same number of pictures. How many pictures are in each room?

Activity 1

Divide 48 into 3 equal groups. Write $48 \div 3$ or $3 \overline{)48}$. Make a model to show how many are in each group.

STEP 1   Show 48 as 4 tens 8 ones. Draw circles to make 3 equal groups.

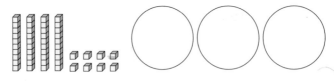

STEP 2   Place an equal number of tens into each group.

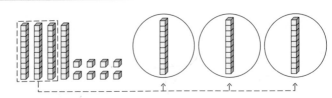

STEP 3   Regroup 1 ten 8 ones as 18 ones. Place an equal number of ones into each group.

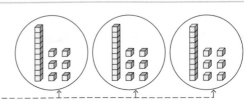

So, there are 16 pictures in each room.

• How many groups did you make?

• How many are in each group?

Try It

Model. Tell how many are in each group.

a. $26 \div 2$ **b.** $42 \div 3$ **c.** $64 \div 4$

We have modeled 26. How many circles should we draw to show $26 \div 2$?

Activity 2

Here is a way to record division. Divide 57 by 2.

STEP 1   Show the model and 2 equal groups.

Record:

2)57

STEP 2   Divide the tens.

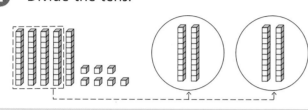

$\begin{array}{r} 2 \\ 2\overline{)57} \\ -4 \\ \hline 1 \end{array}$ 2 tens in each group
4 tens used
1 ten left

STEP 3   Regroup. Divide the ones.

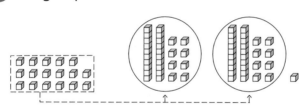

$\begin{array}{r} 28\ r1 \\ 2\overline{)57} \\ -4\downarrow \\ \hline 17 \\ -16 \\ \hline 1 \end{array}$ 8 ones in each group

16 ones used
1 one left
So, 57 ÷ 2 = 28 r1.

Read: 57 divided by 2 equals 28 remainder 1.

▶ **Practice and Problem Solving**

Make or draw a model. Record and solve.

1. 35 ÷ 2   2. 45 ÷ 3   3. 49 ÷ 4   4. 47 ÷ 2   5. 72 ÷ 6

6. 7)88   7. 3)56   8. 6)78   9. 5)66   10. 4)72

11. **? What's the Error?** Emily made this model for 3)42.
Describe her error. Draw a correct model.

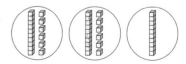

Mixed Review and Test Prep

12. $\begin{array}{r} 1{,}909 \\ -\ \ 287 \end{array}$ (p. 40)

13. $\begin{array}{r} 136 \\ \times\ \ 9 \end{array}$ (p. 196)

14. 9)81 (p. 144)

15. $\begin{array}{r} 2{,}368 \\ +7{,}416 \end{array}$ (p. 40)

16. **TEST PREP** What time is 2 hours 45 minutes later than 9:35? (p. 120)

   A 10:15   **B** 11:35   **C** 12:20   **D** 1:45

3 Division Procedures

▶ **Learn**

BALE OF TURTLES   A turtle hatchery has 96 turtle eggs. A worker put 7 eggs in each tank. How many tanks were used for the turtle eggs? How many eggs were left over?

Example
Divide 96 by 7. Write 96 ÷ 7 or 7)96.

STEP 1

Divide the 9 tens.

$$1$$
7)96   Divide. 9 ÷ 7
−7   Multiply. 1 × 7
2   Subtract. 9 − 7
       Compare. 2 < 7

The difference, 2, must be less than the divisor, 7.

STEP 2

Bring down the 6 ones.

1
7)96
−7↓
26

STEP 3

Divide the 26 ones.

13 r5
7)96   Divide. 26 ÷ 7
−7↓   Multiply. 3 × 7
26   Subtract. 26 − 21
−21   Compare. 5 < 7
5

Write the remainder next to the quotient.

So, the hatchery used 13 tanks for the turtle eggs. There were 5 eggs left over to be put in another tank.

MATH IDEA   The order of division is as follows: divide, multiply, subtract, compare, and bring down. This order is repeated until the division is complete.

Technology Link

More Practice: Use **Mighty Math Calculating Crew,** *Nick Knack,* Level P.

• When you compare, what must you do if the difference is equal to or greater than the divisor?

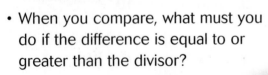

Divide and Check

What if the hatchery put 5 turtle eggs in each tank? How many tanks would be used for the 96 eggs if 6 eggs were put in each tank?

Solve the problems by dividing. Then check your answers.

To check your answer, you can compute:
(divisor × quotient) + remainder = dividend.

Examples

A 5 eggs in each tank

```
  19 r1      CHECK
5)96          19      quotient
 -5↓         × 5      divisor
  46          95
 -45         + 1      remainder
   1          96      dividend
```

So, the hatchery would use 19 tanks with 5 eggs in each tank and 1 egg left over.

B 6 eggs in each tank

```
  16         CHECK
6)96          16      quotient
 -6↓         × 6      divisor
  36          96
 -36         + 0      remainder
   0          96      dividend
```

So, the hatchery would use 16 tanks with 6 eggs in each tank and no eggs left over.

• Why is the remainder zero in Example B?

▶ Check

1. **Tell** what different remainders are possible when dividing any dividend by 5 and when dividing by 6.

Divide and check.

2. 4)58   3. 3)65   4. 5)84   5. 5)79   6. 7)99

7. 39 ÷ 2   8. 84 ÷ 6   9. 62 ÷ 4   10. 95 ÷ 8   11. 55 ÷ 3

Write the check step for each division problem.

12. 78 ÷ 6 = 13   13. 93 ÷ 7 = 13 r2   14. 52 ÷ 3 = 17 r1   15. 64 ÷ 5 = 12 r4

16. Compare each remainder in Exercises 2–5 with the divisor. Why is the remainder always less than the divisor?

LESSON CONTINUES

Divide and check.

17. $4\overline{)84}$ **18.** $4\overline{)51}$ **19.** $7\overline{)52}$ **20.** $2\overline{)46}$

21. $3\overline{)89}$ **22.** $5\overline{)67}$ **23.** $8\overline{)90}$ **24.** $3\overline{)76}$

25. $7\overline{)81}$ **26.** $6\overline{)93}$ **27.** $2\overline{)65}$ **28.** $8\overline{)91}$

29. $4\overline{)56}$ **30.** $4\overline{)59}$ **31.** $6\overline{)88}$ **32.** $5\overline{)69}$

33. $93 \div 8$ **34.** $73 \div 4$ **35.** $94 \div 3$ **36.** $87 \div 5$

Write the check step for each division problem.

37. $57 \div 4 = 14$ r1 **38.** $85 \div 7 = 12$ r1 **39.** $39 \div 3 = 13$ **40.** $82 \div 7 = 11$ r5

Write the division problem for each check.

41. $(12 \times 6) + 3 = 75$ **42.** $(25 \times 3) + 2 = 77$

43. $(14 \times 6) + 3 = 87$ **44.** $(13 \times 5) + 0 = 65$

45. $(14 \times 3) + 1 = 43$ **46.** $(15 \times 4) + 2 = 62$

ALGEBRA Let d = divisor and q = quotient. Find the value of each variable.

47. $97 \div 5 = 19$ r2
$(q \times d) + 2 = 97$

48. $97 \div 7 = 13$ r6
$(q \times d) + 6 = 97$

49. $97 \div 9 = 10$ r7
$(q \times d) + 7 = 97$

50. $97 \div 6 = 16$ r1
$(q \times d) + 1 = 97$

51. $97 \div 8 = 12$ r1
$(q \times d) + 1 = 97$

52. $97 \div 4 = 24$ r1
$(q \times d) + 1 = 97$

USE DATA For 53–56, use the graph.

53. Ms. Juanita put the Thursday and Friday volunteers into equal groups for training classes. Each class had 8 members. How many classes did Ms. Juanita need? Explain.

54. **REASONING** The Monday and Tuesday volunteers meet for lunch. Each table seats 4 people. How many tables are needed for the volunteers?

55. **? What's the Question?** The answer is 25 volunteers.

56. Write a problem that requires division, using the data in the graph.

Oceanside Volunteers

57. Jeremy has 37 pictures of his classmates. He wants to arrange them into groups of 6. How many groups can he make? How many pictures will be left over?

58. Mandy is playing a card game with 4 of her friends. There are 52 cards in a deck. She deals the same number of cards to each player. How many cards are left over?

Mixed Review and Test Prep

59. 1,650 (p. 40)
$$\begin{array}{r} 1,650 \\ -\ \ 938 \\ \hline \end{array}$$

60. 1,879 (p. 40)
$$\begin{array}{r} 1,879 \\ +2,548 \\ \hline \end{array}$$

61. 27 (p. 224)
$$\begin{array}{r} 27 \\ \times 19 \\ \hline \end{array}$$

62. 4)65 (p. 246)

63. Which numbers are missing in the pattern? 12, 24, ■, 48, 60, ■, 84
(p. 148)

64. $20 \times 15 = 300$, so $20 \times 16 = $ ■.
(p. 224)

65. **TEST PREP** If 4 boxes have 60 books, how many books are in each box?
(p. 250)

   A 14   **C** 64

   B 15   **D** 240

USE DATA For 66, use the graph.

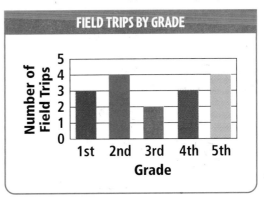

FIELD TRIPS BY GRADE

66. **TEST PREP** The principal at Meade School is planning 2 times as many field trips next year. How many field trips are being planned? (p. 194)

   F 17   **G** 20   **H** 30   **J** 32

PROBLEM SOLVING Thinker's Corner

REASONING Miranda has a collection of jungle cats and sea animals. She has 13 jungle cats and 11 sea animals. She wants to display them on some shelves.

• Each shelf will have the same number of items.

• No shelf can have all jungle cats or all sea animals.

Decide how many shelves Miranda needs. Use models or draw a diagram to help you.

  1. How many shelves will Miranda need?

  2. How many jungle cats and sea animals will be on each shelf?

  3. Draw a diagram to show your answer.

Problem Solving Strategy
Predict and Test

PROBLEM   There were 81 students visiting a museum. When they were placed in equal groups, 3 students were left over. If groups had between 10 and 15 students, how many groups were formed? How many students were in each group?

UNDERSTAND

- What are you asked to find?
- What information will you use?
- Is there any information you will not use? If so, what?

PLAN

- What strategy can you use to solve the problem?
  You can use the strategy *predict and test*.

SOLVE

- How can you use the strategy to solve the problem?
  Predict divisors. Divide to test each prediction.

NUMBER OF GROUPS		THE REMAINDER	
Predict	**Test**	**Compare to 3**	**Does it check?**
5	$81 \div 5 = 16$ r1	$1 < 3$	No
6	$81 \div 6 = 13$ r3	$3 = 3$	Yes
7	$81 \div 7 = 11$ r4	$4 > 3$	No
8	$81 \div 8 = 10$ r1	$1 < 3$	No
9	$81 \div 9 = 9$	$0 < 3$	No
10	$81 \div 10 = 8$ r1	$1 < 3$	No

So, there were 6 groups with 13 students in each group.

CHECK

- How can you check that your answer is correct?

Problem Solving Practice

Predict and test to solve.

1. **What if** there were 86 students touring the museum? After equal groups of between 15 and 20 were formed, 1 student was left over. How many groups were formed? How many students were in each group?

2. Eileen spent $16.50 at the gift shop. She bought 2 gifts. One of them cost $2.50 more than the other. How much was each gift?

PROBLEM SOLVING STRATEGIES

Draw a Diagram or Picture
Make a Model or Act It Out
Make an Organized List
Find a Pattern
Make a Table or Graph
▶ **Predict and Test**
Work Backward
Solve a Simpler Problem
Write an Equation
Use Logical Reasoning

Suppose you have 35 clay pots. You place an equal number of pots on tables and put 2 pots on the floor. Each table can hold up to 15 pots.

3. How many tables are holding the pots?

   A 2   **C** 15
   B 3   **D** 35

4. How many pots are on each table?

   F 10   **H** 12
   G 11   **J** 14

Mixed Strategy Practice

USE DATA For 5–7, use the schedule.

5. It takes Alex 30 minutes to go to or from her home and camp. If she arrives for the start of classes and completes them all, how much time has elapsed when she returns home?

6. **? What's the Question?** The elapsed time is 2 hours and 15 minutes.

ART CAMP SCHEDULE

CERAMICS - 9:00 AM - 10:30 AM
PAINTING - 10:30 AM - 11:30 AM
LUNCH - 11:30 AM - 12:30 PM
PHOTOGRAPHY - 12:30 PM - 1:45 PM
WEARABLE ART - 1:45 PM - 3:00 PM

7. In which of the 4 classes do the campers spend the most time? the least time?

8. **Write About It** If 5 students share some sheets of paper equally, would there ever be more than 4 sheets left over? Explain how you know.

9. Eric drew this table. What is the pattern?

number	15	22	25	30
answer	60	88	100	120

10. Before the game, four soccer players ran a practice drill. Sue finished after Jeremy. Emma finished before Jeremy but after Mark. Who finished first?

Mental Math: Division Patterns

▶ **Learn**

EXTREME FUN Best Sports Shop orders 1,800 skateboards for 3 of its stores to share equally. How many skateboards will each store get?

Example

Find $1,800 \div 3$.

Use basic facts and patterns to find quotients mentally.

dividend		divisor		quotient
18	÷	3	=	6
180	÷	3	=	60
1,800	÷	3	=	600
↑↑ two zeros			two zeros ↑↑	

So, each store will get 600 skateboards.

MATH IDEA As the number of zeros in the dividend increases, the number of zeros in the quotient also increases.

More Examples

Ⓐ $72 \div 9 = 8$ Think: $8 \times 9 = 72$
   $720 \div 9 = 80$
   $7,200 \div 9 = 800$

Ⓑ $40 \div 8 = 5$ Think: $5 \times 8 = 40$
   $400 \div 8 = 50$
   $4,000 \div 8 = 500$

• In Example B, why is there one more zero in the dividend than in the quotient?

▶ **Check**

1. **Tell** how many zeros are in the quotient $72,000 \div 9$ and in the quotient $40,000 \div 8$.

Use a basic division fact and patterns to write each quotient.

2. a. $560 \div 7$
   b. $5,600 \div 7$

3. a. $540 \div 6$
   b. $5,400 \div 6$

4. a. $200 \div 5$
   b. $2,000 \div 5$

5. a. $8,000 \div 8$
   b. $80,000 \div 8$

▶ **Practice and Problem Solving**

Use a basic division fact and patterns to write each quotient.

6. a. $270 \div 3$
   b. $2,700 \div 3$

7. a. $630 \div 9$
   b. $6,300 \div 9$

8. a. $300 \div 5$
   b. $3,000 \div 5$

9. a. $5,400 \div 9$
   b. $54,000 \div 9$

Divide mentally. Write the basic division fact and the quotient.

10. $450 \div 9$
11. $210 \div 7$
12. $160 \div 8$
13. $180 \div 9$

14. $6,300 \div 7$
15. $3,000 \div 6$
16. $1,800 \div 9$
17. $3,200 \div 8$

18. $2,800 \div 4$
19. $3,600 \div 9$
20. $15,000 \div 3$
21. $48,000 \div 6$

 **ALGEBRA** Write the value of *n*.

22. $420 \div 7 = n$

23. $n \div 9 = 30$

24. $350 \div n = 50$

25. $4,800 \div 8 = n$

26. $24,000 \div n = 8,000$

27. $72,000 \div 9 = n$

USE DATA Use the pictograph for 28–31.

28. Each can of tennis balls holds 3 balls. How many cans of tennis balls does the shop have?

29. How many more baseballs and basketballs are there than golf balls?

30. ✎ Write a problem about the data in the pictograph that can be solved by using a basic division fact and patterns.

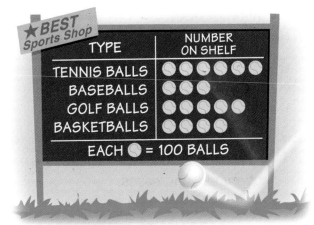

31. What if the shop receives a new shipment of 2,000 balls, made up of an equal number of each type of ball? How many of each type of ball are now in the shop?

Mixed Review and Test Prep

32. $\begin{array}{r} 45 \\ \times 27 \\ \hline \end{array}$ (p. 224)

33. $\begin{array}{r} 985 \\ +1,307 \\ \hline \end{array}$ (p. 40)

34. Round 61,879 to the nearest hundred.
(p. 26)

35. Round 206,714 to the nearest thousand. (p. 26)

36. TEST PREP How many faces does a cube have?

A 4   **B** 5   **C** 6   **D** 8

6 Estimate Quotients

▶ **Learn**

BIG EATER Suzi, a baby elephant, ate 528 pounds of food in one week. About how much did she eat in one day?

Estimate. 528 ÷ 7

Think of numbers close to the dividend 528 that can be easily divided by 7. Numbers that are easy to compute mentally are called **compatible numbers**.

Use division facts for 7 and patterns to find nearby compatible numbers for 528.

490 ÷ 7   **Think:** 528 is between 490 and
560 ÷ 7   560. So, use these compatible
            numbers as dividends.

To estimate, Kendra used 490 and Jamal used 560.

Both 70 and 80 are reasonable estimates.

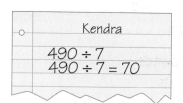

Suzi ate about 70 pounds of food each day.

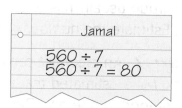

Suzi ate about 80 pounds of food each day.

MATH IDEA Compatible numbers can be used to estimate quotients.

• Shanique estimated 346 ÷ 7 by rounding the dividend to 300. Is 300 a good choice? Explain.

Quick Review

1. 4 × 7   2. 4 × 70

3. 4 × 700   4. 280 ÷ 4

5. 2,800 ÷ 4

VOCABULARY

compatible numbers

1. **Find** a compatible number for 331 when the divisor is 4 and you want to estimate the quotient.

Choose the letter of the best estimate.

2. $267 \div 7$ **a.** 3 or 4   **b.** 5 or 6   **c.** 30 or 40

3. $348 \div 5$ **a.** 6 or 7   **b.** 40 or 50   **c.** 60 or 70

► **Practice and Problem Solving**

Choose the letter of the best estimate.

4. $242 \div 3$ **a.** 6 or 7   **b.** 8 or 9   **c.** 80 or 90

5. $372 \div 6$ **a.** 60 or 70   **b.** 50 or 60   **c.** 6 or 7

6. $4,000 \div 9$ **a.** 400 or 500   **b.** 300 or 400   **c.** 40 or 50

Estimate by using compatible numbers.

7. $5\overline{)408}$ 8. $9\overline{)590}$ 9. $4\overline{)137}$ 10. $6\overline{)492}$

11. $8\overline{)444}$ 12. $9\overline{)660}$ 13. $8\overline{)713}$ 14. $9\overline{)6,800}$

15. $9\overline{)3,150}$ 16. $5\overline{)232}$ 17. $3\overline{)2,720}$ 18. $7\overline{)4,570}$

Use estimation to tell which quotient is greater.

19. $190 \div 3$ or $165 \div 4$ 20. $475 \div 8$ or $365 \div 5$ 21. $349 \div 5$ or $703 \div 8$

22. $555 \div 7$ or $303 \div 6$ 23. $777 \div 8$ or $888 \div 7$ 24. $2,000 \div 7$ or $999 \div 2$

25. An elephant family group usually has about 8 members. There are 135 elephants. Estimate the number of family groups.

26. **? What's the Error?** An elephant can eat 3,080 pounds of food in one week. Mike estimated that an elephant eats about 600 pounds of food in one day. Describe his error. Give a more reasonable estimate.

Mixed Review and Test Prep

27. $13,238$ (p. 40)   28. $126 \div 9$ (p. 250)
    $-8,179$

29. Find the product. (p. 150)
    $6 \times 4 \times 3$

30. Find the value of n in the equation
    $18 - 9 = n + 2$. (p. 60)

31. **TEST PREP** Jim started his walk at 11:15 A.M. and stopped at 12:45 P.M. for a 30-minute lunch. Then he walked home. He arrived home at 2:00 P.M. How long did he walk? (p. 120)

    A 1 hr 45 min   **C** 2 hr 30 min
    B 2 hr 15 min   **D** 2 hr 45 min

Review/Test

✔ CHECK VOCABULARY AND CONCEPTS

Choose the best term from the box.

compatible numbers
quotient
remainder

1. A _?_ is the amount left over when a number cannot be divided evenly. (p. 246)

2. Numbers that make it easier to estimate mentally are called _?_. (p. 258)

For 3–6, think of how to model 47 ÷ 3. (pp. 246–249)

3. How many equal groups are needed to model the divisor?

4. Draw the base-ten blocks needed to show the dividend.

5. How many are in each group?

6. How many are left over?

✔ CHECK SKILLS

Divide and check. (pp. 250–253)

7. $3\overline{)23}$ 8. $4\overline{)33}$ 9. $2\overline{)28}$ 10. $6\overline{)72}$

11. $9\overline{)71}$ 12. $5\overline{)85}$ 13. $4\overline{)65}$ 14. $5\overline{)69}$

15. $7\overline{)91}$ 16. $6\overline{)79}$ 17. $8\overline{)98}$ 18. $7\overline{)90}$

19. $57 \div 5$ 20. $89 \div 7$ 21. $85 \div 6$ 22. $91 \div 2$

Divide mentally. Write the basic division fact and the quotient. (pp. 256–257)

23. $150 \div 5$ 24. $210 \div 3$ 25. $360 \div 6$ 26. $4,500 \div 9$ 27. $5,600 \div 8$

Estimate by using compatible numbers. (pp. 258–259)

28. $136 \div 3$ 29. $228 \div 4$ 30. $257 \div 8$ 31. $492 \div 6$

✔ CHECK PROBLEM SOLVING

Solve. (pp. 254–255)

32. There are 93 students. After equal groups are formed, there are 5 students left over. How many groups are formed? How many students are in each group?

33. Lynn has $73 to spend on CDs. She buys several CDs at the same price and has $3 left over. How many CDs does Lynn buy? How much does each CD cost?

Standardized Test Prep

Decide on a plan.
See item **9.**
Write a number sentence and solve it for the first step. Think how you can make equal-size teams.

Also see problem 4, p. H63.

For 1–9, choose the best answer.

1. $29 \div 4$

A 7 r1   **C** 7

B 9 r3   **D** 9

2. $9\overline{)74}$

F 8 r2   **H** 9 r1

G 8 r1   **J** 7 r4

3. 511×35

A 4,088   **C** 17,885

B 16,775   **D** NOT HERE

4. $541 + 1,980 + 37$

F 2,478   **H** 9,010

G 2,558   **J** NOT HERE

5. $1,600 \div 4$

A 4   **C** 400

B 40   **D** 4,000

6. Jason left on a trip to the beach at 9:10 A.M. He arrived at 1:35 P.M. How long did it take him to get to the beach?

F 3 hours 35 minutes

G 3 hours 25 minutes

H 4 hours 35 minutes

J 4 hours 25 minutes

7. What is the value of n for $n \div 6 = 700$?

A 42   **C** 4,200

B 420   **D** 42,000

8. Which is the best estimate for this quotient?

$7,654 \div 9$

F 80   **H** 8,000

G 800   **J** 80,000

9. There are 79 children who signed up to play baseball. Each team needs 9 players. How many more children are needed so each team will have enough players?

A 3   **C** 1

B 2   **D** 0

Write What You Know

10. A roller coaster has 12 cars. Each car holds 3 people. Fifty people are waiting in line for the next ride. How many of the people in line will **not** get on the next ride? Show your work.

11. There are 23 students in Jeremy's class. Tell how they could be divided into 4 groups to do an activity. Explain your answer.

Divide by 1-Digit Divisors

The Hawaiian Islands celebrate Lei (LAY) Day on May 1. Leis are a tradition in Hawaii. They are used to welcome visitors and to celebrate special events. Leis are usually worn around the neck but can also be worn on the head, ankle, or wrist.

PROBLEM SOLVING Use the table to find how many boxes are needed to ship each type of lei. How many boxes in all will be shipped?

DATA LINK

HAWAIIAN LEIS

Type of Lei	Number Ordered	Number per Box
Flower	225	3
Head	120	2
Satin	144	2
Seed	174	6
Shell	65	5
Silk flower	135	9

CHECK WHAT YOU KNOW

Use this page to help you review and remember
important skills needed for Chapter 14.

✔ USE MULTIPLICATION FACTS AND PATTERNS (For Intervention, see p. H16.)

Copy and complete.

1. ■ × 1 = 3
■ × 10 = 30
■ × 100 = 300

2. ■ × 3 = 12
■ × 30 = 120
■ × 300 = 1,200

3. ■ × 2 = 10
5 × 20 = ■
5 × ■ = 1,000

4. 8 × ■ = 24
■ × 30 = 240
8 × 300 = ■

5. 6 × 4 = ■
6 × 40 = ■
6 × 400 = ■

6. 5 × ■ = 25
5 × ■ = 250
■ × 500 = 2,500

7. ■ × 6 = 42
7 × 60 = ■
7 × ■ = 4,200

8. 3 × 9 = ■
3 × ■ = 270
3 × ■ = 2,700

9. 8 × ■ = 40
■ × 50 = 400
8 × 500 = ■

10. 9 × ■ = 36
9 × ■ = 360
■ × 900 = 3,600

11. ■ × 3 = 21
7 × 30 = ■
7 × ■ = 2,100

12. 6 × 7 = ■
6 × 70 = ■
6 × 700 = ■

✔ PLACE VALUE (For Intervention, see p. H17.)

Copy and complete.

13. 578 = 5 hundreds 7 _?_ 8 _?_

14. 936 = 9 _?_ 3 _?_ 6 _?_

15. 3,825 = 3 _?_ 8 _?_ 2 _?_ 5 _?_

16. 4,396 = 4 thousands ■ hundreds ■ tens ■ ones

17. 6 tens 3 ones = 5 tens ■ ones

18. ■ tens 5 ones = 4 tens 15 ones

19. 4 tens 8 ones = 3 tens ■ ones

20. 8 tens ■ ones = 7 tens 12 ones

21. 7 tens ■ ones = 6 tens 14 ones

22. ■ tens 3 ones = 8 tens 13 ones

Write the value of the blue digit.

23. 394

24. 3,678

25. 75,104

26. 2,786

27. 15,073

28. 149,754

29. 77,642

30. 4,682

Place the First Digit

▶ **Learn**

FOLDED FORMS   There are 265 pieces of origami paper to be shared equally among 5 art classes. How many pieces will each class get?

One Way   Estimate to place the first digit in the quotient. Divide 265 by 5. Write $5\overline{)265}$.

STEP 1

Estimate.
Think:

$\dfrac{50}{5\overline{)250}}$ or $\dfrac{60}{5\overline{)300}}$

■ So, the first digit is $5\overline{)265}$ in the tens place.

STEP 2

Divide the 26 tens.

$\begin{array}{r} 5 \\ 5\overline{)265} \\ -25 \\ \hline 1 \end{array}$ Divide. $5\overline{)26}$
Multiply. 5×5
Subtract. $26 - 25$
Compare. $1 < 5$

STEP 3

Bring down the 5 ones.
Divide the 15 ones.

$\begin{array}{r} 53 \\ 5\overline{)265} \\ -25\downarrow \\ \hline 15 \\ -15 \\ \hline 0 \end{array}$ Divide. $5\overline{)15}$
Multiply. 5×3
Subtract. $15 - 15$
Compare. $0 < 5$

So, each class will get 53 pieces of origami paper.

Another Way   Use place value to place the first digit in the quotient. Divide 253 by 4. Write $4\overline{)253}$.

STEP 1

Use place value to place the first digit in the quotient. Look at the hundreds.

$4\overline{)253}$ $2 < 4$, so look at the tens.

■ $25 > 4$, so use $4\overline{)253}$ 25 tens. Place the first digit in the tens place.

STEP 2

Divide the 25 tens.

$\begin{array}{r} 6 \\ 4\overline{)253} \\ -24 \\ \hline 1 \end{array}$ Divide. $4\overline{)25}$
Multiply. 4×6
Subtract. $25 - 24$
Compare. $1 < 4$

STEP 3

Bring down the 3 ones.
Divide the 13 ones.

$\begin{array}{r} 63\ r1 \\ 4\overline{)253} \\ -24\downarrow \\ \hline 13 \\ -12 \\ \hline 1 \end{array}$ Divide. $4\overline{)13}$
Multiply. 4×3
Subtract. $13 - 12$
Compare. $1 < 4$

1. **Explain** how you can use estimation to help you decide whether $903 \div 3 = 31$ is reasonable.

Tell where to place the first digit. Then divide.

2. $6\overline{)38}$ 3. $4\overline{)81}$ 4. $2\overline{)182}$ 5. $5\overline{)85}$ 6. $6\overline{)771}$

► **Practice and Problem Solving**

Tell where to place the first digit. Then divide.

7. $6\overline{)45}$ 8. $3\overline{)125}$ 9. $5\overline{)558}$ 10. $7\overline{)371}$ 11. $3\overline{)634}$

Divide.

12. $2\overline{)87}$ 13. $4\overline{)189}$ 14. $5\overline{)170}$ 15. $6\overline{)378}$ 16. $3\overline{)801}$

17. $6\overline{)318}$ 18. $4\overline{)239}$ 19. $5\overline{)678}$ 20. $8\overline{)488}$ 21. $7\overline{)917}$

22. $462 \div 9$ 23. $694 \div 5$ 24. $969 \div 6$ 25. $998 \div 8$ 26. $891 \div 9$

27. Mrs. Williams bought supplies for the art class project. She bought glitter for $2.49, lace for $4.89, and paint for $5.30. How much change did she receive from $20.00?

28. A total of 144 origami figures were on display. The same number of figures were on each of 6 tables. Did each table have 244 figures, 24 figures, or 2 figures? Explain.

29. For homework, the students in Mr. Charlton's history class each turned in a notebook with 48 pages. There are 24 students. How many pages will Mr. Charlton have to read?

30. 📓 **Write About It** Explain how you can decide where to place the first digit in a division problem.

Mixed Review and Test Prep

31. $3 \times 3 \times 2$ (p. 150)

32. $600 - 79$ (p. 36)

33. Find the product of 16 and 4.

34. Find the sum. $706 + 552$ (p. 36)

35. **TEST PREP** The play started at 7:15 P.M. It lasted 90 minutes and there was a 15 minute intermission. What was the ending time? (p. 120)

A 8:15 P.M.   C 9:00 P.M.

B 8:45 P.M.   D 9:15 P.M.

Divide 3-Digit Numbers

▶ **Learn**

ISLAND ADVENTURE The Aloha Tour Company divided 178 flyers for guided tours equally among 3 resorts. How many flyers did each resort get?

Example

Divide 178 by 3. Write $3\overline{)178}$.

STEP 1

Estimate to place the first digit in the quotient.

Think: $\dfrac{50}{3\overline{)150}}$ or $\dfrac{60}{3\overline{)180}}$

■ Place the first $3\overline{)178}$ digit in the tens place.

STEP 2

Divide the 17 tens.

$$
\begin{array}{r}
5 \\
3\overline{)178} \\
-15 \\
\hline
2
\end{array}
$$
Divide.
Multiply.
Subtract.
Compare.

STEP 3

Bring down the 8 ones. Divide the 28 ones.

$$
\begin{array}{r}
59 \ r1 \\
3\overline{)178} \\
-15\downarrow \\
\hline
28 \\
-27 \\
\hline
1
\end{array}
$$
Divide.
Multiply.
Subtract.
Compare.

STEP 4

To check, multiply the quotient by the divisor. Then add the remainder.

$$
\begin{array}{r}
2 \\
59 \\
\times \ 3 \\
\hline
177 \\
+ \ 1 \\
\hline
178
\end{array}
$$
The answer is correct.

So, each resort will get 59 flyers, with 1 flyer left over. Since 59 is between 50 and 60, the answer is reasonable.

More Examples

A
$$
\begin{array}{r}
37 \ r4 \\
5\overline{)189} \\
-15 \\
\hline
39 \\
-35 \\
\hline
4
\end{array}
$$

B
$$
\begin{array}{r}
193 \\
2\overline{)386} \\
-2 \\
\hline
18 \\
-18 \\
\hline
06 \\
- \ 6 \\
\hline
0
\end{array}
$$

C
$$
\begin{array}{r}
171 \ r3 \\
4\overline{)687} \\
-4 \\
\hline
28 \\
-28 \\
\hline
07 \\
- \ 4 \\
\hline
3
\end{array}
$$

- Why are there 3 digits in the quotients of Examples B and C?

1. **Explain** how you know whether the quotient in Example B on page 266 is reasonable.

Divide and check.

2. $4\overline{)187}$ 3. $2\overline{)453}$ 4. $5\overline{)592}$ 5. $7\overline{)241}$ 6. $6\overline{)687}$

► **Practice and Problem Solving**

Divide and check.

7. $6\overline{)178}$ 8. $4\overline{)472}$ 9. $7\overline{)241}$ 10. $9\overline{)709}$ 11. $3\overline{)470}$

12. $5\overline{)337}$ 13. $2\overline{)372}$ 14. $8\overline{)697}$ 15. $4\overline{)168}$ 16. $6\overline{)749}$

17. $7\overline{)224}$ 18. $9\overline{)530}$ 19. $4\overline{)617}$ 20. $5\overline{)386}$ 21. $8\overline{)944}$

22. $186 \div 3$ 23. $247 \div 8$ 24. $546 \div 5$ 25. $302 \div 2$ 26. $614 \div 6$

 **ALGEBRA** Find the value of $448 \div n$ for each value of n.

27. $n = 2$ 28. $n = 3$ 29. $n = 4$ 30. $n = 5$ 31. $n = 6$

32. **Mental Math** Which is greater, $345 \div 2$ or $345 \div 3$?

33. Which is greater, $2\overline{)452}$ or $4\overline{)452}$? How much greater?

USE DATA For 34–37, use this information.

The Island Tours theater seats 45 people. There are 210 people on a tour.

34. A guide wants the tourists to be in groups of 8. Will this work, or will there be tourists left over?

35. **What if** the theater could seat 55 people? Would all of the tourists be able to watch a movie in 4 showings? Explain.

36. The tour guide has $1,890 for the tickets. If each ticket costs $7, how much money will she have left?

37. ✏ Write a problem about Island Tours that requires dividing three-digit numbers.

Mixed Review and Test Prep

38. $7,500 - 896$ (p. 42)

39. $2,300 \times 4$ (p. 200)

40. $6\overline{)83}$ (p. 248)

41. $7\overline{)37}$ (p. 246)

42. **TEST PREP** What is the value of n? $n + 4 = 15$ (p. 64)

A $n = 60$ **C** $n = 15$

B $n = 19$ **D** $n = 11$

Zeros in Division

▶ Learn

POSTAGE DUE A postal carrier makes 432 stops in 4 hours. How many stops is this every hour?

Example

Divide 432 by 4. Write 4)432.

STEP 1	STEP 2	STEP 3	STEP 4
Estimate to place the first digit in the quotient. **Think:** $\frac{100}{4)400}$ or $\frac{200}{4)800}$ ■ So, place the 4)432 first digit in the hundreds place.	Divide the 4 hundreds. $\begin{array}{r} 1 \\ 4)\overline{432} \\ -4 \\ \hline 0 \end{array}$	Bring down the 3 tens. Divide the 3 tens. $\begin{array}{r} 10 \\ 4)\overline{432} \\ -4\downarrow \\ \hline 03 \\ -\ 0 \\ \hline 3 \end{array}$ 4 > 3, so write a 0 in the quotient.	Bring down the 2 ones. Divide the 32 ones. $\begin{array}{r} 108 \\ 4)\overline{432} \\ -4 \\ \hline 03 \\ -\ 0 \\ \hline 32 \\ -32 \\ \hline 0 \end{array}$

So, the postal carrier makes 108 stops every hour.

More Examples

A
$\begin{array}{r} 101\ r1 \\ 5)\overline{506} \\ -5 \\ \hline 00 \\ -\ 0 \\ \hline 06 \\ -\ 5 \\ \hline 1 \end{array}$

B
$\begin{array}{r} 130 \\ 6)\overline{780} \\ -6 \\ \hline 18 \\ -18 \\ \hline 00 \\ -\ 0 \\ \hline 0 \end{array}$

C
$\begin{array}{r} 104\ r6 \\ 7)\overline{734} \\ -7 \\ \hline 03 \\ -0 \\ \hline 34 \\ -28 \\ \hline 6 \end{array}$

• Explain what would happen in Example A if you did not write a zero in the tens place in the quotient.

Placing Zeros in the Quotient

The fourth graders collected stamps. They put 4 stamps on each page of an album.

The students in Lee Ann's class collected 240 stamps. How many pages did they need?

Look at Lee Ann's paper. Lee Ann divided 240 by 4.

Lee Ann

$$\begin{array}{r} 6 \\ 4\overline{)240} \\ -24 \\ \hline 0 \end{array}$$

• Describe her error. Write the division correctly.

The students in Craig's class collected 412 stamps. How many pages did they need?

Look at Craig's paper. Craig divided 412 by 4.

Craig

$$\begin{array}{r} 13 \\ 4\overline{)412} \\ -4 \\ \hline 12 \\ -12 \\ \hline 0 \end{array}$$

• Describe his error. Write the division correctly.

 MATH IDEA Estimate to decide how many digits should be in the quotient so you do not forget to include zeros.

▶ Check

1. **Tell** what estimate Craig could have made.

2. **Explain** how the estimate could help Craig remember to write the 0 in the quotient.

Write the number of digits in each quotient.

3. $4\overline{)406}$　　　**4.** $7\overline{)610}$　　　**5.** $5\overline{)309}$　　　**6.** $4\overline{)804}$　　　**7.** $5\overline{)650}$

Divide and check.

8. $5\overline{)800}$　　　**9.** $9\overline{)308}$　　　**10.** $3\overline{)609}$　　　**11.** $3\overline{)305}$　　　**12.** $5\overline{)407}$

LESSON CONTINUES ▶

Write the number of digits in each quotient.

13. $8\overline{)818}$ **14.** $6\overline{)510}$ **15.** $3\overline{)207}$ **16.** $4\overline{)600}$ **17.** $5\overline{)405}$

Divide and check.

18. $6\overline{)40}$ **19.** $8\overline{)60}$ **20.** $5\overline{)70}$ **21.** $7\overline{)80}$ **22.** $9\overline{)97}$

23. $5\overline{)405}$ **24.** $4\overline{)240}$ **25.** $6\overline{)243}$ **26.** $7\overline{)636}$ **27.** $8\overline{)706}$

28. $7\overline{)308}$ **29.** $6\overline{)230}$ **30.** $4\overline{)580}$ **31.** $5\overline{)306}$ **32.** $4\overline{)803}$

33. $4\overline{)260}$ **34.** $7\overline{)605}$ **35.** $8\overline{)900}$ **36.** $3\overline{)620}$ **37.** $9\overline{)951}$

38. $402 \div 7$ **39.** $362 \div 9$ **40.** $760 \div 3$ **41.** $860 \div 8$ **42.** $603 \div 6$

43. $361 \div 3$ **44.** $247 \div 8$ **45.** $654 \div 5$ **46.** $421 \div 2$ **47.** $642 \div 6$

Write $+$, $-$, $\times$, or $\div$ for each ●.

48. $(35 \bullet 5) \bullet 5 = 2$ **49.** $(9 \bullet 8) \bullet 4 = 18$ **50.** $(36 \bullet 4) \bullet 3 = 27$

USE DATA For 51–53 and 57, use the table.

51. Manuel mailed 3 boxes. One box weighed 5 pounds. Another box weighed 7 pounds, and the third box weighed 18 pounds. How much did it cost Manuel to send the boxes?

Shipping Costs	
Weight	**Cost**
1 lb - 5 lb	$0.74 each lb
6 lb - 10 lb	$0.69 each lb
11 lb - 15 lb	$0.57 each lb
16 lb - 20 lb	$0.49 each lb

52. Luisa has $7.00. Does she have enough money to mail a box that weighs 13 pounds? How much will it cost to mail the box?

53. Randy mailed a 4-pound box. Kelly mailed a 10-pound box. How much less did Randy's box cost to ship than Kelly's?

54. Speedy Delivery shipped 520 boxes in 5 days. It shipped the same number of boxes each day. How many boxes were shipped each day?

55. ❓ **What's the Error?** Describe the error and then show the correct way to divide.

$$\begin{array}{r} 12 \text{ r}3 \\ 6\overline{)615} \\ -6 \\ \hline 15 \\ -12 \\ \hline 3 \end{array}$$

56. ✏️ **Write About It** Explain how you can decide how many digits will be in a quotient.

57. ❓ **What's the Question?** Lorraine has $15.00 and is mailing two 12-pound boxes. The answer is $1.32.

58. Yoko takes the train to the city and back 4 times each month. She travels a total of 376 miles. How far away is the city?

59. Sami has 960 beads in 8 different colors to make jewelry. He has the same number of each color. How many beads of each color does Sami have?

Mixed Review and Test Prep

For 60–61, write in standard form. (p. 6)

60. seven hundred eleven thousand, forty-five

61. sixty-four thousand, nine hundred fifty-two

62. Write the number that is 10,000 less than two hundred fifty thousand, one hundred sixteen. (p. 44)

63. 6,983 (p. 40) **64.** 7,682 (p. 40)
     −2,094   +6,749

65. $4 \times \blacksquare = 30 - 2$ (p. 164)

66. 400 (p. 36) **67.** 732 (p. 36)
     −283   −165

68. TEST PREP What is the value of eight $10 bills, six $1 bills, 17 dimes, and 9 pennies?
  A $8.79   **C** $86.79
  B $9.86   **D** $87.79

69. TEST PREP Nelson bought 8 pounds of oranges. He paid with $5.00 and got $1.96 change. How much did 1 pound of oranges cost? (p. 250)
  F 29¢   **H** 38¢
  G 31¢   **J** 42¢

PROBLEM SOLVING LiNKUP ... to Social Studies

On April 3, 1860, the Pony Express began carrying mail between Missouri and California. The riders travelled about 150 miles each day—almost twice as far as a day's travel by stagecoach.

USE DATA For 1–4, use the information above.

1. St. Joseph, Missouri, is about 1,500 miles from San Francisco, California. About how many days did it take the Pony Express to travel this distance?

2. A Pony Express rider left Sacramento at 2:45 A.M. and arrived at Placerville at 6:40 A.M. How long did it take the rider to travel the 45 miles?

3. One of the longest non-stop rides was made by Buffalo Bill. He rode for about 22 hours at about 15 miles per hour. He used 21 different horses. About how many miles did he ride?

4. The Pony Express carried a total of 34,753 pieces of mail in 308 trips of 2,000 miles each. How many miles did the Pony Express riders travel?

EXTRA PRACTICE page H45, Set C

4 Choose a Method

Quick Review

1. 3)63
2. 2)424
3. 5)315
4. 4)884
5. 6)1,806

▶ **Learn**

ROOM AND BOARD   For a national gymnastics competition, 6 hotels reserved 1,650 rooms. The rooms were divided equally among the hotels. How many rooms were reserved in each hotel?

Choose a method of computation that will be useful for the numbers given.

Use Paper and Pencil

Divide 1,650 by 6. Write 6)1,650.

STEP 1	STEP 2	STEP 3	STEP 4
Estimate to place the first digit in the quotient.	Divide the 16 hundreds.	Bring down the 5 tens. Divide the 45 tens.	Bring down the 0 ones. Divide the 30 ones.

STEP 1

Estimate to place the first digit in the quotient.

Think:

$$\frac{200}{6)1,200} \text{ or } \frac{300}{6)1,800}$$

$$\frac{\blacksquare\_\_\_}{6)1,650}$$ So, place the first digit in the hundreds place.

STEP 2

Divide the 16 hundreds.

```
    2
6)1,650
 -1 2
    4
```

STEP 3

Bring down the 5 tens. Divide the 45 tens.

```
   27
6)1,650
 -1 2
   45
  -42
    3
```

STEP 4

Bring down the 0 ones. Divide the 30 ones.

```
  275
6)1,650
 -1 2
   45
  -42
   30
  -30
    0
```

So, 275 rooms were reserved in each hotel.

Ⓐ Use Mental Math

$2,480 \div 4 = \blacksquare$

Think: $2,400 \div 4 = 600$
$80 \div 4 = 20$

So, $2,480 \div 4 = 620$.

Ⓑ Use a Calculator

$7,812 \div 5 = \blacksquare$

| 7 | 8 | 1 | 2 | ÷R |

| 5 | = | $=1562^{R2}$ |

So, $7,812 \div 5 = 1,562$ r2.

MATH IDEA   You can divide by using paper and pencil, a calculator, or mental math. Look at the numbers in the problem before you choose a method.

1. **Explain** how looking at the numbers in a problem can help you choose a method to find the quotient.

Divide.

2. $5\overline{)1{,}440}$
3. $3\overline{)2{,}329}$
4. $8\overline{)1{,}712}$
5. $7\overline{)5{,}030}$
6. $6\overline{)4{,}091}$

▶ Practice and Problem Solving

Divide. Write the method you used to compute.

7. $4\overline{)3{,}420}$
8. $3\overline{)\$83.70}$
9. $6\overline{)\$91.20}$
10. $8\overline{)7{,}200}$
11. $2\overline{)1{,}382}$

12. $5\overline{)185}$
13. $7\overline{)5{,}643}$
14. $9\overline{)398}$
15. $3\overline{)483}$
16. $6\overline{)737}$

17. $4\overline{)236}$
18. $2\overline{)3{,}892}$
19. $6\overline{)\$52.14}$
20. $8\overline{)408}$
21. $9\overline{)\$61.47}$

22. $7\overline{)3{,}273}$
23. $3\overline{)240}$
24. $4\overline{)\$78.04}$
25. $7\overline{)3{,}961}$
26. $5\overline{)200}$

27. At the gymnastics competition, there were 276 entries. There was an equal number of entries in 4 different contests. How many entries were in each contest?

28. At the tumbling event, 3 gymnasts each performed for 4 minutes, and one gymnast performed for 3 minutes. How many minutes did the gymnasts perform in all?

USE DATA For 29–30, use the table.

29. Thao wants to buy 3 pounds of bananas and 2 pounds of coconut. He has $10.00. Does he have enough money? How much money does Thao need in all?

30. Holly's mother bought 2 pounds of each fruit on the list to make trail mix. How much did she spend on dried fruit?

DRIED FRUIT	
Fruit	**Cost for Each Pound**
Raisins	$0.97
Bananas	$1.49
Coconut	$2.36
Apples	$3.24

Mixed Review and Test Prep

31. 238 (p. 196)
   $\times\ \ 6$

32. $5{,}000$ (p. 42)
   $-2{,}198$

33. $21{,}000 \div 7$ (p. 256)

34. Reshanda works 27 hours each week at the video store. She earns $6.50 an hour. How much money does Reshanda earn in 3 weeks? (p. 234)

35. **TEST PREP** Jen has 3 quarters, 2 dimes, 4 nickels, and 1 penny in her pocket. Which combination of coins is possible if Jen pulls 4 coins from her pocket?

   A 4 pennies

   B 3 dimes and 1 nickel

   C 2 pennies and 2 quarters

   D 2 quarters and 2 nickels

EXTRA PRACTICE page H45, Set D

Problem Solving Skill
Interpret the Remainder

UNDERSTAND > PLAN > SOLVE > CHECK

WHAT'S LEFT?   Portia and Preston are planning for the school carnival.

Examples

A   Drop the remainder and increase the quotient by 1.

They need 250 cans of soda for the food booth. A carton holds 6 cans of soda. How many cartons of soda will they need to buy?

$$\begin{array}{r} 41 \text{ r}4 \\ 6\overline{)250} \\ -24 \\ \hline 10 \\ -6 \\ \hline 4 \end{array}$$

Since 41 cartons do not hold 250 cans, increase the quotient by 1.

So, they need to buy 42 cartons of soda.

B   Drop the remainder.

They have a 250-foot roll of paper to make posters for the carnival. They will cut the roll into 3-foot posters. How many posters will they have?

$$\begin{array}{r} 83 \text{ r}1 \\ 3\overline{)250} \\ -24 \\ \hline 10 \\ -9 \\ \hline 1 \end{array}$$

Drop the remainder. The remainder is not enough for another 3-foot poster.

So, they will have a total of 83 posters.

C   Use the remainder as the answer.

Portia made 126 cookies to sell. She divided the cookies into packages of 4 and gave the leftover cookies to her brother. How many cookies did she give her brother?

$$\begin{array}{r} 31 \text{ r}2 \\ 4\overline{)126} \\ -12 \\ \hline 06 \\ -4 \\ \hline 2 \end{array}$$

Use the remainder as your answer.

So, she gave her brother 2 cookies.

MATH IDEA   When you solve a division problem that has a remainder, the way you interpret the remainder depends on the situation in the problem.

Talk About It

- Why isn't 41 cartons of soda the correct answer to the first question?

- Why is the remainder dropped to answer the second question?

Problem Solving Practice

Solve. Then write *a*, *b*, or *c* to tell how you interpreted the remainder.

a. increase the quotient by 1

b. drop the remainder

c. use the remainder as the answer

1. An 85-inch piece of wire needs to be cut into 9-inch lengths. How many 9-inch lengths will there be?

2. Dave must pack 55 bottles of juice. Boxes for the bottles hold 8 bottles. How many boxes are needed?

3. Lena has a 50-foot roll of ribbon. She cuts the ribbon into 9-foot pieces to make bows. How many pieces will she have?

4. Dora needs to buy pages for her photo album. Each page holds 6 photos. How many pages will she need for 94 photos?

5. Jan's Pillow Factory stuffs each pillow with 3 pounds of duck feathers. She has 67 pounds of feathers. How many pounds of feathers will be left over?

6. The Pool Supply Shop had 179 outdoor games. It shipped the same number of games to each of 18 stores. How many outdoor games were left over at the shop?

Mixed Applications

USE DATA For 7–9, use the bar graph.

7. If 900 tickets were sold at the carnival, how many tickets did not get used?
 A 20 **B** 40 **C** 120 **D** 140

8. If each ticket cost 50¢, what was the total spent on snacks?
 F $50 **G** $80 **H** $100 **J** $120

9. Games and snacks each cost 2 tickets. Sodas were 1 ticket each, and each ride cost 4 tickets. What was the most popular choice?

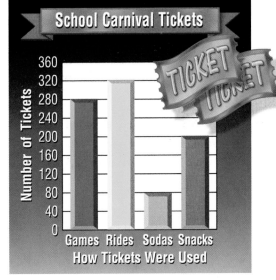

10. The theater has 23 rows of seats. Each row has 15 seats. There are 20 more seats in the balcony. How many seats are in the theater?

11. Nick bought two magazines for $18.50. The difference in the cost of the magazines was $2.50. How much did each magazine cost?

12. Write a problem in which the solution requires that you increase the quotient by 1.

LESSON
6 Find the Mean

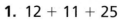

▶ Learn

ON A LEASH Janet earns money walking her neighbor's dogs. Over 4 weeks she earned $4.75, $3.60, $8.25, and $5.00. What is the mean, or average, amount Janet earned each week?

The **mean**, or average, is the number found by dividing the sum of a set of numbers by the number of addends.

Example

STEP 1

Add the amounts she earned.

$4.75
$3.60   4 addends
$8.25
+ $5.00
$21.60 ← sum

STEP 2

Divide the sum by the number of addends.

```
                 $5.40  ← mean
number → 4)$21.60  ← sum
of addends  −20
             16
            −16
             00
            − 0
              0
```

So, the mean, or average, amount Janet earned was $5.40 per week.

- Janet walked the dogs a fifth week and was paid $6.80. How would you find the mean, or average, amount she earned?

Technology Link

More Practice: Use E-Lab, *Finding the Mean.*

www.harcourtschool.com/elab2002

▶ Check

1. **Tell** how much Janet would earn in a month if she was paid $5.40 in each of 4 weeks.

Write the division problem for finding the mean. Then find the mean.

2. 15, 12, 16, 13   3. 358, 460, 733, 197   4. $26.35, $47.83, $62.29

Write the division problem for finding the mean. Then find the mean.

5. 46	**6.** 649	**7.** $0.75	**8.** $15.89	**9.** 4,279	**10.** 1,474
55	153	$0.99	$26.49	2,835	2,820
93	70	$0.29	$53.11	2,351	325
61	429	$0.85	$38.99		2,784
70	334				1,942

Find the mean.

11. 95; 88; 84; 85; 93; 89   **12.** 60; 85; 74; 79; 67   **13.** $3.75; $1.98; $6.75

14. 419; 343; 267; 74; 212   **15.** 598; 1,046; 822   **16.** 4,641; 2,912; 1,816

17. 3,970; 2,753; 1,128   **18.** 2,795; 897; 3,649   **19.** 2,176; 4,212; 1,289

Find the missing number.

20. 8, 10, 12, ▮   Mean is 11.   **21.** ▮, 12, 15, 17, 21   Mean is 15.

USE DATA For 22–24, use the table.

22. Find the average number of points per show the collies won this season.

23. Find the average number of points per show the spaniels won this season.

24. Write < or > to compare the average number of points for collies to the points for spaniels.

DOG SHOW POINT TOTALS		
Show	**Collies**	**Spaniels**
1	45	71
2	56	35
3	79	45
4	32	61
5	44	71
6	56	53

25. REASONING For 5–19, look at the least and greatest numbers in the set. Compare these numbers to the mean of the set. What do you notice?

Mixed Review and Test Prep

Write in order from least to greatest.
(p. 20)

26. 56; 34; 92; 86; 2; 45

27. 12,543; 12,453; 12,354

28. 903 (p. 36)   **29.** 513 (p. 228)
    +745   × 17

30. TEST PREP Jan has 84 points in a game. Mike has 28 fewer points than Jan. How many points does Mike have? (p. 152)

A 56   **C** 102

B 90   **D** 112

Review/Test

✓ CHECK VOCABULARY AND CONCEPTS

Choose the best term from the box.

> mean
> remainder
> quotient
> divisor

1. The _?_, or average, is the number found by dividing the sum of a set of numbers by the number of addends. (p. 276)

2. The way you interpret the _?_ depends upon the situation in the problem. (p. 274)

3. You can estimate or use place value to place the first digit in a _?_. (p. 264)

Tell where to place the first digit.
Then divide. (pp. 264–265)

4. $9\overline{)75}$ 5. $6\overline{)143}$ 6. $4\overline{)346}$ 7. $3\overline{)194}$

8. $5\overline{)275}$ 9. $473 \div 6$ 10. $534 \div 5$ 11. $935 \div 3$

✓ CHECK SKILLS

Divide. (pp. 266–273)

12. $809 \div 7$ 13. $299 \div 8$ 14. $124 \div 3$ 15. $234 \div 4$

16. $569 \div 5$ 17. $831 \div 7$ 18. $971 \div 8$ 19. $325 \div 3$

20. $419 \div 4$ 21. $453 \div 5$ 22. $816 \div 9$ 23. $3,289 \div 5$

24. $\$26.28 \div 6$ 25. $3,957 \div 2$ 26. $4,864 \div 6$ 27. $\$35.20 \div 8$

Find the mean. (pp. 276–277)

28. 13; 19; 22 29. 45; 32; 56; 95

30. 132; 265; 437 31. 1,091; 64; 214; 583

✓ CHECK PROBLEM SOLVING

Solve. (pp. 274–275, 276–277)

32. Tyrone and Jimmy want to build a tree fort. They need 153 feet of lumber. If the lumber comes in 8-foot lengths, how many pieces of lumber do they need?

33. At the bake sale, Ling collected $15.32 the first hour, $28.50 the second hour, and $34.00 the third hour. Find the average amount per hour Ling collected.

⭐Standardized Test Prep

 Decide on a plan.
See item **10**.

Write a number sentence and solve it for the first step. Think about how to be sure there are enough pages for all of the photos.

Also see problem **4**, p. H63.

For 1–10, choose the best answer.

1. $815 \div 3$
   A 270 r2   **C** 27 r5
   B 271 r2   **D** 27 r2

2. $6\overline{)921}$
   F 153 r3   **H** 154 r3
   G 15 r2   **J** 15 r3

3. Mary was at the beach for 7 days. Each day she collected 4 shells. She gave her sister 3 shells. Which expression can be used to find how many shells Mary has?
   A $(7 \times 4) - 3$ **C** $(7 \times 4) + 3$
   B $7 \times (4 - 3)$ **D** $(7 + 4) - 3$

4. How is four million, eight thousand, twenty three written in standard form?
   F 40,080,230   **H** 4,008,023
   G 4,080,023   **J** 4,800,023

5. How many digits are in the quotient?
   $9\overline{)803}$
   A 1   **B** 2   **C** 3   **D** 4

6. $4\overline{)3,286}$
   F 82 r6   **H** 821 r2
   G 821   **J** 921 r2

For 7–8, use the stem-and-leaf plot.

Test Grades

Stem	Leaves
7	6 6 8
8	0 8
9	0 0 0 7

7. What is the median test score?
   A 76   **B** 80   **C** 88   **D** 97

8. What is the mean of the test scores?
   F 90   **G** 88   **H** 86   **J** 85

9. $\$32.90 \div 5$
   A $65.80   **C** $6.80
   B $6.58   **D** NOT HERE

10. Henry is putting photos into an album. He has 69 photos. If each page holds 6 photos, how many pages will he need?
   F 10   **G** 11   **H** 12   **J** 13

Write What You Know

11. Crystal rode back and forth four times to her sister's house last week. Her bicycle odometer showed that she rode 25 miles in all. Explain how to estimate the number of miles Crystal lives from her sister.

12. The five workers at Band Corporation earn $500, $600, $600, $850, and $1,100 per week. Which is greater, the mean salary or the median salary? By how much? Tell how you know.

Divide by 2-Digit Divisors

Thousands of years ago, the Chinese used kites to send messages, to spy on their enemies, and even to fish! People in Europe started making their own kites about 700 years ago. Today, kites are used mostly for fun.

PROBLEM SOLVING
Use the information in the table to find out how many kites you could make with the materials listed.

KITE MAKING		
Materials Needed	**Amount You Have**	**Amount for 1 Kite**
Straws	1 box of 100	24 straws
String	2 rolls, 600 in. each	4 pieces 45 in. long 4 pieces 25 in. long
Posterboard	5 pieces	1 piece
Tissue paper	10 sheets	2 sheets

Use this page to help you review and remember important skills needed for Chapter 15.

✓ VOCABULARY

Choose the best term from the box.

| compatible |
| inverse |
| quotient |
| remainder |

1. The number other than the remainder that is the answer when dividing is called the ? .

2. The amount left over when you find a quotient is called the ? .

3. Numbers that are easy to compute mentally are ? numbers.

✓ DIVIDE WITH REMAINDERS (For Intervention, see p. H17.)

Use the model to find the quotient and remainder.

4.    5.

6.    7.

Find the quotient and remainder.

8. 37 ÷ 5   **9.** 55 ÷ 12   **10.** 76 ÷ 9

11. 23 ÷ 2   **12.** 41 ÷ 4   **13.** 93 ÷ 6

✓ USE COMPATIBLE NUMBERS (For Intervention, see p. H18.)

Rewrite each expression, using compatible numbers.

14. 38 ÷ 6   **15.** 51 ÷ 7   **16.** 42 ÷ 8

17. 71 ÷ 9   **18.** 47 ÷ 4   **19.** 38 ÷ 5

Estimate each quotient, using compatible numbers.

20. 7)59   **21.** 5)48   **22.** 9)37

23. 6)25   **24.** 8)61   **25.** 7)54

Division Patterns to Estimate

▶ Learn

FLYING HIGH   A kite-flying contest had 753 entries from 18 schools. Each school sent in about the same number of entries. About how many entries were there from each school?

You can use basic facts and multiples of 10 to find the estimate.

Estimate.   Look for a pattern.

753 ÷ 18   8 ÷ 2 ← **Think:** Use the basic fact 8 ÷ 2.

↓   ↓   80 ÷ 20 = 4

800 ÷ 20   800 ÷ 20 = 40

8,000 ÷ 20 = 400

So, there were about 40 entries from each school.

Examples

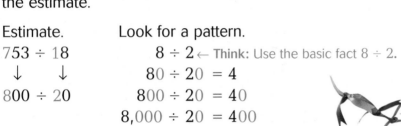

A   6 ÷ 3 = 2 ← basic fact

60 ÷ 30 = 2

600 ÷ 30 = 20

6,000 ÷ 30 = 200

B   18 ÷ 6 = 3 ← basic fact

180 ÷ 60 = 3

1,800 ÷ 60 = 30

18,000 ÷ 60 = 300

C   40 ÷ 5 = 8 ← basic fact

400 ÷ 50 = 8

4,000 ÷ 50 = 80

40,000 ÷ 50 = 800

MATH IDEA   Basic facts and a pattern can help you estimate quotients.

▶ Check

1. **Tell** without dividing how many zeros are in the quotient 900,000 ÷ 30.

Write the numbers you would use to estimate the quotient. Then estimate.

    2. 103 ÷ 11   **3.** 479 ÷ 39

    4. 636 ÷ 81   **5.** 544 ÷ 93

Write the numbers you would use to estimate the quotient. Then estimate.

6. $99 \div 18$ **7.** $450 \div 51$ **8.** $623 \div 82$ **9.** $523 \div 47$

Estimate.

10. $94 \div 11$ **11.** $82 \div 21$ **12.** $97 \div 33$ **13.** $187 \div 31$

14. $313 \div 59$ **15.** $498 \div 55$ **16.** $813 \div 91$ **17.** $478 \div 59$

Copy and complete the tables.

	DIVIDEND		DIVISOR		QUOTIENT
18.	80	÷	40	=	▨
19.	800	÷	40	=	▨
20.	8,000	÷	40	=	▨
21.	80,000	÷	40	=	▨

	DIVIDEND		DIVISOR		QUOTIENT
22.	50	÷	50	=	▨
23.	▨	÷	50	=	10
24.	5,000	÷	50	=	▨
25.	▨	÷	50	=	10,000

26. There are 125 people signed up for the class on making kites. Each group will have 13 people. About how many groups will be formed?

27. Angel arrived at the park at 10:45 A.M. He watched the kite-flying contest and left after 2 hours 35 minutes. At what time did he leave the park?

28. The Pittsfield Mets gave away about 12,080 baseball caps in July. About how many caps did they give away each day?

Technology Link

To learn more about Division Patterns, watch the Harcourt Math Newsroom Video *Fingerprints.*

29. A car dealer sold 58 blue cars, 37 green cars, 63 gold cars, and 42 red cars. How many cars did the dealer sell?

30. 📖 **Write About It** Explain how you can use basic facts and a pattern to estimate quotients.

Mixed Review and Test Prep

Write each number in two other ways. (p. 6)

31. $2,000 + 500 + 40 + 3$

32. seven thousand, sixty

Find the sum or difference. (p. 40)

33. $\begin{array}{r} 2,421 \\ +1,247 \\ \hline \end{array}$ **34.** $\begin{array}{r} 3,631 \\ -1,485 \\ \hline \end{array}$

35. **TEST PREP** Brenden's team scored 101 points, 107 points, 95 points, and 113 points in their last 4 games. What was the mean number of points scored? (p. 276)

A 100   **C** 106
B 104   **D** 108

Model Division

▶ **Explore**

TEA TIME Ann's Gift Shop has 65 teacups to put on display. Each rack holds 31 cups. How many racks will be filled with teacups? How many teacups will be left over?

Activity 1
Divide. 65 ÷ 31

Make a model to divide with a two-digit divisor.

STEP 1	**STEP 2**	**STEP 3**
Show 65 as 6 tens 5 ones.	Make 1 group of 31.	Make 2 groups of 31. Count how many ones are left over.
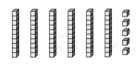	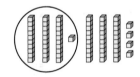	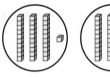 65 ÷ 31 = 2 r3

So, Ann needs 2 racks. There will be 3 teacups left over.

• **Explain** how you can check the quotient and remainder.

Try It

Use base-ten blocks to solve each division problem.

a. 89 ÷ 22   **b.** 76 ÷ 14

c. 92 ÷ 18   **d.** 64 ÷ 12

I have one group of 22. How many more groups of 22 can I make to show 89 ÷ 22?

Activity 2 Here is a way to record division. Divide 65 by 21.

STEP 1 Write the problem 65 ÷ 21.

Model

Record

$21\overline{)65}$

STEP 2 Estimate. 65 ÷ 21

Think: 60 ÷ 20 = 3

Try 3 groups of 21.

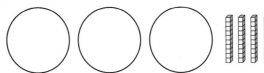

$\dfrac{3}{21\overline{)65}}$

STEP 3 Make 3 groups of 21.

Multiply. 3 × 21 = 63
Count how many ones are left over.
Subtract. 65 − 63
Compare. 2 < 21

$\begin{array}{r} 3\ r2 \\ 21\overline{)65} \\ -63 \\ \hline 2 \end{array}$

▶ **Practice and Problem Solving**

Make a model to divide.

1. $17\overline{)58}$ **2.** $18\overline{)77}$ **3.** $35\overline{)108}$ **4.** $41\overline{)129}$

Divide. You may use base-ten blocks.

5. 259 ÷ 51   **6.** 158 ÷ 25   **7.** 237 ÷ 35   **8.** 301 ÷ 29

9. Mrs. Ching has 8 boxes of 6 cups. She can display 14 cups on each shelf. How many shelves does Mrs. Ching need for her teacups?

10. REASONING There were 6 groups, with 3 tens blocks and 1 ones block in each. What division equation can you write to tell the size of each group?

11. **?** **What's the Error?** Silvia made this model for 59 ÷ 14. Describe her error. Draw the correct model.

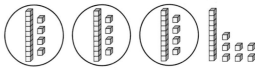

Mixed Review and Test Prep

Compare. Write <, >, or = for each ●. (p. 18)

12. 6,209 ● 6,290   **13.** 4,873 ● 4,783

Find the elapsed time. (p. 120)

14. start: 6:30 A.M.   **end:** 2:15 P.M.

15. 36 × 185 (p. 228)

16. TEST PREP (47 + 7) ÷ (3 × 3) (p. 158)

A 9   C 6

B 8   D 4

Division Procedures

Quick Review

1. 3)96   2. 5)144

3. 7)219   4. 6)512

5. 9)136

▶ Learn

LUNAR PHASES It takes about 27 days for the moon to revolve around the Earth. How many times does the moon revolve around the Earth in 365 days?

Example

Divide. 365 ÷ 27   or   27)365

STEP 1	**STEP 2**	**STEP 3**
Estimate to place the first digit in the quotient.	Divide 36 tens. Write a 1 in the tens place in the quotient.	Bring down the 5 ones. Divide the 95 ones.
Think:		

STEP 1

Estimate to place the first digit in the quotient.

Think:

$$\begin{array}{cc} 10 & 12 \\ 30)\overline{300} & 30)\overline{360} \end{array}$$

So, place the first digit in the tens place.

STEP 2

Divide 36 tens. Write a 1 in the tens place in the quotient.

```
     1
27)365   Multiply. 27 × 1
  −27    Subtract. 36 − 27
    9    Compare. 9 < 27
```

STEP 3

Bring down the 5 ones. Divide the 95 ones.

```
      13 r14
27)365
  −27↓
    95   Multiply. 27 × 3
   −81   Subtract. 95 − 81
    14   Compare. 14 < 27
```

Write the remainder.

365 ÷ 27 = 13 r14
So, the moon revolves around the Earth 13 times in 365 days.

• What would happen if the difference in Step 2 was greater than the divisor?

• How can you check to see if the answer in Step 3 is correct?

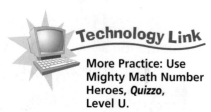

Technology Link

More Practice: Use Mighty Math Number Heroes, *Quizzo*, Level U.

▶ Check

1. **Explain** how an estimate is useful when you divide two-digit numbers.

Divide.

2. 32)972   3. 25)582   4. 9)286   5. 17)365

► Practice and Problem Solving

Divide.

6. $13\overline{)246}$ **7.** $20\overline{)483}$ **8.** $12\overline{)148}$ **9.** $15\overline{)258}$

10. $11\overline{)201}$ **11.** $54\overline{)612}$ **12.** $21\overline{)825}$ **13.** $34\overline{)749}$

14. $32\overline{)676}$ **15.** $27\overline{)543}$ **16.** $41\overline{)784}$ **17.** $53\overline{)582}$

18. $17\overline{)310}$ **19.** $36\overline{)721}$ **20.** $62\overline{)894}$ **21.** $74\overline{)945}$

Write the division problem for each check.

22. $(31 \times 25) + 5 = 780$ **23.** $(22 \times 40) + 8 = 888$ **24.** $(14 \times 52) + 40 = 768$

USE DATA For 25–27, use the map.

25. The Bensons and the Reeds met in Chicago, Illinois. The Bensons drove from Minneapolis, and the Reeds drove from Louisville. After a week's stay they drove home. Who drove farther? How much farther?

26. If the average speed of travel was 60 mph, about how long would it take to drive from Chicago to Minneapolis? from Chicago to Louisville?

27. ☀ **? What's the Question?** Felix and his family drove from Philadelphia to Chicago and back for a vacation. The answer is 1,476 miles.

28. The moon's greatest distance from Earth is 251,927 miles and its least distance is 225,745 miles. How much closer is the moon to Earth from its least distance?

Mixed Review and Test Prep

Find the sum or difference. (p. 44)

29. $\begin{array}{r} 83,204 \\ +\ 1,423 \\ \hline \end{array}$ **30.** $\begin{array}{r} 63,128 \\ -47,129 \\ \hline \end{array}$

31. $20 - (3 \times 2)$ (p. 158)

32. $12 \times 5 \times 7$ (p. 150)

33. **TEST PREP** Joshua travels 23 miles to work and 23 miles home 5 days a week. How many miles does he travel in 4 weeks? (p. 152)

A 920 miles   **C** 720 miles

B 890 miles   **D** 690 miles

Correcting Quotients

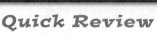

▶ **Learn**

THE BAND MARCHES ON! There are 184 band members going to the Cherry Blossom Festival in Washington, D.C. School policy requires a chaperone for every 21 students. How many chaperones are needed?

Since a chaperone is needed for every 21 students, divide 184 by 21.

Example

Divide. $184 \div 21$ or $21\overline{)184}$
                     ↓   ↓

Estimate. $180 \div 20 = 9$

STEP 1	STEP 2	STEP 3
Try your estimate, 9. $21 \times 9 = 189$	Try 8.	Subtract to find the remainder.
$\begin{array}{r} 9 \\ 21\overline{)184} \\ -189 \end{array}$ Since $189 > 184$, the estimate is too high.	$21 \times 8 = 168$ $\begin{array}{r} 8 \\ 21\overline{)184} \\ -168 \end{array}$	$\begin{array}{r} 8\ r16 \\ 21\overline{)184} \\ -168 \\ \hline 16 \end{array}$ Compare. $16 < 21$

So, 9 chaperones are needed. There are 8 chaperones, each in charge of 21 students, and 1 more chaperone is needed for the 16 remaining students.

More Examples

A Divide. $244 \div 27$
Estimate. $240 \div 30 = 8$

$\begin{array}{r} 8 \\ 27\overline{)244} \\ -216 \\ \hline 28 \end{array}$ Since $28 > 27$, the estimate is too low.

$\begin{array}{r} 9\ r1 \\ 27\overline{)244} \\ -243 \\ \hline 1 \end{array}$

B Divide. $319 \div 84$
Estimate. $320 \div 80 = 4$

$\begin{array}{r} 4 \\ 84\overline{)319} \\ -336 \end{array}$ Since $336 > 319$, the estimate is too high.

$\begin{array}{r} 3\ r67 \\ 84\overline{)319} \\ -252 \\ \hline 67 \end{array}$ $67 < 84$

Check Estimates

To check if your estimate is too high, too low, or just right, multiply it by the divisor. Then compare the product to the dividend. Subtract if possible.

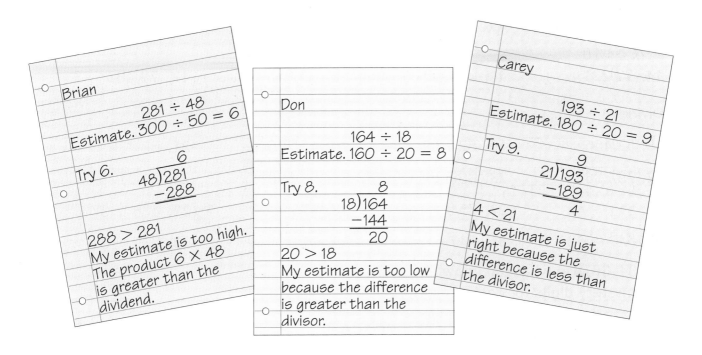

Brian

$281 \div 48$

Estimate. $300 \div 50 = 6$

Try 6.

$$48\overline{)281}$$ with 6 above
$$-288$$

$288 > 281$

My estimate is too high. The product 6×48 is greater than the dividend.

Don

$164 \div 18$

Estimate. $160 \div 20 = 8$

Try 8.

$$18\overline{)164}$$ with 8 above
$$-144$$
$$20$$

$20 > 18$

My estimate is too low because the difference is greater than the divisor.

Carey

$193 \div 21$

Estimate. $180 \div 20 = 9$

Try 9.

$$21\overline{)193}$$ with 9 above
$$-189$$
$$4$$

$4 < 21$

My estimate is just right because the difference is less than the divisor.

MATH IDEA Sometimes your choice for the first digit of a quotient is not correct. You can correct it by increasing or decreasing the first digit.

• What can you do to correct a quotient that is too low? too high?

Technology Link

More Practice: Use *Mighty Math Calculating Crew*, Nick Knack Super Trader, Levels J and T.

▶ Check

1. **Explain** how Brian and Don should correct the quotients shown on their papers above.

Write *too high, too low,* or *just right* for each estimate. Then divide.

2. $27\overline{)257}$ with 8 above

3. $43\overline{)362}$ with 9 above

4. $86\overline{)536}$ with 6 above

5. $51\overline{)462}$ with 9 above

Divide.

6. $15\overline{)146}$

7. $31\overline{)236}$

8. $75\overline{)536}$

9. $67\overline{)436}$

LESSON CONTINUES ▶

▶ Practice and Problem Solving

Write *too high*, *too low*, or *just right* for each estimate.
Then divide.

10. 6 / $18)\overline{108}$

11. 5 / $35)\overline{215}$

12. 9 / $85)\overline{798}$

13. 3 / $85)\overline{367}$

14. 10 / $22)\overline{221}$

15. 9 / $45)\overline{369}$

16. 25 / $36)\overline{853}$

17. 5 / $79)\overline{487}$

Divide.

18. $52)\overline{456}$

19. $82)\overline{736}$

20. $45)\overline{236}$

21. $62)\overline{336}$

22. $79)\overline{238}$

23. $86)\overline{528}$

24. $68)\overline{596}$

25. $81)\overline{377}$

26. $511 \div 42$

27. $754 \div 15$

28. $875 \div 23$

29. $488 \div 37$

30. $647 \div 53$

31. $747 \div 24$

32. $911 \div 33$

33. $939 \div 84$

For 34–36, use what you know about division to find
the mystery digits for each problem.

> **Mystery Digits**
> 0, 1, 2, 3, 4, 7

34. ■■ / $28)\overline{364}$

35. 26 / ■■$)\overline{702}$

36. 5■ / 3■$)\overline{1,700}$

USE DATA For 37–40, use the table.

37. The band director wants the members to march in rows of 14. How many rows will the band have? How many members will not be in a row of 14?

38. The clarinet and the flute players need to march together. If the band marches in rows of 14, how many rows of clarinet and flute players will there be?

39. The percussion and trumpet players march in rows with an equal number of players in each row. How many rows could there be? How many players in each row?

40. ✎ **Write a problem** about a marching band, using division.

The Marching Band

Group	Number of Members
Clarinets	24
Flutes	20
Trumpets	12
Percussion	16
Tubas	4
Flags	20

41. Paul had 10 school books, 45 fiction books, 35 nonfiction books, and 78 science fiction books. He gave away 10 fiction books and 15 science fiction books. How many books does he have left?

42. ☀ **What's the Error?** First Susan estimated $322 \div 42$ as 8. Then she corrected the quotient by writing 9. Describe her error and tell what she should have written.

290

43. Ann wants to use 22 beads on each bracelet. How many beads will she need for 33 bracelets?

44. Phillip put $5 in a machine that gives change in quarters. How many quarters should Phillip get?

Mixed Review and Test Prep

Find the sum or difference. (p. 40)

45. 2,690
        +7,421

46. 5,682
        −1,900

Find the product. (p. 150)

47. $2 \times (3 \times 4)$

48. $4 \times (5 \times 2)$

49. **TEST PREP** $9 \times n = 72$. What is n?
(p. 144)

A 5   **B** 6   **C** 7   **D** 8

50. **TEST PREP** A group of 52 boys, 67 girls, and 4 teachers are taking a bus trip. If each bus holds 48 people, how many buses will they need? (p. 286)

F 2   **G** 3   **H** 4   **J** 5

PROBLEM SOLVING LiNKUP ... to Reading

STRATEGY · SYNTHESIZE INFORMATION

When a problem presents a lot of information, it is helpful to synthesize, or combine, the related facts. You can group the related facts in a table.

Use the graph and the table to find how many rolls Charlie will have of each type of coin.

Synthesize the information by grouping the related facts.

1. How many coins are in a roll of dimes? In a roll of quarters?

2. Copy and complete the table.

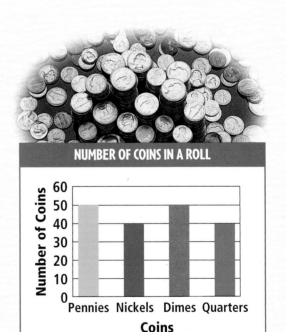

NUMBER OF COINS IN A ROLL

COIN	CHARLIE'S COINS	NUMBER OF COINS IN A ROLL	NUMBER OF ROLLS
Pennies	845	50	▨
Nickels	164	40	▨
Dimes	495	▨	▨
Quarters	282	▨	▨

EXTRA PRACTICE page H46, Set C

Problem Solving Skill
Choose the Operation

Quick Review

1. 12×5 2. $72 \div 8$

3. $17 + 43$ 4. $133 - 21$

5. $144 \div 12$

UNDERSTAND ▷ PLAN ▷ SOLVE ▷ CHECK ▷

MOVING PICTURES Mr. Regis made some notes about his animation classes. Use his notes to solve the problems.

Study the table. Decide how the numbers are related. Then solve each problem.

Add • Join groups of different sizes

Subtract • Take away or compare groups

Multiply • Join equal-size groups

Divide • Separate into equal-size groups
• Find how many in each group

A How many students are taking animation classes at the art school?

B For Tuesday's class, there are 25 sheets of drawing paper for each student. How many sheets of paper are there in all?

C How many more drawings did Maria's group make than Jerry's group?

D For every second of animation, 12 drawings are needed. How long will Tran's animation last? Will there be any drawings left over? If so, how many?

Art School Sign-Up

Monday	95
Tuesday	123
Thursday	78
Saturday	107

Animation Drawings

Group	Drawings
Jerry	501
Maria	810
Greg	642
Tran	756

Talk About It

• How can you decide which operation or operations to use for each problem?

• Solve Problems A–D.

• What operation or operations did you use in each of Problems A–D?

▶ Problem Solving Practice

Solve. Name the operation or operations you used.

1. Eric took pictures on his vacation. He took 18 pictures in Germany, 28 in France, 13 in Spain, and 11 in Portugal. How many pictures did he take?

2. Eric took 5 rolls of film with him to Europe. He can take 24 pictures with each roll of film. How many pictures can he take?

Mary took 96 pictures last year. If each roll of film had 24 pictures, how many rolls of film did she use?

3. What operation would you use to solve the problem?
   A multiplication   **C** addition
   B division   **D** subtraction

4. How many rolls of film did she use?
   F 5   **G** 4   **H** 3   **J** 2

Mixed Applications

USE DATA For 5–7, use the graph.

5. In 2000, Jon visited all the pottery, jewelry, and crafts booths. What was the total number of booths he visited?

6. Mr. Marcel spent 10 minutes talking to each booth owner in 1999. How many minutes did Mr. Marcel spend talking to the booth owners? How many hours and minutes is this?

7. In which year were there more booths at the Carson City Art Festival? How many more?

8. There were 265,980 people at a parade in 1999 and 298,125 people at a parade in 2000. Were there more people at the parade in 1999 or 2000? How many more?

9. **REASONING** There are 45 people at a meeting. Twice as many women as men are at the meeting. How many women are there? How many men?

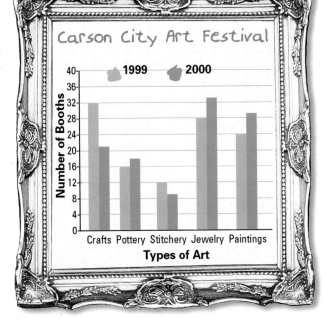

Review/Test

✓ CHECK VOCABULARY AND CONCEPTS

Choose the best term from the box.

divisor
pattern
quotient
remainder

1. Basic facts and a _?_ can help you estimate quotients. (p. 282)

2. Multiply the quotient by the _?_ and add the _?_ to the product to check the quotient and remainder of a division problem. (p. 281)

✓ CHECK SKILLS

Write the numbers you would use to estimate the quotient. Then estimate. (pp. 282–283)

3. $623 \div 22$ **4.** $294 \div 19$ **5.** $761 \div 37$

6. $82 \div 21$ **7.** $385 \div 48$ **8.** $649 \div 52$

Divide. (pp. 284–287)

9. $19\overline{)98}$ **10.** $25\overline{)237}$ **11.** $32\overline{)453}$ **12.** $43\overline{)518}$

13. $27\overline{)394}$ **14.** $34\overline{)619}$ **15.** $17\overline{)123}$ **16.** $56\overline{)767}$

Write _too high, too low,_ or _just right_ for each estimate. Then divide. (pp. 288–291)

17. $48\overline{)288}^{\;5}$ **18.** $35\overline{)175}^{\;4}$ **19.** $37\overline{)260}^{\;7}$ **20.** $71\overline{)513}^{\;8}$ **21.** $52\overline{)468}^{\;8}$

✓ CHECK PROBLEM SOLVING

Solve. Name the operation or operations you used. (pp. 292–293)

22. Jeremy keeps his baseball cards in an album. Each sheet in the album holds 18 cards. How many sheets will he need for 234 new cards?

23. Mr. Davis is writing a 450-page book. He has written 13 pages each day for 25 days. How many more pages does he need to write?

24. Shelly drove a total of 230 miles to work and back in 5 days. How many miles is it from home to work?

25. The market is selling 2 pounds of mixed fruit for $4.50. How much will 1 pound cost? 10 pounds?

Standardized Test Prep

 TIP! **Check your work.**
See item **4.**

Check your division by multiplying the divisor by the quotient and adding the remainder. It should match the dividend. Be sure you answered the question asked.

Also see problem **7,** p. H65.

For 1–9, choose the best answer.

1. Phil worked on his school project from 6:45 P.M. until 8:20 P.M. How long did Phil work on his project?

A 1 hr 25 min   **C** 2 hr 25 min

B 1 hr 35 min   **D** 3 hr 5 min

2. What is the value of p for $p = 6 \times n$ if $n = 3$?

F 2   **G** 3   **H** 9   **J** NOT HERE

3. The table shows how many books were checked out of the library on each of 3 days.

DAY	NUMBER OF BOOKS	CUMULATIVE FREQUENCY
Monday	42	42
Tuesday	▪	78
Wednesday	30	108

How many books were checked out on Tuesday?

A 72   **B** 66   **C** 36   **D** 30

4. What is the remainder for 63 ÷ 19?

F 3   **G** 4   **H** 6   **J** 9

5. 653 ÷ 32

A 20 r13   **C** 21 r19

B 20 r1   **D** 30 r11

6. Mary bought 18 rolls of film. Each roll could take 36 photos. Which operation would be best to find the total number of photos she could take?

F addition

G subtraction

H multiplication

J division

7. 162 ÷ 25

A 5 r12   **C** 7 r3

B 6 r12   **D** 8 r2

8. André sorted his trading cards into 4 groups with 12 cards in each group. He had 3 cards left over. How many trading cards did André have?

F 51   **H** 36

G 48   **J** 19

9. $17\overline{)400}$

A 20   **C** 23 r2

B 23   **D** 23 r9

Write What You Know

10. From June 10 to June 24, there were 840 computers sold at Cyberland. What was the average number sold per day? Explain what your answer means.

11. A bus can hold 32 people. There are 110 people planning a trip to the Grand Canyon. How many buses will they need? Explain.

Patterns with Factors and Multiples

The *Fibonacci* [fee•boh•NAH•chee] *Sequence* is a famous number pattern. One place the *Fibonacci Sequence* can be found is in the spirals formed by the seeds in a sunflower head.

PROBLEM SOLVING   Which of the numbers in this Fibonacci Sequence are prime? composite? neither prime nor composite?

DATA LINK

	NUMBERS IN THE FIBONACCI SEQUENCE											
1st	2nd	3rd	4th	5th	6th	7th	8th	9th	10th	11th	12th	13th
1	1	2	3	5	8	13	21	34	55	89	144	233

Use this page to help you review and remember
important skills needed for Chapter 16.

✓ VOCABULARY

Choose the best term from the box.

array
factors
Grouping Property
product

1. A 2×3 arrangement of tiles can be called an __?__.

2. In $7 \times 3 = 21$, the numbers 7 and 3 are __?__.

3. The __?__ of Multiplication states that when the grouping of factors is changed, the product remains the same.

✓ MULTIPLICATION FACTS (For Intervention, see p. H12.)

Find the product.

4. $\begin{array}{r} 5 \\ \times 2 \\ \hline \end{array}$
5. $\begin{array}{r} 4 \\ \times 8 \\ \hline \end{array}$
6. $\begin{array}{r} 6 \\ \times 2 \\ \hline \end{array}$
7. $\begin{array}{r} 7 \\ \times 5 \\ \hline \end{array}$
8. $\begin{array}{r} 9 \\ \times 7 \\ \hline \end{array}$

9. $\begin{array}{r} 9 \\ \times 6 \\ \hline \end{array}$
10. $\begin{array}{r} 4 \\ \times 5 \\ \hline \end{array}$
11. $\begin{array}{r} 6 \\ \times 0 \\ \hline \end{array}$
12. $\begin{array}{r} 8 \\ \times 5 \\ \hline \end{array}$
13. $\begin{array}{r} 7 \\ \times 8 \\ \hline \end{array}$

14. 4×3
15. 2×9
16. 2×3
17. 7×1

Find the missing factor.

18. $1 \times \blacksquare = 10$
19. $\blacksquare \times 3 = 9$
20. $\blacksquare \times 3 = 27$
21. $12 \times \blacksquare = 24$

22. $6 \times \blacksquare = 42$
23. $\blacksquare \times 4 = 28$
24. $12 \times \blacksquare = 60$
25. $\blacksquare \times 9 = 63$

✓ FACT FAMILIES (For Intervention, see p. H16.)

Write the missing equation in the fact family.

26. $5 \times 4 = 20$
$4 \times 5 = 20$
$20 \div 4 = 5$

27. $9 \times 3 = 27$
$27 \div 3 = 9$
$27 \div 9 = 3$

28. $2 \times 7 = 14$
$7 \times 2 = 14$
$14 \div 7 = 2$

29. $6 \times 7 = 42$
$42 \div 7 = 6$
$42 \div 6 = 7$

Write the fact family for each set of numbers.

30. 2, 4, 8
31. 5, 7, 35
32. 6, 6, 36
33. 6, 9, 54

Factors and Multiples

Quick Review

1. $5 \times \blacksquare = 50$

2. $4 \times \blacksquare = 44$

3. $5 \times \blacksquare = 60$

4. $\blacksquare \times \blacksquare = 25$

5. $\blacksquare \times \blacksquare = 49$

VOCABULARY

multiple

▶ Learn

IT'S ON THE TABLE Look at the multiplication table. Find the factors 6 and 4. Then, follow the *column* and the *row* for the factors to find the product 6×4.

factor　　factor
↓　　　↓
$6 \quad \times \quad 4 = 24$

- What other factors of 24 can you find on the multiplication table?

You can also use a multiplication table to find some of the multiples of 4. A **multiple** is the product of a given number and another whole number. Look at the row or the column for 4. Find the multiples of 4.

So, 4, 8, 12, 16, 20, 24, 28, 32, 36, 40, 44, and 48 are all multiples of 4.

×	1	2	3	4	5	6	7	8	9	10	11	12
1	1	2	3	4	5	6	7	8	9	10	11	12
2	2	4	6	8	10	12	14	16	18	20	22	24
3	3	6	9	12	15	18	21	24	27	30	33	36
4	4	8	12	16	20	24	28	32	36	40	44	48
5	5	10	15	20	25	30	35	40	45	50	55	60
6	6	12	18	24	30	36	42	48	54	60	66	72
7	7	14	21	28	35	42	49	56	63	70	77	84
8	8	16	24	32	40	48	56	64	72	80	88	96
9	9	18	27	36	45	54	63	72	81	90	99	108
10	10	20	30	40	50	60	70	80	90	100	110	120
11	11	22	33	44	55	66	77	88	99	110	121	132
12	12	24	36	48	60	72	84	96	108	120	132	144

▶ Check

1. **Explain** why some products appear several times on the multiplication table.

List the factors you can find on the table for each product.

2. 63　　　　**3.** 36　　　　**4.** 18　　　　**5.** 54

Use the multiplication table to find multiples for each number.

6. 3　　　　**7.** 5　　　　**8.** 6　　　　**9.** 8

List the factors you can find in the table on page 298
for each product.

10. 64   **11.** 72   **12.** 20   **13.** 56

14. 40   **15.** 24   **16.** 28   **17.** 35

Use the multiplication table to find multiples for each number.

18. 2   **19.** 9   **20.** 11   **21.** 7

22. 4   **23.** 12   **24.** 6   **25.** 10

Use what you know about multiplication.
Find as many factors as you can for each product.

26. 10   **27.** 9   **28.** 12   **29.** 48

30. 72   **31.** 30   **32.** 36   **33.** 28

USE DATA For 34–36, use the table.

34. About how much money will Todd
need to buy 6 angelfish?

35. Rachel bought 2 goldfish each for her
3 best friends. How much money did
she spend?

36. Tyler got $5.00 for his birthday. Can
he buy 1 guppy and 2 swordtails?
Explain.

TROPICAL FISH SALE	
Angelfish	$3.09
Guppy	$1.89
Goldfish	$0.79
Swordtail	$1.19

37. NUMBER SENSE The product is
32. One factor is 2 times the other
factor. What are the factors?

38. ? What's the Error? Brian writes
6, 12, 18, 24, 30 as the factors
of 6. Describe his error. Write the
correct answer.

39. Write About It Ross says he found
the factors for 42 in the multiplication
table. Explain how he could find the
factors for 42.

Mixed Review and Test Prep

40. Find $25 - y$ for $y = 4$. (p. 168)

41. $12 + \blacksquare = 9 + 3 + 8$ (p. 66)

42. What is the mode of the data 2, 2, 4,
5, 7, 8, 6, 6, 5, 4, 3, 2? (p. 86)

43. Marie left for a walk at 6:35 A.M. and
returned at 8:10 A.M. How long did
Marie walk? (p. 120)

44. TEST PREP 10 decades = ? (p. 126)

A 110 years   **C** 50 years

B 1 century   **D** 10 years

2 Factor Numbers

Quick Review

Compare. Write <, >, or = for each ●.

1. 5×6 ● 5×7

2. 2×12 ● 3×8

3. 4×8 ● 3×12

4. 8×8 ● 4×12

5. 6×10 ● 5×12

▶ **Learn**

BREAK IT UP Many whole numbers can be broken down into factors in different ways.

$12 = 3 \times 4$ $\quad$ $12 = 2 \times 6$ $\quad$ $12 = 1 \times 12$

Use arrays to show the relationships between factors and products.

Activity

Use arrays to break down 72 into factors.

MATERIALS: grid paper, scissors

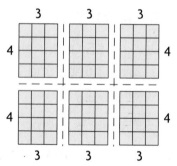

STEP 1

Outline a rectangle that has 72 squares.

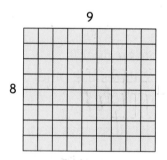

This array shows 8×9.
$72 = 8 \times 9$

STEP 2

Cut the rectangle into equal parts to find other factors.

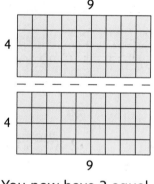

You now have 2 equal arrays.
Each array is 4×9.
$72 = 2 \times (4 \times 9)$

STEP 3

Cut apart each array into equal parts to find more factors.

You have 6 equal arrays.
Each array is 4×3.
$72 = 6 \times (4 \times 3)$

Some of the ways to break down 72 are 8×9, $2 \times (4 \times 9)$, and $6 \times (4 \times 3)$.

• What other arrays can you make to break down 72?

Remember

An *array* is an arrangement of objects in rows and columns.

1. **Tell** how an array model can help you understand that whole numbers can be broken down in different ways.

Write an equation for the arrays shown.

2. 42
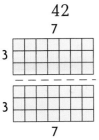

3. 54
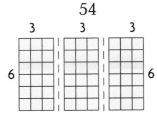

4. 36
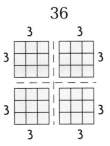

▶ **Practice and Problem Solving**

Write an equation for the arrays shown.

5.
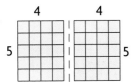

6.
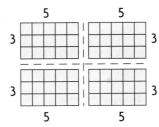

7.
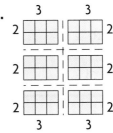

Copy the array model. Show two ways to break apart the model.

8.
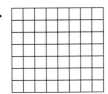

9.
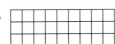

10.
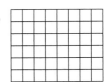

11. Sara wants to make a rectangular design with 24 square tiles. What array can she make? Draw a picture and write an equation for the array.

12. **REASONING** You have bags with 4, 6, 9, 12, 16, 24, and 25 tiles. Which of these bags of tiles can be used to make a square? Explain.

Mixed Review and Test Prep

13.   457 (p. 44)
    12,305
  +   462

14.   6,000 (p. 42)
    −   470

15. Doug's soccer practice starts at 3:45 P.M. and ends at 6:15 P.M. How long is Doug's soccer practice? (p. 120)

16. Compare: 75,526 ● 75,562. Write <, >, or =. (p. 18)

17. **TEST PREP** Kevin earned test scores of 85, 90, 85, 78, 86, 93, and 92. Which is Kevin's median test score? (p. 86)

   A 85   **B** 86   **C** 87   **D** 90

EXTRA PRACTICE page H47, Set B

LESSON

3 Prime and Composite Numbers

▶ **Learn**

ALL IN A ROW Julio has 5 model train engines that he wants to arrange in equal rows. How many ways can he arrange them?

HANDS ON **Activity 1**

Make all the arrays you can with 5 tiles to show all the factors of the number 5.

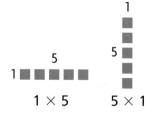

1
5 ▪▪▪▪▪
1 × 5

5 ▪
  ▪
  ▪
  ▪
  ▪
5 × 1

So, Julio can arrange the train engines in 2 ways: 1 row of 5 train engines or 5 rows with 1 train engine each.

The number 5 has two factors, 1 and 5. A **prime number** has exactly two factors, 1 and the number itself. So, 5 is a prime number.

HANDS ON **Activity 2**

Lizette has 12 pots of flowers for her box garden. How many ways can she arrange them in equal rows? Make all the arrays you can with 12 tiles to show all the factors of the number 12.

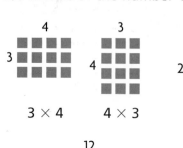

3 × 4   4 × 3   2 × 6   6 × 2

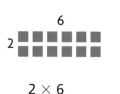
1 ▪▪▪▪▪▪▪▪▪▪▪▪   1 × 12

12 × 1

So, Lizette can arrange her pots in 6 different ways.

The factors of 12 are 1, 2, 3, 4, 6, and 12. A **composite number** has more than two factors. So, 12 is a composite number.

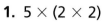

Quick Review

1. $5 \times (2 \times 2)$
2. $2 \times (2 \times 4)$
3. $2 \times (3 \times 3)$
4. $(2 \times 5) \times 6$
5. $(3 \times 3) \times 5$

VOCABULARY

prime number

composite number

Arrays and Factors

The number 1 is neither prime nor composite since it has only one factor, 1.

 MATH IDEA You can tell from an array whether a number is prime or composite.

HANDS ON

Activity 3

MATERIALS: square tiles

Use square tiles to make all the arrays you can for the numbers 2–11. Make a table like the one below to show the arrays and factors for each number.

Number	Arrays	Factors	Prime or Composite?
2		1, 2	prime
3		1, 3	prime
4		1, 2, 4	composite
5			

- Look at your table. For each number, count the arrays and the factors. How are they related?

Technology Link

More Practice: Use E-Lab, *Prime and Composite Numbers.*

www.harcourtschool.com/ elab2002

▶ Check

1. **Describe** how you can tell whether a number is prime or composite by looking at the factor column of your table.

Make arrays to find the factors. Write *prime* or *composite* for each number.

2. 16   **3.** 27   **4.** 13   **5.** 19   **6.** 21

LESSON CONTINUES

Make arrays to find the factors. Write *prime* or *composite*
for each number.

7. 31   **8.** 20   **9.** 15   **10.** 29   **11.** 45

12. 17   **13.** 18   **14.** 37   **15.** 22   **16.** 24

Write *prime* or *composite* for each number.

17. 23   **18.** 36   **19.** 47   **20.** 63   **21.** 50

22. 33   **23.** 144   **24.** 132   **25.** 121   **26.** 28

27. 26   **28.** 81   **29.** 41   **30.** 35   **31.** 43

Rodney has some books to stack on a library table. Each
stack must have an equal number of books. How many ways
can he stack the books found in each box? List the ways.

32.

8 books

33.

11 books

34.

27 books

35.

16 books

36. ❓ **What's the Error?** Mike listed
the first five prime numbers as 1, 2, 3,
5, 7. Describe his error. Write the
correct answer.

37. ❓ **What's the Question?** Irene
listed the factors 1, 2, 4, 8, 16,
and 32.

38. REASONING Josh is 2 years older than Brianna. Sally is
5 years younger than Josh. Josh is 12 years old. How
old are Brianna and Sally?

39. On Monday Ross drew 3 pictures, on Tuesday he drew
twice as many, and on Wednesday he drew twice as many
pictures as he drew on Tuesday. How many pictures did
he draw in all?

40. Rosa scored about 90 points on each test. If she took
20 tests, about how many points did she score?

41. In August, Erin has dance lessons every third day. Her first lesson is on August 3. On what dates in August are her other lessons? Which date contains a prime number?

42. MENTAL MATH Chelsea bought 10 beads to make a bracelet. She paid $0.12 for each bead. If she gave the sales clerk $2.00, how much change did she receive?

Mixed Review and Test Prep

43. 4,380 (p. 40)
+2,647

44. 6,007 (p. 42)
−2,150

Find the sum or difference. (p. 44)

45. 472,804 + 216,865

46. 28,586 − 10,399

Compare. Write < or > for each ●. (p. 18)

47. 3,840 ● 3,804

48. 6,790 ● 6,970

49. REASONING Diane had 2 quarters, a dime, and a nickel. She spent 25¢ for a cookie and 36¢ for an apple. How much money does she have left?

(p. 58)

For 50–51, use the graph. (p. 100)

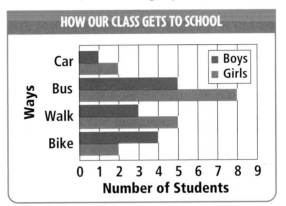

HOW OUR CLASS GETS TO SCHOOL

50. TEST PREP How many more girls ride a bus than ride a bike to school?

A 5 **B** 6 **C** 10 **D** 15

51. TEST PREP In all, how many students are in the class?

F 15 **G** 20 **H** 25 **J** 30

PROBLEM SOLVING LiNKUP... to History

Eratosthenes was a mathematician. He invented a method of sifting out the composite numbers leaving only primes. It is called the Sieve of Eratosthenes.

1	2	3	4	5	6	7	8	9	10
11	12	13	14	15	16	17	18	19	20
21	22	23	24	25	26	27	28	29	30
31	32	33	34	35	36	37	38	39	40
41	42	43	44	45	46	47	48	49	50

Copy the table onto graph paper.

1. Cross out 1. It is neither prime nor composite.

2. Circle 2, 3, 5, and 7. They are all prime. How do you know?

3. Cross out all the multiples of 2, 3, 5, and 7. What kind of numbers are they? Explain.

4. Circle the numbers that are not crossed out. How many factors does each have? What kind of numbers are they?

5. Continue the table to 100. List all the prime numbers.

6. REASONING Explain why 2 is the only prime number that is even.

EXTRA PRACTICE page H47, Set C

▶ **Learn**

FACTOR FACTORY   Any composite number can be written as a product of prime factors. Use a **factor tree** to find the prime factors. **Prime factors** are all the prime numbers in the factor tree.

Quick Review

Write *prime* or *composite.*

1. 23   **2.** 57

3. 11   **4.** 28

5. 44

VOCABULARY

factor tree

prime factors

Example

A factor tree for 20 is shown.

STEP 1   Find any two factors of 20.

$$20$$
$$10 \times 2$$
$$5 \times 2 \times 2$$

STEP 2   Continue factoring until only prime factors are left.

Record the factors from least to greatest.
So, $20 = 2 \times 2 \times 5$.

Remember

Two or more numbers can be multiplied in any order. The product is the same.
$$5 \times 2 = 10$$
$$2 \times 5 = 10$$

More Examples

Ⓐ

$$30$$
$$10 \times 3$$
$$2 \times 5 \times 3$$

So, $30 = 2 \times 3 \times 5$.

Ⓑ

$$36$$
$$9 \times 4$$
$$3 \times 3 \times 2 \times 2$$

So, $36 = 2 \times 2 \times 3 \times 3$.

▶ **Check**

1. Tell how you could check your answer for Example B.

Write each as a product of prime factors.

2. 75   **3.** 42   **4.** 18   **5.** 80   **6.** 56

Write each as a product of prime factors.

7. 9   **8.** 15   **9.** 24   **10.** 99   **11.** 8

12. 25   **13.** 22   **14.** 40   **15.** 21   **16.** 90

17. 44   **18.** 16   **19.** 48   **20.** 12   **21.** 84

22. 14   **23.** 60   **24.** 10   **25.** 120   **26.** 210

$\frac{a+b}{c}$ ALGEBRA   Write the missing factor.

27. $50 = 2 \times 5 \times \blacksquare$ **28.** $45 = \blacksquare \times \blacksquare \times 5$

29. $150 = \blacksquare \times 3 \times 5 \times \blacksquare$ **30.** $81 = 3 \times 3 \times 3 \times \blacksquare$

USE DATA For 31, use the table.

31. John noticed that the numbers of items stacked on shelves are prime and composite numbers. Which numbers of items are prime? composite?

32. **? What's the Error?** Kathy wrote 28 as a product of the prime factors 4 and 7. Describe her error. Write the correct answer.

33. Write About It Anton says the prime factors of a composite number always include at least two different numbers. Do you agree? Explain.

Item	Number
Toy Cars	9
Basketballs	7
Baseballs	13
Baseball Gloves	21

Mixed Review and Test Prep

For 34, use the calendar.

34. Carmon went to his friend's house for 2 weeks and then spent 5 days with his grandmother. He returned home from his vacation on April 25. When did he leave to go on his trip? (p. 126)

APRIL						
Sun	Mon	Tue	Wed	Thu	Fri	Sat
	1	2	3	4	5	6
7	8	9	10	11	12	13
14	15	16	17	18	19	20
21	22	23	24	25	26	27
28	29	30				

35. $\begin{array}{r} 567 \\ \times\ 23 \end{array}$ (p. 228)   **36.** $\begin{array}{r} 1{,}789 \\ \times\ \ \ 65 \end{array}$ (p. 230)

37. Order from least to greatest.
1,267,898; 1,278,987; 1,456,892; 1,987 (p. 20)

38. **TEST PREP** James delivers papers in the morning 7 days a week. His route is 26 miles long. How many miles does he travel in 4 weeks? (p. 196)

A 104 miles   **C** 286 miles

B 182 miles   **D** 728 miles

Problem Solving Strategy
Find a Pattern

PROBLEM   In art class, the students are making beaded jewelry. Janis is using red and blue beads to make a necklace. What colors will the next six beads be?

Quick Review

Find the missing numbers in the pattern.

1. 5, 10, 15, ■, ■, 30

2. 6, 12, 18, ■, ■, 36

3. 2, 10, 3, 15, 4, 20,
   ■, ■, 6, 30

4. 12, 24, ■, 48, ■, 72

5. 2, 3, 6, 7, 14, 15, 30,
   ■, ■, 63

UNDERSTAND

- What are you asked to find?
- What information will you use?
- Is there any information you will not use? Explain.

PLAN

- What strategy can you use to solve the problem?
  You can *find a pattern* in the colors of the beads

SOLVE

- How can you use the strategy to solve the problem?
  You can write the number of each red bead to see the pattern.

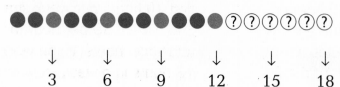

↓	↓	↓	↓	↓	↓
3	6	9	12	15	18

So, the next six beads will be blue, blue, red, blue, blue, red.

The numbers of the red beads are all multiples of 3. The next multiples of 3 are 15 and 18, so the 15th and 18th beads will be red.

Technology Link

More Practice: Use Mighty Math Number Heroes, *Quizzo,* **Level K.**

CHECK

- What other strategy could you use?

► Problem Solving Practice

PROBLEM SOLVING STRATEGIES

Draw a Diagram or Picture
Make a Model or Act It Out
Make an Organized List
► **Find a Pattern**
Make a Table or Graph
Predict and Test
Work Backward
Solve a Simpler Problem
Write an Equation
Use Logical Reasoning

Find a pattern to solve.

1. **What if** Janis replaced the first blue bead with a red bead? Would the pattern be correct? Explain.

2. Continue the pattern.

   1 1
   1 2 1
   1 3 3 1
   1 4 6 4 1
   ■ ■ ■ ■ ■ ■

Look at the pattern of dots.

3. How many dots should be in the next drawing?
   A 12   **C** 14
   B 13   **D** 15

4. Find the missing numbers in the pattern: 1, 3, 6, 10, ■, ■, 28.
   F 12, 15   **H** 15, 21
   G 15, 19   **J** 25, 27

5.  Edith has a puzzle for her classmates. When she says 12, the answer is 24. When she says 15, the answer is 27. When she says 20, the answer is 32. What is the pattern?

6. **REASONING** Tami wrote the following numbers on the board: 1, 2, 6, 30, 210, ■, ■. What are the next two numbers in the pattern? What is the pattern?

⌐ Mixed Strategy Practice ⌐

7. Jake and his family drove for 35 minutes to get to the park. They walked in the park for 1 hour and 45 minutes and spent 20 minutes having lunch. They left the park at 1:15 P.M. At what time did they leave home?

8. Marina earned $6 per hour for the first 40 hours per week she worked. She earned $9 per hour for any additional hours. If she worked 43 hours in a week, how much did she earn?

9. A computer lab has 8 rows of computers. Each row has 4 computers. There are also 4 computers grouped in the center of the room. How many computers are in the lab?

10. Matthew rode his bicycle to and from school for 20 days. He lives 3 miles from school. How many miles did he ride?

Problem Solving Strategy

Review/Test

✓ CHECK VOCABULARY AND CONCEPTS

Choose the best term from the box.

factor
multiple
prime number
composite number

1. A _?_ has exactly two factors, 1 and the number itself. (p. 302)

2. A product of a given number and another whole number is a _?_. (p. 298)

3. A _?_ has more than two factors. (p. 302)

✓ CHECK SKILLS

Use what you know about multiplication. Find as many factors as you can for each product. (pp. 298–299)

4. 8   **5.** 6   **6.** 48   **7.** 28   **8.** 15

Write an equation for the arrays shown. (pp. 300–301)

9.   **10.**   **11.**

Write *prime* or *composite* for each number. (pp. 302–305)

12. 9   **13.** 49   **14.** 11   **15.** 39

Write each as a product of prime factors. (pp. 306–307)

16. 28   **17.** 42   **18.** 50   **19.** 54   **20.** 81

Write the missing factor. (pp. 306–307)

21. $25 = 5 \times \blacksquare$ **22.** $30 = 2 \times \blacksquare \times 5$ **23.** $60 = 2 \times \blacksquare \times 3 \times \blacksquare$

✓ CHECK PROBLEM SOLVING

Solve. (pp. 308–309)

24. Monica does 50 sit-ups every day. How many sit-ups will she do in 30 days?

25. What are the two missing numbers in the pattern? 2, 3, 5, 7, 11, 13, $\blacksquare$, $\blacksquare$, 23

Standardized Test Prep

Look for important words.
See item **2.**

An important word is **not.** Three of the answer choices are multiples of 6. You are to find the one that is **not.**

Also see problem **2**, p. H62.

For 1–12, choose the best answer.

1. Which is the best estimate for this product?

   98
   × 31

   A 300   **C** 3,000
   B 2,700   **D** 30,000

2. Which is **not** a multiple of 6?

   F 16   **G** 18   **H** 24   **J** 30

3. Which is a prime number?

   A 11   **B** 12   **C** 14   **D** 15

4. Which is **not** a factor of 24?

   F 3   **G** 6   **H** 8   **J** 9

5. Which names 42 as a product of prime factors?

   A 2 × 2 × 7   **C** 2 × 4 × 7
   B 3 × 3 × 7   **D** 2 × 3 × 7

Find the missing factor.

6. 70 = 2 × 5 × ■

   F 5   **H** 7
   G 6   **J** 8

7. Which is **not** equivalent to 36?

   A 9 × 4   **C** 8 × 4
   B 6 × 6   **D** 3 × 12

8. What is the median temperature?

   Temperature

Stem	Leaves
6	1 5
7	2 4 5 5
8	2

   F 82   **G** 75   **H** 74   **J** 65

9. What is 372,394 rounded to the nearest ten thousand?

   A 378,000   **C** 380,000
   B 370,000   **D** 400,000

10. (2 × 4) + (2 × 2)

    F 12   **H** 32
    G 24   **J** NOT HERE

11. The numbers 32, 35, 38, 41, 44 follow a counting pattern. What is the next number in the pattern?

    A 45   **B** 46   **C** 47   **D** 48

12. Josh noticed a pattern in the wrapping paper he bought. Each row of figures repeated a baseball, a soccer ball, a football, and a tennis ball, in that order. Which kind of ball would be the 24th in the row?

    F baseball   **H** football
    G soccer ball   **J** tennis ball

Write What You Know

13. Find a number that has 2, 3, 4, 5, and 6 as factors. Show the method you used to find the number.

14. Find the first ten multiples of 6. Describe any patterns you notice.

PROBLEM SOLVING
MATH DETECTIVE

Case of Unknown Numbers

Sheryl, the star mathematics detective, has two cases to solve.
She has gathered the clues but needs your help.

Use your reasoning powers to help solve each case.
Good luck!

Case 1

THE CASE OF THE DISAPPEARING DIVIDEND

Clue 1: The dividend is an odd number between 200 and 250.

Clue 2: When the dividend is divided by 5, the remainder is 0.

Clue 3: The sum of the digits of the dividend is 9. The disappearing dividend is ■.

Case 2

THE CASE OF THE ABSENT AVERAGE

Clue 1: There are four test scores.

Clue 2: The difference between the lowest and highest score is 20 points.

Clue 3: The sum of the two middle scores is 164.

Clue 4: The lowest score is 72. The absent average is ■.

Think It Over!

- 📓 Write About It Explain how you solved each case.

- **STRETCH YOUR THINKING** Write clues that you could give someone to find the quotient $742 \div 7$.

Challenge

Divisibility Rules

Margarette wants to know if she can find out if $5{,}766 \div 3$ has a remainder without having to divide.

Divisibility rules can help you tell whether a number is **divisible** by another number. Divisible means "no remainder after division."

The table below shows divisibility rules for several numbers.

Number	Rule	Examples
2	Even numbers are divisible by 2.	246, 678, 454
3	If the sum of the digits is divisible by 3, then the number is divisible by 3.	$5{,}766 \rightarrow 5 + 7 + 6 + 6 = 24$ 24 is divisible by 3. So, 5,766 is divisible by 3.
5	If the last digit is a 5 or 0, the number is divisible by 5.	5,255; 12,755
9	If the sum of the digits is divisible by 9, then the number is divisible by 9.	$9{,}909 \rightarrow 9 + 9 + 0 + 9 = 27$ 27 is divisible by 9. So, 9,909 is divisible by 9.
10	If the last digit is a 0, the number is divisible by 10.	100; 1,110; 2,050

• Tell if each number is divisible by 2, 3, 5, 9, or 10. Some numbers may be divisible by more than one number.

Try It

Tell whether the number is divisible by 2, 3, 5, 9, or 10.

1. 2,130   **2.** 6,452   **3.** 1,968,085   **4.** 9,876

5. 73,821   **6.** 1,009,980   **7.** 3,459   **8.** 27,000

9. ✏️ **Write About It** Explain how you know that if a number is divisible by 2 and 3, then the number is also divisible by 6.

10. **REASONING** If a number is divisible by 5, is it also divisible by 10? Explain.

Study Guide and Review

VOCABULARY

1. The ? is the number found by dividing the sum of a set of numbers by the number of addends. (p. 276)

2. Sometimes a number cannot be divided evenly. The amount left over is called the ? . (p. 274)

3. A ? has exactly two factors, 1 and the number itself. (p. 302)

> prime number
> composite number
> mean
> remainder

STUDY AND SOLVE

Chapter 13

Divide 2-digit numbers.

```
  16 r3      • Divide.                    Check:
4)67         • Multiply.                      16
 − 4         • Subtract.                     × 4
  27         • Compare difference             64
 − 24          with divisor.                 + 3
   3         • Bring down.                    67
             • Repeat as needed.
```

Divide and check. (pp. 246–253)

4. 6)72   5. 4)95

6. 3)76   7. 2)58

8. 5)83   9. 7)96

10. 8)97   11. 4)79

12. 5)69   13. 3)48

Chapter 14

Divide 3- and 4-digit numbers.

```
  39 r3     • Use order of division to    Check:
6)237         find the quotient.              39
 − 18       • Repeat the order until         × 6
  57          the division is complete.     234
 − 54                                       +  3
   3                                        237
```

Divide and check. (pp. 264–273)

14. 6)383   15. 2)$1.54

16. 3)218   17. 5)4,128

18. 7)2,943   19. 8)$24.48

20. 4)1,862   21. 9)5,097

Find the mean of a set of data.

Find the mean.

• Add. $18 + $29 + $110 + $145 + $173 = $475

• Divide. $475 ÷ 5 = $95.

So, the mean is $95.

Find the mean. (pp. 276–277)

22. 185; 410; 518; 619

23. 13; 25; 29; 35; 48; 52; 64

24. $98; $93; $89; $85; $75

Chapter 15

Divide with 2-digit divisors.

```
     21 r11
16)347
  − 32
     27
   − 16
     11
```
- Divide.
- Multiply.
- Subtract.
- Compare difference with divisor.
- Bring down.
- Repeat as needed.

Divide. (pp. 284–291)

25. $17\overline{)361}$ **26.** $64\overline{)904}$

27. $24\overline{)583}$ **28.** $52\overline{)813}$

29. $31\overline{)843}$ **30.** $47\overline{)666}$

31. $73\overline{)952}$ **32.** $28\overline{)898}$

Chapter 16

Write whether the number is prime or composite.

A **prime number** has exactly two factors.

A **composite number** has more than two factors.

A. 5 has only two factors, 5 and 1. It is prime.

B. 6 has more than 2 factors, 1, 2, 3, and 6. It is composite.

Write *prime* or *composite* for each number. (p. 302–305)

33. 11   **34.** 27   **35.** 52

36. 72   **37.** 17   **38.** 81

39. 132   **40.** 49   **41.** 31

42. 29   **43.** 72   **44.** 91

PROBLEM SOLVING PRACTICE

Solve. (pp. 254-255, 274–275, 292–293, 308-309)

45. There are 78 flowers. After equal bunches are formed, 3 flowers are left over. How many bunches were formed? How many flowers in each bunch?

46. This is Hugo's number pattern. 3, 6, 12, 24, 48, 96, ▪, ▪, ▪ Describe his pattern. Find the missing numbers.

47. Mia is making costumes. She has 108 yards of cloth. Each costume uses 5 yards of cloth. How many costumes can Mia make? How did you interpret the remainder?

48. The hardware store sold 126 hinges. There were 3 hinges in each package. How many packages of hinges did the hardware store sell? What operation did you use?

PERFORMANCE ASSESSMENT

TASK A • SUMMER VACATION

The Turner family lives in Minneapolis, Minnesota. They are planning to take a family trip this summer. Each family member named a city to visit. The cities and their distances from Minneapolis are shown in the table.

CITY	DISTANCE FROM MINNEAPOLIS (in miles)	MILES PER DAY	ESTIMATED AVERAGE SPEED (in mph)
Columbus, OH	778	▪	▪
Detroit, MI	686	▪	▪
Pittsburgh, PA	886	▪	▪
Topeka, KS	504	▪	▪

a. The Turners will travel for 2 days to the city they choose. They will travel the same number of miles on each day of their trip. Copy the table and use this information to complete the "Miles per Day" column.

b. The Turners plan to drive 7 or fewer hours each day. Explain how you could use compatible numbers to estimate the average speed (miles per hour) they will have to drive each day. Then write your estimates in the table.

TASK B • TILE TRIALS

Ryan and Kendra are using square tiles to make rectangular designs for tabletops. Ryan has 20 tiles. Kendra has 19 tiles.

a. How many different 20-tile rectangles can Ryan make? Draw and label the different rectangles.

b. Kendra could make only one rectangle that used all 19 of her tiles. Explain why this is so.

c. Kendra decided to use more tiles. Decide on a number of tiles she can have. Then find out how many different rectangles can be made with the total number she has. Draw and label the rectangles.

Technology Linkup

Calculator • Remainders

Ms. Lee wrote these problems on the board:
$1{,}130 \div 125 = \blacksquare$, $680 \div 75 = \blacksquare$, and $518 \div 57 = \blacksquare$.
Linda says they all have the same quotient.
Greg says they do not. Who is right?

You can use a calculator to prove that both Linda and Greg are right.

Linda's Method	Greg's Method
Use the $\div R$ to divide.	Use the $\div$ to divide.
Find each quotient.	**Find each quotient.**
a. $\boxed{1}\ \boxed{1}\ \boxed{3}\ \boxed{0}\ \boxed{\div R}$ $\boxed{1}\ \boxed{2}\ \boxed{5}\ \boxed{=}$ =9 ^{R5}	a. $\boxed{1}\ \boxed{1}\ \boxed{3}\ \boxed{0}\ \boxed{\div}$ $\boxed{1}\ \boxed{2}\ \boxed{5}\ \boxed{=}$ = 9.04
b. $\boxed{6}\ \boxed{8}\ \boxed{0}\ \boxed{\div R}\ \boxed{7}\ \boxed{5}$ $\boxed{=}$ =9 ^{R5}	b. $\boxed{6}\ \boxed{8}\ \boxed{0}\ \boxed{\div}\ \boxed{7}\ \boxed{5}$ $\boxed{=}$ =9.0666667
c. $\boxed{5}\ \boxed{1}\ \boxed{8}\ \boxed{\div R}\ \boxed{5}\ \boxed{7}$ $\boxed{=}$ =9 ^{R5}	c. $\boxed{5}\ \boxed{1}\ \boxed{8}\ \boxed{\div}\ \boxed{5}\ \boxed{7}$ $\boxed{=}$ =9.0877193

What do you notice about the whole number
parts of Linda's and Greg's quotients?

Practice and Problem Solving

Use $\boxed{\div R}$ and $\boxed{\div}$ to find the quotient in two ways.

1. $356 \div 95$ **2.** $685 \div 103$ **3.** $958 \div 621$ **4.** $1{,}090 \div 55$

5. $4{,}118 \div 32$ **6.** $6{,}217 \div 33$ **7.** $4{,}092 \div 915$ **8.** $9{,}371 \div 208$

Multimedia Math Glossary www.harcourtschool.com/mathglossary

9. Vocabulary Look up *remainder* in the Multimedia Math
Glossary. Look at the example, explain which method
above you would use to show the remainder.

Big Cat Country, at the St. Louis Zoo, is home to tigers, lions, leopards, pumas, and jaguars.

◄ **Snow leopard cubs**

PROBLEM SOLVING ON LOCATION

at the Zoo

BIG CATS AT THE ST. LOUIS ZOO

There are nearly 5,000 animals at the St. Louis Zoo. Many of the animals are endangered species. In fact, all of the animals in Big Cat Country, one of the many favorite attractions at the zoo, are endangered.

1. Snow leopard cubs Shikari and Kashih were born on Thursday, May 28, 1998. On what day of the week were the cubs 30 days old?

2. Snow leopards range from 47 to 59 inches in length. How long is this in feet and inches? (HINT: There are 12 inches in one foot.)

3. A snow leopard at another zoo is fed 93 pounds of food during the month of January. How much food is that per day?

4. What if a snow leopard walks 3 kilometers each day? How far would it walk in 2 weeks? in 3 weeks?

KOALAS

You can also see koalas at the St. Louis Zoo. Koalas spend a lot of time in trees, where they sit comfortably on their thickly padded tails. Eucalyptus leaves are a koala's main diet.

1. **What if,** 10,000 eucalyptus seedlings are raised each year? How many seedlings are raised in 10 years?

2. Koalas sleep about 110 hours a week. About how many hours do they sleep per day?

3. A koala eats about 300 pounds of eucalyptus leaves in a year.

   a. About how many pounds of eucalyptus leaves does a koala eat in one month?

   b. Does a koala eat more than or less than one pound of leaves in a day? Explain how you know.

4. **STRETCH YOUR THINKING** Of the 705 species of animals at the St. Louis Zoo, 66 are endangered. About how many times as many species are not endangered as are endangered species? Explain how you found your answer.

CHAPTER 17 Plane Figures

The Ambassador Bridge connects Detroit, Michigan, and Windsor, Ontario. It is the busiest bridge in North America joining two different countries.

PROBLEM SOLVING Tell what kind of line relationships and angles you see in the picture of the bridge.

CHECK WHAT YOU KNOW

Use this page to help you review and remember
important skills needed for Chapter 17.

✓ IDENTIFY ANGLES (For Intervention, see p. H25.)

Tell if each angle is a *right* angle, *greater than* a right angle,
or *less than* a right angle.

1.

2.

3.

4.

5.

6.

✓ COMPARE FIGURES (For Intervention, see p. H26.)

Are the figures the same size and shape? Write *yes* or *no*.

7.

8.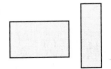

9.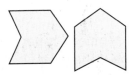

10.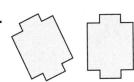

11.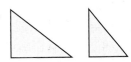

12.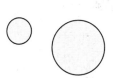

✓ IDENTIFY SYMMETRIC FIGURES (For Intervention see. p. H26.)

Is the blue line a line of symmetry? Write *yes* or *no*.

13.

14.

15.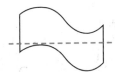

16.

17.

18.

Lines, Rays, and Angles

Quick Review

1. 96 ÷ 18   2. 115 ÷ 32

3. 643 ÷ 52   4. 406 ÷ 21

5. 395 ÷ 46

▶ **Learn**

GEOMETRY EVERYWHERE! These geometric ideas and terms are used to describe the world around us.

- Study the terms and definitions. Then look at the photo. Find as many examples as you can of the terms listed below. Describe how the definitions of the terms match figures in the photo.

VOCABULARY

point	angle
line	vertex
ray	right angle
plane	acute angle
line segment	obtuse angle

Term and Definition	Draw It	Read It	Write It
A **point** names a location on an object or in space.	• A	point A	point A
A **line** is a straight path of points that goes on and on in both directions. It has no endpoints.	 K L	line KL	$\overleftrightarrow{KL}$
A **line segment** is part of a line. It has two endpoints.	 K L	line segment KL	$\overline{KL}$
A **ray** is part of a line. It has one endpoint and goes on and on in one direction.	 K L	ray KL	$\overrightarrow{KL}$
A **plane** is a flat surface of points, with no end. A plane is named by at least three points in the plane.	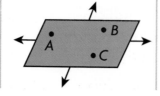 • B A • C	plane ABC	plane ABC

Types of Angles

Term and Definition	Draw It	Read It	Write It
Two rays with the same endpoint form an **angle**. The endpoint is called the **vertex**.	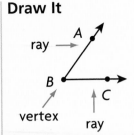	angle *ABC* angle *CBA* angle *B* NOTE: The vertex is always the middle letter or the single letter that names the angle.	∠*ABC* ∠*CBA* ∠*B*

 Activity

Make an angle using a sheet of paper. Fold the paper twice to make an angle like this. The angle you made is called a right angle.

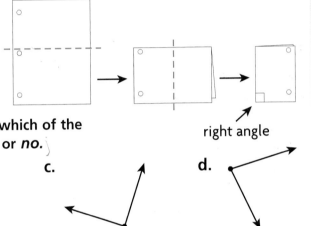

right angle

Use the right angle you made to find out which of the following are also right angles. Write *yes* or *no*.

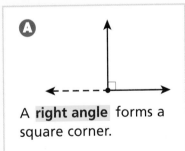

a.

b.

c.

d.

Examples   The size of an angle depends on the size of the opening between the rays. Here are some types of angles.

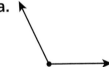

A

A **right angle** forms a square corner.

B

An **acute angle** is an angle that measures less than a right angle.

C

An **obtuse angle** is an angle that measures greater than a right angle.

- What types of angles do you see in the photo on page 320? Describe where they are.

▶ **Check**

1. **Give** some examples of line segments that you see in your classroom.

LESSON CONTINUES ▶

Name a geometric term that describes each.

2. corner of a page   **3.** sharp tip of a pencil   **4.** a beam of light

Practice and Problem Solving

Name a geometric term that describes each.

5. a flagpole

6. an open laptop computer

7. a parking lot

8. tip of a tack

9. side edge of door

10. corner where wall meets floor

Draw and label an example of each.

11. line *BC*

12. line segment *PQ*

13. point *G*

14. obtuse angle *RST*

15. right angle *M*

16. ray *XY*

What kind of angle is each? Write *right, acute,* or *obtuse*.

17.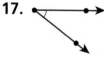

18.

19.

20.

21.

22.

Draw each line segment with the given length.

23. $\overline{BC}$, 3 cm   **24.** $\overline{AE}$, 2 in.   **25.** $\overline{JK}$, $3\frac{1}{2}$ in.   **26.** $\overline{RS}$, 6 cm

27. Look at the figure below. Name the figure three different ways.

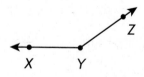

28. REASONING Use the corner of a sheet of paper to prove or disprove that the three angles in the letter *M* are right angles.

M

29. **? What's the Error?** Wanda said that the letter *W* had two angles. What error did she make?

30. **Write About It** Explain the differences between a line, a ray, and a line segment.

31. Name an object in your classroom that is like a line segment.

32. Give a time when the hands on a clock represent each type of angle: acute, obtuse, and right.

33. Suchada had some money. First she spent $15 at the gift shop. Later her father gave her $10 to buy lunch, but she only spent $7. At the end of the day, Suchada had $10. How much money did she have when she began her day?

Mixed Review and Test Prep

34. 500 − 490 (p. 40)

35. 703 − 585 (p. 40)

36. 6,435 + 797 + 285 (p. 40)

37. Jim ordered 3 pizzas. Each pizza had 8 slices. How many slices of pizza were there altogether? (p. 144)

38. Write 45 as a product of prime factors. (p. 306)

39. $(3 + 4) \times 2$ (p. 158)

40. ■ ÷ 8 = 16 (p. 266)

41. Kim opened a carton of 12 eggs. She put 4 eggs in each of 2 bowls. How many eggs were left in the carton? Write the expression you used. (p. 158)

42. **TEST PREP** What is the value of the blue digit in 45,678,342? (p. 8)
  - **A** 800
  - **B** 8,000
  - **C** 80,000
  - **D** 800,000

43. **TEST PREP** What is 745,864 rounded to the nearest ten thousand? (p. 26)
  - **F** 700,000
  - **G** 740,000
  - **H** 745,000
  - **J** 750,000

PROBLEM SOLVING LiNKUP...to Art

When architects design houses or buildings, they draw different views by using points, planes, line segments, and angles.

This drawing shows only the front view of this building.

Use the drawing and name each of the following.

1. line segment
2. right angle
3. obtuse angle
4. acute angle
5. point
6. plane

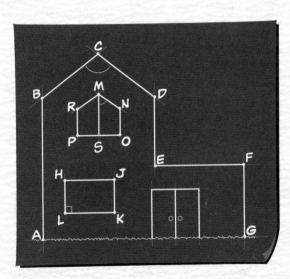

EXTRA PRACTICE page H48, Set A

Line Relationships

▶ Learn

FOLLOW THE LINES Look at the term and definition for each line relationship. Find these same relationships on the road map.

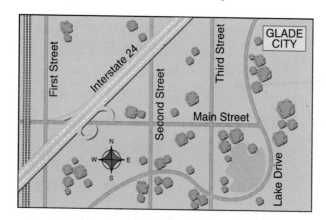

Term and Definition	Draw It	Read It	Write It
Intersecting lines are lines that cross each other. They form four angles.	obtuse, acute, obtuse, acute (A B C D E)	Line *AE* intersects line *DC* at point *B*.	$\overleftrightarrow{AE}$ intersects $\overleftrightarrow{DC}$ at point *B*.
Parallel lines are lines that never intersect.	A E / D C	Line *AE* is parallel to line *DC*.	$\overleftrightarrow{AE} \parallel \overleftrightarrow{DC}$
Perpendicular lines are lines that intersect to form four right angles.	B C A E D	Line *AE* is perpendicular to line *DC*.	$\overleftrightarrow{AE} \perp \overleftrightarrow{DC}$

- Which term identifies the relationship between Third Street and Second Street on the map?

▶ Check

1. Name two streets on the map that are perpendicular. Name two streets that are intersecting.

Name any line relationship you see in each figure. Write
intersecting, *parallel*, or *perpendicular lines*.

2.

3.

4.

Practice and Problem Solving

Name any line relationship you see in each figure. Write
intersecting, *parallel*, or *perpendicular lines*.

5.

6.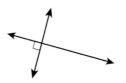

7.

8.

9.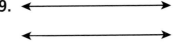

10.

For 11–20, use the drawing at the right.

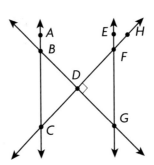

11. Name a point.   **12.** Name two lines.

13. Name a ray.   **14.** Name two parallel lines.

15. Name four line segments that include point *G*.

16. Name two intersecting lines.

17. Name two perpendicular lines.   **18.** Name an acute angle.

19. Name an obtuse angle.   **20.** Name a right angle.

21. **REASONING** Is this statement true or false? "Perpendicular lines are also intersecting lines." Explain your answer.

22. **? What's the Error?** Tricia said that all intersecting lines are perpendicular. Explain her error. Include a drawing with your explanation.

Mixed Review and Test Prep

23. 542×6 (p. 196)

24. $804 \div 5$ (p. 266)

25. The product of two numbers is 45. Their sum is 18. What are the numbers? (p. 148)

26. $96,784 + 8,400$ (p. 44)

27. **TEST PREP** In what place is the 9 in 3,902,817? (p. 8)

A thousands   **C** hundred thousands

B ten thousands   **D** millions

EXTRA PRACTICE page H48, Set B

Congruent Figures and Motion

Quick Review

1. 9
   ×6

2. 4
   ×5

3. 8
   ×7

4. 10
   ×10

5. 8
   ×6

▶ **Learn**

SAME SIZE, SAME SHAPE   Figures that have the same size and shape are **congruent**. Figures do not have to be in the same position to be congruent. You can move a figure to test if two figures are congruent.

VOCABULARY

congruent   turn

slide   similar

transformations flip

congruent

not congruent

 Activity 1

MATERIALS: dot paper, rulers, scissors

STEP 1

Copy each pair of figures on dot paper.

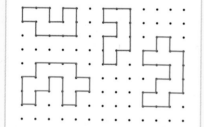

STEP 2

Cut out one in each pair, and move it in any way to check for congruency.

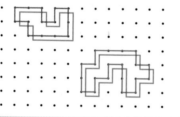

Transformations are different ways to move a figure. Three kinds of transformations are **slide**, **flip**, and **turn**.

Examples

A   A **slide** moves a figure to a new position.

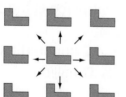

B   A **flip** flips a figure over a line.

C   A **turn** rotates a figure around a point.

Point of Rotation

Similar Figures

When you enlarge or reduce a figure, the new figure is
similar to the original figure. The figures have the same shape
but may have different sizes.

similar   not similar

Activity 2

MATERIALS: centimeter dot paper, rulers, scissors

STEP 1	STEP 2	STEP 3
Draw a 2 cm by 2 cm square on your dot paper.	Enlarge the square by multiplying each side of the original square by 2.	Reduce the square by dividing each side of the original square by 2.

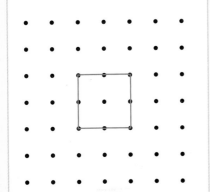 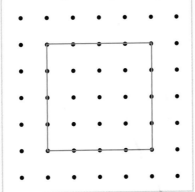 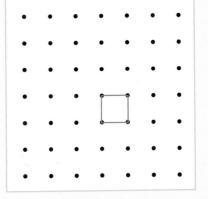

• How are your three squares alike? How are they different?

Technology Link

To learn more about Congruent Figures and Motion, watch the Harcourt Math Newsroom Video *Sydney Opera House.*

 **Check**

1. **Draw** a triangle. Draw another triangle that is similar to it, but not congruent.

Tell how each figure was moved. Write *slide, flip,* or *turn*.

2.

3.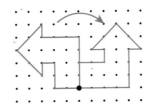

4.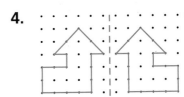

LESSON CONTINUES ▶

Tell whether the two figures are *congruent, similar,*
or *neither.*

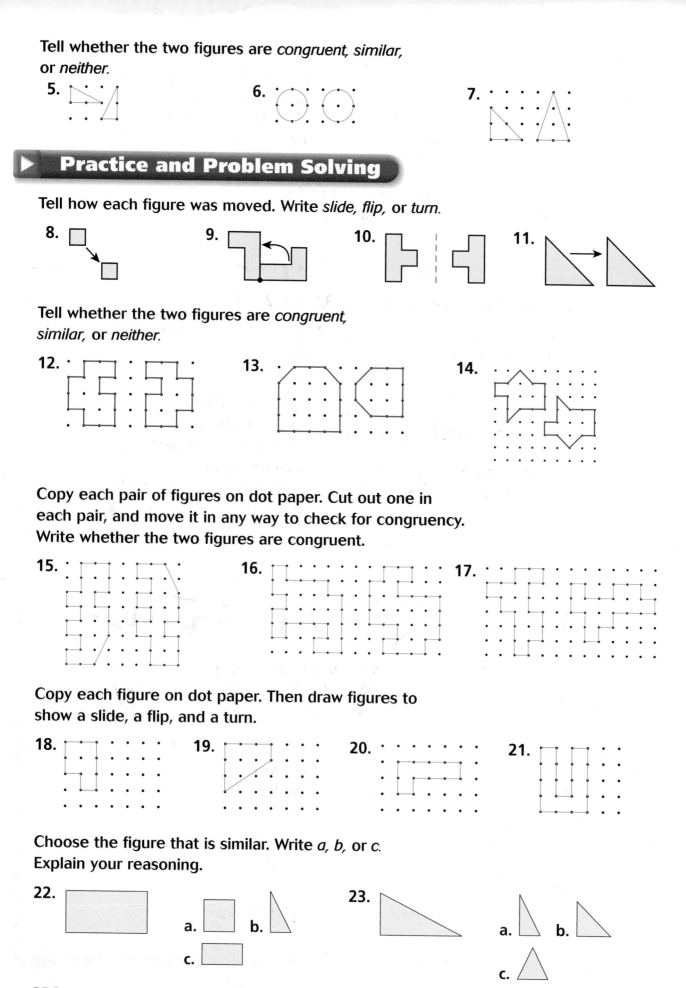

5.

6.

7.

Practice and Problem Solving

Tell how each figure was moved. Write *slide, flip,* or *turn.*

8.

9.

10.

11.

Tell whether the two figures are *congruent,*
similar, or *neither.*

12.

13.

14.

Copy each pair of figures on dot paper. Cut out one in
each pair, and move it in any way to check for congruency.
Write whether the two figures are congruent.

15.

16.

17.

Copy each figure on dot paper. Then draw figures to
show a slide, a flip, and a turn.

18.

19.

20.

21.

Choose the figure that is similar. Write *a, b,* or *c.*
Explain your reasoning.

22.

a.

b.

c.

23.

a.

b.

c.

24. Mr. Jones plays a number game. When he says 7, the answer is 19. When he says 13, the answer is 25. When he says 21, the answer is 33. What is the answer when Mr. Jones says 35? What is his rule?

25. Lydia bought tacks for $3, paper for $2, and stickers for $3. Later, Tom gave Lydia $2. Then Lydia had $5. How much money did Lydia have before she bought the supplies?

26. **What's the Question?** George's teacher drew the figure shown at the right. The answer the students gave was *flip*.

b
p

Mixed Review and Test Prep

27. 118 × 53 (p. 228)   **28.** 625 ÷ 5 (p. 266)

29. Amanda told her mother she would be back in one hour and ten minutes. How many minutes would that be altogether? (p. 120)

30. 7 × 21 (p. 190)   **31.** 17)210 (p. 286)

32. Tonya has 45 more pennies than Cari. If Tonya has 139 pennies, how many pennies does Cari have? (p. 36)

33. 5 × 2 × 5 (p. 150)

34. **TEST PREP** Which is the mean of 88, 84, 75, 60, and 93? (p. 276)
A 80   **B** 84   **C** 87   **D** 90

35. **TEST PREP** Naga bought a notebook for $4.89 and a toy for $12.49. How much change should he get from a $20 bill? (p. 40)
F $1.62   **H** $2.62
G $2.02   **J** $3.72

PROBLEM SOLVING | Thinker's Corner

SELF-SIMILARITY Some figures are called **self-similar**. A figure is self-similar when it is made up entirely of smaller figures that are basically similar to the whole figure, at different sizes.

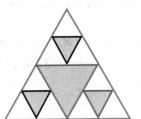

1. Trace and cut out three copies of the figure at the right.

2. Paste them together, corner to corner, on a larger piece of paper. This new figure should look basically like the original triangle, only larger.

3. Form a group with two other students. Paste your three figures together, corner to corner, to form yet a larger triangle, basically like the original.

4. Now look at the result. How is your large figure like the original figure?

Symmetric Figures

▶ **Learn**

MIRROR, MIRROR   A figure can have
rotational symmetry, **line symmetry**, or both.

VOCABULARY

rotational symmetry

line symmetry

Turn this figure around the
center point. It looks the same
at each quarter turn. It has
rotational symmetry.

Fold the figure along a line so that
its two parts match exactly. It has
line symmetry. A figure can have
more than one line of symmetry.

- Which of the above figures has both rotational
symmetry and line symmetry?

HANDS ON

Activity

MATERIALS: paper, straightedge, scissors

STEP 1	**STEP 2**	**STEP 3**
On a sheet of paper, draw an equilateral triangle. Label the center point. Cut it out.	Label the corners of the triangle 1, 2, and 3.	Turn the triangle about its central point. How does the triangle fit back into the space?

- How many times did you turn the triangle?

- What kind of symmetry does this show?

▶ **Check**

1. **Draw** half of a design. Explain how to complete the
design so that the figure has line symmetry.

Tell whether the figure has *rotational symmetry, line symmetry,* or *both.*

2.

3.

4.

5.

▶ Practice and Problem Solving

Tell whether the figure has *rotational symmetry, line symmetry,* or *both.*

6.

7.

8.

9.

10.

11.

Copy each design on dot paper. Complete each design to show line symmetry.

12.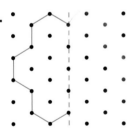

13.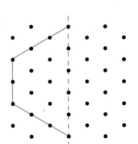

14.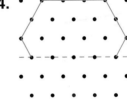

15. The word BOX has a horizontal line of symmetry. Find two other words with line symmetry.

16. REASONING Which capital letters of the alphabet have no line symmetry?

Mixed Review and Test Prep

17. 56×143 (p. 228)

18. $430 + 178$ (p. 40)

19. $112 \div 7$ (p. 266)

20. $(8 \times 9) \div 6$ (p. 158)

21. TEST PREP What motion do you do to Figure A to make it look like Figure B? (p. 326)

A slide   **B** turn   **C** flip   **D** similar

Problem Solving Strategy
Make a Model

PROBLEM Terry is helping decorate a bulletin board about symmetry. He wants to make a large picture of the geometric design at the right. How can he make a model to help him?

UNDERSTAND

- What are you asked to find?
- What information will you use?
- Is there any information you will not use? If so, what?

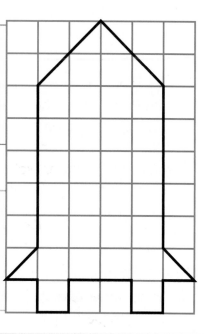

PLAN

- What strategy can you use to solve the problem?

  You can *make a model* of the design that is larger than but similar to the one shown here.

SOLVE

- How can you make a model?

  You can use 1-inch grid paper to enlarge the figure. Copy the picture, square by square, to make a larger picture.

CHECK

- How else might Terry enlarge the picture?

Problem Solving Practice

For 1–2, make a model to solve.

1. Grant wants to reduce the figure below to put it on a postcard. Use 0.5-centimeter grid paper to help him make a smaller picture.

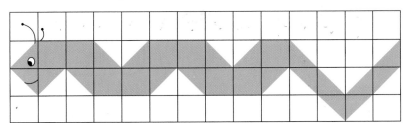

Draw a Diagram or Picture
Make a Model or Act It Out
Make an Organized List
Find a Pattern
Make a Table or Graph
Predict and Test
Work Backward
Solve a Simpler Problem
Write an Equation
Use Logical Reasoning

2. **What if** you want to make a larger picture of the figure at the right to put on a poster? Use 1-cm grid paper to help you make a larger picture.

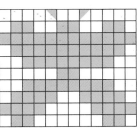

Two squares are 4 units on each side. One square is drawn on 0.5-cm grid paper and the other is on 1-cm grid paper.

3. What is the length of one side of the square on the 0.5-cm grid paper?
   A 20 cm   **C** 2 cm
   B 10 cm   **D** 0.02 cm

4. How many squares on the 0.5-cm grid will fit inside a square on the 1-cm grid without overlapping?
   F 2   **H** 6
   G 4   **J** 8

Problem Solving Strategy

Mixed Strategy Practice

Solve.

5. Lizette had $10.25. She bought 2 play tickets for $3.75 each. How much does Lizette have left?

6. The sum of two numbers is 23. Their difference is 5. What are the numbers?

7. At a concession stand, Luisa bought a drink for $0.75, a hot dog for $1.25, and 3 bags of peanuts for $0.30 each. She received $2.10 in change. How much money did she start with?

8. Jack's band had 56 tickets for a concert. Each of the 7 members received the same number of tickets. How many tickets did each member receive?

Review/Test

✓ CHECK VOCABULARY AND CONCEPTS

Choose the best term from the box.

| congruent |
| angle |
| line |
| similar |

1. Figures that have the same shape but may have different sizes are _?_. (p. 327)

2. Two rays with the same endpoint form an _?_. (p. 321)

3. Figures that have the same size and shape are _?_. (p. 326)

For 4–5, use the drawing at the right.

4. Name two parallel lines. (p. 324)

5. Name an acute angle. (p. 321)

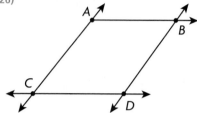

✓ CHECK SKILLS

Name the term that describes each. Write *point, plane, line, line segment,* or *ray*. (pp. 320–323)

6.

7. A

8. •———→

9. •———————•

10.

Tell how each figure was moved. Write *slide, flip,* or *turn*. (pp. 326–329)

11.

12.

Tell whether the two figures in each pair are *congruent, similar,* or *neither*. (pp. 326–329)

13.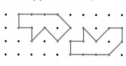

14.

Tell whether each figure has *rotational symmetry, line symmetry,* or *both*. (pp. 330–331)

15.

16.

17.

18.

✓ CHECK PROBLEM SOLVING

Solve. (pp. 332–333)

19. Fred has a map of Kansas in his textbook. There is also a map of Kansas in a dictionary. Are the maps similar? Why or why not?

20. Lucy chose a drawing on 1-cm grid paper to reduce in size. Should she choose 0.5-cm or 2-cm grid paper?

⭐Standardized Test Prep

Eliminate choices.
See item **5.**

Hold your left hand in the same position as the drawing. Move your hand in the same way the figure was moved over the dotted line. Look for the term that describes this motion.

Also see problem 5, p. H64.

For 1–8, choose the best answer.

1. What kind of angle is ∠*LMN*?

   A acute   **C** obtuse
   B right   **D** ray

2. Which expression has a value of 12?
   F $25 - (10 - 3)$ **H** $25 - (10 + 3)$
   G $25 + (10 - 3)$ **J** $25 + (10 + 3)$

3. What is the length of one side of the shaded square?

   A 16 cm
   B 8 cm
   C 4 cm
   D 2 cm

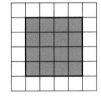

   □ = 0.5 cm

4. 6×42
   F 212   **H** 252
   G 242   **J** 282

5. What term best describes the motion shown by the figure?

   A turn   **C** flip
   B slide   **D** NOT HERE

6. Which geometric term best describes the corner of the picture frame?

   F plane   **H** ray
   G right angle **J** point

7. 4×9
   A 32   **C** 36
   B 35   **D** NOT HERE

8. Which figure is similar to this figure?

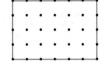

   F   **H**

   G   **J**

Write What You Know

9. Describe any line symmetry you see in the 5 letters below.

   A   **E**   **O**   **P**   **S**

10. Explain the following statement. "Perpendicular lines must intersect, but intersecting lines may not be perpendicular."

Measure and Classify Plane Figures

The Ferris wheel was the sensation of the 1893 World's Fair in Chicago, Illinois. At the time, the wheel was taller than any existing building. The table shows the diameters of some popular amusement wheels in the United States.

PROBLEM SOLVING How do you think the diameter and circumference of each wheel are related? Use this relationship to estimate the unknown circumferences in the table.

SIZES OF AMUSEMENT WHEELS	
Diameter	Circumference (rounded to the nearest foot)
35 ft	110 ft
40 ft	126 ft
50 ft	157 ft
60 ft	188 ft
88 ft	■
105 ft	■
164 ft	■
203 ft	■
250 ft	■

Use this page to help you review and remember
important skills needed for Chapter 18.

✓ VOCABULARY

Choose the best term from the box.

		acute
		obtuse
		right
		ray
		angle

1. An angle with a measure less than a right angle is _?_.

2. An angle with a measure greater than a right angle is _?_.

3. When two rays have the same endpoint, they form an _?_.

4. A part of a line that has one endpoint is a _?_.

✓ CLASSIFY ANGLES (For Intervention, see p. H30.)

Tell if each angle is a *right* angle, *greater than* a right angle,
or *less than* a right angle.

5. **6.** **7.**

8. 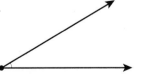 **9.** **10.**

Tell if each angle is *acute, right,* or *obtuse*.

11. 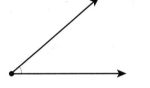 **12.**  **13.**

14.  **15.**  **16.**

HANDS ON

Turns and Degrees

▶ **Explore**

The unit used to measure an angle is a **degree (°)**.

Turning ray $\overrightarrow{CD}$ around the circle makes angles of different sizes. A complete turn around the circle is 360°.

VOCABULARY

degree (°)

MATERIALS 2 strips of paper, paper fastener

Activity Use turns of geostrips to show different angles.

STEP 1	STEP 2	STEP 3
Open the geostrip to form a 90° angle.	Now open the geostrip $\frac{1}{4}$ turn more to make a 180° angle.	Open your geostrip another $\frac{1}{4}$ turn to make a 270° angle.

This is a $\frac{1}{4}$ turn around a circle.	This is a $\frac{1}{2}$ turn around a circle.	This is a $\frac{3}{4}$ turn around a circle.

⚡ **MATH IDEA** Angle measures can be related to a complete turn (360°), a $\frac{3}{4}$ turn (270°), a $\frac{1}{2}$ turn (180°), and a $\frac{1}{4}$ turn (90°).

Try It

Tell whether the rays on the circle show a $\frac{1}{4}$, $\frac{1}{2}$, or $\frac{3}{4}$ turn.

a.

b.

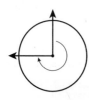

What kind of turn should I make with my geostrips?

▶ Connect

Relate turns and angles measured in degrees to the hands of a clock. Let the hands of a clock represent rays. There are 6° from one minute mark on a clock to the next.

15 minute marks
$6° \times 15 = 90°$

30 minute marks
$6° \times 30 = 180°$

45 minute marks
$6° \times 45 = 270°$

▶ Practice and Problem Solving

Tell whether the rays on the circle show a $\frac{1}{4}$, $\frac{1}{2}$, $\frac{3}{4}$, or full turn.

1.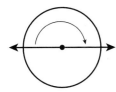

2.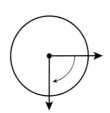

3.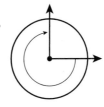

4.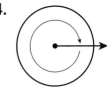

Tell whether the figure has been turned 90°, 180°, 270°, or 360°.

5.

6.

7.

8. REASONING How many 30° angles can be drawn in a circle without overlapping any? Explain.

9. ❓ **What's the Question?** The minute hand on a clock moves from 12 to 4. The answer is 120°.

Mixed Review and Test Prep

10. $1,342 + 908$ (p. 40)

11. $6,428 - 729$ (p. 40)

12. What is 32,704 written in expanded form? (p. 4)

13. What is 48 divided by 2? (p. 248)

14. TEST PREP Which letter has line symmetry? (p. 330)

  A N   **B** M   **C** F   **D** P

HANDS ON Measure Angles

▶ **Explore**

Use a **protractor** to measure the size of the opening of an angle. The scale on a protractor is marked from 0° to 180°.

Activity

Use a protractor to measure angle *ABC*.

STEP 1

Place the center of the protractor on the vertex of the angle.

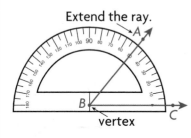

Extend the ray.

vertex

STEP 2

Line up the center point and the 0° mark on protractor with one ray of the angle.

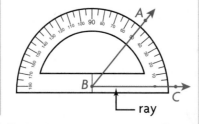

ray

STEP 3

Read the angle measure where the ray passes through the scale.

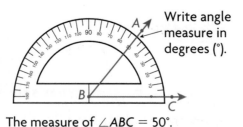

Write angle measure in degrees (°).

The measure of ∠*ABC* = 50°.

VOCABULARY

protractor

MATERIALS protractor, straightedge

Remember

The vertex of an angle is the point where the two rays meet.

Try It

Trace each figure. Use a protractor to measure the angle.

a.

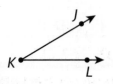

b.

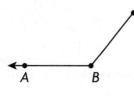

I have the protractor in place. Now, how can I read the measure of the angle?

340

▶ Connect

You can measure angles that are drawn in different positions. Remember to line up the center mark and the 0° mark on the protractor with one of the rays.

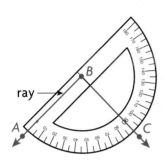

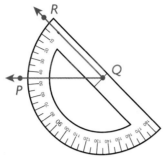

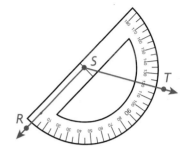

The measure of ∠ABC is 90°.

The measure of ∠PQR is 45°.

The measure of ∠RST is 120°.

MATH IDEA Use a protractor to measure the degrees of an angle by lining up the center mark and the 0° mark with one of the rays.

▶ Practice and Problem Solving

Trace each angle. Then use a protractor to measure the angle.

1.

2.

3.

4.

5. The limbs of a tree form angles with the trunk of the tree. Classify Angles A and B in this tree as acute, obtuse, or right.

6. The measure of ∠XYZ is an odd number. The sum of the two digits is 12. The tens digit is 2 greater than the ones digit. What is the measure of ∠XYZ?

Mixed Review and Test Prep

7. 900 ÷ 30
(p. 282)

8. 1,400 ÷ 200
(p. 282)

9. Write 205,020 in word form. (p. 6)

10. Find the value.
(16 × 12) ÷ 8 (p. 158)

11. **TEST PREP** It takes Pam 40 minutes to walk home from school. If she left school at 11:45 A.M., what time would she get home? (p. 118)

A 12:10 P.M.   C 12:25 P.M.

B 12:20 P.M.   D 12:45 P.M.

Circles

▶ **Learn**

ROUND AND ROUND   A **circle** is a closed figure made up of points that are the same distance from the **center**.

center

Circle *P*

VOCABULARY

circle	diameter
center	radius
chord	compass

Other parts of a circle:

A **chord** is a line segment that has its endpoints on the circle.

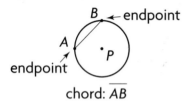

B ← endpoint

A

endpoint

chord: $\overline{AB}$

A **diameter** is a chord that passes through the center.

C   *D*

P

diameter: $\overline{CD}$

A **radius** is a line segment that connects the center of the circle with a point on the circle.

P *K*

radius: $\overline{PK}$

The radius of a circle is half the length of the diameter.

A **compass** is a tool used to construct circles.

HANDS ON

Activity

MATERIALS: compass, ruler

STEP 1

Draw a point to be the center of the circle. Label it with the letter *P*.

STEP 2

Set the compass to the length of the radius you want.

STEP 3

Hold the compass point at point *P*, and move the compass to make the circle.

▶ Check

1. **Explain** how you can find the length of the diameter of a circle that has a radius of 3 inches.

Draw circle *P* with a 2-inch radius. Label each of the following.

2. radius: $\overline{PQ}$ 3. chord: $\overline{JK}$ 4. diameter: $\overline{XY}$

Technology Link

More Practice: Use E-Lab, *Exploring Circles*

www.harcourtschool.com/ elab2002

▶ Practice and Problem Solving

For 5–10, use the drawing and a centimeter ruler. Copy and complete the table.

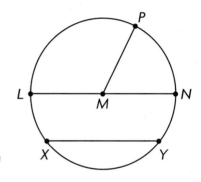

NAME	PART OF CIRCLE	LENGTH IN CM
5. $\overline{MP}$	?	⬛
6. $\overline{XY}$	?	⬛
7. $\overline{MN}$	?	⬛
8. $\overline{LN}$	?	⬛

9. The center of the circle is point _?_.

10. Two points on the circle are _?_ and _?_.

11. Draw a circle. Label the center point *E*. Draw a radius: $\overline{EF}$. Draw a diameter: $\overline{GH}$.

USE DATA For 12–13, use the circle at the right.

12. Name each radius and diameter of the circle shown. Name their lengths in centimeters.

13. ✎ **Write About It** Describe any angles formed by each radius and the diameter.

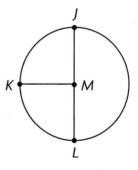

14. **REASONING** Can the length of a chord be greater than the circle's diameter? Explain.

Mixed Review and Test Prep

15. 7×9 (p. 144)   16. 8×8 (p. 144)

17. Find the value. $24 \times (72 \div 9)$ (p. 158)

18. Find the mean of 2, 4, 7, 2, 10. (p. 276)

19. **TEST PREP** Which of the following is true? (p. 18)

A $45{,}126 < 45{,}216$

B $45{,}261 < 45{,}216$

C $45{,}126 > 45{,}621$

D $45{,}216 > 45{,}612$

HANDS ON Circumference

▶ **Explore**

The **circumference** of a circle is the measure of the distance around the circle.

Activity

Use a ruler and a string to find the circumference and diameter of a circular object such as a lid.

STEP 1

Wrap string around a circular object.

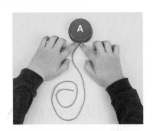

STEP 2

Measure the string in centimeters. The length of the string is the circumference of the circle.

STEP 3

Trace your object. Draw the diameter and then measure it.

Record the circumference and the diameter.

Try It

Use a string and a metric ruler to find the circumference of each object to the nearest centimeter.

a. can   **b.** jar

I have wrapped the string around the cylinder. What do I do next to find the circumference?

▶ Connect

Look at the data for each circle in the table. About how many times as great as the diameter is each circle's circumference?

The circumference of a circle is about 3 times the diameter of the circle. If you know the diameter of a circle, you can estimate its circumference.

Estimate the circumference of a circle with a diameter of 4 cm.

Think: $4 \times 3 = 12$

So, the circumference is about 12 cm.

CIRCLE	DIAMETER	ESTIMATED CIRCUMFERENCE
A	2 cm	about 6 cm
B	3 cm	about 9 cm
C	4 cm	about 12 cm
D	5 cm	about 15 cm

▶ Practice and Problem Solving

Estimate each circumference.

1.

10 cm

2.

12 ft

3.

10 cm

4. Diameter: 38 feet

5. Diameter: 19 miles

6. Diameter: 42 meters

USE DATA For 7–8, use the graph.

7. How much shorter is the diameter of the Mid-size wheel than the diameter of the Ultra wheel?

8. Estimate the circumference of the Standard wheel.

9. **? What's the Error?** The diameter of a circle that Nina measured is 18 inches. Nina estimated the circumference to be about 6 inches. Describe and correct her error.

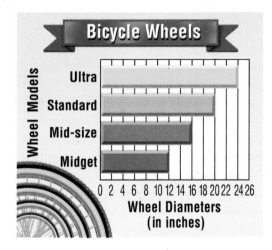

Mixed Review and Test Prep

10. $46 \div 3$ (p. 242) **11.** $79 \div 5$ (p. 242)

12. Write five hundred thousand, thirty in standard form. (p. 6)

13. Write five thousand, two hundred thirty-four in expanded form. (p. 6)

14. **TEST PREP** Which term best describes the line relationship of the hands on the clock? (p. 324)

  A perpendicular
  B parallel
  C obtuse
  D intersecting

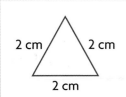 Classify Triangles

▶ **Learn**

TAKE SIDES Triangles can be classified according to the lengths of their sides.

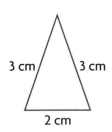 A triangle with 3 sides congruent is an **equilateral triangle** .

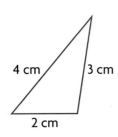 A triangle with only 2 sides congruent is an **isosceles triangle** .

A triangle with no sides congruent is a **scalene triangle** .

Remember

A triangle has three sides and three angles.

side   angle

• What does it mean for sides to be congruent?

▶ **Check**

1. Describe the difference between an equilateral, an isosceles, and a scalene triangle.

Classify each triangle as *isosceles, scalene,* **or** *equilateral.*

2.
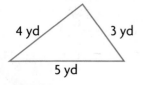
4 yd   3 yd   5 yd

3.
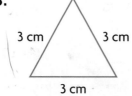
3 cm   3 cm   3 cm

4.
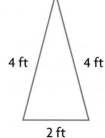
4 ft   4 ft   2 ft

346

▶ Practice and Problem Solving

Classify each triangle. Write *isosceles, scalene,* or *equilateral.*

5.
3 ft
7 ft
9 ft

6.
5 yd
5 yd
5 yd

7.
5 in.
5 in.
4 in.

8.
4 ft
3 ft
2 ft

9.

10.

11.

12.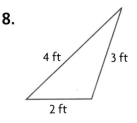

Classify each triangle by the lengths of its sides.
Write *isosceles, scalene,* or *equilateral.*

13. 12 ft, 12 ft, 12 ft

14. 9 cm, 7 cm, 4 cm

15. 13 mm, 13 mm, 8 mm

16. 6 in., 14 in., 14 in.

17. 43 mm, 43 mm, 43 mm

18. 29 yd, 28 yd, 6 yd

Measure the sides of each triangle using a centimeter ruler. Write *isosceles, scalene,* or *equilateral.*

19.

20.

21.

22. I have 3 sides and 3 angles. Only two of my angles are acute. What kinds of triangles could I be?

23. **Write a problem** about a mystery triangle. Give at least three clues that will help identify the triangle.

Mixed Review and Test Prep

24. If 5 shelves hold 60 boxes, how many shelves hold 144 boxes? (p. 222)

25. In June, Irma rode her bicycle 8 miles a day for 15 days. How far did Irma ride her bicycle? (p. 196)

26. $7\overline{)362}$ (p. 266) **27.** $5\overline{)235}$ (p. 266)

28. **TEST PREP** What is the median of this set of numbers? (p. 86)
12, 14, 14, 9, 14, 8, 12
A 14 **B** 12 **C** 8 **D** 4

EXTRA PRACTICE page H49, Set B

6 Classify Quadrilaterals

▶ **Learn**

CLASSIC LINES A figure with 4 sides and 4 angles is called a quadrilateral. There are many kinds of quadrilaterals. They can be classified by their features.

 Activity

- Look at the figures below. What do they all have in common?

- Which have 2 pairs of parallel sides?

- Which have 4 right angles?

- Which have both pairs of opposite sides congruent?

- Which have only 1 pair of parallel sides?

▲ The stained glass lamp shade made by Tiffany Studios reflects light through colored pieces of glass, some of which are shaped like quadrilaterals.

Special Quadrilaterals

There are five special types of quadrilaterals: **parallelogram**, square, rectangle, **rhombus**, and **trapezoid**. Each has different features, and some can be classified in more than one way. Use the diagram to help you identify each type of quadrilateral.

The diagram shows that all rectangles are parallelograms and quadrilaterals.

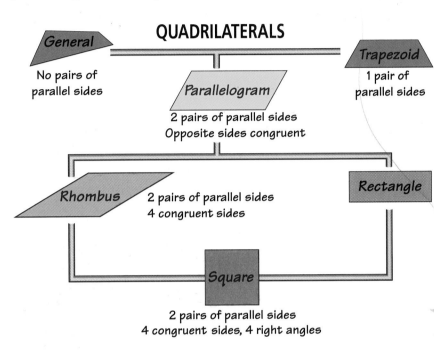

QUADRILATERALS

General
No pairs of parallel sides

Trapezoid
1 pair of parallel sides

Parallelogram
2 pairs of parallel sides
Opposite sides congruent

Rhombus
2 pairs of parallel sides
4 congruent sides

Rectangle
2 pairs of parallel sides
Opposite sides congruent
4 right angles

Square
2 pairs of parallel sides
4 congruent sides, 4 right angles

Technology Link

More Practice: Use Mighty Math Number Heroes, *Geoboard*, Levels P and Q.

- A quadrilateral has 4 congruent sides. What figures could it be?

 **Check**

1. **Compare and contrast** a trapezoid and a parallelogram.

Classify each figure in as many ways as possible. Write *quadrilateral, parallelogram, rhombus, rectangle, square,* **or** *trapezoid.*

2.

3.

4.

Draw an example of each quadrilateral.

5. It has 4 congruent sides and no right angles.

6. Its opposite sides are parallel and it has 4 right angles.

7. It is a trapezoid with 2 congruent sides.

LESSON CONTINUES ▶

Classify each figure in as many ways as possible. Write *quadrilateral, parallelogram, rhombus, rectangle, square,* or *trapezoid.*

8.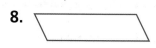

9.

10.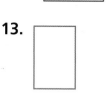

11.

12.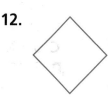

13.

14.

15.

16.

Draw an example of each quadrilateral.

17. It has two pairs of parallel sides and the opposite sides are congruent.

18. The four sides are congruent and there are four right angles.

19. It has one pair of parallel sides.

Choose the figure that does not belong.

20. a.   **b.**    **c.**   **d.**

21. a.    **b.**    **c.**    **d.**

22. REASONING Is a square also a rhombus? Explain how you know.

23. I have four sides and four angles. At least one of my angles is acute. What figures could I be?

24. Draw a trapezoid. What figure would you make if you extended the two non-parallel sides from the endpoints of the shorter side?

25. Write a problem about a mystery quadrilateral. Give at least three clues that will help identify the quadrilateral.

26. At sunrise the temperature was 72°F. At noon it was 19° warmer. It cooled off 12° in the evening. What was the evening temperature?

27. I have 600 rectangles and trapezoids. I have 80 more trapezoids than rectangles. How many of each do I have?

Mixed Review and Test Prep

28. 3×184 (p. 196)

29. $\$7,426 + \$2,915$ (p. 40)

30. $\$5,003 - \$1,879$ (p. 42)

31. ■ $\times 8 = 64$ (p. 144)

32. 7×48 (p. 190)

33. $804 \div ■ = 67$ (p. 286)

34. In September the Reading Club members sold 1,032 movie passes. In October they sold 2,940 passes. In November they sold 125 more than in September. How many passes were sold in all? (p. 40)

35. **TEST PREP** Find the time shown on the clock. (p. 116)
   - **A** 4:08
   - **B** 4:40
   - **C** 8:20
   - **D** 8:40

36. **TEST PREP** There are a total of 42 students on 6 teams. Each team has the same number of students. Which equation can be used to find *n,* the number of students on each team? (p. 170)
   - **F** $n \times 6 = 42$
   - **G** $n \div 6 = 42$
   - **H** $6 \times 42 = n$
   - **J** $n \times 42 = 6$

PROBLEM SOLVING LiNKUP ... to Reading

STRATEGY · CLASSIFY AND CATEGORIZE
When you *classify* information, you group similar information. When you *categorize*, you name the groups that you have classified.

The diagram on page 349 classifies and categorizes information about quadrilaterals.

For 1–6, use the diagram. Tell if the statement is *true* or *false*. If the statement is false, explain why.

1. All rectangles are squares.

3. All squares are rectangles.

5. Some rhombuses are squares.

2. All rhombuses are parallelograms.

4. Some trapezoids are parallelograms.

6. No trapezoids have 2 pairs of parallel sides.

Problem Solving Strategy
Draw a Diagram

UNDERSTAND ⟩ PLAN ⟩ SOLVE ⟩ CHECK

Quick Review

1. 50 − 10

2. 50 − 1

3. 500 − 100

4. 500 − 1

5. 5,000 − 1,001

PROBLEM   Mrs. Stein asked her students to sort these figures into two groups according to the lengths of the sides of each figure. How can the figures be sorted?

A

B

C

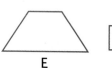
D

E

F

UNDERSTAND

- What are you asked to do?
- What information will you use?
- Is there any information you will not use? Explain.

PLAN

- What strategy can you use to solve the problem?

  Draw a diagram to solve the problem.

SOLVE

- What kind of diagram can you make?

  Make a Venn diagram showing 2 separate circles to sort the figures.

  Label one circle *With 4 Congruent Sides*. Put figures B and D in this circle. Label the other circle *Without 4 Congruent Sides*. Put figures A, C, E, and F in this circle.

Quadrilaterals
With 4 Congruent Sides

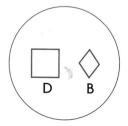
D   B

Without 4 Congruent Sides

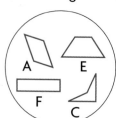
A   E
F   C

CHECK

- How do you know that the answer is correct?

Problem Solving Practice

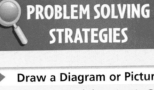

PROBLEM SOLVING STRATEGIES

▶ Draw a Diagram or Picture
  Make a Model or Act It Out
  Make an Organized List
  Find a Pattern
  Make a Table or Graph
  Predict and Test
  Work Backward
  Solve a Simpler Problem
  Write an Equation
  Use Logical Reasoning

Use *draw a diagram* to solve.

1. **What if** Mrs. Stein had asked the students to sort figures *A–F* into two groups, one with 2 pairs of parallel sides and one with fewer than 2 pairs of parallel sides? Draw a diagram that shows those groupings.

2. Sort these numbers into a Venn diagram showing *Divisible by 2* and *Not Divisible by 2:* 2, 3, 4, 6, 8, 9, 10, 12

Look at the Venn diagram at the right.

3. Which label best describes the figures in Circle A?
   A have 3 sides **C** have 5 sides
   B have 4 sides **D** have 6 sides

4. Which label best describes the figures in Circle B?
   F have 3 sides **H** have 5 sides
   G have 4 sides **J** have 6 sides

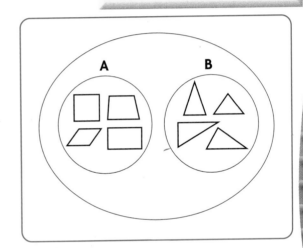

Mixed Strategy Practice

For 5–6, use the stem-and-leaf plot.

5. How many of Jamie's friends were measured?

6. What is the mode?

7. Draw a Venn diagram to sort figures *A–F* on page 352 into these groups: *With Only Right Angles* and *With Other Kinds of Angles*.

9. A wheel has a circumference of 12 inches. It rolled 144 inches. How many complete turns did the wheel make?

Heights of Jamie's Friends (in inches)	
Stem	Leaves
5	7 8 9
6	0 1 1 2 4 5

8. ✎ **Write About It** Describe the rule for this pattern: 1, 2, 4, 8, 16, 32, 64.

10. **REASONING** What is the length of one side of the smallest square that a plate with a 12-inch diameter will fit into?

Problem Solving Strategy

Review/Test

✓ CHECK VOCABULARY AND CONCEPTS

Choose the best term from the box.

1. The unit used to measure an angle is called a _?_. (p. 338)

2. A triangle that has 3 congruent sides is _?_. (p. 346)

> compass
> degree
> equilateral

✓ CHECK SKILLS

Trace each angle. Then use a protractor to measure the angle. (pp. 340–341)

3.
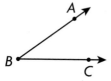

4.
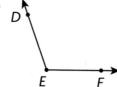

5.
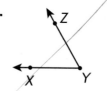

For 6–7, use the drawing. (pp. 342–343)

6. Name a chord.

7. Name a radius.

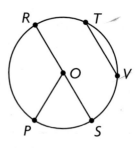

Classify each triangle by the lengths of its sides. Write *isosceles, scalene,* or *equilateral.* (pp. 346–347)

8. 17 ft, 22 ft, 16 ft

9. 5 cm, 5 cm, 5 cm

10. 11 mm, 11 mm, 8 mm

Classify each figure in as many ways as possible. Write *quadrilateral, parallelogram, rhombus, rectangle, square,* or *trapezoid.* (pp. 348–351)

11.

12.

13.

✓ CHECK PROBLEM SOLVING (pp. 352–353)

14. Sort these numbers into a Venn diagram showing *Divisible by 5* and *Not Divisible by 5*: 1, 5, 10, 12, 16, 20, 24.

15. Draw a Venn diagram to sort parallelogram, rhombus, rectangle, and square into two groups: *With Only 2 Congruent Sides* and *With 4 Congruent Sides.*

⭐Standardized Test Prep

Choose the answer.
See item **7**.

Think about the relationship of the diameter of a circle to its circumference. Read the answer choices and relate them to the problem one by one.

Also see problem **6**, p. H64.

For 1–8, choose the best answer.

1. What time does the clock show?

  A 1:40
  B 2:20
  C 4:05
  D 4:10

2. What is the most reasonable measure of the angle formed by the hands of a clock showing 2:00?

  F 10°   **H** 60°
  G 30°   **J** 90°

3. What point represents the center of the circle?

  A *A*   **B** *B*   **C** *C*   **D** *D*

4. Which term best describes the numbers in circle *X*?

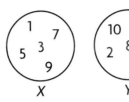

  F even   **H** composite
  G multiples of 3 **J** odd

5. Joe drew an angle that was 4 times the size of a 40° angle. What was the measure of the angle Joe drew?

  A 44°   **C** 180°
  B 160°   **D** 200°

6. What is the value of *t*?

$t \times 15 = 45$

  F $t = 30$ **H** $t = 5$
  G $t = 13$ **J** $t = 3$

7. What would be the measure of the circumference of a circle with a diameter of 5 centimeters?

  A about 6 cm   **C** about 15 cm
  B about 8 cm   **D** about 18 cm

8. $424 \div 4$

  F 12   **H** 106
  G 101   **J** NOT HERE

Write What You Know

9. Point *O* is the center of the circle shown. Points *A* and *B* lie on the circle. Explain why triangle *AOB* is an isosceles triangle.

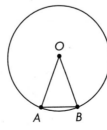

10. Can a figure be both a trapezoid and a parallelogram? Draw each figure and explain.

Paw Prints

Jen's dog, Chester has several play areas in the shape of quadrilaterals. One of these quadrilaterals is Chester's favorite play area. Use the diagram on page 349 and Chester's paw clues below to find out the shape of the play area. Write the name of the quadrilaterals identified in each clue.

Clue 1:

Chester's play area has 4 sides and 4 angles.

Clue 2:

Chester's play area has 2 pairs of parallel sides.

Clue 3:

Chester's play area only has opposite sides that are congruent.

Clue 4:

Chester's play area has 4 right angles.

Think It Over!

- Write About It Explain how you can change Clue 4 to identify a different quadrilateral.

- STRETCH YOUR THINKING Write clues that you can give someone to identify a quadrilateral that you choose.

CASE CLOSED

Challenge

Tessellations

You can arrange polygons so that they completely cover a surface without leaving any gaps between them or overlapping. This arrangement is called a **tessellation**. Some polygons will tessellate.

Will a hexagon tessellate? Make a drawing to solve.

▲ The figures in this tile tessellate.

STEP 1

Cut out a hexagon. Trace it once on a sheet of paper.

STEP 2

Trace the hexagon again in another position so that at least one of its sides touches the first hexagon. Make sure there are no overlaps.

STEP 3

Keep tracing the hexagon until you make a design that covers a surface without gaps or overlaps.

So, the hexagon will tessellate.

Use one shape or more than one shape to make tessellations.

Examples

Ⓐ Tell whether the shapes will form a tessellation.

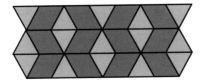

These shapes tessellate.

Ⓑ Tell whether the shape , will form a tessellation.

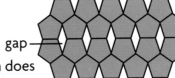

gap ⎯

The pentagon does not fit into this gap.

The pentagon does not tessellate.

Try It

Will these figures tessellate? Tell how you know.

1.
2.
3.

Study Guide and Review

VOCABULARY

Choose a word from the list to complete each sentence.

1. A figure has ? if it can be turned about a point and still look the same in at least two different positions. (page 330)

2. An ? angle is an angle with a measure less than a right angle. (page 321)

3. You can use a ? to measure the size of the opening of an angle. (p. 340)

> acute
> obtuse
> protractor
> rotational symmetry

STUDY AND SOLVE

Chapter 17

Identify lines, rays, angles and line relationships.

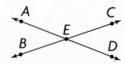

Name 2 acute angles in the figure above.

Both ∠AEB and ∠CED measure less than a right angle.

So, ∠AEB and ∠CED are acute angles.

For 4–7, use the drawing at the left. (pp. 320–323)

4. Name 4 line segments.

5. Name 2 lines.

6. Name a ray.

7. Name 2 obtuse angles.

Chapter 18

Identify congruent and similar figures.

Tell whether the two figures are *congruent, similar,* or *neither.*

Congruent figures have the same size and shape. **Similar figures** have the same shape but may have different sizes.

Since the figures have the same shape, but not the same size, they are similar.

Tell whether the two figures are *congruent, similar,* or *neither.* (pp. 326–329)

8.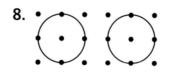

9.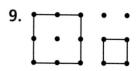

10.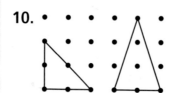

11.

Chapter 18

Identify the parts of a circle and estimate circumference.

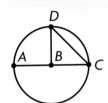

- Point B is the center of the circle.
- Line segment DC is a chord.
- Line segment AC is a diameter.
- Line segment AB is a radius.

For 12–14, use the drawing. (pp. 342–343)

12. The center of the circle is point _?_.

13. The line segment $\overline{LM}$ is a _?_.

14. The diameter $\overline{HK}$ is 10 centimeters long. Name a radius. How long is it?

Classify triangles.

Classify each triangle as *isosceles*, *scalene*, or *equilateral*.

A

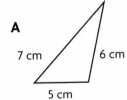

7 cm 6 cm 5 cm

B

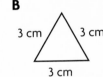

3 cm 3 cm 3 cm

C

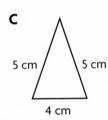

5 cm 5 cm 4 cm

Triangle A is scalene, triangle B is equilateral, and triangle C is isosceles.

Classify each triangle as *isosceles*, *scalene*, or *equilateral*. (pp. 346-347)

15.

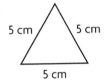

5 cm 5 cm 5 cm

16.

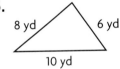

8 yd 6 yd 10 yd

17.

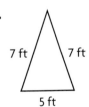

7 ft 7 ft 5 ft

PROBLEM SOLVING PRACTICE

Solve. (pp. 332–333, 352–353)

18. Tracy drew a square with sides of 6 units on 1-cm grid paper. How long would each side of the square be if Tracy used 0.5-grid paper?

19. Sort the numbers 2–20 into a diagram showing *Prime* and *Composite*.

PERFORMANCE ASSESSMENT

TASK A • PICTURE THIS

The Rodriquez family visited the Museum of Modern Art in New York City. They saw plane figures in many of the paintings and drawings.

a. Draw a picture that includes the plane figures in List A.

b. Write a description of your picture using the terms in List B.

c. Tell if your picture has line symmetry or rotational symmetry.

List A
right angle
acute angle
obtuse angle
intersecting lines
parallel lines
perpendicular lines

List B
line segment
ray
plane
vertex
congruent

TASK B • WHAT AM I?

Jamie found this riddle in a book.
    I have 4 congruent sides.
    All of my angles are right angles.
    What am I?
    I am a square.

He wants to have a collection of riddles like this.

a. Write three different riddles for Jamie. Make the answer to each riddle the name of a plane figure.

b. Tell if there is more than one answer to any of your riddles. Explain.

Technology Linkup

Mighty Math Number Heroes • Symmetry

Click on *Geoboard*.

Click 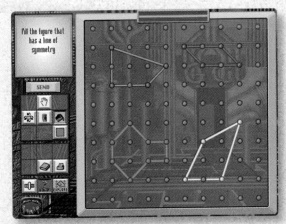. Choose Level M.

- Answer at least 5 questions.
- Draw pictures to record your work.

Practice and Problem Solving

Click **EXPLORE**. Use the Geoboard to make each figure.

Draw a picture of each figure and its line or lines of symmetry.

1. Copy the figure below on the Geoboard. Draw its line of symmetry.

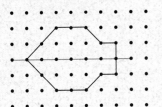

2. Copy the figure below on the Geoboard. Draw 2 lines of symmetry.

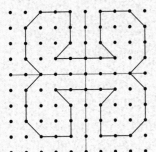

3. Make a figure with no lines of symmetry and a figure with one line of symmetry.

4. Write a problem about a figure with two lines of symmetry. Use the Geoboard. Draw a picture to show your work.

Multimedia Math Glossary www.harcourtschool.com/mathglossary

5. Vocabulary Use the Multimedia Math Glossary to write a paragraph comparing and contrasting a *rhombus,* a *parallelogram,* and a *trapezoid.*

The Ferris wheel at Navy Pier is 150 feet tall and holds up to 240 people.

PROBLEM SOLVING ON LOCATION

in

Chicago

NAVY PIER

Navy Pier is a popular tourist destination in Chicago. It offers many attractions including rides, museums, theaters, and shopping. Built in 1916, Navy Pier was originally a commercial-shipping pier.

1. The screen in a theater at Navy Pier is six stories tall and measures 60 feet by 80 feet. Erika says the screen is shaped like a rectangle. Ian says the screen is shaped like a parallelogram. Who is correct? Explain.

2. At the Smith Museum of Stained Glass Windows, there is a window made up of equilateral, isosceles, and scalene triangles. Draw a picture of what the window might look like. Label each type of triangle.

3. The Ferris wheel at Navy Pier has a diameter of 140 feet. About how long is the circumference? Explain.

4. The diameter of the carousel at Navy Pier is 44 feet. How much greater is the diameter of the Navy Pier Ferris wheel than the carousel?

OTHER CHICAGO SIGHTS

Navy Pier is not the only interesting place to visit in Chicago. Chicago also has an aquarium, a planetarium, and many museums and unique buildings.

1. What line relationships can you identify in the photograph of the John Hancock Center? What types of angles do you see? What types of triangles can you find? Draw a diagram with labels to show your answer.

2. Trace one of the triangles shown on the photograph of the John Hancock Center.
   a. Draw a triangle that is congruent to the triangle you traced.
   b. Make a drawing in which you slide, flip, and turn the triangle. Label each move.

3. Suppose the frame for a painting in the Museum of Contemporary Art is described as having 4 congruent sides, 2 pairs of parallel sides, and no right angles.
   a. Draw a picture to show the shape of the frame.
   b. Write a name for the quadrilateral you drew.

4. The diameter of the circular viewing area at the domed Sky Theater at Adler Planetarium is 68 feet.
   a. What is the radius of the Sky Theater?
   b. About how many feet is it around the edge of the viewing area?

John Hancock Center is one of the 10 tallest buildings in the world. ▶

Understand Fractions

Each golden retriever puppy in the photograph weighs about 9 pounds.

PROBLEM SOLVING
Write a fraction that shows the weight of one puppy compared to the weight of all the puppies. The total weight of all the puppies is $\frac{1}{2}$ the weight of the adult dog. What is the weight of the adult dog?

DATA LINK

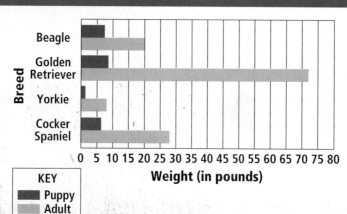

AVERAGE WEIGHTS OF DOGS

Breed

Beagle
Golden Retriever
Yorkie
Cocker Spaniel

0 5 10 15 20 25 30 35 40 45 50 55 60 65 70 75 80
Weight (in pounds)

KEY
Puppy
Adult

Use this page to help you review and remember important skills needed for Chapter 19.

✓ VOCABULARY

Choose the best term from the box.

1. The 4 in $\frac{3}{4}$ is called the __?__ .

2. The 3 in $\frac{3}{4}$ is called the __?__ .

> fraction
> numerator
> denominator

✓ PARTS OF A WHOLE (For Intervention, see p. H18.)

Choose the word to name the equal parts in each whole.

> halves   thirds   fourths   sixths

3.

4.

5.

6.

Write a fraction for each shaded part.

7.

8.

9.

10.

✓ PARTS OF A GROUP (For Intervention, see p. H19.)

Draw a picture for each description.

11. Four out of five stars are shaded.

12. Three out of six circles are shaded.

13. Four out of seven squares are shaded.

14. Two out of four triangles are shaded.

Write a fraction for the shaded part.

15.

16.

17.

18.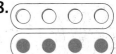

Read and Write Fractions

▶ Learn

BOW WOW BISCUITS Mario wants to cut the dog treats recipe in half. What fraction shows the amount of oats he needs?

A **fraction** is a number that can name a part of a whole. One whole can be divided into 2 equal parts.

each part → $\frac{1}{2}$ ← numerator

total equal parts → ← denominator

Read: one half   **Write:** $\frac{1}{2}$

one divided by two

So, Mario needs $\frac{1}{2}$ cup of oats.

Mario baked 8 treats on one tray. He fed two treats to his puppy. What fraction of the treats were eaten?

A fraction can also name a part of a group.

number eaten → $\frac{2}{8}$ ← numerator

number in the group → ← denominator

Read: two eighths   **Write:** $\frac{2}{8}$

two out of eight

So, $\frac{2}{8}$ of the treats were eaten.

Dog Treats

2 cups flour
6 tablespoons oil
2 eggs, beaten
2 packages yeast
1 teaspoon salt
1 cup oats
1 cup bran
1 cup hot water

Mix ingredients. Spoon onto greased cookie sheet. Bake at 350° for 25 minutes.

▶ Check

1. Tell what fraction of the treats were **not** eaten.

Tell what part is shaded. Tell what part is unshaded.

2. 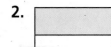   **3.** △ △ △ △   **4.** ☆ ☆ ☆ ☆ ☆ ☆ ☆ ☆ ☆   **5.**

Write a fraction for the shaded part. Write a fraction for the unshaded part.

6.

7.

8.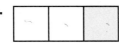

9.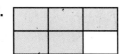

10.

11.

12.

13.

Draw a picture and shade part of it to show the fraction. Write a fraction for the unshaded part.

14. $\frac{3}{5}$ **15.** $\frac{1}{2}$ **16.** $\frac{4}{6}$ **17.** $\frac{6}{8}$ **18.** $\frac{1}{8}$ **19.** $\frac{3}{10}$

20. $\frac{2}{7}$ **21.** $\frac{4}{4}$ **22.** $\frac{2}{3}$ **23.** $\frac{10}{12}$ **24.** $\frac{6}{6}$ **25.** $\frac{1}{5}$

For 26–30, use the figure at the right.
Write a fraction for each part of the figure.

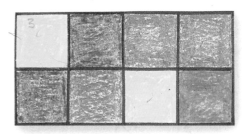

26. green   **27.** red   **28.** yellow

29. not yellow or red

30. red, blue, yellow, or green

31. How much money can Mr. Drew save by buying one large box of dog biscuits for $0.96 rather than the same number of dog biscuits in 2 small boxes for $0.59 each?

Mixed Review and Test Prep

32.   3,456 (p. 40)
    − 2,785

33.   921 (p. 230)
    × 45

34. $4 \times 4 \times 2$ (p. 150)

35. Thomas arrived at the library at 11:45 A.M. He left 1 hour 40 minutes later. At what time did Thomas leave the library? (p. 120)

36. **TEST PREP** Which term best describes a triangle with 2 sides of the same length?
  A equilateral
  B scalene
  C acute
  D isosceles

HANDS ON

Equivalent Fractions

Quick Review

Write the fraction for each.

1. one third 2. five sixths

3. one half 4. four fifths

5. three eighths

▶ **Explore**

Fractions that name the same amount are called **equivalent fractions**.

VOCABULARY

equivalent fractions

MATERIALS fraction bars

Activity

Use fraction bars to model fractions that name $\frac{1}{3}$.

STEP 1

Start with the bar for 1. Then line up the bar for $\frac{1}{3}$.

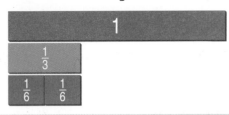

STEP 2

Line up other fraction bars that show the same amount as $\frac{1}{3}$.

Since $\frac{1}{3}$ and $\frac{2}{6}$ name the same amount, they are equivalent fractions.

☀ **MATH IDEA** Fraction bars that are the same length show equivalent fractions.

- **Explain** how you can tell if $\frac{1}{2}$ and $\frac{6}{10}$ are equivalent fractions.

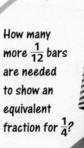

How many more $\frac{1}{12}$ bars are needed to show an equivalent fraction for $\frac{1}{4}$?

Try It

Use models to find an equivalent fraction for each fraction.

a. $\frac{1}{4}$ b. $\frac{3}{4}$ c. $\frac{4}{8}$ d. $\frac{1}{6}$ e. $\frac{8}{10}$

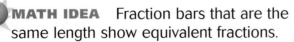

Another way to model equivalent fractions is to use number lines.

Find fractions that are equivalent to $\frac{8}{12}$.

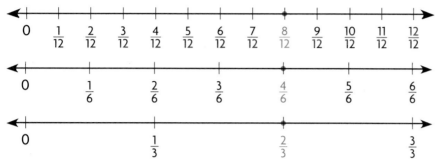

Fractions that line up with $\frac{8}{12}$ are equivalent to $\frac{8}{12}$.

So, $\frac{4}{6}$ and $\frac{2}{3}$ are equivalent to $\frac{8}{12}$.

- What other equivalent fractions do the number lines show? Explain.

Technology Link

More Practice: Use E-Lab, *Equivalent Fractions.*

www.harcourtschool.com/elab2002

▶ **Practice and Problem Solving**

Use fraction bars or number lines to find at least one equivalent fraction for each.

1. $\frac{4}{10}$ 2. $\frac{8}{12}$ 3. $\frac{2}{8}$ 4. $\frac{4}{4}$ 5. $\frac{3}{5}$ 6. $\frac{9}{12}$

7. $\frac{6}{12}$ 8. $\frac{3}{6}$ 9. $\frac{1}{2}$ 10. $\frac{1}{3}$ 11. $\frac{4}{12}$ 12. $\frac{6}{6}$

13. Find a fraction that is equivalent to $\frac{5}{6}$ and has a numerator of 10.

14. Find a fraction that is equivalent to $\frac{6}{12}$ and has a numerator of 3.

15. At the basketball game, every eighth guest will receive a free cap. How many caps will be given away when 123 guests have entered the gym?

16. **REASONING** Write $0.75 as 2 different equivalent fractions of a dollar.

Mixed Review and Test Prep

17. Round 23,714 to the nearest thousand. (p. 26)

18. Find the value of $12 \times (5 - 2)$. (p. 158)

19.   86 (p. 224)
    $\times$ 29

20. $4,806 \div 9 = $ ▧
    (p. 272)

21. **TEST PREP** Which term best describes an angle greater than a right angle?
   A scalene
   B obtuse
   C acute
   D isosceles

Equivalent Fractions

▶ **Learn**

PAMPERED PUP Carrie made 6 dog pillows for a craft fair. Two of the pillows are red. What fraction of the pillows are red?

$\frac{2}{6}$ red pillows
pillows in all

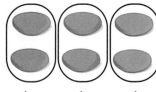

$\frac{1}{3}$ $\frac{1}{3}$ $\frac{1}{3}$

$\frac{1}{3}$ red group
groups in all

$\frac{2}{6}$ and $\frac{1}{3}$ are equivalent fractions. They name the same amount. So, $\frac{1}{3}$ or $\frac{2}{6}$ of the pillows are red.

Example

Find an equivalent fraction for $\frac{5}{6}$.

One Way   Use fraction bars.

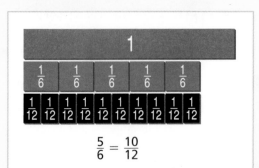

$\frac{5}{6} = \frac{10}{12}$

Another Way   Use multiplication.

Multiply the numerator and denominator by 2.

$\frac{5}{6} = \frac{5 \times 2}{6 \times 2} = \frac{10}{12}$

• What other equivalent fractions do you see in these models?

Multiply or Divide

You can multiply the numerator and the denominator by any number except zero to find equivalent fractions. Sometimes you can divide to find equivalent fractions.

Find equivalent fractions for $\frac{4}{6}$.

One Way Multiply the numerator and denominator by the same number.

> Try 3.
>
> $$\frac{4}{6} = \frac{4 \times 3}{6 \times 3} = \frac{12}{18}$$
>
> So, $\frac{4}{6}$ is equivalent to $\frac{12}{18}$.

Another Way Divide the numerator and denominator by the same number.

> Try 2.
>
> $$\frac{4}{6} = \frac{4 \div 2}{6 \div 2} = \frac{2}{3}$$
>
> So, $\frac{4}{6}$ is equivalent to $\frac{2}{3}$.

If you continue to divide until 1 is the only number that can be divided into the numerator and the denominator evenly, you find the fraction in **simplest form**. So, $\frac{4}{6}$ in simplest form is $\frac{2}{3}$.

Examples

Find the simplest form of $\frac{45}{60}$.

> **Try 5.**   Divide the numerator and denominator by 5.
>
> $$\frac{45}{60} = \frac{45 \div 5}{60 \div 5} = \frac{9}{12}$$
>
> **Next try 3.**   Divide the numerator and denominator by 3.
>
> $$\frac{9}{12} = \frac{9 \div 3}{12 \div 3} = \frac{3}{4}$$

Now the only number that can evenly divide both the numerator and denominator of $\frac{3}{4}$ is 1. So, the simplest form of $\frac{45}{60}$ is $\frac{3}{4}$.

▶ **Check**

1. **Explain** how you can find the simplest form of $\frac{18}{24}$.

Write two equivalent fractions for each picture.

2.

3.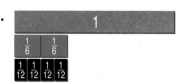

4. ○○○○○
   ○○○○○

LESSON CONTINUES ▶

Write two equivalent fractions for each picture.

5. 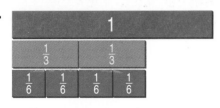   **6.**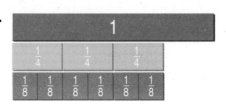

Write two equivalent fractions for each.

7. $\frac{9}{12}$ **8.** $\frac{6}{8}$ **9.** $\frac{2}{10}$ **10.** $\frac{8}{10}$ **11.** $\frac{1}{5}$ **12.** $\frac{4}{4}$

13. $\frac{8}{12}$ **14.** $\frac{4}{10}$ **15.** $\frac{6}{9}$ **16.** $\frac{1}{2}$ **17.** $\frac{2}{4}$ **18.** $\frac{3}{9}$

19. $\frac{5}{6}$ **20.** $\frac{3}{5}$ **21.** $\frac{2}{8}$ **22.** $\frac{7}{12}$ **23.** $\frac{3}{3}$ **24.** $\frac{6}{10}$

Tell whether each fraction is in simplest form. If not, write
it in simplest form.

25. $\frac{2}{4}$ **26.** $\frac{1}{3}$ **27.** $\frac{6}{9}$ **28.** $\frac{3}{4}$ **29.** $\frac{1}{4}$ **30.** $\frac{5}{15}$

31. $\frac{4}{8}$ **32.** $\frac{2}{6}$ **33.** $\frac{6}{8}$ **34.** $\frac{6}{10}$ **35.** $\frac{12}{16}$ **36.** $\frac{2}{3}$

37. $\frac{13}{20}$ **38.** $\frac{9}{21}$ **39.** $\frac{12}{32}$ **40.** $\frac{11}{44}$ **41.** $\frac{10}{25}$ **42.** $\frac{18}{27}$

ALGEBRA   Find the missing numerator or denominator.

43. $\frac{2}{3} = \frac{\blacksquare}{18}$ **44.** $\frac{\blacksquare}{10} = \frac{2}{5}$ **45.** $\frac{6}{8} = \frac{12}{\blacksquare}$ **46.** $\frac{3}{\blacksquare} = \frac{9}{12}$ **47.** $\frac{4}{4} = \frac{\blacksquare}{8}$

48. $\frac{3}{\blacksquare} = \frac{6}{10}$ **49.** $\frac{\blacksquare}{9} = \frac{4}{18}$ **50.** $\frac{3}{7} = \frac{12}{\blacksquare}$ **51.** $\frac{5}{8} = \frac{\blacksquare}{56}$ **52.** $\frac{5}{6} = \frac{20}{\blacksquare}$

53. Find the missing fraction. Explain.

   a. $\frac{1}{2}, \frac{2}{4}, \frac{3}{6}, \frac{4}{8}, \frac{\blacksquare}{\blacksquare}$

   b. $\frac{5}{15}, \frac{4}{12}, \frac{3}{9}, \frac{2}{6}, \frac{\blacksquare}{\blacksquare}$

54. **Write About It** Write a rule that
you can use to tell whether a fraction
is in simplest form.

55. **? What's the Error?** Babs says
that $\frac{3}{4}$ and $\frac{3}{8}$ are equivalent fractions
based on the model she made.
Describe and correct her error.

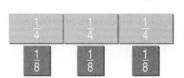

56. James and his friends picked oranges in a grove. James picked 19 oranges, Sal picked 22, Tara picked 12, and Pam picked 11 oranges. If they share the oranges equally, how many oranges will each person get?

57. Raul drank $\frac{1}{4}$ of 12 cans of orange juice. Basil drank the same amount of orange juice as Raul. Find an equivalent fraction to show the amount Basil drank.

58. Sally's dog ate 2 out of 6 dog treats. What fraction of the dog treats was left? Write the fraction in simplest form.

59. **REASONING** Is the product of a 4-digit number and a 2-digit number always a 6-digit number? Explain.

Mixed Review and Test Prep

Which quotient is greater? (p. 266)

60. 160 ÷ 2 or 180 ÷ 3

61. 293 ÷ 5 or 312 ÷ 6

62. **TEST PREP** 8,924 + 3,452 (p. 40)

    A 11,276   **C** 12,276

    B 11,376   **D** 12,376

Find the value of *n*. (p. 286)

63. $n \times 46 = 2,300$

64. **TEST PREP** Which time is 25 minutes before 12:15? (p. 116)

    F 11:40   **H** 12:20

    G 11:50   **J** 12:40

PROBLEM SOLVING LiNKUP ... to Music

In music, one whole note is equivalent to two $\frac{1}{2}$ notes, four $\frac{1}{4}$ notes, or eight $\frac{1}{8}$ notes. The diagram shows how the notes are related.

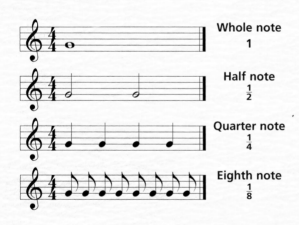

Copy and complete.

1. __?__ $\frac{1}{4}$ note(s) equal two $\frac{1}{2}$ notes.

2. Two $\frac{1}{8}$ notes equal one __?__ note.

3. One whole note equals one $\frac{1}{2}$ note and __?__ $\frac{1}{4}$ note(s).

EXTRA PRACTICE page H50, Set B

Compare and Order Fractions

▶ **Learn**

FOR THE BIRDS At summer camp, Benito got a recipe for wild bird seed food. He is making the recipe. Does the recipe call for a greater amount of sunflower seeds or raisins?

HANDS ON

Activity

MATERIALS: fraction bars

Compare $\frac{1}{2}$ and $\frac{2}{3}$ by using fraction bars.

STEP 1

Start with the bar for 1. Line up the fraction bar for $\frac{1}{2}$.

1
$\frac{1}{2}$

STEP 2

Line up the fraction bars for $\frac{2}{3}$. Compare the two rows of fraction bars. The longer row represents the greater fraction.

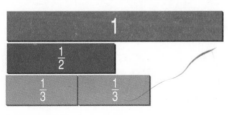

So, $\frac{2}{3} > \frac{1}{2}$. The recipe calls for a greater amount of raisins.

* How can you compare fractions that have the same numerators, but different denominators, such as $\frac{5}{8}$ and $\frac{5}{6}$?

Quick Review

Find an equivalent fraction for each.

1. $\frac{2}{3}$ 2. $\frac{4}{16}$ 3. $\frac{4}{5}$

4. $\frac{14}{28}$ 5. $\frac{6}{8}$

Wild Bird Seed Food

1 cup cornmeal
1 cup shortening
$\frac{1}{2}$ cup sunflower seeds
$\frac{2}{3}$ cup raisins

Mix together all the ingredients. Form into a ball. Chill for 30 minutes or until firm.

Order Fractions

You can use a number line to order three or more fractions.

Activity

Order $\frac{1}{2}$, $\frac{1}{6}$, and $\frac{2}{3}$ from *least* to *greatest*.

Technology Link

More Practice: Use Mighty Math Number Heroes, *Fraction Fireworks*, Levels B and D.

STEP 1

Draw a number line.

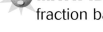

0   1

STEP 2

Place $\frac{1}{2}$, $\frac{1}{6}$, and $\frac{2}{3}$ on the number line.
The fraction closest to 1 is the greatest.
The fraction closest to 0 is the least.

0    $\frac{1}{6}$ $\frac{1}{2}$ $\frac{2}{3}$ 1

$\frac{2}{3}$ is closest to 1, so it is the greatest of these fractions.

$\frac{1}{6}$ is closest to 0, so it is the least of these fractions.

So, the order from least to greatest is $\frac{1}{6}$, $\frac{1}{2}$, $\frac{2}{3}$.

Use $<$, $>$, or $=$ when comparing and ordering fractions.
So, you can write $\frac{1}{6} < \frac{1}{2} < \frac{2}{3}$.

• Order $\frac{1}{2}$, $\frac{1}{6}$, and $\frac{2}{3}$ from *greatest* to *least*, using symbols.

MATH IDEA You can compare and order fractions by using fraction bars and number lines.

▶ Check

1. **Explain** how you can compare fractions that have the same denominators, but different numerators, such as $\frac{2}{5}$ and $\frac{3}{5}$.

Compare the fractions. Write $<$, $>$, or $=$ for each ●.

2. $\frac{2}{5}$ ● $\frac{1}{5}$

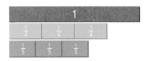

3. $\frac{3}{4}$ ● $\frac{3}{5}$

4. $\frac{2}{6}$ ● $\frac{1}{3}$

LESSON CONTINUES ▶

Compare the fractions. Write <, >, or = for each .
Use the models to help you.

5. $\frac{2}{8}$ ● $\frac{5}{8}$

6. $\frac{4}{8}$ ● $\frac{1}{2}$

7. $\frac{6}{10}$ ● $\frac{2}{3}$

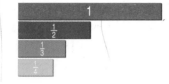

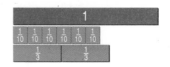

Write <, >, or = for each ●.

8. $\frac{7}{8}$ ● $\frac{3}{8}$

9. $\frac{2}{6}$ ● $\frac{1}{3}$

10. $\frac{5}{8}$ ● $\frac{5}{6}$

11. $\frac{3}{10}$ ● $\frac{3}{4}$

12. $\frac{5}{7}$ ● $\frac{4}{7}$

13. $\frac{8}{16}$ ● $\frac{9}{12}$

14. $\frac{6}{12}$ ● $\frac{7}{14}$

15. $\frac{3}{9}$ ● $\frac{2}{4}$

Order the fractions from *least* to *greatest.* Use the models,
fraction bars, or a number line to help you.

16. $\frac{1}{2}, \frac{1}{3}, \frac{1}{4}$

17. $\frac{3}{5}, \frac{3}{8}, \frac{4}{6}$

18. $\frac{1}{10}, \frac{2}{3}, \frac{4}{12}$

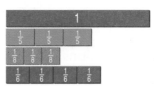
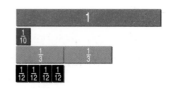

19. $\frac{3}{4}, \frac{3}{6}, \frac{3}{5}$

20. $\frac{7}{8}, \frac{1}{4}, \frac{5}{8}$

21. $\frac{2}{12}, \frac{3}{6}, \frac{2}{3}$

22. $\frac{2}{4}, \frac{9}{12}, \frac{1}{8}$

23. $\frac{2}{4}, \frac{2}{7}, \frac{2}{6}$

24. $\frac{5}{6}, \frac{5}{9}, \frac{5}{10}$

25. $\frac{3}{8}, \frac{4}{16}, \frac{3}{4}$

26. $\frac{5}{6}, \frac{5}{12}, \frac{2}{3}$

Order the fractions from *greatest* to *least.*

27. $\frac{1}{4}, \frac{3}{4}, \frac{2}{4}$

28. $\frac{2}{6}, \frac{1}{12}, \frac{3}{4}$

29. $\frac{1}{4}, \frac{1}{2}, \frac{1}{3}$

30. $\frac{5}{10}, \frac{4}{5}, \frac{2}{5}$

31. **ALGEBRA** Name all possible whole number values
for n when $\frac{1}{2} > \frac{n}{3}$.

USE DATA For 32–33, use the price list and fraction bars.

32. REASONING If Jimmy buys $\frac{1}{2}$ pound of the deluxe mix,
will he pay more or less than $1.35? Explain.

33. Write a problem using the fractions in the price list.

PRICE LIST	
Item	**Price**
$\frac{1}{3}$ pound almonds	$1.95
$\frac{1}{2}$ pound cashews	$3.15
$\frac{1}{4}$ pound deluxe mix	$1.35

34. Rhonda has $\frac{1}{4}$ yard of ribbon. She needs $\frac{3}{8}$ yard more for a costume. How much does she need in all?

35. Darrell had $\frac{1}{3}$ cup of raisins and $\frac{1}{2}$ cup of sunflower seeds left from his bird seed mix. Which is the greater amount?

36. Lee scored 91 points on a math test. On the next test, she scored 13 fewer points. On the third test, she scored 92. What is the difference between Lee's highest and lowest scores?

Mixed Review and Test Prep

For 37–38, write each fraction in simplest form. (p. 369)

37. $\frac{9}{12}$

38. $\frac{5}{10}$

39. $\begin{array}{r} 2,422 \\ \times \quad 4 \end{array}$ (p. 200)

40. $\begin{array}{r} 29,000 \\ - 13,752 \end{array}$ (p. 42)

41. Find the mean for 59, 63, 53, 75, 65, 81, 73. (p. 276)

42. Chen bought 2 bags of popcorn for $1.05 each. Felipe bought 2 drinks for a total of $2.25. Who spent more money? How much more? (p. 162)

43. 64 ÷ 4 (p. 250)　　**44.** 521 ÷ 3 (p. 266)

45. **TEST PREP** What are the missing numbers in the pattern? (p. 36)

23, 38, 53, 68, ■, ■, ■, 128

A 78, 83, 93　　**C** 88, 93, 103
B 83, 98, 113　　**D** 98, 108, 113

46. **TEST PREP** Mr. Edwards drives a total of 25 miles to and from work each day. He works 6 days each week. How many miles is this in 4 weeks? (p. 196)

F 200 miles　　**H** 500 miles
G 400 miles　　**J** 600 miles

PROBLEM SOLVING   THiNKER'S CorNer

FUN WITH FRACTIONS Solve each problem. Then match each answer to its fraction bar. The letters on the fraction bars will spell the answer to the riddle below. The first problem is done for you.

Which fraction is greater?

1. $\frac{3}{4}$, $\frac{4}{6}$　**2.** $\frac{1}{6}$, $\frac{1}{2}$　**3.** $\frac{1}{8}$, $\frac{6}{8}$　**4.** $\frac{1}{2}$, $\frac{3}{4}$　**5.** $\frac{4}{6}$, $\frac{2}{5}$

　[A]　　□　　□　　□　　□

Which fraction is least?

6. $\frac{2}{3}$, $\frac{2}{5}$, $\frac{1}{2}$　**7.** $\frac{2}{3}$, $\frac{3}{4}$, $\frac{1}{2}$　**8.** $\frac{3}{4}$, $\frac{2}{3}$, $\frac{6}{8}$　**9.** $\frac{1}{6}$, $\frac{3}{4}$, $\frac{6}{8}$　**10.** $\frac{1}{2}$, $\frac{2}{5}$, $\frac{1}{8}$

　□　　□　　□　　□　　□

Riddle: When can you hear the band play?

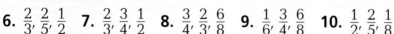

Problem Solving Strategy
Make a Model

PROBLEM   Sammy, Henry, and Jeb are gopher tortoises. They are training for a 1-yard race. In the first week, who ran the greatest distance? the least distance?

TORTOISE TRAINING	
Name	**Yards Run First Week**
Sammy	$\frac{1}{2}$
Henry	$\frac{3}{4}$
Jeb	$\frac{2}{3}$

UNDERSTAND

- What are you asked to do?
- What information will you use?
- Is there information you will not use? If so, what?

PLAN

- What strategy can you use to solve the problem?

  Make a model by using a number line. Locate and mark points to represent $\frac{1}{2}$, $\frac{3}{4}$, and $\frac{2}{3}$ yards.

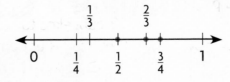

SOLVE

- How can you use the model to solve the problem?

  $\frac{3}{4}$ is closest to 1, so it is the greatest fraction.

  $\frac{1}{2}$ is closest to 0, so it is the least fraction. So, Henry ran the greatest distance and Sammy ran the shortest distance.

CHECK

- How can you decide if your answer is correct?

▶ Problem Solving Practice

Make a model to solve.

1. **What if** Sammy ran $\frac{1}{2}$ yard more the first week? Would he have run more than Henry ran the first week? Explain.

🔍 PROBLEM SOLVING STRATEGIES

Draw a Diagram or Picture
Make a Model or Act It Out
Make an Organized List
Find a Pattern
Make a Table or Graph
Predict and Test
Work Backward
Solve a Simpler Problem
Write an Equation
Use Logical Reasoning

A spinner has 12 equal sections. Two of the sections are blue, 3 sections are yellow, 2 sections are red, and 5 sections are green.

2. What color covers $\frac{1}{4}$ of the spinner?
   A red   **C** green
   B blue   **D** yellow

3. If the red sections are changed to blue, what fraction of the spinner will be blue?
   F $\frac{2}{12}$ **G** $\frac{1}{8}$ **H** $\frac{1}{3}$ **J** $\frac{1}{4}$

Mixed Strategy Practice

4. Ty's tomato sauce has $\frac{1}{4}$ teaspoon basil, $\frac{1}{2}$ teaspoon oregano, and $\frac{1}{8}$ teaspoon pepper. Order the amounts from greatest to least.

5. **REASONING** Explain how you know whether Ty used more or less than 1 teaspoon of ingredients in the tomato sauce.

6. Beatrice read 3 pages of her book on Monday, 6 pages on Tuesday, 12 pages on Wednesday, and 24 pages on Thursday. If this pattern continues, how many pages will she read on Friday?

7. Jon and his sister made cupcakes. They each ate 2 cupcakes. Jon took 12 cupcakes to school and left 8 cupcakes at home. What fraction of the cupcakes did Jon take to school?

8. Cindy is making a square design with 16 tiles. The 4 corner tiles are red, and the rest of the outside border tiles are blue. She puts 4 green tiles in the middle. Show what Cindy's design will look like.

USE DATA For 9–10, use the table.

9. How long will each assembly last?

10. Which assembly will be the longest? the shortest?

MORNING ASSEMBLY SCHEDULE	
Kindergarten and First Grade	8:45–9:15
Second Grade and Third Grade	9:30–10:15
Fourth Grade and Fifth Grade	10:30–11:20

Problem Solving Strategy

6 Mixed Numbers

VOCABULARY

mixed number

▶ Learn

HOP TO IT Susan gave one and one fourth cups of rabbit food to Whiskers, her pet rabbit.

A **mixed number** is made up of a whole number and a fraction. Look at the pictures that represent one and one fourth cups of food.

Read: One and one fourth

Write: $1\frac{1}{4}$

Examples

Write a mixed number for each picture.

A one and one fourth

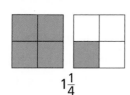

$1\frac{1}{4}$

B two and two thirds

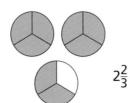

$2\frac{2}{3}$

C three and one sixth

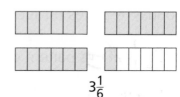

$3\frac{1}{6}$

D two and one half

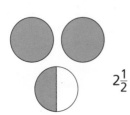

$2\frac{1}{2}$

E three and three fifths

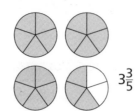

$3\frac{3}{5}$

F one and four sixths

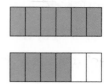

$1\frac{4}{6}$, or $1\frac{2}{3}$

• Look at Example A. How many fourths does it take to make two wholes?

Rename Fractions

Sometimes the numerator of a fraction is greater than the denominator. These fractions have a value greater than 1. They can be renamed as a mixed number.

Jason renamed $\frac{7}{5}$.

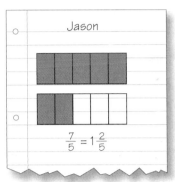

$$\frac{7}{5} = 1\frac{2}{5}$$

Kareem renamed $\frac{8}{3}$.

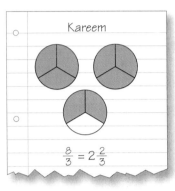

$$\frac{8}{3} = 2\frac{2}{3}$$

Dawn renamed $\frac{10}{4}$.

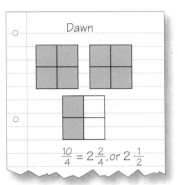

$$\frac{10}{4} = 2\frac{2}{4}, \text{ or } 2\frac{1}{2}$$

Since $\frac{15}{4}$ means $15 \div 4$, you can use division to rename a fraction greater than 1 as a mixed number.

$$
\begin{array}{r}
3 \text{ r}3 \\
\text{denominator} \rightarrow 4\overline{)15} \quad \leftarrow \text{numerator} \\
-12 \\
\hline
3 \quad \leftarrow \text{number of fourths left over}
\end{array}
$$

• In a fraction greater than 1, which is greater—the denominator or the numerator? Explain.

▶ Check

1. Explain how you can tell that a fraction is greater than 1.

Write a mixed number for each picture.

2.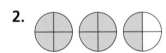

3.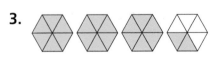

4.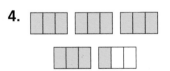

Rename each fraction as a mixed number. You may wish to draw a picture.

5. $\frac{4}{3}$ 6. $\frac{11}{5}$ 7. $\frac{13}{4}$ 8. $\frac{19}{6}$ 9. $\frac{5}{2}$

LESSON CONTINUES ▶

Write a mixed number for each picture.

10.  11.  12.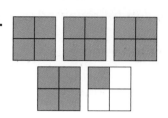

Rename each fraction as a mixed number. You may wish to draw a picture.

13. $\frac{5}{2}$ 14. $\frac{10}{3}$ 15. $\frac{10}{8}$ 16. $\frac{15}{4}$

17. $\frac{10}{6}$ 18. $\frac{21}{8}$ 19. $\frac{19}{3}$ 20. $\frac{25}{9}$

Compare. Write <, >, or = for each ●.

21. $\frac{10}{6}$ ● 3   22. $\frac{13}{4}$ ● $\frac{5}{2}$ 23. $\frac{15}{4}$ ● $\frac{25}{9}$

24. $\frac{17}{8}$ ● $\frac{22}{7}$ 25. $\frac{10}{8}$ ● $\frac{5}{4}$ 26. 6 ● $\frac{37}{7}$

27. $\frac{19}{3}$ ● $\frac{19}{2}$ 28. $3\frac{3}{4}$ ● $6\frac{1}{2}$ 29. $4\frac{2}{6}$ ● $4\frac{1}{3}$

For 30–34, use the figures at the right.

30. Write an expression for the shaded part in the third figure as a fraction.

31. How many whole figures are shaded in the given picture?

32. What fraction and mixed number can you write for the picture?

33. How can you change the model to show 3 wholes?

34. **REASONING** How can you change the model to show $1\frac{4}{6}$?

35. A cup holds 8 ounces of liquid. Mary used 24 ounces of milk to make waffles. How many cups of milk did she use?

36. At snack time, Tim drank $\frac{1}{2}$ cup of juice and Cory drank $\frac{3}{4}$ cup of juice. Who drank more than $\frac{5}{8}$ cup of juice?

37. **ESTIMATION** Mrs. James cut up $\frac{9}{4}$ grapefruit to make juice. Is that closer to 2 or 3 whole grapefruit?

38. ✎ **Write About It** How can you tell when a fraction names a number greater than 1?

39. In 1940, a man set a world record by riding his bike about 200 miles each day for 500 days. About how many miles did he ride?

40. A serving of rabbit food is $1\frac{1}{4}$ cups. If you measure $3\frac{3}{4}$ cups, how many servings do you have?

Mixed Review and Test Prep

41.   3,524 (p. 40)   **42.**   67,004 (p. 44)
   $+ 1,524$ $- 9,386$

43. Round 2,316,790 to the nearest thousand. (p. 26)

Write <, >, or = for each ⬤.

44. 5,236,909 ⬤ 5,237,987 (p. 18)

45. $\frac{3}{8}$ ⬤ $\frac{6}{5}$ (p. 372)

46. $98.02 × 6 (p. 200)

47. **TEST PREP** How many zeros are in the product of 65 and 40,000? (p. 212)
   A 2   **C** 4
   B 3   **D** 5

48. **TEST PREP** Which product is different from the others? (p. 150)
   F $3 × 3 × 2$ **H** $3 × 2 × 4$
   G $2 × 2 × 2 × 3$ **J** $2 × 2 × 6$

PROBLEM SOLVING THiNKER'S CoRNer

RATIO Double Dutch is a jump rope game that is often played on city streets. In tournaments, teams of 4 jumpers do several stunts. Each team of 4 jumpers uses 2 jump ropes.

A **ratio** compares two amounts. There are three ways to write a ratio comparing the number of jumpers to the number of jump ropes they will use:

   4:2 $\frac{4}{2}$ 4 to 2

All of these ratios are read "4 to 2."

The ratio of the number of jumpers to the number of jump ropes will be equivalent to 4:2 no matter how many teams there are.

Write a ratio to compare the number of jumpers to the number of jump ropes. Write each ratio in three ways.

   1. 8 jumpers   **2.** 12 jumpers   **3.** 16 jumpers   **4.** 20 jumpers

EXTRA PRACTICE page H50, Set D

Review/Test

✅ CHECK VOCABULARY AND CONCEPTS

Choose the best term from the box.

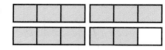

1. Different fractions that name the same amount are called ___?___ . (p. 366)

2. A ___?___ is made up of a whole number and a fraction. (p. 378)

✅ CHECK SKILLS

Write the fraction or mixed number for the shaded part. (pp. 364–365, 378–381)

3.
4.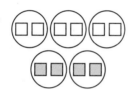
5.

Write two equivalent fractions for each. (pp. 366–371)

6. $\frac{1}{2}$ 7. $\frac{3}{4}$ 8. $\frac{7}{8}$ 9. $\frac{4}{10}$ 10. $\frac{2}{5}$

Write each fraction in simplest form. (pp. 368–371)

11. $\frac{9}{12}$ 12. $\frac{6}{10}$ 13. $\frac{4}{6}$ 14. $\frac{10}{12}$ 15. $\frac{6}{4}$

Write <, >, or = for each ●. (pp. 372–375)

16. $\frac{3}{8}$ ● $\frac{5}{8}$ 17. $\frac{3}{6}$ ● $\frac{4}{8}$ 18. $\frac{2}{3}$ ● $\frac{1}{4}$ 19. $\frac{11}{12}$ ● $\frac{5}{6}$

Rename each fraction as a mixed number. (pp. 378–381)

20. $\frac{7}{2}$ 21. $\frac{8}{3}$ 22. $\frac{16}{3}$ 23. $\frac{13}{5}$

✅ CHECK PROBLEM SOLVING

Make a model to solve. (pp. 376–377)

24. Tom made $\frac{1}{3}$ of his free throws and Maria made $\frac{4}{6}$ of hers. Who made a greater part of his or her free throws?

25. Sue has a set of wrenches. Three of the sizes are $\frac{1}{2}$, $\frac{3}{4}$, and $\frac{5}{8}$ inches. Put the sizes in order from least to greatest.

Standardized Test Prep

Understand the problem.
See item **8**.

The fraction tells what part of the empty box was filled each day. Compare and order the fractions to find the greatest amount filled.

Also see problem **1**, p. H62.

For 1–8, choose the best answer.

1. What fraction names the part of the figure that is shaded?

   A $\frac{3}{5}$ **B** $\frac{5}{3}$ **C** $\frac{3}{8}$ **D** $\frac{8}{3}$

2. What part of this group of balloons has dots?

   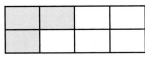

   F $\frac{5}{2}$ **G** $\frac{3}{5}$ **H** $\frac{2}{3}$ **J** $\frac{2}{5}$

3. Which fraction is written in simplest form?

   A $\frac{3}{12}$ **B** $\frac{6}{8}$ **C** $\frac{5}{6}$ **D** $\frac{8}{10}$

4. Which fractions are written in order from *least* to *greatest*?

   F $\frac{1}{4}, \frac{5}{6}, \frac{3}{8}$ **H** $\frac{1}{4}, \frac{3}{8}, \frac{5}{6}$

   G $\frac{3}{8}, \frac{5}{6}, \frac{1}{4}$ **J** $\frac{5}{6}, \frac{1}{4}, \frac{3}{8}$

5. Jarod wrote these fractions that are equivalent to $\frac{1}{2}$.

   $$\frac{2}{4}, \frac{3}{6}, \frac{4}{8}, \frac{5}{10}$$

   He noticed a pattern in his list. If he continues to follow the same pattern, which fraction would he write next?

   A $\frac{6}{11}$ **B** $\frac{6}{12}$ **C** $\frac{7}{10}$ **D** $\frac{7}{14}$

6. Jeremy had 75¢. He gave some money to Tina. Then he had 38¢. Which equation can be used to find the amount Jeremy gave to Tina?

   F $75¢ + 38¢ = a$ **H** $75¢ - a = 38¢$
   G $38¢ - a = 75¢$ **J** $a - 75¢ = 38¢$

7. $5,387 + 2,498$

   A 7,775   **C** 7,875
   B 2,889   **D** NOT HERE

8. Mrs. Reed's class collected newspaper to recycle. Every day the students put their newspapers in a box. The table shows how much of the box they filled each day last week.

MON	TUE	WED	THU	FRI
$\frac{3}{4}$	$\frac{2}{3}$	$\frac{1}{4}$	$\frac{5}{8}$	$\frac{1}{6}$

   On which day did the students fill the greatest part of the box?

   F Monday   **H** Wednesday
   G Tuesday   **J** Thursday

Write What You Know

9. Explain how you know that $\frac{1}{2} < \frac{2}{3}$.

10. Explain why 25¢ is called a *quarter* of a dollar.

Add and Subtract Fractions and Mixed Numbers

How does a printer get all the colors on a page? Mixing colors of ink or paint in different amounts will make other colors.

PROBLEM SOLVING Look at the paint-mixing guide. Copy the table, and find the number of gallons of each new color produced.

PAINT-MIXING GUIDE (in gallons)			
Blue	**Red**	**Yellow**	**New Color**
$\frac{3}{4}$	$1\frac{1}{4}$	0	■ purple
0	$1\frac{1}{2}$	$1\frac{1}{2}$	■ orange
$\frac{1}{3}$	0	$\frac{1}{3}$	■ green

CHECK WHAT YOU KNOW ✓

Use this page to help you review and remember
important skills needed for Chapter 20.

✓ VOCABULARY

Choose the best term from the box.

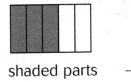

equivalent fractions
simplest form
mixed numbers

1. The numbers $3\frac{2}{6}$ and $5\frac{3}{6}$ are called __?__.

2. The fractions $\frac{3}{6}$ and $\frac{1}{2}$ are called __?__.

✓ COUNT PARTS OF A WHOLE (For Intervention, see p. H18.)

Use the model to complete the fraction.

3.

shaded parts → ▢
—
total equal parts → ▢

4.

shaded parts → ▢
—
total equal parts → ▢

5.

shaded parts → ▢
—
total equal parts → ▢

✓ COUNT PARTS OF A GROUP (For Intervention, see p. H19.)

What part of the group is shaded?

6.

7.

8.

9.

10.

11.

12.

13.

✓ COMPARE PARTS OF A WHOLE (For Intervention, see p. H19.)

Use <, >, or = to compare the shaded areas of
the rectangles.

14.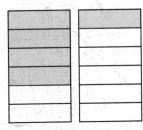

$\frac{4}{6}$ ● $\frac{1}{6}$

15.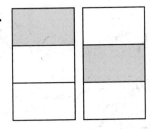

$\frac{1}{3}$ ● $\frac{1}{3}$

16.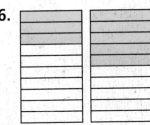

$\frac{3}{10}$ ● $\frac{5}{10}$

Add Like Fractions

▶ **Learn**

A STEP AT A TIME Pamela walked $\frac{1}{4}$ mile to Lori's house. Then, the two girls walked together $\frac{2}{4}$ mile to pottery class. Did Pamela walk more than or less than 1 mile?

Find $\frac{1}{4} + \frac{2}{4}$ and compare to 1.

Like fractions are fractions with the same denominator.

You can add like fractions by using fraction bars.

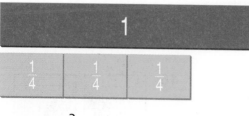

Activity

STEP 1

Line up one of the $\frac{1}{4}$ fraction bars under the bar for 1.

1

$\frac{1}{4}$

STEP 2

Line up two more $\frac{1}{4}$ fraction bars.

Count the $\frac{1}{4}$ fraction bars. $\frac{1}{4} + \frac{2}{4} = \frac{3}{4}$

1

$\frac{1}{4}$	$\frac{1}{4}$	$\frac{1}{4}$

Compare. $\frac{3}{4} < 1$

So, Pamela walked less than 1 mile.

• When using fraction bars, how do you know whether a sum is greater than 1 or less than 1?

VOCABULARY

like fractions

Remember

The *numerator* in a fraction tells how many parts are being considered. The *denominator* tells the total number of equal parts.

$\frac{2}{3}$ ← numerator
← denominator

Add Numerators

What if Pamela walked $\frac{3}{10}$ mile from class to the library and $\frac{7}{10}$ mile home? How far did Pamela walk?

You can add fractions with like denominators by adding the numerators.

Distance to the library   Distance home   Total distance

$$\frac{3}{10} \quad + \quad \frac{7}{10} \quad = \quad \frac{10}{10}$$

← Add the numerators.
← The denominator stays the same.

So, Pamela walked a total of 1 mile.

Examples

A $\frac{2}{10} + \frac{6}{10}$

$\frac{2}{10} + \frac{6}{10} = \frac{8}{10}$

Write the sum in simplest form.

$\frac{8}{10} = \frac{4}{5}$

B $\frac{3}{8} + \frac{5}{8}$

$\frac{3}{8} + \frac{5}{8} = \frac{8}{8}$

Write the sum as a whole number.

$\frac{8}{8} = 1$

C $\frac{2}{6} + \frac{5}{6}$

$\frac{2}{6} + \frac{5}{6} = \frac{7}{6}$

Write the sum as a mixed number.

$\frac{7}{6} = \frac{6}{6} + \frac{1}{6} = 1\frac{1}{6}$

 MATH IDEA To add like fractions, add the numerators. Use the same denominator as the like fractions.

▶ Check

1. **Tell** why you add only the numerators when adding like fractions.

Find the sum.

2.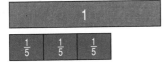

$\frac{2}{5} + \frac{1}{5}$

3.

$\frac{4}{10} + \frac{3}{10}$

4.

$\frac{1}{12} + \frac{4}{12}$

5. $\begin{array}{r} \frac{1}{2} \\ +\frac{1}{2} \\ \hline \end{array}$

6. $\begin{array}{r} \frac{3}{4} \\ +\frac{1}{4} \\ \hline \end{array}$

7. $\begin{array}{r} \frac{4}{8} \\ +\frac{2}{8} \\ \hline \end{array}$

8. $\begin{array}{r} \frac{1}{3} \\ +\frac{1}{3} \\ \hline \end{array}$

LESSON CONTINUES

Find the sum.

9.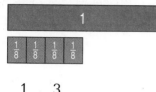

$$\frac{1}{8} + \frac{3}{8}$$

10.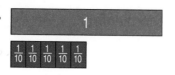

$$\frac{1}{10} + \frac{4}{10}$$

11.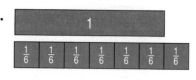

$$\frac{4}{6} + \frac{3}{6}$$

12. $\frac{2}{6} + \frac{4}{6}$

13. $\frac{3}{4} + \frac{2}{4}$

14. $\frac{3}{5} + \frac{2}{5}$

15. $\frac{2}{3} + \frac{2}{3}$

16. $\begin{array}{r} \frac{3}{12} \\ +\frac{5}{12} \\ \hline \end{array}$

17. $\begin{array}{r} \frac{9}{12} \\ +\frac{10}{12} \\ \hline \end{array}$

18. $\begin{array}{r} \frac{3}{10} \\ +\frac{7}{10} \\ \hline \end{array}$

19. $\begin{array}{r} \frac{3}{3} \\ +\frac{1}{3} \\ \hline \end{array}$

Compare. Write <, >, or = for each ●.

20. $\frac{1}{4} + \frac{2}{4} \; ● \; \frac{1}{2}$

21. $\frac{1}{6} + \frac{3}{6} \; ● \; \frac{2}{3}$

22. $\frac{4}{8} + \frac{3}{8} \; ● \; 1$

23. $\frac{1}{5} + \frac{1}{5} \; ● \; \frac{3}{10}$

24. $\frac{1}{3} + \frac{1}{3} \; ● \; \frac{1}{2}$

25. $\frac{1}{12} + \frac{6}{12} \; ● \; \frac{1}{2}$

26. $\frac{4}{10} + \frac{1}{10} \; ● \; \frac{2}{5}$

27. $\frac{2}{8} + \frac{4}{8} \; ● \; \frac{3}{4}$

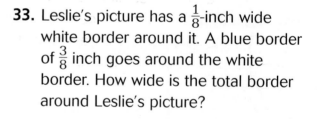 **ALGEBRA**   **For 28–31, find the value of n.**

28. $\frac{7}{10} + \frac{2}{n} = \frac{9}{10}$

29. $\frac{5}{n} + \frac{3}{n} = 1$

30. $\frac{5}{n} + \frac{3}{n} = \frac{8}{12}$

31. $\frac{4}{n} + \frac{3}{n} = \frac{7}{9}$

32. During pottery class, a clay block was sliced into 8 equal pieces. Peter and Jack each got $\frac{1}{4}$ of the clay. What fraction of the total clay block did they get altogether? How many pieces is this?

33. Leslie's picture has a $\frac{1}{8}$-inch wide white border around it. A blue border of $\frac{3}{8}$ inch goes around the white border. How wide is the total border around Leslie's picture?

34. **REASONING** Three identical jars have pottery glaze. Will all the pottery glaze fit into one of the jars? Explain.

35. **?** **What's the Error?** Allen says the sum $\frac{3}{4} + \frac{2}{4}$ is $\frac{5}{8}$. Describe his error. Write the correct answer.

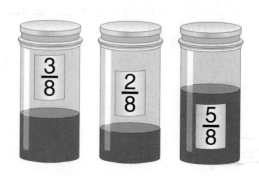

36. Each lap around a track is $\frac{1}{4}$ mile. Leslie walked 1 lap around the track on Saturday and walked 2 laps around the track on Sunday. How far did Leslie walk on Saturday and Sunday, in miles?

37. During pottery class, a clay block was sliced into 9 equal pieces. Wendy, Peter, and Jack each got $\frac{1}{9}$ of the clay. What fraction of the total clay block did they get altogether? How many pieces is this?

Mixed Review and Test Prep

38. Order $\frac{1}{2}$, $\frac{1}{4}$, and $\frac{2}{3}$ from greatest to least. (p. 372)

39. $\begin{array}{r} 983 \\ \times \quad 6 \end{array}$ (p. 196)

40. $\begin{array}{r} 1,820 \\ \times \quad 54 \end{array}$ (p. 230)

41. School starts at 8:15 A.M. and ends at 2:50 P.M. How long is the school day? (p. 120)

42. $\begin{array}{r} 82 \\ \times 51 \end{array}$ (p. 224)

43. $275 \div 8$ (p. 266)

44. **TEST PREP** Which fraction is in simplest form? (p. 368)

A $\frac{2}{10}$ **B** $\frac{3}{6}$ **C** $\frac{2}{3}$ **D** $\frac{6}{8}$

45. **TEST PREP** Evie had 3 packages of pens with 10 pens in each pack. She gave Toby some pens. Choose an expression that shows how many she had left. (p. 162)

F $(3 \times 10) + n$ **H** $(3 + n) \times 10$
G $(3 \times 10) - n$ **J** $(3 - n) \times 10$

PROBLEM SOLVING LiNKUP... to Art

Pueblo Indian artists create pottery by using methods dating back to A.D. 700. Clay from a quarry is dried and then soaked in water and mixed with volcanic material. The artist forms the clay into a pot and paints a design on it. The pot is fired, or baked, in a kiln, a type of oven.

Clay

$\frac{1}{2}$ cup flour	$\frac{2}{4}$ cup water
$\frac{1}{4}$ cup salt	food color
1 tablespoon vegetable oil	$\frac{1}{2}$ teaspoon cream of tartar

Mix the dry ingredients together. Mix wet ingredients together in a separate bowl. Combine the wet and dry ingredients together in a saucepan. Cook on low until the dough thickens to modeling consistency.

For 1–3, use the recipe at the right.

1. If you combined the salt and water, how many cups would you have?

2. **What if** you want to make 4 batches of modeling clay? Write a new recipe showing how much of each ingredient you will need.

3. **Write a problem** that involves adding like fractions. Use the recipe at the right.

Subtract Like Fractions

▶ **Explore**

How can you subtract two fractions with like denominators?

Activity

MATERIALS fraction bars

Use fraction bars to find $\frac{5}{6} - \frac{3}{6}$.

One Way   Take away fraction bars.

STEP 1	STEP 2	STEP 3
Line up 5 of the $\frac{1}{6}$ bars.	Take away 3 of the $\frac{1}{6}$ bars.	Count the $\frac{1}{6}$ bars left.

$\frac{5}{6} - \frac{3}{6} = \frac{2}{6}$

There are two $\frac{1}{6}$ bars left. So, the answer is $\frac{2}{6}$, or $\frac{1}{3}$.

Another Way   Compare two groups of fraction bars.

STEP 1	STEP 2	STEP 3
Line up 5 of the $\frac{1}{6}$ bars.	Line up 3 of the $\frac{1}{6}$ bars.	Compare the bars. Find the difference.

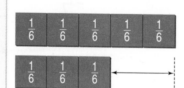

The difference is $\frac{2}{6}$.

So, $\frac{5}{6} - \frac{3}{6} = \frac{2}{6}$, or $\frac{1}{3}$.

Try It

Find the difference.

a. $\frac{7}{8} - \frac{2}{8}$ b. $\frac{8}{12} - \frac{6}{12}$

c. $\frac{5}{10} - \frac{2}{10}$ d. $\frac{3}{4} - \frac{2}{4}$

I am comparing $\frac{7}{8}$ and $\frac{2}{8}$. What is the difference?

> **Connect**

Here is a way to record subtraction for $\frac{7}{8} - \frac{3}{8}$.

Example

To subtract like fractions, first subtract the numerators.
Take away 3 from 7.

$$\frac{7}{8} - \frac{3}{8} = \frac{4}{8}$$

STEP 2

The denominator stays the same.
Write the difference over the denominator.

Write the answer in simplest form.

$$\frac{7}{8} - \frac{3}{8} = \frac{4}{8}$$

$$\frac{4}{8} = \frac{1}{2}$$

MATH IDEA To subtract like fractions, subtract the numerators. Use the same denominator as the like fractions.

> **Practice and Problem Solving**

Use fraction bars to find the difference.

1. $\frac{3}{5} - \frac{1}{5}$ 2. $\frac{8}{10} - \frac{1}{10}$ 3. $\frac{6}{6} - \frac{1}{6}$ 4. $\frac{11}{12} - \frac{5}{12}$

Find the difference.

5.

6.

7.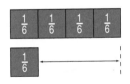

8.

9. In a survey $\frac{4}{10}$ of the students chose an animal as a mascot. The rest chose a cartoon character. Which mascot received more votes? Explain.

Mixed Review and Test Prep

10. List the factors of 36. (p. 298)

11. Write the first five multiples of 7. (p. 298)

12. $595 \div 5$ (p. 266)   13. $438 \div 2$ (p. 266)

14. **TEST PREP** Hannah can fit 32 names on each page of the school directory. She has 416 names. How many pages does she need? (p. 286)

A 12   B 13   C 14   D 17

> More Practice: Use E-Lab, *Subtracting Like Fractions.*
>
> www.harcourtschool.com/elab2002

3 Add and Subtract Mixed Numbers

Quick Review
1. $\frac{1}{4} + \frac{2}{4}$ 2. $\frac{1}{3} + \frac{2}{3}$
3. $\frac{2}{5} + \frac{1}{5}$ 4. $\frac{4}{6} - \frac{1}{6}$
5. $\frac{5}{8} - \frac{3}{8}$

▶ **Learn**

SET THE SCENE Tim is painting a backdrop for the school play. He mixed $2\frac{2}{4}$ gallons of blue paint and $1\frac{1}{4}$ gallons of red paint to make dark purple. How many gallons of dark purple paint did Tim make?

Example

Add. $2\frac{2}{4} + 1\frac{1}{4}$

STEP 1

Add the fractions first.

$$2\frac{2}{4}$$
$$+1\frac{1}{4}$$
$$\overline{\quad \frac{3}{4}}$$

STEP 2

Then add the whole numbers.

$$2\frac{2}{4}$$
$$+1\frac{1}{4}$$
$$\overline{3\frac{3}{4}}$$

So, Tim made $3\frac{3}{4}$ gallons of dark purple paint.

More Examples

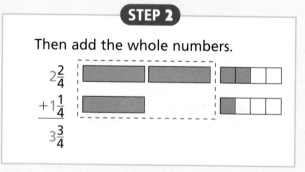

Ⓐ $2\frac{1}{3}$
$+2\frac{1}{3}$
$\overline{4\frac{2}{3}}$

Ⓑ $\frac{3}{5}$
$+1\frac{4}{5}$
$\overline{1\frac{7}{5}}$, or $2\frac{2}{5}$

MATH IDEA When you add mixed numbers with like denominators, add the fractions first, and then add the whole numbers.

Subtract Mixed Numbers

Subtracting mixed numbers with like denominators is similar to adding mixed numbers.

Mrs. Baker has $3\frac{5}{8}$ yards of fabric. She needs $2\frac{1}{8}$ yards of fabric to make a costume. How much fabric will Mrs. Baker have left?

Example

Subtract. $3\frac{5}{8} - 2\frac{1}{8}$

STEP 1

Subtract the fractions first.

$$
\begin{array}{r}
3\frac{5}{8} \\
-2\frac{1}{8} \\
\hline
\frac{4}{8}
\end{array}
$$

STEP 2

Then subtract the whole numbers.

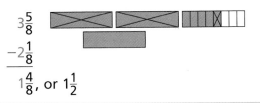

$$
\begin{array}{r}
3\frac{5}{8} \\
-2\frac{1}{8} \\
\hline
1\frac{4}{8}, \text{ or } 1\frac{1}{2}
\end{array}
$$

So, Mrs. Baker will have $1\frac{1}{2}$ yards of fabric left.

More Examples

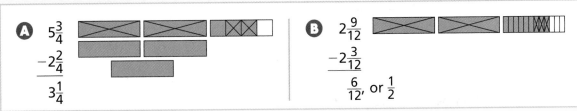

A
$$
\begin{array}{r}
5\frac{3}{4} \\
-2\frac{2}{4} \\
\hline
3\frac{1}{4}
\end{array}
$$

B
$$
\begin{array}{r}
2\frac{9}{12} \\
-2\frac{3}{12} \\
\hline
\frac{6}{12}, \text{ or } \frac{1}{2}
\end{array}
$$

► **Check**

1. **Tell** how you can check whether your answer to a subtraction problem is correct.

Find the sum or difference.

2.
$$
\begin{array}{r}
2\frac{1}{4} \\
+3\frac{3}{4} \\
\hline
\end{array}
$$

3.
$$
\begin{array}{r}
4\frac{3}{5} \\
+1\frac{3}{5} \\
\hline
\end{array}
$$

4.
$$
\begin{array}{r}
2\frac{3}{5} \\
-1\frac{1}{5} \\
\hline
\end{array}
$$

5.
$$
\begin{array}{r}
2\frac{5}{6} \\
-1\frac{3}{6} \\
\hline
\end{array}
$$

Technology Link

More Practice: Use Mighty Math Calculating Crew, *Nautical Number Line*, Level N.

LESSON CONTINUES ▶

Find the sum or difference.

6. $3\frac{1}{3}$
    $+2\frac{1}{3}$

7. $4\frac{3}{4}$
    $-1\frac{1}{4}$

8. $5\frac{1}{4}$
    $+1\frac{2}{4}$

9. $3\frac{5}{6}$
    $-2\frac{2}{6}$

10. $6\frac{10}{12}$
    $-2\frac{4}{12}$

11. $8\frac{1}{6}$
    $+1\frac{5}{6}$

12. $5\frac{8}{10}$
    $-3\frac{3}{10}$

13. $7\frac{2}{9}$
    $+4\frac{8}{9}$

14. $4\frac{3}{6}$
    $+2\frac{4}{6}$

15. $4\frac{2}{2}$
    $-3\frac{1}{2}$

16. $8\frac{7}{10}$
    $+7\frac{2}{10}$

17. $4\frac{6}{8}$
    $-1\frac{2}{8}$

18. $3\frac{5}{6} - 1\frac{4}{6}$

19. $5\frac{3}{8} + 1\frac{5}{8}$

20. $6\frac{5}{8} - 3\frac{2}{8}$

21. $7\frac{9}{12} + 4\frac{2}{12}$

Compare. Write <, >, or = for each ●.

22. $1\frac{2}{8} + 3\frac{3}{8}$ ● $4\frac{1}{2}$

23. $5\frac{3}{5} + 4\frac{2}{5}$ ● 10

24. $7\frac{2}{4} + 9\frac{1}{4}$ ● $16\frac{7}{8}$

25. $4\frac{5}{6} - 2\frac{2}{6}$ ● $2\frac{1}{2}$

26. $12\frac{8}{10} - 8\frac{2}{10}$ ● $4\frac{4}{5}$

27. $12\frac{4}{5} - 7\frac{2}{5}$ ● $5\frac{3}{10}$

28. $5\frac{2}{3} - 3\frac{1}{3}$ ● $8\frac{2}{3}$

29. $6\frac{5}{6} - 1\frac{2}{6}$ ● $5\frac{1}{2}$

30. $4\frac{3}{8} + 6\frac{1}{8}$ ● $10\frac{1}{4}$

ALGEBRA **For 31–33, find the value of *n*.**

31. $3\frac{6}{8} + 5\frac{2}{8} = n$

32. $9\frac{6}{10} + 11\frac{n}{10} = 20\frac{9}{10}$

33. $7\frac{4}{6} - 4\frac{1}{6} = 3\frac{3}{6}$, or $3\frac{1}{n}$

USE DATA For 34–36, use the table.

34. Luisa has $4\frac{2}{4}$ yards of fabric. How many colonial girl and ballerina costumes can she make?

35. Ms. Jenkins has $1\frac{2}{4}$ yards of fabric. How much more fabric does she need to make a lion costume?

36. **? What's the Question?** Two mixed numbers are $1\frac{1}{4}$ and $2\frac{1}{4}$. The answer is $3\frac{1}{2}$.

COSTUME	FABRIC NEEDED
Colonial Girl	$3\frac{1}{4}$ yards
Ballerina	$1\frac{1}{4}$ yards
Lion	$2\frac{3}{4}$ yards

37. Bonnie needs 5 gallons of paint. If she mixes $2\frac{4}{10}$ gallons of blue paint and $2\frac{7}{10}$ gallons of yellow paint, will she have enough paint? Explain.

38. Draw pictures to represent the mixed numbers $3\frac{1}{4}$ and $1\frac{1}{2}$. Explain how to redraw the second mixed number so you could add them together.

Mixed Review and Test Prep

39. $\begin{array}{r} 5{,}463 \\ \times\quad 8 \\ \hline \end{array}$ (p. 200)

40. $\begin{array}{r} \$7{,}492 \\ \times\quad 28 \\ \hline \end{array}$ (p. 230)

41. Mrs. Troy had 6 cans of apple juice. Keisha, Geoffry, and Latika each drank 1 can. Mr. Troy drank 2 cans. What fraction of the apple juice is left? (p. 390)

42. ($\blacksquare \times 6$) − 13 = 35 (p. 158)

43. **TEST PREP** If a water cooler bottle holds 12 gallons of water, how many gallons do 150 bottles hold? (p. 228)
  A 180
  B 1,800
  C 18,000
  D 180,000

44. **TEST PREP** Find the elapsed time from 12:45 P.M. to 5:18 P.M. (p. 120)
  F 3 hr 40 min
  G 4 hr 5 min
  H 4 hr 33 min
  J 5 hr 20 min

PROBLEM SOLVING — THINKER'S CORNER

ESTIMATE WITH FRACTIONS You can use fraction models to estimate whether fractions are greater than or less than $\frac{1}{2}$.

A pizza is cut into eighths. Ty and Ron ate $\frac{3}{8}$ of the pizza. Is the part that is left *greater than* or *less than* $\frac{1}{2}$ of the whole?

Look at the picture. Compare $\frac{5}{8}$ with $\frac{1}{2}$. You can see that $\frac{5}{8} > \frac{1}{2}$.

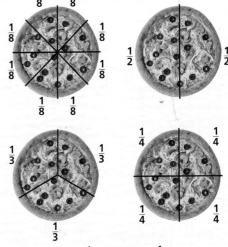

Compare. Write < or > for each **.**

1. $\frac{3}{4} \bullet \frac{1}{2}$ **2.** $\frac{1}{3} \bullet \frac{1}{2}$ **3.** $\frac{3}{8} \bullet \frac{1}{2}$ **4.** $\frac{3}{5} \bullet \frac{1}{2}$

You can round fractions and mixed numbers to the nearest whole number. If the fractional part of a mixed number is equal to or greater than $\frac{1}{2}$ the whole number increases by 1. For example, in $2\frac{5}{8}$, $\frac{5}{8} > \frac{1}{2}$. So, $2\frac{5}{8}$ rounds to 3. If the fraction is less than $\frac{1}{2}$, the whole number stays the same.

Round each mixed number to the nearest whole number. Then estimate the sum.

5. $1\frac{1}{3} + 4\frac{3}{4}$ **6.** $3\frac{1}{2} + 5\frac{1}{5}$ **7.** $2\frac{4}{5} + 2\frac{2}{3}$ **8.** $2\frac{3}{4} + 5\frac{1}{4}$ **9.** $4\frac{1}{3} + 1\frac{2}{3}$ **10.** $3\frac{1}{4} + 6\frac{5}{8}$

Problem Solving Skill
Choose the Operation

GROWING GAINS   Steve and Leon each did a science project on plant growth. They recorded the weekly growth for the plants and made graphs.

Study the data. Then read Problems A and B.

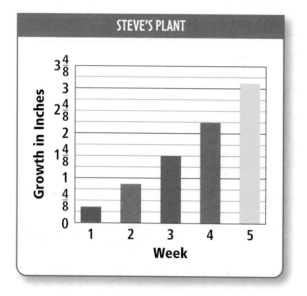

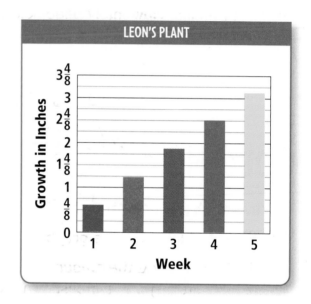

A. How much did Steve's plant grow the first two weeks? Leon's plant?

B. How much more did Leon's plant grow in Week 1 than Steve's plant?

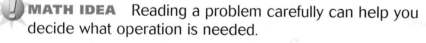 **MATH IDEA**   Reading a problem carefully can help you decide what operation is needed.

Talk About It

• Discuss how you would solve Problems A and B. Then solve.

• What operation would you use to find the amount of growth of Steve's or Leon's plant after 5 weeks?

• What operation would you use to find the difference in the amount of growth of Steve's and Leon's plants after 5 weeks?

► Problem Solving Practice

Write the operation. Then solve each problem.

1. Meí ate $\frac{1}{12}$ of the watermelon and Billy ate $\frac{2}{12}$. What fraction of the watermelon did they eat?

2. Michelle practiced the piano for $\frac{5}{6}$ hour and Seth practiced for $\frac{3}{6}$ hour. How much longer did Michelle practice than Seth?

For a science report, Lloyd says that he wrote $\frac{2}{4}$ page more than Mary and Bob combined. Mary wrote $\frac{3}{4}$ page and Bob wrote $1\frac{1}{4}$ pages. How many pages did Lloyd write?

3. Which expression could you use to solve the problem?

 A $\left(\frac{5}{4} - \frac{3}{4}\right) - \frac{2}{4}$ **C** $\left(\frac{5}{4} - \frac{3}{4}\right) + \frac{2}{4}$

 B $\left(\frac{3}{4} + \frac{5}{4}\right) + \frac{2}{4}$ **D** $\left(\frac{5}{4} + \frac{3}{4}\right) - \frac{2}{4}$

4. Which is NOT an answer to the question?

 F $\frac{10}{4}$ **H** $2\frac{2}{4}$

 G $\frac{6}{4}$ **J** $2\frac{1}{2}$

Mixed Applications

USE DATA For 5–7, use the graph.

Mr. Smith measured the amount of rainfall for five days. The measurements are shown on the graph.

5. What was the difference in the rainfall amount on Friday and Monday?

6. Find the total amount of rainfall for the five days.

7. Was the total rainfall for the first two days greater than the rainfall amount on Friday? Explain.

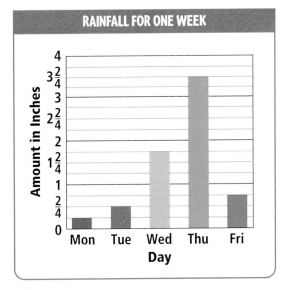

8. Arjun spent 3 of his 8 quarters at the store. He gave a friend 2 quarters. What fraction of his quarters does he have left?

9. ✎ **Write About It** Explain how you know what operation to use when solving a word problem.

10. **REASONING** Find the number that comes next in the sequence. $0, \frac{1}{2}, 1, \frac{3}{2}, 2, \frac{5}{2},$ ■. Tell a rule.

11. Find the value for ▲ that makes the equation true.

 $$▲ \times 3,000 = 360,000$$

Add Unlike Fractions

HANDS ON

VOCABULARY
unlike fractions

▶ Explore

Unlike fractions are fractions with different denominators. You can use fraction bars to add unlike fractions.

Activity

MATERIALS fraction bars

Find $\frac{1}{2} + \frac{1}{4}$.

STEP 1	STEP 2	STEP 3
Model with fraction bars.	Find the like fraction bars that together are equivalent to $\frac{1}{2} + \frac{1}{4}$ in length.	Three $\frac{1}{4}$ fraction bars are equal in length to $\frac{1}{2} + \frac{1}{4}$.

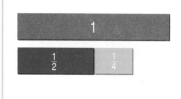

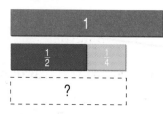

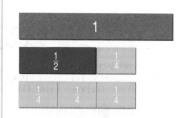

So, $\frac{1}{2} + \frac{1}{4} = \frac{3}{4}$.

- What would your answer be if you used two $\frac{1}{8}$ fraction bars instead of $\frac{1}{4}$ fraction bars? How is $\frac{2}{8}$ related to $\frac{1}{4}$?

Try It

Use fraction bars to find the sum.

a. $\frac{1}{4} + \frac{3}{8}$

b. $\frac{1}{3} + \frac{1}{6}$

c. $\frac{1}{6} + \frac{5}{12}$

d. $\frac{3}{4} + \frac{1}{6}$

How many $\frac{1}{8}$ bars are needed to show an equivalent fraction for $\frac{1}{4}$?

▶ Connect

You can use equivalent fractions to add fractions with unlike denominators.

Activity

Add. $\frac{2}{3} + \frac{1}{2}$

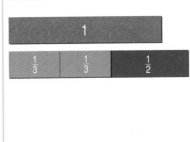

STEP 1

To add unlike fractions, first look at the denominators.

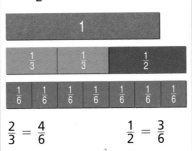

STEP 2

Find the like fraction bars that fit exactly under $\frac{2}{3}$ and $\frac{1}{2}$.

$\frac{2}{3} = \frac{4}{6}$ $\qquad$ $\frac{1}{2} = \frac{3}{6}$

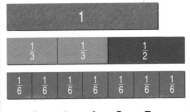

STEP 3

Add the like fractions. Write the answer as a mixed number.

So, $\frac{2}{3} + \frac{1}{2} = \frac{4}{6} + \frac{3}{6} = \frac{7}{6}$, or $1\frac{1}{6}$.

▶ Practice and Problem Solving

Use fraction bars to find like fractions. Write the fractions and the sum.

1.

2.
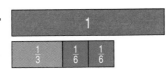

3.
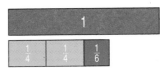

4. $\frac{2}{5} + \frac{1}{2}$ $\qquad$ 5. $\frac{2}{3} + \frac{1}{4}$ $\qquad$ 6. $\frac{1}{12} + \frac{3}{4}$ $\qquad$ 7. $\frac{1}{5} + \frac{3}{10}$

8. **? What's the Error?** Lisa is making a layered bow and has $9\frac{7}{12}$ yards of ribbon. If she needs $5\frac{2}{3}$ yards for the large bow and $2\frac{3}{4}$ yards for the small bow, how many yards will she have left?

Mixed Review and Test Prep

Order from *least* to *greatest*. (p. 372)

9. $\frac{1}{3}, \frac{1}{2}, \frac{1}{10}$ $\qquad$ 10. $\frac{3}{6}, \frac{2}{8}, \frac{3}{9}$

11. $24\overline{)369}$ (p. 286) $\qquad$ 12. $32\overline{)512}$ (p. 286)

13. **TEST PREP** Eric ate $\frac{4}{3}$ pizza. What mixed number represents how much pizza Eric ate? (p. 378)

A $1\frac{4}{3}$ $\qquad$ **B** $1\frac{3}{3}$ $\qquad$ **C** $1\frac{1}{3}$ $\qquad$ **D** 1

Subtract Unlike Fractions

Quick Review

1. $\frac{6}{10} - \frac{2}{10}$

2. $\frac{10}{12} - \frac{3}{12}$ 3. $\frac{6}{14} - \frac{2}{14}$

4. $\frac{10}{15} - \frac{9}{15}$ 5. $\frac{8}{9} - \frac{2}{9}$

▶ **Explore**

Use fraction bars to find $\frac{5}{8} - \frac{1}{4}$.

Activity
MATERIALS fraction bars

STEP 1

Place five $\frac{1}{8}$ fraction bars under a 1 whole bar. Then place one $\frac{1}{4}$ fraction bar under the five $\frac{1}{8}$ bars.

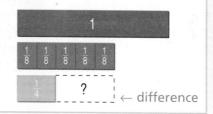
← difference

STEP 2

Compare the bars. Find like fraction bars that fit exactly under the difference $\frac{5}{8} - \frac{1}{4}$.

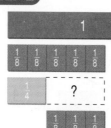

$\frac{5}{8} - \frac{1}{4} = \frac{3}{8}$

So, $\frac{5}{8} - \frac{1}{4} = \frac{3}{8}$.

• **REASONING** Suppose the problem is $\frac{6}{8} - \frac{1}{4}$. What is the difference?

Try It
Use fraction bars to find the difference.

a. $\frac{5}{6} - \frac{1}{3}$ b. $\frac{7}{8} - \frac{1}{4}$

c. $\frac{1}{2} - \frac{1}{4}$ d. $\frac{9}{10} - \frac{2}{5}$

How many $\frac{1}{6}$ fraction bars are needed to show the difference $\frac{5}{6} - \frac{1}{3}$?

▶ Connect

When you subtract unlike fractions, you first find equivalent fractions with like denominators.

Activity

Subtract. $\frac{2}{3} - \frac{1}{4}$

STEP 1	STEP 2	STEP 3
Place two $\frac{1}{3}$ fraction bars under a 1 whole bar. Then place one $\frac{1}{4}$ fraction bar under the two $\frac{1}{3}$ bars. 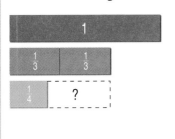	Find fraction bars that are equivalent to $\frac{2}{3}$ and $\frac{1}{4}$. 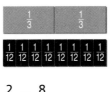 $\frac{2}{3} = \frac{8}{12}$ $\frac{1}{4} = \frac{3}{12}$	Compare the bars. Find the number of $\frac{1}{12}$ bars that fit exactly under the difference $\frac{2}{3} - \frac{1}{4}$. 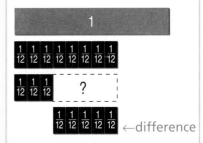 ←difference

So, $\frac{2}{3} - \frac{1}{4} = \frac{8}{12} - \frac{3}{12} = \frac{5}{12}$.

▶ Practice and Problem Solving

Use fraction bars to find like fractions. Write the fractions and the difference.

1. $\frac{3}{4} - \frac{1}{3}$ 2. $\frac{6}{8} - \frac{1}{4}$ 3. $\frac{9}{10} - \frac{3}{5}$ 4. $\frac{7}{12} - \frac{1}{4}$

5. Crystal has $\frac{11}{12}$ cup of sugar to make two cakes. The first cake needs $\frac{2}{3}$ cup of sugar and the second needs $\frac{1}{4}$ cup of sugar. Does she have enough sugar to make both cakes? Explain.

6. Find the distance, in miles, from York to Camden.

Mixed Review and Test Prep

7. List the factors of 7. Is 7 prime or composite? (p. 306)

Rename as mixed numbers. (p. 378)

8. $\frac{11}{4}$ 9. $\frac{10}{3}$ 10. $\frac{8}{5}$

11. **TEST PREP** John and his brother ate $\frac{3}{8}$ of a cake. His sisters ate $\frac{1}{4}$ of the same cake. How much cake was eaten in all? (p. 398)

A $\frac{1}{8}$ B $\frac{3}{4}$ C $\frac{1}{2}$ D $\frac{5}{8}$

Review/Test

✅ CHECK VOCABULARY AND CONCEPTS

Choose the best term from the box.

| like fractions |
| unlike fractions |
| numerators |
| denominators |

1. Fractions with different denominators are ___?___. (p. 398)

2. Fractions that have the same denominator are ___?___. (p. 386)

3. Add like fractions by adding the ___?___ and using the same denominator. (p. 386)

✅ CHECK SKILLS

Find the sum. (pp. 386–389; 392–395)

4. $\frac{2}{8} + \frac{5}{8}$ **5.** $\frac{3}{7} + \frac{4}{7}$ **6.** $\frac{1}{12} + \frac{4}{12}$ **7.** $\frac{2}{10} + \frac{6}{10}$

8. $\begin{array}{r} 3\frac{1}{4} \\ +12\frac{3}{4} \\ \hline \end{array}$ **9.** $\begin{array}{r} 2\frac{2}{3} \\ +1\frac{2}{3} \\ \hline \end{array}$ **10.** $\begin{array}{r} 2\frac{1}{4} \\ +8\frac{2}{4} \\ \hline \end{array}$ **11.** $\begin{array}{r} 7\frac{6}{8} \\ +5\frac{3}{8} \\ \hline \end{array}$

Find the difference. (pp. 390–395)

12. $\frac{5}{8} - \frac{2}{8}$ **13.** $\frac{3}{4} - \frac{1}{4}$ **14.** $\frac{5}{5} - \frac{2}{5}$ **15.** $\frac{10}{12} - \frac{3}{12}$

16. $\begin{array}{r} 5\frac{8}{9} \\ -2\frac{5}{9} \\ \hline \end{array}$ **17.** $\begin{array}{r} 3\frac{3}{5} \\ -\frac{2}{5} \\ \hline \end{array}$ **18.** $\begin{array}{r} 9\frac{5}{8} \\ -7\frac{3}{8} \\ \hline \end{array}$ **19.** $\begin{array}{r} 2\frac{11}{12} \\ -2\frac{5}{12} \\ \hline \end{array}$

Find the sum or difference. (pp. 398–401)

20. $\frac{1}{4} + \frac{1}{3}$ **21.** $\frac{3}{8} + \frac{3}{4}$ **22.** $\frac{7}{8} - \frac{1}{4}$ **23.** $\frac{1}{2} - \frac{2}{5}$

✅ CHECK PROBLEM SOLVING

Write the operation(s). Then solve each problem. (pp. 396–397)

24. Justin ate $\frac{1}{2}$ of the pizza and Julie ate $\frac{1}{3}$ of the pizza. How much of the pizza is left?

25. Ed worked $2\frac{3}{4}$ hours and Tom worked $3\frac{1}{4}$ hours. Find the total time they worked.

Standardized Test Prep

Get the information you need.
See item **9**.

You need to list factors of 100. Find the number that is not in the list.

Also see problem **3**, p. H63.

For 1–9, choose the best answer.

1. Which pair of fractions have like denominators?

A $\frac{2}{3}$ and $\frac{2}{5}$ **C** $\frac{6}{11}$ and $\frac{11}{6}$

B $\frac{1}{3}$ and $\frac{4}{3}$ **D** $\frac{1}{2}$ and $\frac{1}{3}$

2. Beth brought $2\frac{1}{2}$ dozen cookies to the party, and Cary brought $3\frac{1}{2}$ dozen. The party guests ate 4 dozen cookies. Which expression can be used to find how many dozen cookies are left?

F $\left(2\frac{1}{2} + 3\frac{1}{2}\right) - 4$

G $\left(4 + 2\frac{1}{2}\right) - 3\frac{1}{2}$

H $4 - \left(2\frac{1}{2} + 3\frac{1}{2}\right)$

J $3\frac{1}{2} + \left(2\frac{1}{2} + 4\right)$

3. $\frac{2}{3} + \frac{1}{4}$

A $\frac{3}{7}$ **C** $\frac{11}{12}$

B $\frac{3}{12}$ **D** $\frac{3}{4}$

4. $\frac{9}{11} - \frac{4}{11}$

F 0 **H** $\frac{5}{11}$

G $\frac{4}{11}$ **J** $\frac{6}{11}$

5. $4\frac{2}{3} + 6\frac{2}{3}$

A 10 **C** $10\frac{2}{3}$

B $11\frac{2}{3}$ **D** $11\frac{1}{3}$

6. Which expression is **not** equivalent to the others?

F $3 \times (5 + 2)$ **H** $7 \times (2 + 1)$

G $(5 + 9) + 7$ **J** $2 + (1 \times 7)$

7. Jay bought $\frac{3}{4}$ pound of lunch meat. He used $\frac{1}{2}$ pound to make some lunches. How much of the lunch meat is left?

A $\frac{1}{4}$ pound   **C** $\frac{4}{6}$ pound

B $\frac{1}{2}$ pound   **D** $\frac{2}{2}$ pound

8.
$$3\frac{4}{6}$$
$$+ 1\frac{3}{6}$$

F $4\frac{7}{12}$ **H** $4\frac{1}{6}$

G $5\frac{1}{6}$ **J** $5\frac{7}{12}$

9. Which is **not** a factor of 100?

A 3 **C** 5

B 10 **D** 4

Write What You Know

10. Describe how to subtract fractions with unlike denominators. Use your method to show that $\frac{4}{5} - \frac{1}{2} = \frac{3}{10}$.

11. Explain and correct the mistake in this problem.

$$\frac{5}{9} + \frac{2}{9} = \frac{7}{18}$$

Understand Decimals

Swimming became an organized sport in the 1800s. The first swimming event, the 50-meter freestyle, was held at the 1896 Olympic Games.

PROBLEM SOLVING The table lists the winning times at a school swim meet. Which event had the fastest time? the slowest time?

50-METER RACE WINNING TIMES				
	Backstroke (sec)	Breaststroke (sec)	Butterfly (sec)	Freestyle (sec)
Girls	81.81	140.77	78.15	70.23
Boys	82.86	123.86	78.15	65.43

CHECK WHAT YOU KNOW

Use this page to help you review and remember
important skills needed for Chapter 21.

VOCABULARY

Choose the best term from the box.

1. In the number 0.5, the 5 is in the ? place.

2. In the number 0.07, the 7 is in the ? place.

> ones
> tenths
> hundredths

MODEL DECIMALS (For intervention, see p. H20.)

Write the decimal for the shaded part.

3.

4.

5.

6.

7.

8.

9.

10.

RELATE DECIMALS TO MONEY (For intervention, see p. H20.)

Write the name of the coin that is described.

11. 0.1 of a dollar is a ?.

12. 0.01 of a dollar is a ?.

Write a decimal for the money amount.

13. two dollars and fifteen cents

14. three dollars and twelve cents

FRACTIONS WITH DENOMINATORS OF 10 AND 100 (For intervention, see p. H21.)

Write a fraction for each.

15. five tenths

16. nine hundredths

17. fifty-three hundredths

Complete to show equivalent fractions.

18.

$$\frac{6}{10} = \frac{\blacksquare}{100}$$

19.

$$\frac{1}{10} = \frac{\blacksquare}{100}$$

20.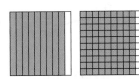

$$\frac{9}{10} = \frac{\blacksquare}{100}$$

Tenths and Hundredths

Quick Review

Draw a model for each.

1. $\frac{3}{10}$ 2. $\frac{7}{10}$ 3. $\frac{9}{10}$

4. $\frac{15}{100}$ 5. $\frac{4}{100}$

VOCABULARY

decimal

BATTER UP! A **decimal** is a number with one or more digits to the right of the decimal point.

Understanding fractions that have a denominator of 10 or 100 will help you understand decimals.

Gina plays on a Little League team. Gina scored 14 of her team's 100 runs this year. In the past three games, she was at bat 10 times and had 6 hits.

Example Gina had 6 hits out of 10 tries at bat.

Model	**Fraction**	**Decimal**
	Write: $\frac{6}{10}$ Read: six tenths	Write: 0.6 Read: six tenths

Gina scored 14 of her team's 100 runs this year.

Model	**Fraction**	**Decimal**
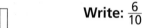	Write: $\frac{14}{100}$ Read: fourteen hundredths	Write: 0.14 Read: fourteen hundredths

Gina got a hit $\frac{6}{10}$, or 0.6, of her times at bat recently and scored $\frac{14}{100}$ or 0.14, of her team's runs this year.

More Examples

A Write: $\frac{5}{10}$, or 0.5

Read: five tenths

B Write: $\frac{25}{100}$, or 0.25

Read: twenty-five hundredths

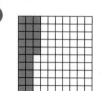

Read and Write Decimals

Decimals, like whole numbers, can be written in standard form, word form, and expanded form. Look at these numbers on the place-value chart.

Examples

Ones	.	Tenths	Hundredths
0	.	8	
0	.	1	2

Standard Form	Word Form	Expanded Form
0.8	eight tenths	8 tenths, or 0.8
0.12	twelve hundredths	12 hundredths, or $0.10 + 0.02$

You can write a decimal for a fraction that has a denominator other than 10 or 100. First write the fraction using a denominator of 10 or 100.

Technology Link

To learn more about fractions and decimals, watch the Harcourt Math Newsroom Video, *Windless Kites*.

Example

What decimal shows the same amount as $\frac{1}{2}$?

$$\frac{1}{2} = \frac{1 \times 5}{2 \times 5} = \frac{5}{10} \qquad \frac{5}{10} = 0.5$$

So, $\frac{1}{2}$ is the same as 0.5.

A number line divided into 100 equal parts can be used to model fractions and decimals that name the same amount in tenths or hundredths.

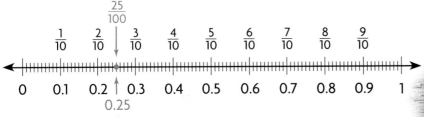

So, the decimal 0.25 shows the same amount as $\frac{25}{100}$, or $\frac{1}{4}$.

LESSON CONTINUES

1. Explain how the tenths model is different from the hundredths model.

Write the decimal and fraction shown by each model or number line.

2.   **3.**   **4.**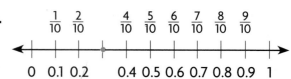

▶ **Practice and Problem Solving**

Write the decimal and fraction shown by each model or number line.

5.   **6.**   **7.**   **8.**

9. 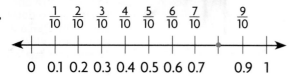   **10.**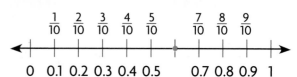

Write each fraction as a decimal.

11. $\frac{8}{10}$ **12.** $\frac{7}{10}$ **13.** $\frac{60}{100}$ **14.** $\frac{25}{100}$ **15.** $\frac{4}{100}$

16. $\frac{32}{100}$ **17.** $\frac{1}{5}$ **18.** $\frac{2}{100}$ **19.** $\frac{2}{4}$ **20.** $\frac{4}{5}$

21. $\frac{5}{10}$ **22.** $\frac{3}{5}$ **23.** $\frac{47}{100}$ **24.** $\frac{7}{100}$ **25.** $\frac{9}{10}$

 **ALGEBRA**   **Find the missing number or digit.**

26. $\frac{\blacksquare}{10} = 0.70$ **27.** $0.\blacksquare5 = \frac{3}{4}$ **28.** $\frac{\blacksquare}{4} = 0.50$ **29.** $\frac{2}{5} = 0.\blacksquare0$ **30.** $\frac{1}{\blacksquare} = 0.25$

Write the decimal two other ways.

31. 0.1 **32.** $0 + 0.4 + 0.07$ **33.** 0.4 **34.** 0 ones $+ 8$ tenths

35. In three weeks a theater sold $9,500 worth of tickets per week. For each of the next 2 weeks, the theater sold $7,200 worth of tickets per week. What were the total ticket sales?

36. 📖 **Write About It** Aline walked $\frac{3}{4}$ mile to school. Dave walked 0.75 mile to school. Aline said she walked farther than Dave. Is she correct? Explain.

37. The graph shows how Pepe spends 100 minutes of baseball practice. He spends $\frac{28}{100}$ of the time running bases. What decimal is this?

38. Write a decimal to show what part of the total time is spent on warm-ups.

39.  Write a problem about the time Pepe spends fielding and throwing.

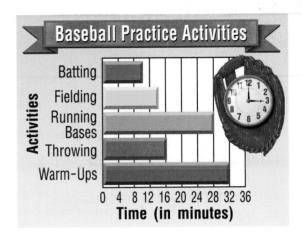

Baseball Practice Activities

Activities: Batting, Fielding, Running Bases, Throwing, Warm-Ups

Time (in minutes)
0 4 8 12 16 20 24 28 32 36

Mixed Review and Test Prep

40. Round 23,425 to the nearest hundred. (p. 26)

41. 398×11 (p. 224)

42. $\$5.21 \times 8$ (p. 200)

43. $\frac{2}{3} + \frac{1}{3}$ (p. 386)

44. 543×23 (p. 228)

45. **TEST PREP** What is $\frac{6}{10}$ in simplest form? (p. 368)

   A $\frac{1}{4}$ **B** $\frac{2}{5}$ **C** $\frac{1}{2}$ **D** $\frac{3}{5}$

46. **TEST PREP** Which number does **not** divide into 210 evenly? (p. 266)

   F 2   **G** 3   **H** 4   **J** 5

PROBLEM SOLVING LiNKUP ... to Science

How is the Earth's crust like an apricot? They are both made of common elements!

Elements are basic substances that form all matter. In the Earth's crust, the elements oxygen, silicon, aluminum, and iron occur in the greatest amounts. Oxygen makes up 0.47, or "forty-seven hundredths" of the elements in the Earth's crust. Twenty-eight hundredths of the Earth's crust is silicon, 0.08 is aluminum, and about 0.05 is iron.

Write each decimal as a fraction.

1. Oxygen: 0.47

2. Aluminum: 0.08

3. Silicon: 0.28

4. Iron: 0.05

▲ The Earth's crust contains oxygen, silicon, aluminum, and iron in greatest amounts.

▲ Apricots contain iron and silicon.

2 Thousandths

Learn

TINY PARTS Thousandths are even smaller parts than hundredths. If one hundredth were divided into ten equal parts, each part would represent one **thousandth**.

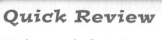
Quick Review
Write each fraction as a decimal.

1. $\frac{4}{10}$ 2. $\frac{34}{100}$ 3. $\frac{1}{10}$

4. $\frac{61}{100}$ 5. $\frac{50}{100}$

VOCABULARY

thousandth

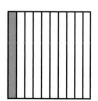 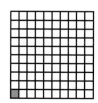 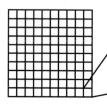 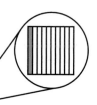

Fraction: $\frac{1}{1}$ $\frac{1}{10}$ $\frac{1}{100}$ $\frac{1}{1,000}$

Decimal: 1   0.1   0.01   0.001

Read: one   one tenth   one hundredth   one thousandth

Examples You can use a place-value chart to help you understand thousandths.

Ones	.	Tenths	Hundredths	Thousandths
0	.	4	6	3
0	.	0	1	9
0	.	0	0	2

Ty Cobb has the all-time highest career batting average of .367. ▼

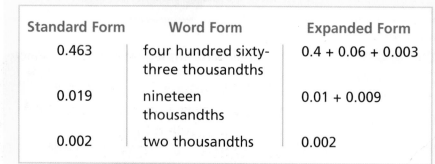

Standard Form	Word Form	Expanded Form
0.463	four hundred sixty-three thousandths	0.4 + 0.06 + 0.003
0.019	nineteen thousandths	0.01 + 0.009
0.002	two thousandths	0.002

• What is the value of the digit 6 in Ty Cobb's career batting average?

1. Tell how many thousandths are in one hundredth. How many thousandths are in one?

Write each decimal as a fraction.

2. 0.095　　　**3.** 0.418　　　**4.** 0.639　　　**5.** 0.002　　　**6.** 0.007

► **Practice and Problem Solving**

Write each decimal as a fraction.

7. 0.005　　　**8.** 0.749　　　**9.** 0.038　　　**10.** 0.001　　　**11.** 0.634

Use a place-value chart to write the value of the digit 5 in each decimal.

12. 0.025　　　**13.** 0.519　　　**14.** 0.153　　　**15.** 0.465　　　**16.** 0.593

Write each decimal in expanded form.

17. 0.034　　　**18.** 0.027　　　**19.** 0.689　　　**20.** 0.193　　　**21.** 0.042

Write each decimal in standard or word form.

22. four thousandths　　　**23.** 0.398　　　**24.** one hundred two thousandths

Complete.

25. 0.36■ = 0.3 + 0.06 + 0.004.

26. 0.903 = nine _?_ three thousandths

27. 0.408 = 0.4 + ■

28. 0.072 = seventy-two _?_

29. A 1-hour piano lesson costs $18. Students receive 1 free lesson for every 8 paid lessons. How much will 36 lessons cost?

30. Pam practiced the piano for $\frac{1}{6}$ hr on Wednesday and $\frac{1}{3}$ hr on Friday. How many minutes did she practice in all?

Mixed Review and Test Prep

31. Write the word form of 0.14. (p. 406)

32. 4 × (25 × 2) (p. 158)

33. 427 ÷ 7 (p. 266)

34. $4\frac{2}{3} + 5\frac{2}{3}$ (p. 392)

35. TEST PREP What is five million, two thousand, eight hundred sixteen written in standard form? (p. 8)

A 502,816　　　**C** 5,020,816

B 5,002,816　　　**D** 5,028,160

HANDS ON

Equivalent Decimals

Quick Review

1. $3.\blacksquare = 3\frac{2}{10}$
2. $2.6 = 2\frac{\blacksquare}{10}$
3. $7\frac{41}{\blacksquare} = 7.41$
4. $8.9 = 8\frac{9}{\blacksquare}$
5. $6.\blacksquare = 6\frac{49}{100}$

▶ **Explore**

Equivalent decimals are decimals that name the same number.

Use models and paper folding to find equivalent decimals. Are 0.2 and 0.20 equivalent decimals?

VOCABULARY

equivalent decimals

MATERIALS tenths and hundredths models; two different-colored markers

Activity

STEP 1

Shade 0.2 of the tenths model and 0.20 of the hundredths model.

0.2

two tenths
2 out of 10

0.20

twenty hundredths
20 out of 100

STEP 2

Fold 0.2 of the tenths model and 0.20 of the hundredths model. Then compare the models.

0.2

fold

0.20
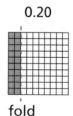
fold

So, 0.2 and 0.20 are equivalent decimals.

Try It

Use a tenths model and a hundredths model. Are the two decimals equivalent? Write *yes* or *no*.

a. 0.50 and 0.6   b. 0.3 and 0.30

c. 0.70 and 0.75   d. 0.8 and 0.80

How do these models show whether or not 0.50 and 0.6 are equivalent?

▶ Connect

Felipe said that $0.30 is 3 tenths of a dollar. Lea said that $0.30 is 30 hundredths of a dollar. Who was correct?

Technology Link

More Practice: Use E-Lab, *Equivalent Decimals.*

www.harcourtschool.com/
elab2002

Example Compare the models.

Felipe used a tenths model to show
$0.30 = 3 tenths.

Each column is
equal to 0.1,
or one tenth,
of a dollar.

0.3 of a dollar

Lea used a hundredths model to show
$0.30 = 30 hundredths.

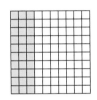

Each square
is equal to
0.01, or one
hundredth,
of a dollar.

0.30 of a dollar

The two models show that 3 tenths of a dollar is equal to 30 hundredths of a dollar. So, both Felipe and Lea are correct.

▶ Practice and Problem Solving

Are the two decimals equivalent? Write *yes* or *no*.

1. 0.7 and 0.70   **2.** 0.04 and 0.4   **3.** 0.9 and 0.09

4. 0.28 and 0.82   **5.** 0.17 and 0.07   **6.** 0.1 and 0.10

Write an equivalent decimal for each. You may use decimal models.

7. 0.8   **8.** 0.7   **9.** 0.90   **10.** 0.2

11. 0.5   **12.** 0.10   **13.** 0.40   **14.** 0.6

15. MENTAL MATH Erin's family plants a garden on 0.5 acre of their land. Write an equivalent decimal for this amount.

16. 📝 **Write About It** Make a model to show that 0.8 and 0.80 are equivalent. Explain your model.

Mixed Review and Test Prep

17. $\frac{4}{5} + \frac{4}{5}$ (p. 386)

18. $5\frac{5}{7} - 2\frac{2}{7}$ (p. 392)

19. $43.25 - $10.80

20. $20.00 - $5.25

21. **TEST PREP** Which number is **not** a factor of 320? (p. 298)

A 4   **B** 6   **C** 8   **D** 10

Chapter 21 **413**

Relate Mixed Numbers and Decimals

▶ **Learn**

ANNA'S BANANAS   Plantains, a variety of banana, grow in Mexico. Anna bought two and three tenths pounds of plantains at the store. How can you write this weight as a mixed number and as a decimal?

Mixed Number: $2\frac{3}{10}$

Decimal: 2.3

Read: two and three tenths

So, write the weight as $2\frac{3}{10}$, or 2.3 pounds.

- **What if** Anna also needs $3\frac{4}{10}$ pounds of peanuts? What decimal will a decimal scale show if she has the correct amount?

▲ There are 300 varieties of bananas worldwide.

Examples

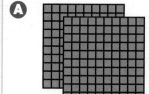

Ⓐ

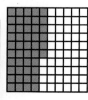

Mixed Number: $2\frac{46}{100}$

Decimal: 2.46

Read: two and forty-six hundredths

Ⓑ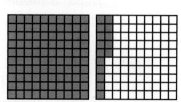

Mixed Number: $1\frac{15}{100}$

Decimal: 1.15

Read: one and fifteen hundredths

- **REASONING** How can you write four and fifty-one thousandths as a mixed number and as a decimal?

Decimal Equivalents

Sandy and Bill brought $2\frac{1}{2}$ pounds of bananas to the class picnic. The decimal equivalent for $2\frac{1}{2}$ can be found by using a number line or a decimal model.

Technology Link

More Practice: Use Mighty Math Calculating Crew, *Nautical Number Line*, Level Q.

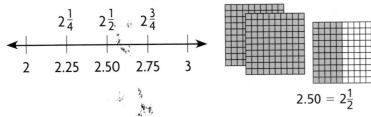

$2.50 = 2\frac{1}{2}$

The number line and the decimal model both show that $2\frac{1}{2}$ and 2.50 name the same amount.

More Examples
You can use decimal models to show other mixed numbers and decimals that are equivalent.

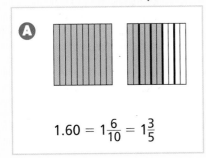

A $1.60 = 1\frac{6}{10} = 1\frac{3}{5}$

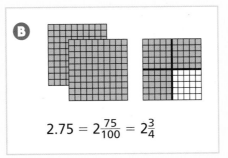

B $2.75 = 2\frac{75}{100} = 2\frac{3}{4}$

• What would the decimal model for 3.50 look like? What mixed number is equivalent to 3.50?

▶ Check

1. **Tell** how you can use a number line or a decimal model to relate mixed numbers and decimals.

Write an equivalent decimal and mixed number for each decimal model. Then write the word form.

2.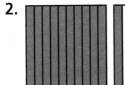

3.

4.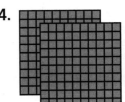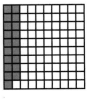

LESSON CONTINUES ▶

Write an equivalent decimal and mixed number for each decimal model. Then write the word form.

5.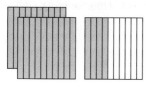

6.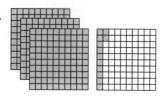

7.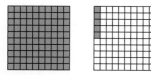

8.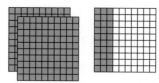

Write an equivalent mixed number or a decimal.

9. 11.50

10. $9\frac{1}{4}$

11. 7.25

12. $4\frac{1}{5}$

13. 7.7

14. $4\frac{2}{5}$

15. 8.06

16. 16.3

17. $27\frac{3}{5}$

18. $6\frac{1}{2}$

Use the number line to write an equivalent mixed number or a decimal for each letter.

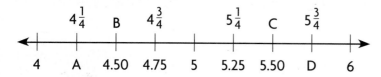

19. C

20. B

21. A

22. D

23. Kerri went to the store with $100.00. She spent $22.00 on a shirt, $12.00 on a hat, and $45.00 on a pair of lawn chairs for her parents. How much money did Kerri have left?

24. Tony needs to buy 6.75 feet of rope to make a rope swing. A worker at the hardware store measured a length of $6\frac{3}{4}$ feet of rope. Did the worker measure enough rope? Explain.

25. Joe's marble collection was $\frac{2}{5}$ blue. Alex's collection was 0.50 blue. Whose collection had a greater portion of blue marbles?

26. Use the number line above to name two mixed numbers and two decimals between 4.25 and 4.50.

27. ⭐ **?** **What's the Error?** Kris wrote the mixed number $2\frac{7}{10}$ in two ways, shown at the right. Describe her error. Write the correct answer.

Kris

two and seven tenths

2.07

28. Tasha has 16 boxes of dolls. Each box has 8 dolls. How many dolls does Tasha have?

29. Sumi ran 10.75 miles. Larry ran $10\frac{4}{5}$ miles. Who ran farther? Explain.

Mixed Review and Test Prep

30. Order from least to greatest: 27,654; 26,654; 27,754; 27,652 (p. 20)

31. Kara recorded these temperatures each day for a week: 65°F, 72°F, 74°F, 68°F, 67°F, 62°F, 65°F. What is the median? (p. 86)

32. List the factors of 21. (p. 298)

33. $2.17 × 5 (p. 200)

34. 3,425 × 9 (p. 200)

35. 26 × 24 (p. 224)

36. 441 ÷ 9 (p. 266)

37. Write $\frac{6}{9}$ in simplest form. (p. 368)

38. **TEST PREP** Which fraction is the greatest? (p. 372)

A $\frac{2}{3}$ **B** $\frac{3}{4}$ **C** $\frac{4}{5}$ **D** $\frac{5}{6}$

39. **TEST PREP** Which is equivalent to 2.4? (p. 414)

F 2.04 **G** $2\frac{4}{10}$ **H** $2\frac{4}{100}$ **J** 2.41

PROBLEM SOLVING Thinker's Corner

PERCENT A sewing store sells 100 buttons on a card. One half of the buttons are green. What percent is this?

You have seen how decimals and fractions are related. Let's explore how they are related to percents.

Percent (%) means "per hundred." So, you can write a percent as a fraction or a decimal.

$$50\% = \frac{50}{100} = 0.50 \qquad\qquad 25\% = \frac{25}{100} = 0.25$$

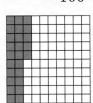

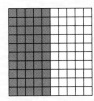

Think: 50 out of 100 squares are shaded.

Think: 25 out of 100 squares are shaded.

So, 50% of the buttons on a card are green.

Write the following decimals or fractions as percents.

1. 0.10 **2.** $\frac{40}{100}$ **3.** 0.80 **4.** $\frac{33}{100}$ **5.** 0.30

Compare and Order Decimals

▶ **Learn**

TUNNEL TRAVEL The Brooklyn-Battery Tunnel in New York is 1.73 miles long. The E. Johnson Memorial Tunnel in Colorado is 1.70 miles long. Which tunnel is longer?

One Way Ken used a number line to compare the decimals.

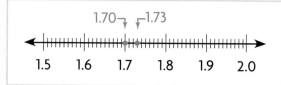

Another Way Sela used place value to compare the decimals.

ONES	.	TENTHS	HUNDREDTHS
1	.	7	3
1	.	7	0

1 = 1   7 = 7   3 > 0

Since 3 > 0, 1.73 > 1.70.

Think: Line up the decimal points. Compare the digits, beginning with the greatest place value.

So, the Brooklyn-Battery Tunnel is longer.

Example

Compare 0.316 and 0.398.

ONES	.	TENTHS	HUNDREDTHS	THOUSANDTHS
0	.	3	1	6
0	.	3	9	8

3 = 3   1 < 9

Since 1 < 9, 0.316 < 0.398.

Remember

On a number line, the numbers to the right are greater than the numbers to the left.

Order Decimals

One Way   Use a number line to order decimals.
Order 9.4, 9.63, and 9.27 from greatest to least.

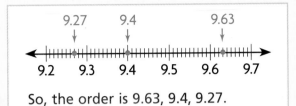

So, the order is 9.63, 9.4, 9.27.

Another Way   Use place value to order decimals. Order
1.23, 0.98, and 1.28 from least to greatest.

STEP 1	STEP 2	STEP 3
Line up the decimal points. Compare the digits in the greatest place.	Compare the tenths.	Compare the hundredths.
1.23 ↓ 0.98 0 < 1 ↓ 1.28	1.23 ↓ 2 = 2 1.28	1.23 ↓ 3 < 8 1.28
Since 0 < 1, 0.98 is the least.	There are the same number of tenths.	So, the order from least to greatest is 0.98, 1.23, 1.28.

- **What if** you wanted to write the decimals from greatest to least?
  How would this change the order?

Example

Order 0.813, 0.6, 0.65 from *least* to *greatest*.

0.813		**Think:** 8 > 6, so 0.813 is the greatest.
↓		
0.6	0.600	0.6 is equivalent to 0.600.
↓	↓	0.65 is equivalent to 0.650.
0.65	0.650	0 < 5, so 0.6 is the least.

The order from least to greatest is 0.6, 0.65, 0.813.

LESSON CONTINUES

1. **Tell** how you can use a number line to help you compare decimals.

Compare. Write <, >, or = for each ⬤.

2. 0.45 ⬤ 0.35 **3.** 0.5 ⬤ 0.7 **4.** 0.03 ⬤ 0.30 **5.** 5.4 ⬤ 5.243 **6.** 1.036 ⬤ 1.308

Use the number line to order the decimals from *least* **to** *greatest.*

2.0	2.1	2.2	2.3	2.4	2.5	2.6	2.7	2.8	2.9	3.0

7. 2.01, 2.10, 2.2, 2.02   **8.** 2.7, 2.67, 2.76, 2.6

Compare. Write <, >, or = for each ⬤.

9. 0.82 ⬤ 0.93 **10.** 0.81 ⬤ 0.18 **11.** 0.5 ⬤ 0.51 **12.** 0.20 ⬤ 0.02

13. 1.0 ⬤ 1.029 **14.** 0.600 ⬤ 0.6 **15.** 2.316 ⬤ 2.63 **16.** 0.74 ⬤ 0.53

Use the number line above to order the decimals from *least* **to** *greatest.*

17. 2.01, 2.11, 2.13, 2.10 **18.** 2.23, 2.45, 2.32, 2.5 **19.** 2.94, 2.49, 2.4, 3.00

Order the decimals from *greatest* **to** *least.*

20. 1.04, 4.11, 0.41, 1.40

21. 0.96, 1.06, 0.9, 1.6

22. 4.08, 4.3, 4.803, 4.038

23. 2.007, 2.714, 2.09, 2.97

24. 0.086, 8.6, 8.069, 0.006

25. 1.703, 1.037, 1.37, 1.073

26. ❓ **What's the Question?** Wes has $4.10 more than June. Debbie has $7.25 less than Wes. June has $5.58. The answer is $3.15.

27. Compare the decimals 0.8 and 0.2 using < or >. Then explain how you can use a number line to find the difference between them.

28. **REASONING** Which of these numbers has the same value as the digit 7 in the number 136.074? 70, 7, 0.7, 0.07, 0.007

29. **REASONING** List all the possible digits for the missing number.

12.34 < 12.■6 < 12.77

USE DATA For 30–32, use the table.

30. What was the time for the fastest runner? What was the time for the slowest runner?
    HINT: The least time is the fastest.

31. Mia also ran the mile. Her time was 6.48 minutes. Order the times from least to greatest.

32. Keisha ran the mile in 6.43 minutes. Compare her time to Jessica's time. Who was faster?

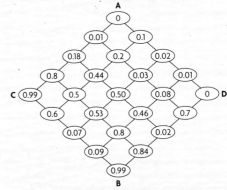

MILE RUN	
Runner	**Time (in minutes)**
Lisa	6.50
Jessica	6.45
Kelly	6.40

Mixed Review and Test Prep

33. Write $\frac{9}{12}$ in simplest form. (p. 368)

34. $3,618 \div 9$ (p. 272)

35. List the factors of 36. (p. 300)

36. 32×100 (p. 228)

37. Write a fraction equivalent to 0.73. (p. 406)

38. Write a decimal equivalent to 5.4. (p. 412)

39. **TEST PREP** Choose the letter of the fraction equivalent to 1.06. (p. 414)

   A $1\frac{6}{10}$ C $1\frac{60}{10}$

   B $1\frac{6}{100}$ D $1\frac{60}{100}$

40. **TEST PREP** Choose the letter of the greatest fraction. (p. 372)

   F $\frac{1}{3}$ H $\frac{1}{2}$

   G $\frac{1}{6}$ I $\frac{1}{8}$

PROBLEM SOLVING | Thinker's Corner

A-MAZE-ING REASONING

MATERIALS: Decimal Maze worksheet

1. On the worksheet, trace a path through the maze from A to B. For each step, move to a number of greater value.

2. On the worksheet, trace a path through the maze from C to D. For each step, move to a number of lesser value.

3. Then, using the blank maze on the worksheet, make your own maze. Try to make your path the only possible way to get across the maze. Give your maze to a partner to solve.

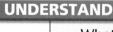
Problem Solving Strategy
Use Logical Reasoning

Quick Review

Which is greater?

1. 0.5 or 0.9 2. 0.12 or 0.21

3. 3.1 or 2.5 4. 5.50 or 5.35

5. 37.0 or 37.10

PROBLEM Miss Epps used a stopwatch to time Max, Jenna, and Dalia in a race. The times were 25.15 sec, 30.50 sec, and 34.10 sec. Jenna was slower than Dalia. A boy came in second. Who received first, second, and third place?

UNDERSTAND

- What are you asked to find?
- What information will you use?
- Is there any information you will not use?

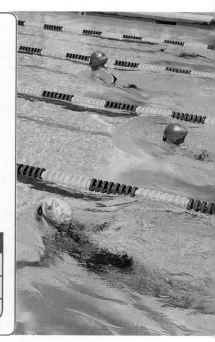

▲ This stopwatch shows thirteen and fifteen hundredths seconds.

PLAN

- What strategy can you use to solve the problem?

  Use *logical reasoning* to determine the order in which the students finished.

SOLVE

- How can you use the strategy?

  Organize what you know in a table. Show all the possibilities.

A A boy came in second.

Max is the only boy, so he must have the middle time. No two people have the same time, so there can be only one *yes* in each row and column.

	25.15	30.50	34.10
Max	NO	YES	NO
Dalia		NO	
Jenna		NO	

B Jenna was slower than Dalia.

Since 34.10 sec is slower than 25.15 sec, Dalia's time must be 25.15 sec.

	25.15	30.50	34.10
Max	NO	YES	NO
Dalia	YES	NO	NO
Jenna	NO	NO	YES

So, Dalia was first, Max was second, and Jenna was third.

CHECK

- How can you check your work?

▶ Problem Solving Practice

1. **What if** after the race, Max, Dalia, and Jenna were thirsty? One person had a sports drink. The winner chose juice, and another person had water. Max does not like sports drinks. What did each person drink?

The temperatures last week were 85°F, 75°F, 77°F, 83°F, and 81°F. Monday was the hottest day, and Thursday was the coolest. Wednesday was cooler than Friday but warmer than Tuesday.

2. On which day was the temperature 83°F?
   - **A** Monday
   - **B** Tuesday
   - **C** Wednesday
   - **D** Friday

3. On which day was the median temperature recorded?
   - **F** Monday
   - **G** Tuesday
   - **H** Wednesday
   - **J** Friday

🔍 PROBLEM SOLVING STRATEGIES

Draw a Diagram or Picture
Make a Model or Act It Out
Make an Organized List
Find a Pattern
Make a Table or Graph
Predict and Test
Work Backward
Solve a Simpler Problem
Write an Equation
▶ **Use Logical Reasoning**

Mixed Strategy Practice

4. Tim earns $4.00 an hour mowing lawns. He worked from 12:00 P.M. to 5:00 P.M. How much did he earn?

5. What two numbers come next in this pattern? Explain.

   32, 28, 23, 17, ▨, ▨

6.

   If this pattern continues, how many squares will be in the 7th figure? Explain.

7.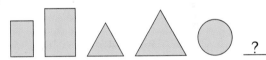

   Look at the pattern. Draw the next figure.

8. Shannon went grocery shopping. She spent $1.54 for grapes, $2.25 for milk, and $0.75 for a newspaper. When Shannon got home, she had $15.46. How much money did Shannon bring to the grocery store?

9. Bruno, David, and Zack are out for breakfast. The waiter brings eggs and bacon, pancakes, and cereal to their table. Bruno needs syrup and David does not like meat. Who gets which order?

10. Write the greatest and least four-digit decimal expressed to hundredths. Do not write zero in the hundredths place.

Review/Test

CHAPTER 21

✓ CHECK VOCABULARY AND CONCEPTS

Choose the best term from the box.

> decimal
> decimal point
> equivalent decimals

1. A number with one or more digits to the right of the decimal point is a _?_. (p. 406)

2. Decimals that name the same number are _?_. (p. 412)

✓ CHECK SKILLS

Write each fraction as a decimal. (pp. 406–411)

3. $\frac{6}{10}$ 4. $\frac{1}{100}$ 5. $\frac{8}{10}$ 6. $\frac{48}{100}$ 7. $\frac{2}{5}$

8. $\frac{4}{1,000}$ 9. $\frac{89}{1,000}$ 10. $\frac{9}{25}$ 11. $\frac{312}{1,000}$ 12. $\frac{176}{1,000}$

Write an equivalent decimal for each. (pp. 412–413)

13. 0.60   14. 0.9   15. 0.4   16. 0.50   17. 0.70

Write an equivalent mixed number or a decimal. (pp. 414–417)

18. $6\frac{79}{100}$ 19. $1\frac{67}{1,000}$ 20. $3\frac{5}{100}$ 21. 8.16   22. 4.002

Compare. Write <, >, or = for each ●. (pp. 418–421)

23. 0.71 ● 0.63   24. 0.56 ● 0.837   25. 2.603 ● 2.61   26. 1.4 ● 1.40

Order the decimals from *least* to *greatest*. (pp. 418–421)

27. 1.23, 2.23, 1.32, 0.89, 2.03   28. 3.06, 3.97, 3.614, 3.8

✓ CHECK PROBLEM SOLVING

Solve. (pp. 422–423)

29. May, Peg, Lon, and Tim each bought a gift. The gifts cost $9.57, $8.64, $9.32, and $8.97. May's gift cost more than Tim's but less than Lon's. Lon's cost more than Peg's. Tim's gift cost $8.97. Name each child and the amount of his or her gift.

30. Four runners ran a mile in 6.52 min, 7.20 min, 6.59 min, and 7.16 min. Elena finished after Lara but before Nick. Jan ran the fastest. List the runners with their times, from first through fourth place.

★Standardized Test Prep

 Eliminate choices.
See item **2**.

First find the answer choices that are ordered. Choose the one ordered from *least* to *greatest*.

Also see problem **5**, p. H64.

For 1–10, choose the best answer.

1. Which fraction is equivalent to 0.02?

   A $\frac{2}{100}$ **C** $\frac{2}{1}$

   B $\frac{2}{10}$ **D** $\frac{2}{1,000}$

2. Which shows the numbers in order from least to greatest?

   F 2.22, 2.42, 2.24, 2.20

   G 2.42, 2.24, 2.22, 2.20

   H 2.22, 2.20, 2.24, 2.42

   J 2.20, 2.22, 2.24, 2.42

3. How is $4\frac{7}{1,000}$ written as a decimal?

   A 4.700   **C** 4.710

   B 4.070   **D** 4.007

4. There are 83 students going on a field trip. Each school bus holds 60 students. How many buses will be needed?

   F 1   **G** 2   **H** 5   **J** 6

5. The ages of all of Roger's aunts and uncles are 38, 24, 37, 29, 42, 34, and 25. What is the median age?

   A 24   **B** 29   **C** 34   **D** 38

6. Which pair shows decimals that are **not** equivalent?

   F 0.38 and 0.038

   G 0.06 and 0.060

   H 0.80 and 0.8

   J 0.4 and 0.40

7. Maria needs $12\frac{3}{4}$ inches of ribbon. Emily needs 12.75 inches. Who needs more ribbon?

   A Emily; 12.75 inches is more than $12\frac{3}{4}$ inches.

   B Maria; $12\frac{3}{4}$ inches is more than 12.75 inches.

   C Neither; the amounts are equivalent.

   D Cannot be determined

8. Which is true?

   F $0.56 > 0.59$

   G $0.75 > 0.31$

   H $0.078 < 0.05$

   J $0.04 = 0.4$

9. $13 + (11 \times 9)$

   A 112   **C** 216

   B 116   **D** NOT HERE

10. Which fraction is equivalent to 10.75?

    F $10\frac{5}{7}$ **H** $10\frac{3}{4}$

    G $10\frac{7}{10}$ **J** $10\frac{1}{2}$

Write What You Know

11. Draw a model to compare 1.7 and 1.70. Explain how the model helps you compare the decimals.

12. Write the decimal equivalent for the shaded part of the model shown. Tell how you found the decimal.

Add and Subtract Decimals

The first weather station in Alaska was established in Anchorage in 1915. The average snowfall in Anchorage is more than 71 inches per year.

PROBLEM SOLVING The table below shows the deepest February snowfalls. What is the range of the record snowfalls shown?

DATA LINK

RECORD FEBRUARY SNOWFALLS IN ANCHORAGE, ALASKA	
Year	**Snowfall (in inches)**
1955	48.50
1956	33.10
1968	26.10
1974	23.30
1978	20.80
1990	23.00
1996	52.10

Owl snow sculpture in Alaska

CHECK WHAT YOU KNOW

Use this page to help you review and remember
important skills needed for Chapter 22.

✔ VOCABULARY

Choose the best term from the box.

1. The numbers 0.9 and 0.90 are ? .

2. The ? shows that 2.3 + 9.7 = 9.7 + 2.3.

> equivalent decimals
> Order Property
> Grouping Property

✔ ROUND MONEY AMOUNTS (For Intervention, see p. H21.)

Round each to the nearest dollar.

3. $5.89 4. $9.28 5. $2.51 6. $7.47

7. $16.47 8. $52.73 9. $20.65 10. $133.33

Round each to the nearest ten cents.

11. $5.22 12. $8.99 13. $1.49 14. $72.19 15. $47.34

16. $24.52 17. $44.39 18. $138.12 19. $210.96 20. $627.53

✔ ADD AND SUBTRACT MONEY (For Intervention, see p. H22.)

Find the sum.

21.	22.	23.	24.	25.
$4.99 +$6.48	$6.00 +$2.89	$6.45 +$0.39	$22.75 +$19.23	$147.50 +$ 32.99

26.	27.	28.	29.	30.
$6.02 +$3.56	$7.41 +$2.80	$8.67 +$4.76	$80.19 +$44.97	$212.45 +$189.20

Find the difference.

31.	32.	33.	34.	35.
$7.99 −$2.32	$9.00 −$3.95	$5.63 −$1.50	$9.50 −$7.68	$6.55 −$4.90

36.	37.	38.	39.	40.
$10.00 −$ 5.20	$20.00 −$13.09	$15.01 −$ 3.41	$100.00 −$ 72.18	$1,000.00 −$ 970.74

Round Decimals

Quick Review

Round each number to the nearest ten.

1. 56 **2.** 84 **3.** 938

4. 4,892 **5.** 15,284

SNOW TREK Lisa and some friends went cross-country skiing. They covered 4.2 miles.

Round 4.2 to the nearest whole number.

One Way Use a number line.

4.2

4   5

4.2 is between 4 and 5, but it is closer to 4.

Another Way Use the rounding rules.

Look at the tenths place.   4.2

Since 2 < 5, the digit 4 stays the same.

So, 4.2 rounded to the nearest whole number is 4.

Remember

Rounding Rules:
- Find the place to which you want to round.
- Look at the digit to its right.
- If that digit is less than 5, the digit in the rounding place stays the same.
- If that digit is 5 or more, the digit in the rounding place is increased by 1.

Examples

A Round 6.48 to the nearest whole number. Look at the number line.

6.48

6   7

6.48 is closer to 6 than to 7. So, 6.48 rounds to 6.

B Round 5.076 to the nearest hundredth.

Use the rounding rules.

Look at the thousandths place.   5.076

Since 6 > 5, the digit 7 is increased by 1.

So, 5.076 rounds to 5.08.

MATH IDEA Decimals can be rounded using a number line or the rounding rules.

▶ Check

1. **Explain** how to use a number line to round 3.4 to the nearest whole number.

Round each number to the place of the blue digit.

2. 2.2   3. 1.8   4. $16.98   5. 7.305   6. 6.327

Round to the nearest tenth. Use the number lines or the rounding rules.

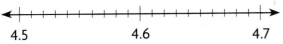

7. 4.50   8. 4.55   9. 4.61   10. 9.67   11. 9.52

▶ Practice and Problem Solving

Round each number to the place of the blue digit.

12. 5.84   13. 3.18   14. $1.43   15. $7.71   16. $36.52

17. 13.68   18. 49.274   19. 27.643   20. $83.54   21. $54.91

Round to the nearest hundredth.

22. 10.076   23. 61.349   24. 5.181   25. 9.413   26. 24.259

27. Round 5.261 and 5.19 to the nearest tenth, and compare.

28. **REASONING** For what digits will 43.9■5 round to 43.9?

29. David bought a jacket on sale for $32.49. To the nearest ten dollars, how much did David's jacket cost?

30. What is two and fifty-one thousandths rounded to the nearest hundredth? to the nearest tenth?

31. **REASONING** James paid $5.82 for a paint set. He told Pete the cost was about $6.00. Was $6.00 a reasonable rounded amount? Explain.

32. **? What's the Question?** Ted has half as many marbles as Joel and 25 fewer marbles than Spencer. Spencer has 60 marbles. The answer is 70 marbles.

Mixed Review and Test Prep

33. Write an equivalent decimal for 8.3. (p. 412)

34. Write 235,617 in expanded form. (p. 6)

35. Round 217,627 to the nearest ten thousand. (p. 26)

36. Find the value of $n + 8$ for $n = 16$. (p. 54)

37. **TEST PREP** Which number is NOT a multiple of 9? (p. 298)

   A 3   **C** 18
   B 9   **D** 90

EXTRA PRACTICE page H53, Set A

Estimate Sums and Differences

▶ Learn

PACK YOUR BAGS!   The table shows the three countries that had the most visitors in 1998. Altogether, about how many people traveled to these countries?

Estimate by rounding to the nearest ten.

$$
\begin{array}{rcl}
70.0 & \to & 70 \\
47.7 & \to & 50 \\
+47.1 & \to & \underline{50} \\
& & 170
\end{array}
$$

• Line up the decimal points.
• Round to the nearest ten.

So, altogether about 170 million travelers visited France, Spain, and the United States in 1998.

A travel magazine showed that in San Francisco, California, each traveler spent an average of $72.58 per day. In Paris, France, it was an amount equal to $81.55. About how much more did each traveler spend in Paris per day?

Estimate by rounding to the nearest dollar.

$$
\begin{array}{rcl}
\$81.55 & \to & \$82 \\
-\$72.58 & \to & \underline{-\$73} \\
& & \$9
\end{array}
$$

• Line up the decimal points.
• Round to the nearest dollar.

So, each traveler spent about $9 more per day in Paris.

• Name some situations in which you might need only an estimated sum or difference.

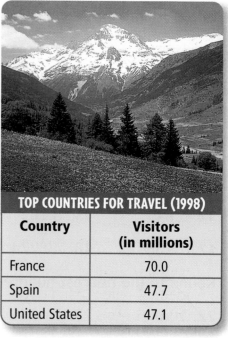

TOP COUNTRIES FOR TRAVEL (1998)

Country	Visitors (in millions)
France	70.0
Spain	47.7
United States	47.1

Spring Meadow, France

▶ Check

1. Explain how you would determine if a total cost of $19.48 is reasonable for a travel book that costs $12.99 and a poster that costs $6.49.

Estimate the sum or difference.

2. $\begin{array}{r} \$24.76 \\ +\$\ 5.21 \end{array}$ **3.** $\begin{array}{r} 5.25 \\ +7.06 \end{array}$ **4.** $\begin{array}{r} 2.314 \\ -1.238 \end{array}$ **5.** $\begin{array}{r} \$7.80 \\ -\$2.07 \end{array}$ **6.** $\begin{array}{r} 17.136 \\ +19.785 \end{array}$

1. **Describe** how regrouping to subtract decimals is like regrouping to subtract whole numbers.

Find the difference. Estimate to check.

2. 0.8
   -0.5

3. 20.82
   $-\ 7.71$

4. 22.3
   -11.9

5. 3.426
   -0.249

6. 2.914
   -1.685

► **Practice and Problem Solving**

Find the difference. Estimate to check.

7. 0.9
   -0.2

8. 6.93
   -0.54

9. 1.6
   -0.8

10. 41.97
   -10.38

11. 52.72
   -21.28

12. 2.453
   -2.386

13. 3.517
   -1.274

14. 4.768
   -2.993

15. 16.702
   $-\ 5.178$

16. 5.083
   -2.226

17. $7.89 - 4.37$

18. $9.84 - 3.87$

19. $8.2 - 6.9$

20. $29.53 - 18.98$

21. $9.124 - 3.076$

22. $4.237 - 2.819$

23. $13.207 - 11.496$

24. $9.532 - 4.107$

For 25–30, find the missing digits.

25. $8.\blacksquare - \blacksquare.9 = 1.6$

26. $\blacksquare.3 - 2.\blacksquare = 1.5$

27. $\blacksquare 6.5 - 3.\blacksquare = 12.8$

28. $7.\blacksquare - \blacksquare.3 = 5.3$

29. $\blacksquare.6 - 4.\blacksquare = 2.5$

30. $\blacksquare.9 - \blacksquare.8 = 7.1$

31. **REASONING** The difference of two 3-digit decimal numbers is 1.34. One number has a 6 in the ones and hundredths places. The other has a 6 in the tenths place. What are the two numbers?

32. **? What's the Error?** Describe the error. Write the correct answer.

   4.6
   -3.9
   1.7

33. Molly has saved $600 for a trip to France with the French Club. She needs twice that amount plus $50 more. How much does the trip cost?

Mixed Review and Test Prep

34. List the factors of 18. (p. 300)

35. Order 5.82, 4.95, and 5.89 from least to greatest. (p. 418)

36. 8×136 (p. 196)

37. $156 \div 4$ (p. 266)

38. **TEST PREP** Find 5×13. (p. 196)
   A 55   **B** 60   **C** 65   **D** 70

5 Add and Subtract Decimals

Quick Review

Compare. Write <, >, or = for each ⬤.

1. 47.5 ⬤ 4.81

2. 3.2 ⬤ 2.3   3. 0.75 ⬤ 0.72

4. 2.8 ⬤ 2.84   5. 9.9 ⬤ 9.90

▶ **Learn**

RAIN, RAIN, GO AWAY! Danny is doing a report on Iowa weather. Des Moines receives about 33 inches of rain each year. Use the table to find the total average rainfall from April through June.

Example 1

Add. 3.21 + 3.96 + 4.18

Estimate. 3 + 4 + 4 = 11

STEP 1

Line up the decimal point and place value of each number.

ones	.	tenths	hundredths

	.		
3	.	2	1
3	.	9	6
+4	.	1	8

STEP 2

Add as you do with whole numbers. Place the decimal point in the sum.

ones	.	tenths	hundredths

	.	1	1
3	.	2	1
3	.	9	6
+4	.	1	8
11	.	3	5

AVERAGE RAINFALL (DES MOINES, IA)

Month	Amount of Rain (in inches)
April	3.21
May	3.96
June	4.18

The answer of 11.35 is close to the estimate of 11, so the answer is reasonable. The total is 11.35 inches.

Example 2

What if Danny wanted to compare the average rainfall for April and June? On average, how many more inches are received in June?

Subtract. 4.18 − 3.21

Estimate. 4 − 3 = 1

ones	.	tenths	hundredths

	.		
³4̷	.	¹¹1̷	8
−3	.	2	1
0	.	9	7

Line up the decimal points.

So, the average rainfall is about 0.97 inch more in June than in April.

Equivalent Decimals

Sometimes one number has more decimal places after the decimal point than the other. Write equivalent decimals with the same number of decimal places before adding or subtracting.

Andrea goes to the store to buy a sun visor that costs $12.56. She gives the cashier $20. How much change should she receive?

Example

Subtract. $20 − $12.56

Estimate. $20 − $13 = $7

STEP 1

Line up the decimal points. Place zeros to the right of the decimal point so each number has the same number of digits after the decimal point.

$$
\begin{array}{r}
\$2\,0\,.\,0\,0 \\
-\$1\,2\,.\,5\,6 \\
\hline
\end{array}
$$

STEP 2

Subtract as you do with whole numbers. Place the decimal point.

$$
\begin{array}{r}
\overset{9\quad9}{\underset{}{1\ \cancel{10}\ \cancel{10}10}} \\
\$2\,\cancel{0}\,.\,\cancel{0}\,\cancel{0} \\
-\$1\,2\,.\,5\,6 \\
\hline
\$\ \ 7\,.\,4\,4 \\
\end{array}
$$

So, Andrea will receive $7.44 in change.

- How much change should Andrea get back if she gave the cashier $20.01? Why do you think Andrea may want to do this?

More Examples

Ⓐ $25 − $16.33

$$
\begin{array}{r}
\overset{14\quad9}{\underset{}{1\ \cancel{4}\ \cancel{10}10}} \\
\$2\,\cancel{5}\,.\,\cancel{0}\,\cancel{0} \\
-\$1\,6\,.\,3\,3 \\
\hline
\$\ \ 8\,.\,6\,7 \\
\end{array}
$$

Ⓑ $32.56 + $57.89

$$
\begin{array}{r}
\overset{1\ \ 1\ \ 1}{} \\
\$3\,2\,.\,5\,6 \\
+\$5\,7\,.\,8\,9 \\
\hline
\$9\,0\,.\,4\,5 \\
\end{array}
$$

Ⓒ 57.68 − 38.567

$$
\begin{array}{r}
\overset{4\ 17\quad7\ 10}{\underset{}{5\ \cancel{7}\,.\,6\,8\,\cancel{0}}} \\
-3\,8\,.\,5\,6\,7 \\
\hline
1\,9\,.\,1\,1\,3 \\
\end{array}
$$

LESSON CONTINUES ▶

1. Explain why zeros are sometimes placed to the right of the decimal point of numbers.

Find the sum or difference. Estimate to check.

2. $21 − $10.20   **3.** 5.4 + 0.39   **4.** 13 + 9.12   **5.** 15.03 − 9.647

▶ **Practice and Problem Solving**

Find the sum or difference. Estimate to check.

6. 9.5
    +2.52

7. 6.4
    −2.26

8. 3
    −1.39

9. 3.8
    +4.073

10. 21.28
     − 8

11. 43.8
    + 1.73

12. 7
    −3.18

13. 5.3
    −2.87

14. 56.123
     − 8

15. 16.2
    + 9.5

16. 19
    − 4.37

17. 79.142
    − 3.861

18. 7.9
    −2.58

19. 74.68
    + 8.3

20. 43
    − 6.507

21. $6.99 + $2.09   **22.** 1.3 − 0.4   **23.** 18.7 − 5.941   **24.** $65 − $30.50

25. 56.83 − 0.67   **26.** $1.34 + $12.09   **27.** 41.36 − 7.89   **28.** 69.4 + 7.802

29. 4.5 + 19 + 6.032   **30.** 14.4 + 19 + 7.74   **31.** $45 + $31.50 + $20

Find the missing number.

32. $2.51 − 0.8 = \blacksquare$ **33.** $2.32 − 1.6 = \blacksquare$ **34.** $\blacksquare − 0.90 = 0.2$

35. $3.02 − \blacksquare = 1.31$ **36.** $0.9 + 2.25 = \blacksquare$ **37.** $\blacksquare + 0.52 = 1.12$

38. $4.76 − \blacksquare = 2.93$ **39.** $15.86 + 3.79 = \blacksquare$ **40.** $\blacksquare − 6.5 = 21.89$

41. Find two 3-digit decimals whose difference is 7.09.

42. Find three 3-digit decimals whose sum is 16.5.

43. **? What's the Error?** Maria added 3.16, 1.04, and 0.07 and got a sum of 42.7. Describe her error. Write the correct answer.

44. Write a problem using addition or subtraction in which the answer is $9.21.

45. Alex went to the store with $10 and left the store with $5.98. He bought milk, eggs, and bread. The milk was $1.49 and the eggs were $0.89. How much was the bread?

46. Trevor scored 2 home runs in a college baseball game. Use the diagram to find how many feet Trevor ran to score the 2 home runs.

47. Will is walking to the ball field, which is 2.3 kilometers from his home. He has walked 0.8 kilometer. How much farther must he walk?

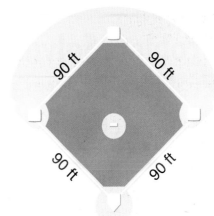

Mixed Review and Test Prep

48. $\frac{5}{6} - \frac{1}{6}$ (p. 390)

49. $40 \times 8,000$ (p. 210)

50. 9×367 (p. 196)

51. $\frac{3}{5} + \frac{1}{10}$ (p. 398)

52. Round 9.81 to the nearest whole number. (p. 428)

53. Find the value of $(17 - 8) \times 7$ (p. 158)

54. **TEST PREP** What is $456 \div 7$? (p. 266)

   A 46 r1   **B** 65 r1   **C** 65   **D** 65 r4

55. **TEST PREP** Which is equivalent to 9.1? (p. 414)

   F $\frac{9}{10}$ **G** $\frac{91}{100}$ **H** $9\frac{1}{10}$ **J** $9\frac{9}{10}$

PROBLEM SOLVING Thinker's Corner

PYRAMID POWER Pyramids are built one stone at a time. This pyramid uses one addition at a time to find the value of the stone at the top. The method for finding the missing numbers on this pyramid is shown below.

A = 3.7 + 4.02, or 7.72   **B** = 4.02 + 6.8, or 10.82   **C** = 6.8 + 9.1, or 15.9

D = the sum of A and B;   **E** = the sum of B and C;   **F** = the sum of D and E;
      7.72 + 10.82, or 18.54   10.82 + 15.9, or 26.72   18.54 + 26.72, or 45.26

Use addition to find the missing numbers.

1.

2.

	F		
	E	D	
B	A	C	
17.8	4.6	13.51	12.9

Use subtraction to find the missing numbers.

3.

82.7		
13.8	C	
5.9	B	D
A	2.4	E	F

4.

85.18		
45.09	A	
C	B	18.10
D	14.31	E	F

Problem Solving Skill
Evaluate Reasonableness of Answers

Quick Review

Write each mixed number as a decimal.

1. $3\frac{1}{10}$ **2.** $1\frac{28}{100}$

3. $9\frac{7}{10}$ **4.** $4\frac{2}{5}$

5. $7\frac{1}{4}$

ROAD TRIP   Mrs. Pate drives a bus three days a week. The table shows the Virginia cities she went to last week. How many miles did Mrs. Pate drive last week if she made one round trip to each city?

Which is a more reasonable answer?

a. Joe got an answer of 516.6 miles.
b. Jane got an answer of 1,033.2 miles.

MILES FROM RICHMOND, VA	
Charlottesville	70.5
Danville	162.7
Norfolk	91.4
Roanoke	192.0

MATH IDEA   If you estimate before solving a problem, then you can compare your answer to the estimate. If your answer is close to your estimate, then your answer is reasonable.

Estimate by rounding to the nearest 10.

$$70.5 + 162.7 + 91.4 + 192.0$$

$$\downarrow \qquad \downarrow \qquad \downarrow \qquad \downarrow$$

$$70 + 160 + 90 + 190 = 510 \text{ miles}$$

$$510 + 510 = 1,020 \text{ miles}$$ Think: 510 miles is the estimate for one-way trips.

You can add to see that Jane answered the problem correctly.

$$70.5 + 162.7 + 91.4 + 192.0 = 516.6 \text{ miles}$$
$$516.6 + 516.6 = 1,033.2 \text{ miles}$$

1,033.2 miles is close to the estimate of 1,020 miles.

So, Jane's answer is reasonable. Mrs. Pate drove 1,033.2 miles last week.

Talk About It

• Why is it helpful to estimate to see if an answer is reasonable?

• Why is Joe's answer not reasonable for the problem?

1. Mark ran 3 miles on Friday, 4.4 miles on Saturday, and 3.5 miles on Sunday. Which of the following is reasonable? Explain.
   a. Mark ran a total of 10.9 miles.
   b. Mark ran a total of 6.5 miles.

2. Joe is buying a notebook that costs $2.25, an eraser that costs $0.99, and a pen that costs $1.25. Which of the following is reasonable? Explain.
   a. Joe's supplies cost $11.74.
   b. Joe's supplies cost $4.49.

Tim has 3 hours to drive to his grandmother's house, which is 129.5 miles away. He drove 52.3 miles the first hour and 47.7 miles the second hour.

3. Which is the best estimate of the number of miles Tim drove in 2 hours?
   A 75 miles   **C** 130 miles
   B 100 miles   **D** 180 miles

4. Which is a reasonable answer for the number of miles Tim must travel in the last hour?
   F 10 miles   **H** 60 miles
   G 30 miles   **J** 230 miles

Mixed Applications

5. Jodie swam 2.8 miles on Monday. She swam 0.5 mile more on Tuesday than Monday. If she swam a total of 8 miles Monday through Wednesday, how many miles did she swim on Wednesday?

USE DATA For 7–10, use the table of Amy Chow's 2000 Olympics All-Around Finals scores.

7. What was Amy's total score?

8. Write Amy's scores in order from least to greatest.

9. On which event did Amy receive her highest score? her lowest score?

10. Find the difference between Amy's highest and lowest scores.

11. 📓 **Write About It** Why is estimating a good way to check the reasonableness of your answer?

6. Katie completed the first half of a race in 8.6 minutes and the second half in 11.1 minutes. Is it reasonable to say that Katie finished the race in about 20 minutes?

AMY CHOW'S SCORES	
Event	**Result**
Vault	9.443
Bars	9.737
Beam	9.225
Floor	9.187

Review/Test

✅ CHECK VOCABULARY AND CONCEPTS

Choose the best term from the box.

1. When adding or subtracting decimals, first line up the _?_ . (pp. 432, 434)

2. 0.6 and 0.60 are _?_ since they are different names for the same amount. (p. 437)

> decimal points
> equivalent decimals
> tenths

✅ CHECK SKILLS

Round each number to the place of the blue digit. (pp. 428–429)

3. 8.294　　　　**4.** $4.68　　　　**5.** $9.76　　　　**6.** 3.49

Round to the nearest hundredth. (pp. 428–429)

7. 7.695　　　　**8.** 3.504　　　　**9.** 1.635　　　　**10.** 14.839

Estimate the sum or difference. (pp. 430–431)

11. $\begin{array}{r} 1.29 \\ +3.46 \end{array}$　　**12.** $\begin{array}{r} \$5.98 \\ +\$9.04 \end{array}$　　**13.** $\begin{array}{r} 16.104 \\ +87.259 \end{array}$　　**14.** $\begin{array}{r} 14.337 \\ -\ 4.652 \end{array}$

Find the sum or difference. Estimate to check. (pp. 432–439)

15. 5.7 + 8.4　　　　**16.** 9.61 + 0.81　　　　**17.** 16.309 + 9.743

18. $3.67 − $0.59　　　**19.** 23.107 − 5　　　　**20.** 78.41 − 42.83

21. 1.94 + 0.8　　　　**22.** 42.51 + 22.4　　　　**23.** 20 − 8.684

✅ CHECK PROBLEM SOLVING

Solve. (pp. 440–441)

24. Dave earned $8.00 from his garage sale. The items for sale were a chair for $2.50, a radio for $1.25, 3 shirts for $2.75, and an old television set for $5.25. Is it reasonable to say that Dave sold all of the items? Explain.

25. Joanna has $5.50. She needs to buy eggs for $0.98, bread for $1.19, and milk for $2.69. Is it reasonable to say that Joanna has enough money to buy all of these items? Explain.

⭐Standardized Test Prep

Choose the answer.
See item **6**.

If your answer doesn't match one of the choices, check your computation. If your computation is correct, mark the letter for NOT HERE.

Also see problem **6**, p. H64.

For 1–12, choose the best answer.

1. What is 48.183 rounded to the nearest tenth?

   A 48.1   **C** 49.0

   B 48.2   **D** 50.0

2. What is the median of the data set?

     8.3, 9.2, 8.6, 8.7, 9.1

   F 8.3   **G** 8.6   **H** 8.7   **J** 9.2

3. Which shows 84 written as factors?

   A $6 \times 4 \times 2$ **C** $4 \times 3 \times 6$

   B $7 \times 2 \times 6$ **D** $3 \times 4 \times 8$

4. Which number makes this equation true?

$$0.87 + \blacksquare = 1.68$$

   F 0.81   **H** 1.51

   G 1.81   **J** 2.55

5. $289 \div 57$

   A 4 r2   **C** 5 r1

   B 4 r9   **D** 5 r4

6. $15 - 6.834$

   F 9.166   **H** 8.266

   G 8.000   **J** NOT HERE

7. What is the value of n for $8.7 - 6.9 = n$?

   A 1.86   **B** 1.8   **C** 1.6   **D** 1.46

8.   14.89
   + 24.235

   F 38.125   **H** 38.025

   G 39.125   **J** NOT HERE

9. Which expression has a value of 17.3?

   A $9.1 + 9.2$ **C** $8.6 + 9.1$

   B $9.2 + 8.3$ **D** $8.6 + 8.7$

10. Sandy bought a shirt for $24.75. What is that amount to the nearest dollar?

   F $25.00   **H** $24.70

   G $24.80   **J** $24.00

11.   64.77
   − 32.98

   A 12.21   **C** 32.69

   B 31.79   **D** NOT HERE

12.   89
   $\times$ 35

   F 3,125   **H** 3,115

   G 3,315   **J** NOT HERE

Write What You Know

13. Tanya gave the clerk 3 nickels and 5 quarters to pay for a book that cost $1.38. List 2 ways she could pay using other coins and bills and still get the same amount of change. Explain.

14. Ellen bought a soccer ball for $15.69 and a soccer uniform for $39.19. She gave the clerk $60.00. Estimate the amount of change she should receive. Tell how you estimated.

PROBLEM SOLVING
MATH DETECTIVE

Mystery Numbers

REASONING Use what you know about adding and subtracting fractions and decimals to break each case. Use the clues to help you test your reasoning powers. Find the mystery numbers.

Case 1

Clue 1: Add my two mystery fractions, and you get $1\frac{1}{12}$.

Clue 2: Subtract my two mystery fractions, and you get $\frac{5}{12}$.

The mystery fractions are $\frac{\square}{\square}$ and $\frac{\square}{\square}$.

Case 2

Clue 1: Add my two mystery fractions, and you get $1\frac{3}{8}$.

Clue 2: Subtract my two mystery fractions, and you get $\frac{1}{8}$.

The mystery fractions are $\frac{\square}{\square}$ and $\frac{\square}{\square}$.

Case 3

Clue 1: Add my two mystery decimals, and you get 1.8.

Clue 2: Subtract my two mystery decimals, and you get 0.

The mystery decimals are ▪ and ▪.

Case 4

Clue 1: Add my three mystery decimals, and you get 4.5.

Clue 2: Subtract the greatest and least of my mystery decimals, and you get 2.

The mystery decimals are ▪, ▪, and ▪.

Think It Over!

- Write About It Explain how you solved each case.

- **STRETCH YOUR THINKING** Find three different pairs of fractions that have a sum of $1\frac{7}{8}$. For each pair, find the difference.

Challenge

Circle Graphs

A **circle graph** shows data as parts of a whole circle.

The circle graph shows each part of the total budget that Mrs. Snider spent for art supplies. What fraction of her total art supply budget did Mrs. Snider spend for paper?

Since the part labeled "paper" takes up about one half of the whole circle, Mrs. Snider spent about one half of her total budget for paper.

- **What if** Mrs. Snider spent $80 for supplies? How much did she spend for paper?

Art Supplies Bought

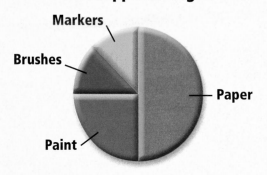

Talk About It

- Tell which two supplies together account for about one fourth of the total budget. Explain.

- On what supply did Mrs. Snider spend about as much as she spent for paint, brushes, and markers?

Try It

USE DATA For 1–5, use the circle graph.

1. Did Karl spend more for games or for magazines? Explain.

2. Name two things for which Karl spent the same amount of money.

3. **REASONING** If Karl's allowance was $24, how much money did he put into savings? How much did he spend for gifts?

4. If Karl spent $2 for snacks this week, how much was his allowance?

5. ✎ **Write a problem** about data that can be shown in a circle graph divided into 10 equal parts.

Karl's Spending

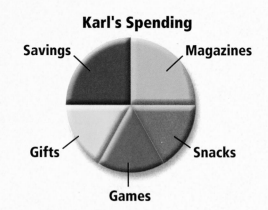

Study Guide and Review

VOCABULARY

Choose the best term from the box.

1. A number that names a part of a whole is a _?_ (p. 364)

2. Fractions that name the same amount are _?_. (p. 366)

3. If you continue to divide the numerator and the denominator of a fraction by the same number until 1 is the only number that can be evenly divided into both, you find the fraction in _?_. (p. 368)

4. Fractions with different denominators are _?_. (p. 398)

5. A number with one or more digits to the right of a decimal point is a _?_. (p. 406)

> decimal
> fraction
> equivalent fractions
> unlike fractions
> simplest form
> mixed numbers

STUDY AND SOLVE

Chapter 19

Find equivalent fractions.

Find equivalent fractions by multiplying or dividing.

Multiply the numerator and denominator by 2.	Divide the numerator and denominator by 6.
$\frac{6}{12} = \frac{6 \times 2}{12 \times 2} = \frac{12}{24}$	$\frac{6}{12} = \frac{6 \div 6}{12 \div 6} = \frac{1}{2}$
So, $\frac{6}{12}$ is equivalent to $\frac{12}{24}$.	So, $\frac{6}{12}$ is equivalent to $\frac{1}{2}$.

Write an equivalent fraction for each. (pp. 366–371)

6. $\frac{2}{3}$ 7. $\frac{3}{4}$ 8. $\frac{2}{5}$

9. $\frac{1}{6}$ 10. $\frac{4}{8}$ 11. $\frac{6}{10}$

12. $\frac{6}{9}$ 13. $\frac{4}{12}$ 14. $\frac{10}{12}$

15. $\frac{7}{14}$ 16. $\frac{6}{8}$ 17. $\frac{4}{6}$

Chapter 20

Add and subtract fractions and mixed numbers.

Subtract.

$2\frac{3}{5} - 1\frac{1}{5}$

$\begin{array}{r} 2\frac{3}{5} \\ -1\frac{1}{5} \\ \hline 1\frac{2}{5} \end{array}$

- Subtract the fractions first.
- Then subtract the whole numbers.

Find the sum or difference. (pp. 386–395)

18. $\begin{array}{r} \frac{7}{12} \\ +\frac{6}{12} \\ \hline \end{array}$ 19. $\begin{array}{r} \frac{7}{8} \\ -\frac{3}{8} \\ \hline \end{array}$ 20. $\begin{array}{r} 9\frac{6}{9} \\ -4\frac{3}{9} \\ \hline \end{array}$

21. $\begin{array}{r} 1\frac{3}{5} \\ +7\frac{2}{5} \\ \hline \end{array}$ 22. $\begin{array}{r} 2\frac{3}{8} \\ +3\frac{1}{8} \\ \hline \end{array}$ 23. $\begin{array}{r} 5\frac{4}{6} \\ -4\frac{3}{6} \\ \hline \end{array}$

Chapter 21

Compare and order decimals.

Tell which number is greater, 3.45 or 3.46. Begin with the digits to the left of the decimal point.

ONES	.	TENTHS	HUNDREDTHS
3	.	4	5
3	.	4	6

3 = 3   4 = 4   6 > 5

Since 6 > 5, 3.46 > 3.45.

• Line up the decimal points.

• Compare the digits, beginning with the greatest place value.

Compare. Write < , >, or = for each ●. (pp. 418–421)

24. 3.65 ● 3.54

25. 6.89 ● 7.32

26. 1.059 ● 1.127

Order the decimals from *least* to *greatest*.

27. 3.48, 3.79, 3.02

28. 8.43, 5.62, 8.47

29. 1.928, 1.849, 1.959

Chapter 22

Add and subtract decimals.

Subtract. 4.53 − 3.47

$$\begin{array}{r} \overset{4\ 13}{4.\cancel{5}\cancel{3}} \\ -\ 3.4\ 7 \\ \hline 1.0\ 6 \end{array}$$

• Line up the decimal points.

• Subtract.

• Place the decimal point.

Find the sum or difference. Estimate to check. (pp. 432–439)

30. 2.4 + 5.1   **31.** 5.87 − 4.72

32. 9.62 + 3.79   **33.** 2.76 − 1.68

34. 6.273 − 3.864   **35.** 3.825 + 7.519

PROBLEM SOLVING PRACTICE

Solve. (pp. 376–377, 396–397, 422–423, 440–441)

36. Heidi, Debbie, Frank, and Carrie painted pictures. Carrie painted a sun. Someone painted a house. A boy painted the farm. Heidi did not paint the cat. Who painted which picture?

37. The gymnastics team had the following scores: 8.75, 9.20, 8.60, and 9.55. Is it reasonable to think that the total score was 50? Explain.

38. Jim and Brianna each ate $\frac{5}{12}$ of the pizza. What fraction of the pizza is left? Name the operation or operations used.

39. Ling has a set of measuring cups. Three of the sizes are $\frac{3}{4}$, $\frac{1}{3}$, and $\frac{1}{2}$ cups. Put the sizes in order from least to greatest.

PERFORMANCE ASSESSMENT

TASK A • SHARE YOUR PIZZA

Curtis, Joe, Sara, and Laura have a coupon for one free pizza at the Pizza Palace. They all want the same number of slices.

a. Tell which size pizza they should order. Explain your answer.

b. Write a fraction that names how much of the pizza will be left after each person eats 1 slice. Write an equivalent fraction for that amount.

c. Should the friends order the same size pizza if Laura decides she does not want any pizza? Explain.

PIZZA PALACE MENU

Size	Servings
Personal	4 slices
Small	6 slices
Medium	8 slices
Large	12 slices
Jumbo	16 slices

TASK B • ON THE EDGE

Tina earns extra money in the spring by putting lawn edging around her neighbors' gardens. The table shows how many feet of edging she will need for each neighbor.

The garden center sells edging in rolls of 10.5 feet and 15 feet.

a. Tina decided to buy the 10.5-foot rolls of edging. How many rolls will she need to buy? Explain how you know.

b. How could Tina buy the edging in different-sized rolls to have the least amount left over?

c. Suppose Tina sends you to the garden center to buy 22 feet of edging. How many of each size roll would you buy?

EDGING FOR NEIGHBORS	
Mrs. Jones	9 ft
Mr. Morgan	6.5 ft
Mr. Rodriguez	11.25 ft
Ms. O'Donnell	8.75 ft

Technology Linkup

Mighty Math Number Heroes • Add and Subtract Fractions

Click on *Fraction Fireworks*.

Click . Choose Level L, M, or T.

• Answer at least 5 questions.

• Draw pictures to record your work.

Practice and Problem Solving

Add or subtract. Then click **EXPLORE**. Make each sum or difference. Draw pictures to record your work.

1. $\frac{1}{4} + \frac{1}{4}$

2. $\frac{5}{12} + \frac{3}{12}$

3. $\frac{4}{9} + \frac{4}{9}$

4. $\frac{2}{7} + \frac{3}{7}$

5. $\frac{9}{10} - \frac{3}{10}$

6. $\frac{7}{11} - \frac{4}{11}$

7. $\frac{7}{8} - \frac{5}{8}$

8. $\frac{5}{12} - \frac{4}{12}$

9. $\frac{3}{10} + \frac{3}{5}$

10. $1\frac{1}{4} + \frac{1}{8}$

11. $\frac{1}{3} + \frac{1}{4}$

12. $\frac{1}{2} + \frac{1}{5}$

13. $\frac{5}{6} - \frac{2}{3}$

14. $\frac{7}{12} - \frac{1}{6}$

15. $\frac{2}{3} - \frac{1}{4}$

16. $\frac{4}{5} - \frac{1}{2}$

Solve.

17. **STRETCH YOUR THINKING** Of the 6 kittens at the rescue center, $\frac{2}{3}$ are males. How many kittens are females? How do you know?

18. **REASONING** How is adding $\frac{1}{2} + \frac{1}{10}$ different from adding $\frac{1}{2} + \frac{1}{5}$?

Multimedia Math Glossary www.harcourtschool.com/mathglossary

19. **Vocabulary** Locate *equivalent fractions* in the Multimedia Glossary. Write an addition problem using unlike fractions. Explain how you found the sum and show the equivalent fractions used.

The Grand Canyon is 18 miles wide from rim to rim.

PROBLEM SOLVING ON LOCATION

at

National Parks

GRAND CANYON NATIONAL PARK

The Grand Canyon was formed about 5 million years ago. The Grand Canyon National Park, in Arizona, was established as a national park on February 26, 1919. The park has about 5 million visitors each year.

1. The Carlsons are spending 7 days in Grand Canyon National Park. They want to spend 1 day in Grand Canyon Village, $\frac{1}{2}$ day at Yavapai Observation Station, and $\frac{1}{2}$ day hiking. What fraction of their 7 days will they use on these activities? Explain how you know.

2. Sue and her family are deciding which trails to hike. Bill Hall Trail is $3\frac{2}{5}$ miles long, Dripping Springs Trail is $1\frac{1}{2}$ miles long, and Bright Angel Trail is $7\frac{7}{10}$ miles long. If Susan and her family hike all of these trails, how far will they hike in all?

3. South Kaibab Trail, which leads from the South Rim down to the Colorado River, is 12.8 miles round trip. North Kaibab Trail, which leads from the North Rim down to the Colorado River, is 29 miles round trip. Which trail is longer? How much longer is it?

BRYCE CANYON NATIONAL PARK

Bryce Canyon National Park, in Utah, was established as a national park on February 25, 1928. The park covers 35,835 acres and has over 1.5 million visitors each year.

AVERAGE RAINFALL AND SNOWFALL IN BRYCE CANYON (in inches)											
Jan	Feb	Mar	Apr	May	Jun	Jul	Aug	Sep	Oct	Nov	Dec
1.7	1.4	1.4	1.2	0.8	0.6	1.4	2.2	1.4	1.4	1.2	1.6

USE DATA For 1–6, use the table.

1. What is the total average rainfall and snowfall for the first three months of the year? for the last three months?

2. Is the average rainfall and snowfall amount greater for January to June or July to December?

3. What is the difference between the greatest and least average monthly rainfall and snowfall?

4. During what month does the most rainfall and snowfall occur? Explain how you know.

5. How many months average less than one inch of rainfall and snowfall? What fraction of the year is this?

6. Write the average rainfall and snowfall for May as a fraction. Write the average rainfall and snowfall for April as a mixed number.

▲ These "hoodoos" at Bryce Canyon National Park were shaped millions of years ago by water, ice, and gravity.

Customary Measurement

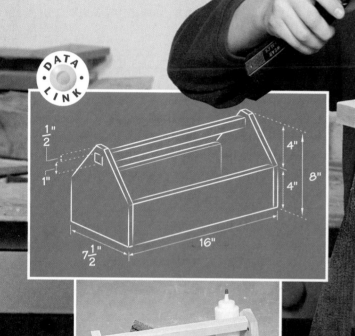

DATA
LINK

$\frac{1}{2}$"

1"

4"

4"

8"

16"

$7\frac{1}{2}$"

Measurements were first used thousands of years ago. Often, units were based on the size of a human arm or finger.

PROBLEM SOLVING   The drawing above is a plan for a tool box. The measurements are in inches. How tall is the finished box?

CHECK WHAT YOU KNOW

Use this page to help you review and remember
important skills needed for Chapter 23.

✓ MEASURE TO THE NEAREST INCH AND HALF-INCH (For Intervention, see p. H22.)

Measure the length of each to the nearest inch.

1.

2.

Measure the length of each to the nearest $\frac{1}{2}$ inch.

3.

4.

✓ MULTIPLICATION (For Intervention, see p. H23.)

Describe each pattern. Write the next number in each pattern.

5. 3, 6, 9, 12, ▪

6. 12, 24, 36, 48, ▪

7. 4, 8, 12, 16, ▪

Multiply.

8. 16
× 3

9. 36
× 3

10. 12
× 4

11. 16
× 4

12. 12
× 7

13. 12
× 3

14. 1,760
× 3

15. 5,280
× 2

16. 12
× 12

17. 36
× 9

✓ DIVISION (For Intervention, see p. H16.)

Divide.

18. 27 ÷ 3

19. 36 ÷ 12

20. 33 ÷ 3

21. 60 ÷ 12

22. 32 ÷ 16

23. 48 ÷ 16

24. 72 ÷ 3

25. 84 ÷ 12

Choose the Appropriate Unit

VOCABULARY

linear units	inch (in.)
foot (ft)	yard (yd)
mile (mi)	

▶ **Learn**

MEASURE UP!   Mike and his dad are building storage cabinets for a tool shed. They want to know the length, width, and height of the shed. What is the most reasonable unit of measure they could use?

Linear units are used to measure length, width, height, and distance. The customary system of measurement is used in the United States.

Examples

Ⓐ An **inch (in.)** is about the length of your thumb from the first knuckle to the tip.

1 in.

Smaller objects, such as a pencil or a nail, are measured in inches.

Ⓑ A **foot (ft)** is about the length of a sheet of paper.

A person's height or the length of a room is measured in feet.

Ⓒ A **yard (yd)** is about the length of a baseball bat.

1 yd

The length of a football field or one lap around a track is measured in yards.

Ⓓ A **mile (mi)** is about the distance you can walk in 20 minutes.

1 mi

The distance a person travels in a car is measured in miles.

So, Mike and his dad could measure the shed's length, width, and height in feet.

• What tools can you use to find linear measurements?

1. **Explain** how you can decide what linear unit to use.

Choose the most reasonable unit of measure. Write _in., ft, yd,_ or _mi._

2. The width of a book is about 8 _?_ .

3. The length of a car is about 10 _?_ .

4. Yesterday Jordan ran 3 _?_ .

Name the greater measurement.

5. 2 yd or 2 mi

6. 16 ft or 16 in.

7. 32 in. or 32 yd

▶ **Practice and Problem Solving**

Choose the most reasonable unit of measure. Write _in., ft, yd,_ or _mi._

8. The distance from New York to Los Angeles is 2,794 _?_ .

9. The door of your classroom is about 1 _?_ wide.

10. The length of a goldfish is about 3 _?_ .

11. The length of the playground is 22 _?_ .

12. The width of my notebook is 10 _?_ .

13. The height of a desk is about 2 _?_ .

Name the greater measurement.

14. 400 ft or 400 yd

15. 10 in. or 10 ft

16. 10 yd or 10 mi

17. 10 yd or 10 ft

18. 128 in. or 128 yd

19. 30 in. or 30 ft

20. Liz took a trip from Baltimore, Maryland, to Denver, Colorado. Which unit of measure would you use to describe the distance between these two cities?

21. The Nile River is 4,160 miles long. The Mississippi River is 2,348 miles long. How many miles longer than the Mississippi is the Nile?

22. **? What's the Question?** Lea stated that 50 inches is more reasonable for her height.

23. ✎ **Write a problem** about choosing a customary unit of measure to solve a problem.

24. Estimate in yards the distance from your desk to the classroom door. Choose a tool and unit, and measure the distance. Record your estimate and measurement.

Mixed Review and Test Prep

25. $\frac{3}{7} + \frac{4}{7}$ (p. 386)

26. $\frac{7}{8} - \frac{3}{8}$ (p. 390)

27. $4\frac{3}{8} + 7\frac{1}{8}$ (p. 392)

28. $512 \div 8$ (p. 262)

29. **TEST PREP** Which shows $\frac{13}{20}$ as a decimal? (p. 406)

A 0.13   **B** 0.20   **C** 0.65   **D** 0.85

Measure Fractional Parts

▶ Learn

LEAF LESSON   Mary collected and classified leaves for her science project. Part of her assignment was to measure each leaf to the nearest $\frac{1}{4}$ inch and $\frac{1}{8}$ inch.

Fractional units, such as $\frac{1}{2}$ inch, $\frac{1}{4}$ inch, and $\frac{1}{8}$ inch, are used to measure lengths that are between two whole units.

Measuring to the nearest fractional unit is like rounding a number.

Examples

Ⓐ Measure to the nearest $\frac{1}{4}$ inch.

The length is closer to $2\frac{2}{4}$ in. than to $2\frac{1}{4}$ in.
So, the leaf's length is about $2\frac{2}{4}$ in., or $2\frac{1}{2}$ in.

Ⓑ Measure to the nearest $\frac{1}{8}$ inch.

The length is closer to $1\frac{7}{8}$ in. than to $1\frac{6}{8}$ in.
So, the leaf's length is about $1\frac{7}{8}$ in.

Technology Link

To learn more about measurement, watch the Harcourt Math Newsroom Video, *Deep Worker Subs.*

Activity

MATERIALS: 5 objects, ruler, yardstick

OBJECT	ESTIMATE	ACTUAL MEASUREMENT TO THE NEAREST:			
		1 in.	$\frac{1}{2}$ in.	$\frac{1}{4}$ in.	$\frac{1}{8}$ in.
?	■	■	■	■	■

STEP 1

Make a table like the one shown. Estimate the lengths of 5 objects to the nearest inch, and record the estimates in your table.

STEP 2

Measure the length of each object to the nearest inch, $\frac{1}{2}$ inch, $\frac{1}{4}$ inch, and $\frac{1}{8}$ inch. Record the measurements in your table.

- For each object, which measurement is the closest to the actual length?

MATH IDEA You can measure length to the nearest $\frac{1}{2}$ inch, $\frac{1}{4}$ inch, or $\frac{1}{8}$ inch. The unit of measure used depends on your reason for measuring. Measurements using smaller units are closer to the actual length.

▶ Check

1. **Explain** how to find the $\frac{1}{2}$-inch marks and the $\frac{1}{4}$-inch marks on a ruler.

Estimate to the nearest inch. Then measure to the nearest $\frac{1}{4}$ inch.

2.

3.

Estimate to the nearest inch. Then measure to the nearest $\frac{1}{8}$ inch.

4.

5.

Order the measurements from *least* to *greatest*.

6. $2\frac{1}{2}$ in., $1\frac{7}{8}$ in., $2\frac{3}{8}$ in., $2\frac{3}{4}$ in.

7. $5\frac{1}{4}$ in., $5\frac{1}{8}$ in., $5\frac{1}{2}$ in.

LESSON CONTINUES ▶

Estimate to the nearest inch. Then measure to the nearest $\frac{1}{4}$ inch.

8.

9.

Estimate to the nearest inch. Then measure to the nearest $\frac{1}{8}$ inch.

10.

11.

Order the measurements from *least* to *greatest*.

12. $1\frac{3}{4}$ in., $2\frac{3}{8}$ in., 1 in., $2\frac{1}{2}$ in.

13. $\frac{3}{4}$ in., $\frac{2}{8}$ in., $\frac{1}{2}$ in., $\frac{7}{8}$ in.

14. $2\frac{5}{8}$ in., $1\frac{3}{4}$ in., $2\frac{1}{2}$ in., 2 in.

15. $\frac{3}{8}$ in., $\frac{1}{2}$ in., 1 in., $\frac{1}{4}$ in.

Order the measurements from *greatest* to *least*.

16. $\frac{5}{8}$ in., $1\frac{1}{4}$ in., $1\frac{1}{8}$ in., 1 in.

17. $2\frac{7}{8}$ in., 2 in., $2\frac{1}{2}$ in., $2\frac{3}{4}$ in.

18. $3\frac{1}{2}$ in., $2\frac{5}{8}$ in., $3\frac{3}{8}$ in., $3\frac{3}{4}$ in.

19. $6\frac{1}{2}$ in., 6 in., $6\frac{1}{8}$ in., $6\frac{3}{4}$ in.

USE DATA For 20–21, use the picture.

20. Find the length of the rock to the nearest inch and nearest $\frac{1}{2}$ inch.

21. Find the length of the fossilized leaf in the rock to the nearest inch.

22. Mindy and Leslie bought gifts at the Natural History Museum store. Leslie spent $4.29 more than Mindy. Together they spent $15.91. How much did Mindy spend?

23. Your class wants to take a tour of the Natural History Museum at 1:15 P.M. It is 11:38 A.M. now. How long is it until the tour begins?

24. **? What's the Error?** Max says that his measurement of $2\frac{3}{8}$ in. is closer to 2 in. than to $2\frac{1}{2}$ in. Describe and correct his error.

25. **Write About It** Could you measure an object to the nearest $\frac{1}{4}$ in. and get a measurement that is a whole number? Explain.

26. Which of the following things would you measure in fractional units—the ribbon for a package, the length of a wallpaper border, the length of a nail, the length of an ant, or the length of a leash for a dog?

27. Ken's project included a maple leaf $5\frac{3}{4}$ in. long, a dogwood leaf $5\frac{1}{2}$ in. long, and a poplar leaf $5\frac{3}{8}$ in. long. Order the lengths from shortest to longest.

28. Find the average height, to the nearest inch, of 6 students who are 52 in., 57 in., 54 in., 49 in., 55 in., and 58 in. tall.

Mixed Review and Test Prep

29. Write 31 pennies as a fraction of a dollar and as a decimal. (p. 406)

30. $6\frac{2}{3} - 2\frac{1}{3}$ (p. 392)

31. Write 23 minutes before twelve as it would appear on a digital clock. (p. 116)

32. $7 \times (8 \times 10)$ (p. 208)

33. Round 214,568 to the nearest hundred thousand. (p. 26)

34. Write $\frac{10}{12}$ as a fraction in simplest form. (p. 368)

35. Write $\frac{242}{100}$ as a decimal. (p. 420)

36. **TEST PREP** 12×345 (p. 224)

   A 3,000   **C** 4,060

   B 3,600   **D** 4,140

37. **TEST PREP** $816 \div 34$ (p. 282)

   F 20   **G** 24   **H** 28   **J** 30

PROBLEM SOLVING LiNKUP ... to Reading

STRATEGY • USE GRAPHICS AIDS A map scale shows how distance on a map compares to actual distances. Look at the scale on this map. The scale shows $\frac{1}{4}$ inch = 25 miles.

Use the scale to find the actual distance from Akron to Dover.

Measure the length between the cities on the map. The length is $\frac{1}{2}$ inch. Since there are two $\frac{1}{4}$-inch lengths in $\frac{1}{2}$ inch, multiply 25 by 2.

So, 2 lengths $\times$ 25 miles = 50 miles.

Use the scale to find the actual distances.

  1. Canton to Dover

  3. Lima to Dayton

  2. Columbus to Marion

  4. Columbus to Wheeling

Algebra: Change Linear Units

▶ **Learn**

TALL ENOUGH? In order to ride the roller coaster, Kyle must be at least 48 inches tall. Kyle is 5 feet tall. Is he tall enough to ride the roller coaster?

Example 1   5 feet = ■ inches

A foot is a larger unit than an inch. When you change larger units to smaller units, multiply.

feet	×	inches in 1 foot	=	total inches
↓		↓		↓
5	×	12	=	60

> **Equivalent Measures**
> 1 foot (ft) = 12 inches (in.)
> 1 yard (yd) = 3 feet
> 1 mile (mi) = 5,280 feet

60 inches is greater than 48 inches.
So, Kyle is tall enough to ride the roller coaster.

Example 2   How many yards are 10,560 feet?

Think: 10,560 feet = ■ yards

When you change smaller units to larger units, divide.

You can divide 10,560 by 3 with a calculator.

$$1\ 0\ 5\ 6\ 0\ ÷$$

$$10560 ÷ 3 =$$
$$3520$$

3 Enter =

You must be as tall as my tail to ride the roller coaster.

So, 10,560 feet is the same as 3,520 yards.

You can use an equation to change units. Since feet are smaller units than yards, divide.

Example 3   Use $y = f ÷ 3$ to complete the table.

Feet, f	Yards, y
24	■
30	■
36	■

$y = f ÷ 3$
$y = 24 ÷ 3$ $y = 8$
$y = 30 ÷ 3$ $y = 10$
$y = 36 ÷ 3$ $y = 12$

Feet, f	Yards, y
24	8
30	10
36	12

1. **Choose** whether you would multiply or divide to change miles to yards. Explain.

Complete. Tell whether you multiply or divide.

2. 36 in. = ▊ ft

3. 2 mi = ▊ ft

4. 5 yd = ▊ ft

▶ **Practice and Problem Solving**

Complete.

5. 8 ft = ▊ in.

6. 1,200 ft = ▊ yd

7. 2 mi = ▊ yd

8. 48 in. = ▊ ft

9. 432 in. = ▊ yd

10. 3 mi = ▊ ft

USE DATA For **11–12**, write an equation that can be used to complete each table. Copy and complete the table.

11.

Feet, f	2	3	4	5	7	9
Inches, i	▊	36	▊	60	84	▊

12.

Inches, i	36	72	108	144	252	324
Yards, y	1	2	▊	▊	▊	9

Compare. Write <, >, or = for each ●.

13. 144 in. ● 5 yd

14. 1,820 yd ● 1 mi

15. 525 ft ● 6,300 in.

16. 22 in. ● 2 ft

17. 16 yd ● 45 ft

18. 6 ft ● 72 in.

19. Admission prices to the park are $16.95 for children under 10 and $22.95 for ages 10 and over. The Sigmans paid $79.80 for tickets. How many of each type did they buy?

20. **REASONING** To make a bench for their playhouse, the children need a board 80 inches long. They have a board that is $6\frac{1}{2}$ feet long. Is the board long enough? Explain.

21. The *Raven*, a roller coaster at Holiday World in Indiana, requires riders to be 46 inches tall. If Fernando is 4 feet tall, can he ride the *Raven*? Explain.

22. Felicia bought 3 yards of fabric for $3.59 per yard. She gave the clerk $20.00. How much change did she receive?

Mixed Review and Test Prep

23. $4 \times (25 + 4)$ (p. 158)

Compare. Write <, >, or = for each ●.

24. 0.75 ● 0.79 (p. 414)

25. $5\frac{3}{4}$ ● $\frac{23}{4}$ (p. 378)

26. $0.25 + ▊ = 0.35$ (p. 436)

27. **TEST PREP** Which unit would you use to measure the length of a room? (p. 452)

A in.　　B ft　　C mi　　D lb

HANDS ON

Capacity

▶ Explore

The **capacity** of a container is the amount it can hold when it is filled. The customary units for measuring capacity are **cup**, **pint**, **quart**, and **gallon**.

VOCABULARY

capacity	cup (c)
pint (pt)	quart (qt)
gallon (gal)	

MATERIALS 1-cup, 1-pint, 1-quart, and 1-gallon containers, water

Activity

How many cups fill a pint? a quart? a gallon?

STEP 1

Copy the table. Estimate the number of cups needed to fill each container. Record your estimate.

CONTAINER	NUMBER OF CUPS	
How many cups are in each?	**Estimate**	**Actual**
Pint	▦	▦
Quart	▦	▦
Gallon	▦	▦

STEP 2

Fill the measuring cup with water to the 1-cup line.

STEP 3

Pour the water into each of the containers. In your table, record the actual number of cups.

Try It

Use measuring containers.

a. How many pints fill a gallon container?

b. How many cups fill 3 pint containers?

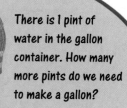

There is 1 pint of water in the gallon container. How many more pints do we need to make a gallon?

▶ Connect

You can change units of capacity from one unit to another. Use the table of measures and multiply or divide.

Examples

Ⓐ How many quarts are in 2 gallons?

$$2 \text{ gal} = \blacksquare \text{ qt}$$

gallons	quarts in 1 gallon	quarts
↓	↓	↓
2	× 4	= 8

So, 2 gallons equals 8 quarts.

Ⓑ How many pints are equal to 6 cups?

$$6 \text{ c} = \blacksquare \text{ pt}$$

cups	cups in 1 pint	pints
↓	↓	↓
6	÷ 2	= 3

So, 6 cups equals 3 pints.

▶ Practice and Problem Solving

Copy and complete the tables. Change the units.

1.

Cup	2	6	14
Pint	■	■	■

2.

Pint	8	24	■
Gallon	■	■	8

3.

Pint	2	■	10
Quart	■	4	■

Choose the unit of capacity. Write a, b, or c.

4.
  a. cup
  b. pint
  c. gallon

5.
  a. gallon
  b. cup
  c. quart

6.
  a. quart
  b. cup
  c. pint

7. Alice bought 5 half gallons of orange juice, 2 quarts of pineapple juice, and 2 gallons of lemonade to make fruit punch. How many gallons of drink did she buy altogether?

8. **ᵃ⁺ᵇ⁄ᶜ ALGEBRA** Use the equation $c = 16 \times g$ to find the number of cups in 7 gallons.

c = number of cups
g = number of gallons

Mixed Review and Test Prep

9. 4 ft = ■ in. (p. 458) **10.** 6 yd = ■ ft (p. 458)

11. Write 0.87 in word form. (p. 406)

12. 4,008 − 1,992 (p. 42)

13. **TEST PREP** Which is equal to 36? (p. 302)
  A $2 \times 2 \times 3$ **C** $2 \times 3 \times 4$
  B $2 \times 2 \times 3 \times 3$ **D** $2 \times 2 \times 4$

 HANDS ON

Weight

VOCABULARY

weight   ounce (oz)

pound (lb)   ton (T)

MATERIALS spring scale

▶ Explore

How would you describe the weight of a toy car? a real car?

The customary units for measuring weight are ounce (oz), pound (lb), and ton (T).

One toy car weighs about 1 oz.   One remote control car weighs about 1 lb.

One real car weighs about 1 T.

Activity

Choose 5 classroom objects that you can weigh with your scale.

STEP 1

Make a table. Estimate the weight of each object in ounces or pounds. Record the object and estimated weight.

STEP 2

Weigh each object. Record each actual weight in your table.

Try It

Choose *ounce* or *pound* for each object. Then measure the weight.

a. a stapler   **b.** box of crayons

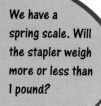

We have a spring scale. Will the stapler weigh more or less than 1 pound?

▶ Connect

You can multiply or divide to change one unit of weight to another.

Weight Equivalents
16 oz = 1 lb
2,000 lb = 1 T

Examples

A How many ounces are in 7 pounds?

$$7 \text{ lb} = \blacksquare \text{ oz}$$

pounds ounces in 1 pound ounces
↓   ↓   ↓
7   ×   16   =   112

So, 7 pounds equals 112 ounces.

B How many tons are equal to 4,000 pounds?

$$4,000 \text{ lb} = \blacksquare \text{ T}$$

pounds pounds in 1 ton tons
↓   ↓   ↓
4,000 ÷ 2,000 = 2

So, 4,000 pounds equals 2 tons.

▶ Practice and Problem Solving

Choose the more reasonable measurement.

1. 15 oz or 15 lb

2. 3 lb or 3 T

3. 14 oz or 14 lb

Complete.

4. 3 lb = ■ oz

5. 5 T = ■ lb

6. 160 oz = ■ lb

For 7–10, change to pounds.

7. 32 oz

8. 2 T

9. 3 T

10. 96 oz

11. REASONING Tamika weighed 7 lb 6 oz at birth. Her weight doubled in three months. How much did she weigh at the end of three months?

12. ❓ **What's the Error?** Chris needs 2 lb of snack mix. He bought two 8-oz packages. Describe his error. Tell what Chris should have bought.

13. Which weight is less, 3,000 lb or 2 T? Explain.

14. Which weight is greater, 5 lb or 88 oz?

Mixed Review and Test Prep

15. Round 213,467 to the nearest ten. (p. 26)

16. $5\frac{4}{5} - 3\frac{1}{5}$ (p. 392)

17. $\begin{array}{r} 4,306 \\ -2,987 \end{array}$ (p. 40)

18. 5 ft = ■ in. (p. 458)

19. ⭐ **TEST PREP** How many inches are in 8 yards? (p. 458)

A 24 in. **B** 96 in. **C** 288 in. **D** 324 in.

Problem Solving Strategy
Compare Strategies

PROBLEM Carol's mom is making punch for 8 children. If she mixes 5 pints of pineapple juice and 2 quarts of orange juice, will each child be able to have 2 cups of punch?

UNDERSTAND

- What are you asked to find?
- What information will you use?
- Is there any information you will not use?

PLAN

- What strategy can you use?
  Often you can use more than one strategy to solve a problem. Use *draw a picture* or *make a table*.

SOLVE

- How will you solve the problem?

Draw a Picture Show how to find the total cups of punch.

Pineapple Juice	Orange Juice
1 pt = 2 c	1 qt = 2 pt

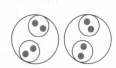

5 pt = 10 c	2 qt = 4 pt
	4 pt = 8 c

10 + 8 = 18, or 18 cups

Make a Table Show the relationships between cups, pints, and quarts.

1 qt = 2 pt   1 pt = 2 c

PINEAPPLE JUICE					
pints	1	2	3	4	5
cups	2	4	6	8	10

ORANGE JUICE		
quarts	1	2
pints	2	4
cups	4	8

5 pints of pineapple juice makes 10 cups, and 2 quarts of orange juice makes 8 cups.

10 + 8 = 18, or 18 cups

Carol's mom needs 16 cups of punch. Since 18 > 16, there will be enough punch for each child to have 2 cups.

CHECK

- What other strategy could you use?

► Problem Solving Practice

Choose a strategy to solve. Explain your choice.

PROBLEM SOLVING STRATEGIES

▶ Draw a Diagram or Picture
Make a Model or Act It Out
Make an Organized List
Find a Pattern
▶ Make a Table or Graph
Predict and Test
Work Backward
Solve a Simpler Problem
Write an Equation
Use Logical Reasoning

1. **What if** Carol's mom mixed 2 quarts of orange juice and 1 quart of pineapple juice? Would there be enough for each child to have 2 cups of punch? Explain.

2. Deborah is making punch. One cup of concentrated juice makes 1 quart of punch. If Deborah buys 3 pints of concentrated juice, how many quarts of punch can she make?

Luisa is sewing doll clothes and uses lace for the trim. Each dress needs 1 foot of lace. Luisa buys 2 yards of lace.

3. **ALGEBRA** Which equation represents the number of dresses Luisa can trim?

   y = yards, d = number of dresses

   A $y = 3 + d$ **C** $d = 3 + y$
   B $d = 3 \div y$ **D** $d = y \times 3$

4. How many doll dresses can Luisa trim?
   F 5   **H** 11
   G 6   **J** 12

Mixed Strategy Practice

5. Theo caught a fish that was 2 feet long. Mitch caught a fish that was 5 inches shorter than the fish Theo caught. How long was the fish that Mitch caught?

USE DATA For 7–8, use the graph.

7. Maria kept a record of how many miles she walked around a track each school day. It takes 4 laps around the track to equal 1 mile. How many laps around the track did Maria walk in one week?

8. **? What's the Error?** Anna says that the distance Maria walked for the week was over 16,000 yards. Describe and correct her error.

6. The test scores for four students were 85, 93, 77, and 92. Jackie's score was 8 points higher than Mark's but 7 points lower than Mary's. Nick had the highest score. What was each student's test score?

Miles Walked Each Day

Review/Test

✓ CHECK VOCABULARY AND CONCEPTS

Choose the best term from the box.

| capacity |
| cup |
| foot |
| linear |
| mile |
| ounce |
| pound |
| ton |
| yard |

1. The ? of a container is the amount it can hold when it is filled. (p. 460)

2. Units used to measure length, width, height, or distance are ? units. (p. 452)

3. The customary units for weight are ?, ?, and ?. (p. 462)

✓ CHECK SKILLS

Choose the most reasonable unit of measure.
Write *in., ft, yd,* or *mi.* (p. 452–453)

4. The height of a desk is about 3 ?.

5. The length of a bus route is about 10 ?.

6. The width of a door is about 36 ?.

Name the greater measurement. (pp. 452–453, 460–461, 462–463)

7. 300 yd or 300 ft

8. 3 c or 3 pt

9. 4 T or 4 lb

10. 64 oz or 3 lb

11. Measure to the nearest $\frac{1}{4}$ inch.
(pp. 454–457)

12. Measure to the nearest $\frac{1}{8}$ inch.
(pp. 454–457)

Complete. (pp. 458–459)

13. 2 mi = ■ yd

14. 2 yd = ■ in.

15. 48 in. = ■ ft

16. 2 lb = ■ oz

17. 3 T = ■ lb

18. 4 pt = ■ qt

✓ CHECK PROBLEM SOLVING

Choose a strategy to solve. Explain your choice. (pp. 464–465)

19. There are 22 students in the class. If each student will drink 1 cup of punch, how many pints are needed?

20. Karl's grandmother needs 2 yards of lace for her quilt. Lace costs $3 per foot. How much will she pay?

Standardized Test Prep

Decide on a plan.
See item **9.**

You need to know the relationships among cups, quarts, and gallons. Change the units for quarts and gallon to cups.

Also see problem **5**, p. H63.

For 1–10, choose the best answer.

1. Which unit would be best to measure the length of your arm?

A inch   **C** yard
B pound   **D** kilometer

2. Which is the greatest length?

F 5 ft   **H** 48 in.
G 3 yd   **J** 8 ft

3. Use your ruler to measure this leaf to the nearest $\frac{1}{4}$ inch.

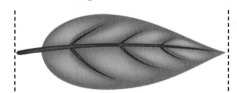

A $1\frac{3}{8}$ in.   **C** $2\frac{1}{8}$ in.
B $3\frac{1}{8}$ in.   **D** $2\frac{1}{4}$ in.

4. 12,986 + 4,603

F 17,579   **H** 16,589
G 17,589   **J** NOT HERE

5. Which lengths are in order from *least* to *greatest*?

A $\frac{1}{4}$ in.; $\frac{7}{8}$ in.; $1\frac{3}{4}$ in.; $1\frac{3}{8}$ in.
B $\frac{7}{8}$ in.; $\frac{1}{4}$ in.; $1\frac{3}{8}$ in; $1\frac{3}{4}$ in.
C $\frac{7}{8}$ in.; $\frac{1}{4}$ in.; $1\frac{3}{4}$ in.; $1\frac{3}{8}$ in.
D $\frac{1}{4}$ in.; $\frac{7}{8}$ in.; $1\frac{3}{8}$ in.; $1\frac{3}{4}$ in.

6. How many feet are in 4 miles?

F 21,120   **H** 20,000
G 20,800   **J** 15,000

7. Which is the greatest amount?

A 5 quarts   **C** 8 pints
B 3 gallons   **D** 17 cups

8. Which decimal is equivalent to $\frac{3}{12}$?

F 0.20   **H** 0.25
G 0.48   **J** 0.32

9. Mrs. Johnson bought 3 quarts of grape juice and 1 gallon of orange juice. How many 1-cup servings of juice did she buy?

A 12   **C** 28
B 18   **D** 30

10. 137 × 25

F 3,425   **H** 3,195
G 2,195   **J** NOT HERE

Write What You Know

11. Tanya's height to the nearest inch is 4 feet 3 inches tall. What could be her actual height? Explain your answer.

12. Find the number of ounces in three tons. Tell why ounces would not be the best unit to use to measure heavy objects.

Metric Measurement

Grizzly Bear

Cardinal

Raccoon

Ostrich

When animals walk through soft soil, they often leave tracks. People can look at the tracks and tell what kind of animal has been there.

PROBLEM SOLVING Look at the drawings of the animal tracks. Which animal has the largest track? the smallest? Measure your foot. How does it compare to the tracks?

DATA LINK

ANIMAL TRACKS		
	Ostrich	14 cm
	Grizzly Bear	18 cm
	Cardinal	3 cm
	Raccoon	8 cm

CHECK WHAT YOU KNOW

Use this page to help you review and remember important skills needed for Chapter 24.

✔ VOCABULARY

Choose the best term from the box.

1. The length of a living room is best measured using ?.

2. The length of a pencil is best measured using ?.

> centimeters
> kilometers
> meters

✔ MEASURE TO THE NEAREST CENTIMETER (For Intervention, see p. H23.)

For 3–8, measure the length of each object to the nearest centimeter.

3.

4.

5.

6.

7.

8.

✔ MULTIPLY BY 10, 100, AND 1,000 (For Intervention, see p. H24.)

Find the value of *n*.

9. $12 \times 10 = n$

10. $6 \times 1{,}000 = n$

11. $100 \times 16 = n$

12. $15 \times 1{,}000 = n$

13. $270 \times 10 = n$

14. $35 \times 100 = n$

15. $18 \times n = 1{,}800$

16. $60 \times n = 600$

17. $n \times 1{,}000 = 36{,}000$

18. $n \times 10 = 10{,}000$

19. $17 \times n = 1{,}700$

20. $1{,}000 \times 32 = n$

Linear Measure

► Learn

HOW LONG? HOW SHORT? The western pygmy-blue butterfly is one of the smallest butterflies. Measure the width of its wingspan to the nearest centimeter using a metric ruler.

The wingspan of this butterfly is between 2 and 3 centimeters. To the nearest centimeter, the wingspan is 2 centimeters.

You can use different metric units to measure length or distance.

VOCABULARY

centimeter (cm)

decimeter (dm)

meter (m)

kilometer (km)

centimeters

A **centimeter (cm)** is about the width of your index finger.

A **decimeter (dm)** is about the width of an adult's hand.

A **meter (m)** is about the distance from one hand to the other when you stretch out your arms.

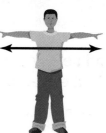

HANDS ON

Activity 1

MATERIALS: centimeter ruler, meterstick

- Estimate and measure the lengths of 5 objects in your classroom to the nearest centimeter, decimeter, or meter. Record the estimates and measures in a table.
- How did you estimate the length of each object?

Object	Unit of Measure	Estimate	Measurement
1.			
2.			
3.			

470

Greater Lengths

Activity 2

MATERIALS: centimeter ruler

Now measure your math book.

- Choose a unit of measure. Record all the linear measurements of your math book to the nearest centimeter or decimeter.

- Compare your results to your classmates' results. Are the measurements of your math book the same? If not, why not?

Mrs. Chen's class is taking a bus to the zoo. She wants to know how far the zoo is from the school. What unit of measure should she use?

Larger distances and lengths can be measured in kilometers. A **kilometer (km)** is about the length of 10 football fields.

So, Mrs. Chen would measure the distance from the school to the zoo in kilometers.

MATH IDEA The centimeter (cm), decimeter (dm), meter (m), and kilometer (km) are metric units of length or distance.

▶ Check

1. **Explain** how you decide what unit of measure to use when measuring an object.

Use a centimeter ruler or a meterstick to measure each item. Write the measurement and unit of measure used.

2. height of a classmate
3. length of a stapler
4. length of a classroom

Choose the most reasonable unit of measure. Write *cm,* *dm, m,* **or** *km*.

5.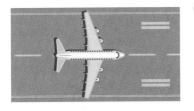

The length of an airport runway is about 3 _?_.

6.

The length of a baseball bat is 1 _?_.

7.

The length of a leaf is 5 _?_.

LESSON CONTINUES

▶ Practice and Problem Solving

Use a centimeter ruler or a meterstick to measure each item. Write the measurement and unit of measure used.

8. length of a pencil

9. width of a poster

10. length of a chalkboard

Choose the most reasonable unit of measure. Write *cm, dm, m,* or *km.*

11.

The height of the table is 76 ? .

12.

The length of the earthworm is 10 ? .

13.

The length of the car is 4 ? .

14. distance between two cities

15. height of a maple tree

16. length of a ballpoint pen

Compare. Write <, >, or = for each ●.

17. 5 cm ● 5 dm

18. 100 dm ● 100 km

19. 25 km ● 25 m

20. 98 cm ● 98 m

21. 13 m ● 13 dm

22. 87 m ● 87 km

For 23–24, use the ruler to find the distance.

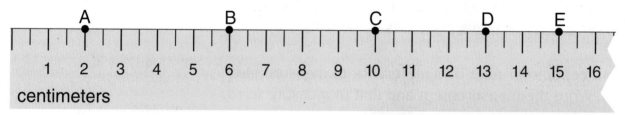

23. between point A and point B

24. between point B and point D

25. Hudson ran a 5-kilometer race. Use the table to find how many minutes he ran in all. If the race began at 7:35 A.M., when did he cross the finish line?

26. REASONING Cary's broken ruler begins at the 16-cm mark. If she draws a 7-cm line beginning at the 16-cm mark, at which mark will she stop?

27. ✎ Write a problem that can be solved by using a centimeter ruler.

HUDSON'S RACE	
Kilometer	**Time (each kilometer)**
1 and 2	5 min 30 sec
3, 4, and 5	6 min

28. Five students are 149 cm, 151 cm, 139 cm, 152 cm, and 144 cm tall. What is their average height?

29. Look at the pattern. Write the rule and complete. 12 m, 24 m, 36 m, ▨ m, ▨ m, ▨ m

Mixed Review and Test Prep

30. Evaluate $5 \times (2 + 3)$. (p. 158)

31. Find the mean of 14, 16, 17, 17, 18, 19, 22, 23, 25. (p. 276)

32. What are the factors of 36? (p. 300)

33. Multiply 437 by 6. (p. 196)

34. $408 \div 7$ (p. 266)

35. $\begin{array}{r} 721 \\ \times\ 36 \end{array}$ (p. 228) **36.** $\begin{array}{r} 3.12 \\ +\ 0.98 \end{array}$ (p. 432)

37. Order 0.08, 0.73, 0.9, and 0.12 from least to greatest. (p. 418)

38. **TEST PREP** How many inches are in 3 feet? (p. 458)

A 9 in.   **C** 32 in.

B 12 in.   **D** 36 in.

39. **TEST PREP** Which fraction is equivalent to $\frac{8}{12}$? (p. 368)

F $\frac{2}{6}$ **G** $\frac{2}{4}$ **H** $\frac{4}{8}$ **J** $\frac{2}{3}$

PROBLEM SOLVING   Thinker's Corner

REASONING You can use a meterstick to model decimal numbers. The meter shows the whole number. The decimeters and centimeters are fractional parts of the meter.

meter (m) 1.0 meter	decimeter (dm) *Deci* means "tenths."	centimeter (cm) *Centi* means "hundredths."
 meter	1 decimeter = 0.1, or $\frac{1}{10}$, meter.	1 centimeter = 0.01, or $\frac{1}{100}$, meter.

Write 3.26 meters as meters + decimeters + centimeters.

$\frac{2}{10}$ meter = 2 decimeters $\frac{6}{100}$ meter = 6 centimeters

So, 3.26 meters = 3 meters + 2 decimeters + 6 centimeters.

Complete.

1. $\frac{4}{10}$ m = ▨ dm

2. $\frac{23}{100}$ m = ▨ cm

3. $\frac{9}{100}$ m = ▨ cm

4. 4.52 m = 4 m + ▨ dm + ▨ cm

5. 0.51 m = ▨ m + ▨ dm + ▨ cm

Algebra: Change Linear Units

▶ **Learn**

ANIMAL ENGINEER In five hours, a mole can dig a tunnel 500 decimeters long. How many centimeters long would the tunnel be?

Think: 500 dm = ■ cm

A decimeter is a larger unit than a centimeter. When you change larger units to smaller units, you multiply.

Length of tunnel in decimeters		Centimeters in 1 decimeter		Length of tunnel in centimeters
↓		↓		↓
500	×	10	=	5,000

So, the tunnel would be 5,000 centimeters long.

▲ Moles live mainly underground and dig new tunnels daily. Each tunnel is about 5 cm wide.

How many meters long would the tunnel be?

Think: 500 dm = ■ m

A decimeter is a smaller unit than a meter. When you change smaller units to larger units, you divide.

Equivalent Measures

1 decimeter = 10 centimeters
1 meter = 100 centimeters
1 meter = 10 decimeters
1 kilometer = 1,000 meters

Length of tunnel in decimeters		Decimeters in 1 meter		Length of tunnel in meters
↓		↓		↓
500	÷	10	=	50

So, the tunnel would be 50 meters long.

▶ **Check**

1. **Tell** whether you would multiply or divide to change meters to centimeters.

Complete.

2. 5 m = ■ dm   3. 4 dm = ■ cm   4. 16 m = ■ cm

Complete.

5. 500 dm = ▓ m

6. 35 m = ▓ cm

7. 10 dm = ▓ m

8. 10 m = ▓ dm

9. 8,000 m = ▓ km

10. 2 km = ▓ m

Write the correct unit.

11. 20 dm = 200 _?_

12. 400 cm = 4 _?_

13. 6 _?_ = 60 dm

14. 14 m = 1,400 _?_

15. 500 _?_ = 50 m

16. 8 _?_ = 80 cm

Compare. Write <, >, or = for each ●.

17. 8 m ● 400 cm

18. 40 cm ● 10 dm

19. 900 m ● 9 km

20. 2,000 cm ● 20 m

21. 10 dm ● 1,000 m

22. 1 km ● 500 m

Order from least to greatest.

23. 3 dm; 120 cm; 18 dm; 1m

24. 2,010 m; 2 km; 2,100 dm

For 25–26, use this information.
A beaver's dam can reach 300 meters long.
Some beavers dig canals so they can move
sticks to their dams. These canals are about
40 cm deep and 50 cm wide and can be
210 m long.

25. Write the lengths of a beaver canal
and a beaver dam in centimeters.
How many centimeters is this in all?

26. Write the width and depth of a beaver
canal in decimeters.

27. The top of one beaver's lodge is
130 centimeters above the water. Is
this more than or less than 2 meters?
Explain.

28. **?** **What's the Error?** Marco is 2 m
tall. He wrote his height as 2,000
centimeters. Describe his error. Write
the correct height.

Mixed Review and Test Prep

29. 198.42 − 39.75 (p. 434)

30. Write an expression for 2 greater than
some number. Use the variable *n*. (p. 60)

31. $\frac{3}{4} + \frac{5}{8}$ (p. 398)

32. 2 ft = ▓ in. (p. 458)

33. **TEST PREP** Choose the most
reasonable measurement for the
height of the front door of a house.
(p. 452)

A 8 inches   **C** 8 yards

B 8 feet   **D** 8 miles

Capacity

HANDS ON

Quick Review

1. ■ dm = 10 cm

2. 100 cm = ■ dm

3. ■ dm = 1 m

4. ■ cm = 3 m

5. 2 km = ■ m

▶ **Explore**

To stay healthy, most people should drink over $1\frac{1}{2}$ liters of water every day.

A **milliliter (mL)** and a **liter (L)** are metric units of capacity.

1 mL

1 L

1,000 mL = 1 L

VOCABULARY

milliliter (mL)   **liter (L)**

MATERIALS 1-liter container, metric measuring cup, small plastic cup, bucket, tall plastic cup

What is the capacity of a plastic cup? of a bucket? You can measure the capacity of a container by using water.

Activity

STEP 1

Copy the table. Write an estimate of the capacity of each container in your table.

CONTAINER	ESTIMATE	ACTUAL
small cup		
tall cup		
bucket		

STEP 2

Measure the actual capacity of each container by using the measuring cup or the 1-liter container. Record the actual measurements in your table.

Try It

Choose *mL* or *L* for each. Then measure the capacity.

a. small spoon   **b.** drink pitcher

c. small cup   **d.** trash can

Which should we use to measure the capacity of the spoon?

► Connect

Sometimes you need to change liters to milliliters. When you change larger units to smaller units, multiply.

Examples

Ⓐ 2 L = ▨ mL

Think: There are 1,000 mL in 1 L.

2 L = 2 × 1,000 = 2,000 mL
So, 2 L = 2,000 mL.

Ⓑ 23 L = ▨ mL

Think: There are 1,000 mL in 1 L.

23 L = 23 × 1,000 = 23,000 mL
So, 23 L = 23,000 mL.

► Practice and Problem Solving

Choose the more reasonable unit of measure. Write *mL* or *L*.

1. a raindrop

2. water in a pool

3. juice in a punch bowl

Choose the best estimate. Write *a, b,* or *c*.

4.
   a. 25 mL
   b. 250 mL
   c. 25 L

5.
   a. 2 mL
   b. 20 mL
   c. 2 L

6.
   a. 5 mL
   b. 500 mL
   c. 5 L

7.
   a. 6 mL
   b. 600 mL
   c. 6 L

Change to milliliters.

8. 3 L

9. 5 L

10. 10 L

11. 42 L

12. Eric drinks 8 glasses of water each day. Each glass contains 300 milliliters. Anna drinks 2 liters of water each day. Who drinks more water? How much more?

13. ✎ **Write About It** Explain how two different containers can have the same capacity. Give an example.

Mixed Review and Test Prep

14. 82 × 63 (p. 224)

15. 216 ÷ 12 (p. 286)

16. $\frac{9}{10} - \frac{2}{10}$ (p. 390)

17. 0.78 − 0.43 (p. 434)

18. **TEST PREP** Which shows $\frac{12}{100}$ written as a decimal? (p. 406)

   A 1,200   **B** 12   **C** 1.2   **D** 0.12

HANDS ON

Mass

Quick Review

1. 3 × 1,000

2. 10 cm = 1 _?_

3. 1,000 × 2

4. 100 _?_ = 1 m

5. 5,000 ÷ 1,000

▶ Explore

Matter is what all objects are made of. **Mass** is the amount of matter in an object. Metric units of mass are the **gram (g)** and the **kilogram (kg)**.

The mass of a dollar bill or a small paper clip is about 1 gram.

The mass of a baseball bat is about 1 kilogram.

VOCABULARY

mass   gram (g)

kilogram (kg)

_____

MATERIALS balance, gram and kilogram masses

Activity

Use a balance scale to find the mass of other objects.

STEP 1

Find the mass of each of 5 objects. Make a table. Estimate the mass of each object in grams or kilograms. Record the object and the estimated mass in the table.

STEP 2

Place each object on the scale. Record the actual measurements in your table.

• Explain how to order the items you measured from least to greatest in mass.

Try It

Choose *g* or *kg* for each mass. Then measure the mass.

  a. the mass of a stapler

  b. the mass of this book

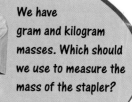

We have gram and kilogram masses. Which should we use to measure the mass of the stapler?

▶ Connect

You can use multiplication to change kilograms to grams.

Mass Equivalent
1 kg = 1,000 g

Examples

A How many grams are equivalent to 2 kilograms?

Think: There are 1,000 g in 1 kg.

2 kg = 2 × 1,000 = 2,000 g
So, 2,000 g are equivalent to 2 kg.

B How many grams are equivalent to 8 kilograms?

Think: There are 1,000 g in 1 kg.

8 kg = 8 × 1,000 = 8,000 g
So, 8,000 g are equivalent to 8 kg.

▶ Practice and Problem Solving

Choose the more reasonable measurement.

1.

1 g or 1 kg

2.

20 g or 200 g

3.

10 g or 10 kg

4.

10 kg or 1,000 kg

Change to grams.

5. 4 kg

6. 7 kg

7. 3 kg

8. 10 kg

USE DATA For 9–10, use the bar graph.

9. Find the vegetable whose sales increased from Day 1 to Day 2. How many more kilograms were sold on Day 2?

10. Find the total mass of the vegetables sold on the two days.

11. **ALGEBRA** Write an equation that can be used to change kilograms to grams. Use *g* for the number of grams and *k* for the number of kilograms.

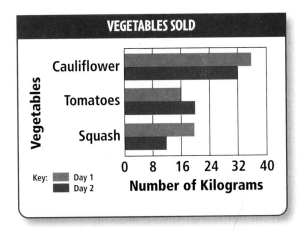

Mixed Review and Test Prep

12. 6,580 (p. 40)
    −2,461

13. 18 (p. 224)
    ×35

Find the missing number.

14. ■ + 0.67 = 2.75 (p. 436)

15. 3,000 − ■ = 1,284 + 254 (p. 42)

16. TEST PREP There are 256 cookies to be placed on 8 platters at the bake sale. How many cookies will be on each platter? (p. 266)

A 22   C 32
B 23   D 37

Problem Solving Strategy
Draw a Diagram

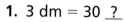

Quick Review

1. 3 dm = 30 _?_

2. 4 _?_ = 400 cm

3. 100 cm = 10 _?_

4. 15 m = 1,500 _?_

5. 800 cm = 8 _?_

PROBLEM   Darrell and Charles each caught a fish while deep-sea fishing. The two fish have a mass of 10 kilograms. The mass of Charles's fish is 1 kilogram more than twice the mass of Darrell's fish. What is the mass of each fish?

UNDERSTAND

• What are you asked to find?

• What information will you use?

• Is there any information you will not use? If so, what?

PLAN

• What strategy can you use to solve the problem?
   You can *draw a diagram*.

SOLVE

• How can you use the strategy to solve the problem?
   Draw a diagram showing the relationship between the masses of the two fish.

| Charles's | ▓ kg | ▓ kg | 1 kg | } Total mass |
| Darrell's | ▓ kg | | | is 10 kg. |

Technology Link

More Practice: Use E-Lab, Mass.

www.harcourtschool.com/
elab2002

Subtract 1 kg from 10 kg to find the sum of the three equal parts.   10 − 1 = 9

Divide the sum by 3 to find the value of each part.   9 ÷ 3 = 3

Each part is 3 kg, so Darrell's fish has a mass of 3 kg and Charles's fish has a mass of 3 + 3 + 1, or 7 kg.

CHECK

• What other strategy could you use?

▶ Problem Solving Practice

Draw a diagram to solve.

1. Val, Brent, and Don caught a total of 24 fish. Val caught 4 more fish than Brent. Don caught twice as many fish as Brent. How many fish did each catch?

PROBLEM SOLVING STRATEGIES

▶ Draw a Diagram or Picture
Make a Model or Act It Out
Make an Organized List
Find a Pattern
Make a Table or Graph
Predict and Test
Work Backward
Solve a Simpler Problem
Write an Equation
Use Logical Reasoning

Matt's dog is 15 centimeters longer than Michael's dog. Derek's dog is 9 centimeters greater than twice the length of Michael's dog. Michael's dog is 3 decimeters long.

2. What should you do to change the unit of length of Michael's dog from decimeters to centimeters?
  A divide by 100
  B divide by 10
  C multiply by 100
  D multiply by 10

3. What are the lengths of Matt's, Michael's, and Derek's dogs in centimeters?
  F 18 cm; 30 cm; 15 cm
  G 45 cm; 30 cm; 69 cm
  H 45 cm; 30 cm; 60 cm
  J 180 cm; 30 cm; 150 cm

Mixed Strategy Practice

USE DATA For 4–5, use the table.

4. The Martins went fishing for the weekend. They rented a fishing boat for 4 hours. They also rented 3 fishing poles and bought one bucket of bait. How much money did the Martins spend?

Lakeside Fish Camp
Open Daily at 10 AM

Boat Rental.............. $25 per hour
Fishing Pole Rental... $4.25 per day
Bucket of Bait.......... $5.75 each

5. The Kells arrived at the Lakeside Fish Camp when it opened. They spent $4\frac{1}{2}$ hours fishing, 30 minutes eating lunch, and 90 minutes driving home from the camp. At what time did they get home?

6. Jay and Norma work as fishing guides. They each work every other day. If Norma works this Saturday, who will work on the Saturday 4 weeks from this Saturday? Who will work on the Saturday 7 weeks from now?

7. Mario folded a sheet of paper in half several times. When he opened it up, there were 32 rectangles. How many times did Mario fold the paper?

8. ✎ **Write a problem** that can be solved by drawing a diagram. Exchange problems with a classmate and solve.

Review/Test

✔ CHECK VOCABULARY AND CONCEPTS

Choose the best term from the box.

| milliliter |
| mass |
| decimeter |
| meter |

1. Metric units of capacity are _?_ and liter. (p. 476)

2. The amount of matter in an object is the _?_. (p. 478)

3. The width of an adult's hand is about a _?_. (p. 470)

✔ CHECK SKILLS

Complete. (pp. 474–475; 476–477; 478–479)

4. 6 L = ■ mL   5. ■ cm = 8 m   6. 30 m = ■ cm   7. 854 dm = ■ cm

8. 20 kg = ■ g   9. 12 m = ■ dm   10. 90 dm = ■ m   11. ■ g = 2 kg

Compare. Write <, >, or = for each ●. (pp. 474–475)

12. 16 cm ● 12 dm   13. 4 m ● 40 cm   14. 90 dm ● 10 m

15. 2,000 m ● 2 km   16. 300 dm ● 3 km   17. 720 cm ● 72 dm

Choose the most reasonable unit of measure. Write *cm, dm, m, km, mL, L, g,* or *kg*. (pp. 470–473; 476–477; 478–479)

18. the mass of a pin

19. the height of a door

20. the capacity of a mug

21. the mass of a brick

✔ CHECK PROBLEM SOLVING

Solve. (pp. 474–475; 476–477; 480–481)

22. George bought a liter of milk. He used 480 milliliters to make pudding. How many milliliters of milk does George have left?

23. Aiesha's poster is 62 centimeters long. Enrico's poster is 7 decimeters long. Whose poster is longer? How much longer?

24. The mass of a package of cookies is 60 grams. The mass of a box of crackers is 3 grams more than twice the mass of the package of cookies. What is the mass of the box of crackers?

25. Vince is building a table. Each leg will be 75 centimeters long. He has a piece of wood that is 3 meters long. Can he make 4 table legs? Explain.

Standardized Test Prep

Check your work.

See item **3.**

The question asks how long the segment is from **A** to **B**, not from the beginning of the ruler. Check the position of the ruler in the drawing and the units shown on the ruler.

Also see problem **7**, p. H65.

For 1-11, choose the best answer.

1. Which names the greatest mass?

  A 0.50 kg   **C** 1,500 g

  B 5 kg   **D** 550 g

2. Which is longer than 19 meters?

  F 19 cm   **H** 19 km

  G 19 dm   **J** 19 g

3. How long is the segment from A to B?

  A 5 cm **B** 5 dm **C** 6 cm **D** 6 dm

4. How many milliliters are in 5 liters?

  F 5   **H** 500

  G 50   **J** 5,000

5. How many kilograms are in 7,000 grams?

  A 7   **B** 70   **C** 700   **D** 7,000

6. 75 × 800

  F 60,000   **H** 600

  G 6,000   **J** NOT HERE

7. Which lists the lengths in order from least to greatest?

  A 2 m; 4 dm; 16 cm

  B 2 m; 16 cm; 4 dm

  C 4 dm; 16 cm; 2 m

  D 16 cm; 4 dm; 2 m

8. 673 ÷ 13

  F 59 r6   **H** 51 r10

  G 59 r4   **J** 51 r1

9. Which fraction is equivalent to 0.40?

  A $\frac{1}{4}$ **B** $\frac{3}{8}$ **C** $\frac{3}{4}$ **D** $\frac{2}{5}$

10. Which is a reasonable height for a tree?

  F 7 cm **G** 7 dm **H** 7 m   **J** 7 km

11. Which is a reasonable amount for a glass of milk?

  A 30 mL   **C** 30 L

  B 300 mL   **D** 300 L

Write What You Know

12. Jeff's shoe is 8 centimeters longer than Mary's shoe. Brenda's shoe is a decimeter shorter than Jeff's. Mary's shoe is 22 centimeters long. Explain how to find the length of Brenda's shoe.

13. Louisa wants to find the number of centimeters in 12 kilometers. She says that you can use multiplication to find the number of centimeters. Is she right? Tell how you know.

Perimeter and Area of Plane Figures

More thoroughbred horses are raised in Kentucky than in any other state in the United States. More than 8,000 thoroughbreds are born each year. Trainers work with them in outdoor corrals.

PROBLEM SOLVING The table gives sizes for some corrals. What are the perimeter and area of each corral?

HORSE CORRAL SIZES

Length	Width	Number of Horses
12 ft	12 ft	1
16 ft	16 ft	1
12 ft	24 ft	2
50 ft	50 ft	4
50 ft	100 ft	6

CHECK WHAT YOU KNOW

Use this page to help you review and remember
important skills needed for Chapter 25.

✓ SORT POLYGONS (For Intervention, see p. H27.)

   A   B   C   D   E   F

For 1–4, use the polygons.

1. Which have 3 sides?

2. Which have 4 sides?

3. Which has 4 equal sides?

4. Which have right angles?

✓ IDENTIFY PLANE FIGURES (For Intervention, see p. H27.)

Write the name of each figure.

5.

6.

7.

8.

✓ FIND PERIMETER (For Intervention, see p. H28.)

Count to find the perimeter of each figure.

9.

10.

11.

12.

✓ FIND AREA (For Intervention, see p. H28.)

Count to find the area of each figure.

13.

14.

15.

16.

17.

18.

19.

20.

Perimeter of Polygons

Quick Review

1. $7 + 8 + 14$

2. $14 + 22 + 9 + 31$

3. $52 + 37 + 51$

4. $62 + 18 + 27 + 43$

5. $45 + 34 + 21 + 27$

▶ **Learn**

THE SIDE STORY   A polygon is a closed plane figure with straight sides. Polygons are named by the number of sides or number of angles they have. Here are some polygons.

triangle	quadrilateral	pentagon	hexagon	octagon
3 sides 3 angles	4 sides 4 angles	5 sides 5 angles	6 sides 6 angles	8 sides 8 angles

VOCABULARY

polygon	octagon
triangle	regular polygon
quadrilateral	perimeter
pentagon	formula
hexagon	

The hexagon and octagon above are special polygons because all the sides in each figure are the same length.

Each is called a regular polygon. Any polygon can be regular if all its sides are the same length.

The triangle, quadrilateral, and pentagon shown above have some sides that are different lengths. These are NOT regular polygons.

Examples

These polygons are regular.

triangle

quadrilateral

pentagon

hexagon

These polygons are not regular.

triangle

quadrilateral

pentagon

hexagon

MATH IDEA   Polygons can be many different shapes. The number of sides a figure has tells what kind of polygon it is.

Find Perimeter of Polygons

Lizzie is putting a new fence around Dancer's corral. What is the perimeter?

You can find the perimeter by finding the distance around the corral.

A **perimeter** is the distance around a polygon. You can use a **formula**, or mathematical rule, to find perimeter.

Dancer's corral has 5 sides, so you can use a formula to find the perimeter of a pentagon.

P = sum of the length of the sides

So, a perimeter formula for a pentagon is

$P = a + b + c + d + e.$

$P = 14 + 17 + 13 + 15 + 16$

$P = 75$

The length of each side is represented by a variable.

So, the perimeter is 75 m.

Dancer's Corral

▶ Check

1. Draw and label a polygon that has a perimeter of 15 units.

Name the polygon and tell if it is *regular* or *not regular*.

2.

3.

4.

5.

Find the perimeter.

6. 14 m, 7 m, 15 m

7. 12 yd, 4 yd, 4 yd, 12 yd

8. 4 in., 6 in., 4 in., 5 in., 1 in., 4 in.

9. 4 ft, 8 ft, 4 ft, 3 ft, 6 ft, 2 ft

Name the polygon and tell if it is *regular* or *not regular*.

10.

11.

12.

13.

Find the perimeter.

14.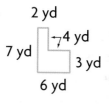
8 ft
4 ft   4 ft
8 ft

15.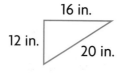
16 in.
12 in.   20 in.

16.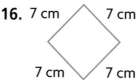
7 cm   7 cm
7 cm   7 cm

17.
4 in.
4 in.   4 in.
4 in.   4 in.
4 in.   4 in.
4 in.

18.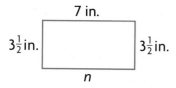
2 yd
4 yd
7 yd   3 yd
6 yd

19.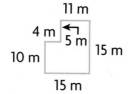
11 m
4 m   5 m
10 m   15 m
15 m

20.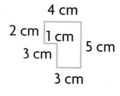
4 cm
2 cm   1 cm
3 cm   5 cm
3 cm

21.
15 yd   11 yd
7 yd
21 yd   24 yd
32 yd

Using the perimeter given, find the unknown length, *n*.

22. $P = 21$ in.

7 in.
$3\frac{1}{2}$ in. $3\frac{1}{2}$ in.
n

23. $P = 92$ cm

23 cm
11.5 cm   11.5 cm
n   11.5 cm
23 cm

24. **REASONING** An equilateral triangle and a square each have a perimeter of 24 inches. Draw each figure.

25. ? **What's the Question?** A quadrilateral has sides with lengths of 4 feet, 5 feet, 8 feet, and 10 feet. The answer is 27 feet.

26. **Draw** three quadrilaterals, one with 4 equal sides, one with 2 pairs of equal sides, and one with no equal sides.

27. ✎ **Write About It** Explain how you would write a formula for the perimeter of any polygon.

28. At lunch, Jaco walked around the perimeter of a field at the right. Use the measurements in the drawing to find the perimeter of the field. What is the perimeter?

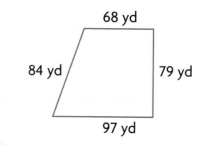
68 yd
84 yd   79 yd
97 yd

29. The perimeter of a triangle is 24 inches. Two sides of the triangle measure 10 inches and 6 inches. What is the length of the third side?

30. Draw two regular polygons that have the same shape but are different in both size and position.

31. Janice has only dimes. She spends 6 dimes on juice, and has 30¢ left over. How many dimes did Janice have before buying juice?

32. Dennis had 42 marbles. He gave some away and now has 14 marbles. Write an equation that can be used to find how many marbles he gave away.

Mixed Review and Test Prep

33. $\frac{3}{8} + \frac{1}{4}$ (p. 398)   **34.** $\frac{2}{3} - \frac{1}{6}$ (p. 400)

35. $356 \div 67$ (p. 200)

36. List 6 multiples of 3. (p. 298)

37. Round 2.35 to the nearest tenth. (p. 428)

38. Write 24,607 in expanded form. (p. 6)

39. **TEST PREP** Mike bought a flashlight for $6.95 and a CD for $10.99. How much change did he get from a $20 bill? (p. 434)

A $17.95   **C** $9.01
B $13.05   **D** $2.06

40. The bandstand in the park has 5 sides. What is another name for a polygon with 5 sides? (p. 486)

A quadrilateral   **C** hexagon
B pentagon   **D** octagon

PROBLEM SOLVING | THINKER'S CORNER

REASONING A long time ago, a Chinese man named Tan had a square tile. The tile fell and broke into seven pieces like the ones shown here. The pieces are called tans.

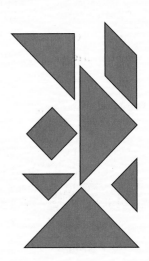

1. Help Tan put his tile back together. Use the tangram pattern and cut out the pieces. Use all of the pieces to make one square. You can flip, slide, and turn the pieces.

Rearrange your tangram pieces to make the pictures below.

2.

3.

Estimate and Find Perimeter

▶ Learn

MEASURING MATH You can use what you know about units of length to estimate perimeter.

Activity

MATERIALS: centimeter ruler

- Look at the cover of your math book. Estimate the perimeter of your math book in centimeters.

- Now measure the perimeter with your ruler. How does your estimate compare to your measurement?

There are special formulas that you can use to find the perimeter of a rectangle and a square.

Polygon	Perimeter	Formula
rectangle	Perimeter = length + width + length + width	$P = l + w + l + w$
	or	or
	Perimeter = 2 × length + 2 × width	$P = (2 \times l) + (2 \times w)$
square	Perimeter = side + side + side + side	$P = s + s + s + s$
	or	or
	Perimeter = 4 × side	$P = 4 \times s$

Examples

A

7 in. / 5 in. / 5 in. / 7 in.

Use the formula.

$P = (2 \times l) + (2 \times w)$
$P = (2 \times 7) + (2 \times 5)$
$P = 14 + 10$
$P = 24$

The perimeter is 24 inches.

B

3 ft / 3 ft

Use the formula.

$P = 4 \times s$
$P = 4 \times 3$
$P = 12$

The perimeter is 12 feet.

▶ Check

1. **Give** an estimate in inches for the perimeter of your desk. Then measure the perimeter and compare the measurement to your estimate.

Use a formula to find the perimeter.

2.
8 m
5 m 5 m
8 m

3. 6 ft
6 ft

4. 12 in.
3 in.

5. 9 cm
9 cm

▶ Practice and Problem Solving

Use a formula to find the perimeter.

6. 12 yd
9 yd 9 yd
12 yd

7. 11 cm
11 cm

8. 2 yd
4 yd

9. 13 m
13 m

10. 15 ft
9 ft

11. 24 yd
6 yd

12. 8 cm
4 cm

13. 20 cm
20 cm

14. Mai walked around the perimeter of the playground 3 times. The rectangular playground is 25 meters long and 15 meters wide. How far did Mai walk?

15. **REASONING** The perimeter of a rectangle is 26 centimeters. The length is 8 centimeters. What is the width?

16. **? What's the Error?** Gabe says the perimeter of the figure at the right is 10 inches. Describe his error. Find the correct perimeter.

7 in.

3 in.

Mixed Review and Test Prep

17. 1,096 (p. 200)
    × 8

18. 24)372 (p. 286)

19. Subtract 1.38 from 3.07. (p. 434)

20. Find the median for this set of numbers.
    60, 62, 64, 64, 64, 68, 68, 68, 68
    (p. 86)

21. **TEST PREP** Traci wants to buy a stereo that costs $130. She saves $14 each week. How many weeks will it take Traci to save enough money to buy the stereo? (p. 286)

    A 8 weeks   **C** 10 weeks
    B 9 weeks   **D** 11 weeks

EXTRA PRACTICE page H56, Set A

Estimate and Find Area

VOCABULARY

area

▶ **Learn**

ROW BY ROW Mr. Jones is putting 1-square-foot tiles on the shower wall of his bathroom. Estimate the number of tiles he needs to cover the wall.

Think: How many squares in each row?   about 6 squares
How many rows are needed?   about 8 rows

$6 \times 8 = 48$

So, Mr. Jones needs about 48 tiles to cover the area of the shower wall.

Area is the number of square units needed to cover a surface.

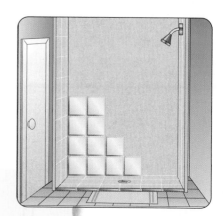

Example

One Way You can count square units to find the area.

STEP 1	**STEP 2**	**STEP 3**
Using a centimeter ruler and grid paper, draw a rectangle 7 cm long and 5 cm wide.	Count the number of squares. 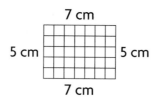	Record your answer in square units. $A = 35$ sq cm

Another Way You can also use a formula.

The formula for the area of a rectangle is
Area = length × width or $A = l \times w$.

Use the formula to find the area.
$A = l \times w$
$A = 7 \times 5$
$A = 35$
So, the area is 35 sq cm.

7 cm

5 cm   5 cm

7 cm

• Use the formula to find the area of a rectangle with a length of 8 cm and a width of 6 cm.

Divide Figures into Parts

What is the area of this figure?

It is easier to find the area of some figures by dividing them into rectangles first.

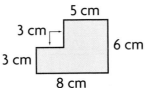

Example

STEP 1

Divide the figure into two rectangles.

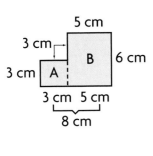

STEP 2

Find the area of each part.

Figure A

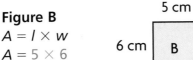

$A = l \times w$
$A = 3 \times 3$
$A = 9$

The area is 9 sq cm.

Figure B

$A = l \times w$
$A = 5 \times 6$
$A = 30$

The area is 30 sq cm.

STEP 3

Add the areas together to find the total area of the figure.

$9 + 30 = 39$

So, the total area is 39 sq cm.

• **Discuss** why you can use the formula $A = s \times s$ for the area of a square.

Technology Link

More Practice: Use E-Lab, *Exploring Area.*

www.harcourtschool.com/elab

▶ Check

1. **Draw** this figure and show one way to divide it into rectangles so you can find the area.

Find the area.

2.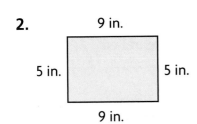

3.

4.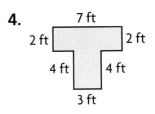

LESSON CONTINUES ▶

Find the area.

5.
9 yd
3 yd   3 yd
9 yd

6.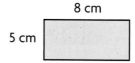
6 m
3 m   3 m
6 m

7.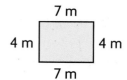
7 m
4 m   4 m
7 m

8.
4 in.
4 in.   4 in.
4 in.

9.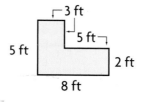
8 cm
5 cm

10.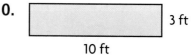
3 ft
10 ft

11.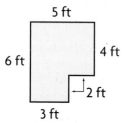
5 ft
6 ft   4 ft
2 ft
3 ft

12.
3 ft
5 ft
5 ft   2 ft
8 ft

13.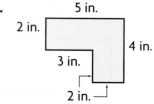
5 in.
2 in.   4 in.
3 in.
2 in.

**Use a centimeter ruler to measure each figure.
Find the area and perimeter.**

14.

15.

USE DATA For 16–17, use the bar graph.

16. The length of a twin and a full blanket is 86 inches long. What is the difference in the area of a full blanket and a twin blanket?

17. The length of a queen and a king blanket is 94 inches long. What is the difference in the area of a king blanket and a queen blanket?

18. **? What's the Question?** The school yard has an area of 90 sq ft set aside for hopscotch. The length of the section is 10 ft. The answer is 9 ft.

19. **Write About It** One bulletin board is a rectangle. Another is a square. Explain how 16 square pictures could cover either one.

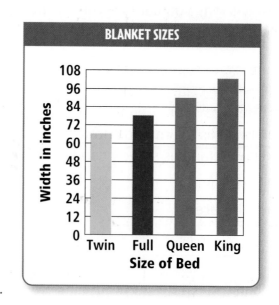

BLANKET SIZES

Width in inches

108
96
84
72
60
48
36
24
12
0
   Twin   Full   Queen   King
Size of Bed

20. Josh is covering a rectangular table top with tiles. There will be 12 rows. Each row will have 8 tiles. How many tiles will Josh use?

21. Andrea's living room is 4 yards wide. The perimeter is 20 yards. What is the area?

22. Mr. Robledo is putting carpet in a bedroom that measures 14 ft by 12 ft. How many square feet of carpet does he need?

23. Mrs. Page's dining room measures 16 ft by 13 ft. She wants to tile the room using 1-square-foot tiles. How many tiles does she need?

Mixed Review and Test Prep

24. 7,218 (p. 40)
  −2,643

25. 5,207 (p. 40)
  −1,958

26. $3\frac{3}{8} + 7\frac{1}{8}$ (p. 392)

27. $2\frac{2}{5} + 4\frac{1}{5}$ (p. 392)

Find the value of each expression. (p. 54)

28. $(36 - 12) + 8$

29. $36 - (12 + 8)$

30. Write an equivalent decimal for 0.7.
(p. 412)

31. **TEST PREP** Which number shows $\frac{30}{6}$ in simplest form? (p. 368)

A $\frac{1}{6}$ **B** 5 **C** 24 **D** 180

32. **TEST PREP** The movie Jada wants to see starts at 6:50 P.M. and lasts for 1 hour 40 minutes. At what time should Jada's mother pick her up from the movie? (p. 120)

F 7:30 P.M. **H** 8:30 P.M.
G 7:40 P.M. **J** 8:40 P.M.

PROBLEM SOLVING LiNKUP ... to Architecture

Architects design buildings and manage the construction. To do this, they draw floor plans. This floor plan shows the rooms for a house.

For 1–4, use the floor plan to find the area of the room.

1. Kitchen
2. Living Room
3. Family Room
4. Bedroom 3

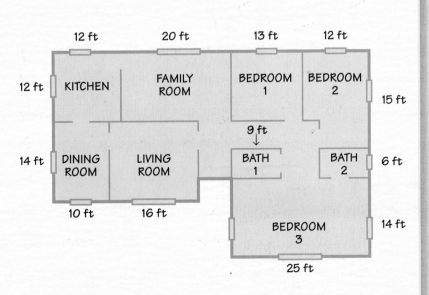

EXTRA PRACTICE page H56, Set B

Relate Area and Perimeter

▶ **Learn**

AREA VS. PERIMETER Two figures can have the same area but different perimeters, or different areas but the same perimeter.

 Activity 1 Find the area and perimeter of each figure.

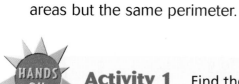

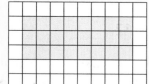

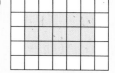

- Compare and contrast figures A and B.
- Compare and contrast figures C and D.

Activity 2 Show a rectangle that has the same area as rectangle A but a different perimeter.

8 ft

A

6 ft

Area = 48 sq ft
Perimeter = 28 ft

One possible rectangle:

12 ft

4 ft

Area = 48 sq ft
Perimeter = 32 ft

▶ **Check**

1. **Draw** two figures that have different perimeters but the same area. Use grid paper.

2. **Draw** two figures that have different areas but the same perimeter. Use grid paper.

For 3–8, find the area and perimeter of each figure.
Then draw another figure that has the same perimeter
but a different area.

3.

4.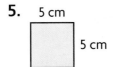
10 cm
4 cm ☐ 4 cm
10 cm

5. 5 cm
☐ 5 cm

6. 9 in.
4 in. ☐

7. 9 cm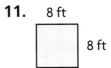
10 cm ☐ 10 cm
9 cm

8.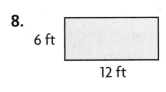
6 ft ☐
12 ft

For 9–12, find the area and perimeter of each figure.
Then draw another figure that has the same area but
a different perimeter.

9.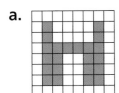

10. 9 cm
☐ 3 cm

11. 8 ft
☐ 8 ft

12. 7 m
8 m ☐ 8 m
7 m

For 13–14, use figures a–d.

13. Which of the figures below have the
same area but different perimeters?

14. Which of the figures below have the
same perimeter but different areas?

a. 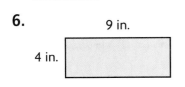 **b.** **c.** **d.**

15. ✏ Write a problem about the
square floor of a tree house that is
8 feet on each side.

16. **REASONING** The area of a rectangular
carpet is 24 square feet. Give
4 possible perimeters of the carpet.

17. 47×22 (p. 224)   **18.** 71×18 (p. 224)

19. $938 \div 31$ (p. 286)

20. Find the value of the expression
$37 - (3 \times 9)$ (p. 158)

21. ⭐ **TEST PREP** Cindy had 416 trading
cards. She gave 72 to her sister and
got 37 for her birthday. How many
cards did she have then? (p. 234)
A 109   **B** 344   **C** 379   **D** 381

Relate Formulas and Rules

▶ **Learn**

COVER IT UP! Handy's Hardware store sells wallpaper by the roll. Each roll covers an area of 96 square feet and is 2 feet wide. What is the length of the roll?

A formula is a special kind of rule. To find the length, think about the formula for the area of a rectangle.

Formula: area = length × width

$$A = l \times w$$ Replace A with 96 and
$$96 = l \times 2$$ w with 2.
$$96 \div 2 = l$$ Since division is the inverse
$$48 = l$$ of multiplication, divide 96 by 2.

So, the length of the roll of wallpaper is 48 feet.

Quick Review

Find the missing factor.

1. $4 \times \blacksquare = 56$

2. $4 \times \blacksquare \times 2 = 40$

3. $4 \times \blacksquare = 36$

4. $\blacksquare \times 2 \times 3 = 42$

5. $24 = 2 \times 4 \times \blacksquare$

Examples

Ⓐ Find the width. The area is 24 square feet.

$$A = l \times w$$
$$24 = 8 \times w$$
$$24 \div 8 = w$$
$$3 = w$$

8 ft

?

The width is 3 feet.

Ⓑ Find the length. The area is 56 square yards.

$$A = l \times w$$
$$56 = l \times 7$$
$$56 \div 7 = l$$
$$8 = l$$

?

7 yd

The length is 8 yards.

Find an Unknown Side

You can also use a formula to find the length of a side of a triangle if you know the perimeter and the lengths of two of the sides.

Handy's Hardware store also sells triangular signs. Each sign has a perimeter of 62 inches. One side of the triangle is 20 inches long, and another side is 22 inches long. How long is the third side?

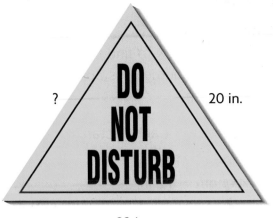

?   20 in.

22 in.

Formula: Perimeter $= a + b + c$

$$P = a + b + c$$
$$62 = 20 + 22 + c$$
Replace P with 62, a with 20, and b with 22.

$$62 = 42 + c$$
$$62 - 42 = c$$
Since subtraction is the inverse of addition, subtract 42

$$20 = c$$

So, the third side is 20 inches long.

 MATH IDEA   You can use formulas to find unknown lengths when some lengths and the area or perimeter are known.

Examples

A   Find the unknown length.
The perimeter is 24 meters.

8 m   ?   6 m

$$P = a + b + c$$
$$24 = 6 + 8 + c$$
$$24 = 14 + c$$
$$24 - 14 = c$$
$$10 = c$$
The length is 10 meters.

B   Find the unknown length.
The perimeter is 42 centimeters.

?   14 cm   14 cm

$$P = a + b + c$$
$$42 = 14 + 14 + c$$
$$42 = 28 + c$$
$$42 - 28 = c$$
$$14 = c$$
The length is 14 centimeters.

LESSON CONTINUES ▶

1. **Explain** how to find the length of a side of a square whose perimeter is 36 inches. What is the length of the side?

Complete for each rectangle.

2. Area = 48 sq in.
Length = 12 in.
Width = ▦ in.

3. Area = 63 sq mi
Width = 9 mi
Length = ▦ mi

4. Area = 56 sq in.
Length = 7 in.
Width = ▦ in.

▶ Practice and Problem Solving

Complete for each rectangle.

5. Area = 100 sq in.
Length = 25 in.
Width = ▦ in.

6. Area = 13 sq ft
Width = 1 ft
Length = ▦ ft

7. Area = 54 sq mi
Length = 9 mi
Width = ▦ mi

8. Area = 84 sq m
Width = 7 m
Length = ▦ m

9. Area = 110 sq cm
Length = 22 cm
Width = ▦ cm

10. Area = 150 sq ft
Width = 10 ft
Length = ▦ ft

Find the unknown length.

11. 9 yd
?
Area = 45 sq yd

12. 9 ft
?
Area = 108 sq ft

13. 10 cm
?
Area = 70 sq cm

14. 12 ft   8 ft
?
Perimeter = 35 ft

15. 24 m   22 m
?
Perimeter = 72 m

16. 15 in.
?
12 in.
Perimeter = 36 in.

Write the formula you would use to find the area or perimeter of each figure. Then find the unknown length.

17. 6 in.
?
Area = 24 sq in.

18. 4 cm   ?
3 cm
Perimeter = 12 cm

19. ?
6 m
Area = 84 sq m

20. Ivonne wants to know how many quilt blocks are in her quilt. It is 12 blocks long and 6 blocks wide. How many blocks are in her quilt?

21. How could you find the length of a rectangle when you know that its perimeter is 48 inches and its width is 4 inches? What is the length?

Mixed Review and Test Prep

22. $4\frac{2}{5} + 3\frac{4}{5}$ (p. 392)　　**23.** $7\frac{3}{8} - 3\frac{1}{8}$ (p. 392)

24. List the first six multiples of 6. (p. 298)

25. $\begin{array}{r} 7,802 \\ +2,008 \end{array}$ (p. 40)　　**26.** $\begin{array}{r} 4,500 \\ -1,905 \end{array}$ (p. 40)

27. Joan has $\frac{3}{8}$ yard of ribbon. She needs $\frac{3}{4}$ yard to make a costume. How much more ribbon does she need? (p. 400)

28. **TEST PREP** Find the value of $(9 + 3) \times (24 - 8)$. (p. 158)
A 96　**B** 154　**C** 192　**D** 432

29. **TEST PREP** Suzy's Seashell Shop has 144 shells equally divided into 3 baskets. Suzy divides 24 more shells into the baskets. If each basket has the same number, how many shells are in each basket? (p. 234)
F 44　**G** 56　**H** 68　**J** 72

PROBLEM SOLVING LINKUP... to Reading

STRATEGY • OBSERVING RELATIONSHIPS

Observing relationships can help you solve problems. Look at flag of Washington state. About how many square units of the flag are covered by the design?

To estimate the area covered by the circle:

- Find how many partial squares are more than $\frac{1}{2}$ covered by the design. Divide the number of partial squares by 2.
- Add the number of whole squares to the number of partial squares to estimate the area.

So, you can estimate the area of a circle by observing relationships of grid squares.

Record:
$(12 \div 2)$　+　9　=　15
　↑　　　　　↑　　　　↑
partial　　whole　　estimated
units　　　units　　　area in
　　　　　　　　　　square units

Estimate the area of each circle.

1.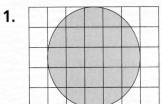

2.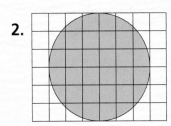

Problem Solving Strategy
Find a Pattern

PROBLEM   Mr. Jiminez wants to build different-size storage buildings, so he drew plans for the buildings. Some of the buildings will be twice as long as others but with the same width. He wants to know how the areas of the buildings are related. How do the areas change?

JIMINEZ STORAGE BUILDINGS

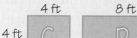

4 ft   8 ft
3 ft   A   B   3 ft
4 ft   8 ft
4 ft   C   D   4 ft
4 ft   8 ft
5 ft   E   F   5 ft
4 ft   8 ft
6 ft   G   H   6 ft

UNDERSTAND
- What do you know?
- What are you asked to find?

PLAN
- What strategy can you use to solve the problem?
  You can *find a pattern* to solve the problem.

SOLVE
- How can you use *find a pattern* to solve the problem?
  Look for a pattern in the areas.

	LENGTH	WIDTH	AREA		LENGTH	WIDTH	AREA
Building A	4	3	12	Building C	4	4	16
Building B	8	3	24	Building D	8	4	32
Building E	4	5	20	Building G	4	6	24
Building F	8	5	40	Building H	8	6	48

The areas change from 12 to 24, 16 to 32, 20 to 40, and 24 to 48.

So, as the length of one side doubles, the area also doubles.

CHECK
- Explain how you can check your answer.

Problem Solving Practice

Use *find a pattern* **and solve.**

1. **What if** Mr. Jiminez wants to know how the areas of his storage buildings change when both sides are doubled? Make a table to show how the areas change. Explain.

2. Mr. Jiminez also wants to know what happens to the perimeters of his storage buildings when both sides are doubled. Make a table to show how the perimeters change. Explain.

PROBLEM SOLVING STRATEGIES

Draw a Diagram or Picture
Make a Model or Act It Out
Make an Organized List
▶ Find a Pattern
Make a Table or Graph
Predict and Test
Work Backward
Solve a Simpler Problem
Write an Equation
Use Logical Reasoning

Rectangular swimming pools come in different sizes. Use the table to find how the perimeters of the swimming pools are related.

3. How do the widths of the swimming pools change?
   A decrease by 20 ft
   B decrease by 30 ft
   C increase by 10 ft
   D increase by 20 ft

SWIMMING POOL SIZES

Pool	Length	Width	Perimeter
A	12 ft	20 ft	64 ft
B	12 ft	30 ft	84 ft
C	12 ft	40 ft	104 ft
D	12 ft	50 ft	124 ft

4. What pattern do you see in the perimeters of the swimming pools?
   F increase by 20 feet   **G** double in size   **H** decrease by 10 ft   **J** decrease by half

Mixed Strategy Practice

Use any strategy to solve.

5. Gene, Erin, and Axel collect coins. Gene has 115 more coins than Erin. Erin has 72 fewer coins than Axel. Axel has 185 coins. How many coins does Gene have?

6. Three stickers cost 15¢. Six stickers cost 30¢ and 9 stickers cost 45¢. How much will 15 stickers cost? How much will 30 stickers cost?

7. Mark travels 14.75 miles to soccer practice. First, he travels 5.25 miles to pick up John, then 3.75 miles to pick up Harold, and finally 2.25 miles to pick up Frank. How many miles does he have left to travel to soccer practice?

8. ✎ **Write About It** Explain how to find the unknown length.

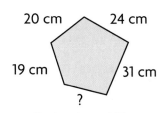

20 cm   24 cm
19 cm   31 cm
?
Perimeter = 115 cm

Problem Solving Strategy

Review/Test

✓ CHECK VOCABULARY AND CONCEPTS

Choose the best term from the box.

| area |
| formula |
| perimeter |
| polygon |

1. The distance around a figure is its __?__. (p. 487)

2. The number of square units needed to cover a surface is its __?__. (p. 492)

3. A closed plane figure with straight sides is a __?__. (p. 486)

✓ CHECK SKILLS

Find the perimeter. (pp. 486–491)

4.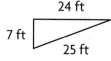
24 ft
7 ft
25 ft

5.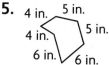
4 in. 5 in.
4 in. 5 in.
6 in. 6 in.

Find the area. (pp. 492–495)

6.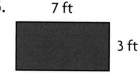
7 ft
3 ft

7.
4 ft
4 ft

Find the area and perimeter of each figure. Then draw another figure that has the same perimeter but a different area. (pp. 496–497)

8.

9.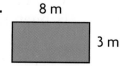
8 m
3 m

10.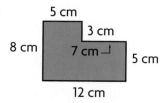
5 cm
3 cm
8 cm
7 cm
5 cm
12 cm

Complete for each rectangle. (pp. 498–501)

11. Area = 300 sq in.

    Width = 15 in.

    Length = ▇ in.

12. Area = 175 sq m

    Length = 25 m

    Width = ▇ m

13. Area = 208 sq ft

    Width = 13 ft

    Length = ▇ ft

✓ CHECK PROBLEM SOLVING

Solve. (pp. 502–503)

14. Mr. Walker's front porch is 6 feet by 8 feet. He wants to double the length and the width of his porch. How will the area of the porch change?

15. A stone wall has 28 stones on the first layer, 24 stones on the second layer, and 20 stones on the third layer. How many stones are on the fifth layer?

Standardized Test Prep

 Decide on a plan.

See item **4.**

You can divide the figure into two rectangles and then use a formula to answer the problem.

Also see problem **4**, p. H63.

For 1–8, choose the best answer.

1. What is the perimeter?

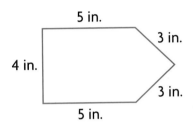

5 in.

4 in.

3 in.

3 in.

5 in.

A 10 in. **C** 18 in.

B 15 in. **D** 20 in.

2.

PLAY AREA PERIMETERS			
	Length (ft)	Width (ft)	Perimeter (ft)
Area A	8	15	46
Area B	10	15	50
Area C	12	15	54

If the length of a play area increases by 2 feet, how does the perimeter change?

F It decreases by 4 ft.

G It increases by 5 ft.

H It increases by 4 ft.

J It doubles.

3. Which term best describes this angle?

A acute **C** obtuse

B right **D** ray

4. What is the area of this figure?

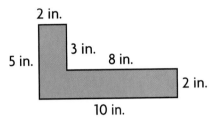

2 in.

5 in. 3 in. 8 in.

2 in.

10 in.

F 50 sq in. **H** 26 sq in.

G 28 sq in. **J** NOT HERE

5. How is $8\frac{4}{10}$ written as a decimal?

A 84 **C** 8.10

B 8.4 **D** 8.04

6. A rectangle has an area of 35 square feet. The length is 7 feet. What is the width?

F 4 ft **H** 10 ft

G 5 ft **J** 28 ft

7. 768 ÷ 3

A 223 **C** 256

B 243 **D** NOT HERE

8. What is the mean score for 85, 96, 79, and 92?

F 80 **H** 86

G 85 **J** 88

Write What You Know

9. The perimeter of a rectangle is 20 cm and its width is 4 cm. How can you find the length?

10. Show that two rectangles can have an area of 36 square feet but different perimeters.

Solid Figures and Volume

The Rose Center in New York City is the new home of the Hayden Planetarium. The Hayden Planetarium is a sphere that appears to be suspended inside a glass cube. The cube, itself sits on a one-story high granite base.

PROBLEM SOLVING The graph shows the volume of the cube, the sphere, and the base. Would the combined volume of the sphere and base fit inside the cube? Explain your thinking.

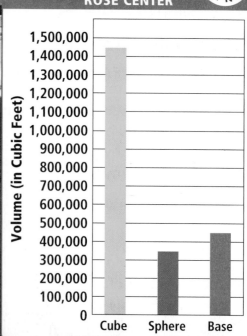

ROSE CENTER

CHECK WHAT YOU KNOW

Use this page to help you review and remember
important skills needed for Chapter 26.

✓ IDENTIFY SOLID FIGURES (For Intervention, see p. H29.)

Choose a name from the list for each solid shape.

cube	cone	cylinder
sphere	square pyramid	rectangular prism

1.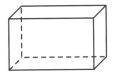

2.

3.

4.

5.

6.

7.

8.

9.

✓ DRAW PLANE FIGURES (For Intervention, see p. H29.)

Draw the plane figure named.

10. circle   **11.** triangle   **12.** square   **13.** rectangle

14. pentagon   **15.** octagon   **16.** hexagon

✓ MULTIPLY THREE FACTORS (For Intervention, see p. H30.)

Find the product.

17. $4 \times 2 \times 2 = \blacksquare$ **18.** $5 \times 2 \times 1 = \blacksquare$ **19.** $4 \times 5 \times 5 = \blacksquare$

20. $7 \times 2 \times 3 = \blacksquare$ **21.** $8 \times 5 \times 2 = \blacksquare$ **22.** $14 \times 5 \times 2 = \blacksquare$

23. $4 \times 4 \times 6 = \blacksquare$ **24.** $2 \times 11 \times 3 = \blacksquare$ **25.** $9 \times 9 \times 1 = \blacksquare$

26. $25 \times 7 \times 0 = \blacksquare$ **27.** $3 \times 2 \times 12 = \blacksquare$ **28.** $4 \times 25 \times 8 = \blacksquare$

Faces, Edges, and Vertices

Quick Review

Tell the number of sides.

1. square

2. triangle

3. rectangle

4. pentagon

5. hexagon

VOCABULARY

two-dimensional

three-dimensional

▶ **Learn**

ANOTHER DIMENSION   Polygons have only length and width, so they are **two-dimensional** figures.

Solid figures have length, width, and height, so they are **three-dimensional** figures.

Study these solid figures. Look for polygons that are faces of each solid figure.

cube

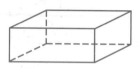
rectangular prism

triangular prism

triangular pyramid

square pyramid

▲ Pyramids at Giza, Egypt

• Name the plane figures found in the faces of each solid above.

Look carefully at the faces of the square pyramid below. The faces of a square pyramid are triangles and a square.

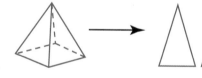

Technology Link

More Practice: Use
Mighty Math
Calculating Crew,
Dr. Gee's Lab,
Levels H, K, and M.

Some solid figures have curved surfaces and no faces.

cylinder

cone

sphere

Activity

Find the number of faces, edges, and vertices of the solid figures in the table below.

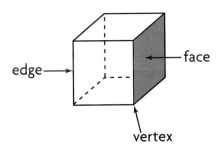

edge → ← face

vertex

Copy and complete the table.

FIGURE	NAME OF FIGURE	NUMBER OF FACES	NUMBER OF EDGES	NUMBER OF VERTICES
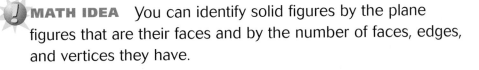	Cube	6	12	8
	Rectangular prism	▨	▨	▨
	Triangular prism	▨	▨	▨
	Triangular pyramid	▨	▨	▨
	Square pyramid	▨	▨	▨

▲ David Smith's sculpture sits in the UCLA Sculpture Gardens.

💡**MATH IDEA** You can identify solid figures by the plane figures that are their faces and by the number of faces, edges, and vertices they have.

▶ Check

1. Name three solid figures that have no faces or straight edges.

Name a solid figure that is described.

2. 9 edges

3. 5 vertices

4. more than 5 faces

5. 8 vertices, 12 edges, 6 faces

LESSON CONTINUES ▶

Name a solid figure that is described.

6. some or all triangular faces

7. some or all rectangular faces

8. curved surfaces

9. 4 vertices

Which solid figure do you see in each?

10.
11.
12.
13.

Write the names of the plane figures that are the faces of each solid figure.

14.
15.
16.
17.

Copy the drawings. Circle each vertex, outline each edge in red, and shade one face in yellow.

18.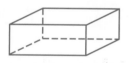
19.
20.
21.

Write the names of the faces and the number of each kind of face of the solid figure.

22. cube

23. rectangular prism

24. square pyramid

25. triangular prism

26. triangular pyramid

27. I am a solid figure with 5 faces, 8 edges, and 5 vertices. What am I?

28. REASONING What plane figure is always found in a pyramid?

29. I am a solid figure with 3 rectangular faces and 2 triangular faces. What am I?

30. I am a solid figure with no edges, no vertices, and a curved surface. What am I?

31. ✎ **Write a problem** in which the answer is a cube.

32. ☀ **What's the Question?** The answer is 4 triangular faces.

33. Laura hung 7 drawings of cats in her room. She hung 10 drawings of dogs. Laura has 1 dog and 2 cats. She has 8 more drawings to hang. How many drawings does Laura have in all?

34. Brandon and Joe played a number game. Joe chose a number and subtracted 3. Then, he added 7 and subtracted 5. The answer was 27. What number did Joe choose?

Mixed Review and Test Prep

Write whether each number is prime or composite. (p. 302)

35. 28   **36.** 31

37. 59   **38.** 56

List the factors of each number. (p. 300)

39. 6   **40.** 15

41. 18   **42.** 20

43. Holly is baking a double batch of cookies. For a single batch, $2\frac{1}{3}$ cups of flour are needed. How much flour does Holly need for the double batch? (p. 392)

44. **TEST PREP** Jane rode her bike 4 miles from her house to school. How many yards did she ride her bike? (HINT: 1 mile = 1,760 yards) (p. 458)

A 440 yards   **C** 6,940 yards

B 1,760 yards   **D** 7,040 yards

45. **TEST PREP** One ant is $\frac{1}{2}$ inch long and a second ant is $\frac{1}{4}$ inch long. How much longer is the first ant than the second ant? (p. 372)

F $\frac{1}{4}$ inch   **H** $\frac{3}{4}$ inch

G $\frac{1}{2}$ inch   **J** 1 inch

PROBLEM SOLVING LiNKUP ... to Social Studies

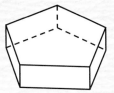

A pentagonal prism is another kind of solid figure.

One of the largest office buildings in the world is shaped like a pentagonal prism. The building is called the Pentagon, and it is located just outside of Washington, D.C. The Pentagon was built in 1943 and is the headquarters of the United States military.

1. How many faces does a pentagonal prism have?

2. What shapes are the faces of a pentagonal prism?

3. How many edges does a pentagonal prism have? How many vertices?

4. How many of each kind of face are on a pentagonal prism?

Patterns for Solid Figures

Quick Review

Tell the number of faces for each.

1. cube

2. square pyramid

3. triangular prism

4. rectangular prism

5. triangular pyramid

▶ Learn

SHAPE OF THINGS A net is a two-dimensional pattern of a three-dimensional figure. It can be folded to make a model of a solid figure.

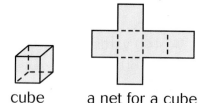

cube   a net for a cube

You can cut apart a three-dimensional box to make a two-dimensional pattern, or net.

VOCABULARY

net

HANDS ON

Activity

MATERIALS: empty container, such as a cereal box; scissors and tape

STEP 1 Cut along some of the edges until the box is flat. Be sure that each face is connected to another face by at least one edge.

STEP 2 Trace the flat shape on a piece of paper. This shape is a net of the box.

STEP 3 Rebuild the net into a three-dimensional box. Use tape to hold it together.

• How is the net for a rectangular prism different from the net for a cube?

Technology Link

More Practice: Use Mighty Math Calculating Crew, *Dr. Gee's Lab*, Levels A, B, C, D and E.

▶ Check

1. **Explain** how the net for a square pyramid is different from the net for a rectangular prism.

2. **Draw** the net for a triangular pyramid.

Would the net make a cube? Write *yes* or *no*.

3.

4.

5.

6.

Write the letter of the figure that is made with each net.

7.

8.

9.

10.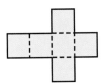

a.

b.

c.

d.

Draw a net that can be cut and folded to make a model of the solid figure shown.

11.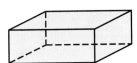

12.

13.

14.

For 15–16, use the pattern.

15. Juanita drew the pattern at the right. What solid figure can she make?

16. Ernest folded Juanita's pattern into a model of a solid figure. How many faces, edges, and vertices did the model have?

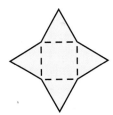

17. **?** **What's the Error?** Nina says the net at the right can be folded to make a triangular prism. Explain Nina's error. Then draw a net for a triangular prism.

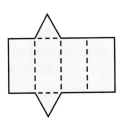

Mixed Review and Test Prep

18. 4×11 (p. 148)

19. 7×12 (p. 148)

20. $60 \div 5$ (p. 142)

21. $99 \div 9$ (p. 144)

22. **TEST PREP** What is $2\frac{1}{4}$ written as a decimal? (p. 406)

A 2.14

C 2.50

B 2.25

D 2.75

Estimate and Find Volume of Prisms

Quick Review

1. $4 \times 4 \times 4$
2. $6 \times 2 \times 3$
3. $4 \times 3 \times 2$
4. $5 \times 5 \times 2$
5. $3 \times 6 \times 5$

▶ **Learn**

HIGH VOLUME The measure of the space that a solid figure occupies is called **volume**. Volume is measured in **cubic units**.

1 cubic unit

VOCABULARY

volume

cubic units

Use what you know about unit cubes to estimate volume. About how many cubes would you need to completely fill a cereal box?

You can find volume in two ways.

One Way Count the number of cubic units as you build a 6-cube × 2-cube × 3-cube rectangular prism.

Another Way Multiply the length, width, and height of the rectangular prism to find the volume in cubic units.

STEP 1

Find the length. Count the number of cubes in one row.

The length is 6 cubes. ⟵ 6 ⟶

STEP 2

Find the width. Count the number of rows in one layer.

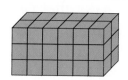

The width is 2 cubes.

STEP 3

Find the height. Count the number of layers in the prism.

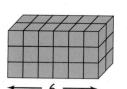

3

The height is 3 cubes.

STEP 4

Multiply length × width × height to find the volume.

6 cubes × 2 cubes × 3 cubes = 36

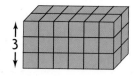

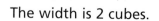

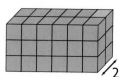

So, the volume is 36 cubic units.

• Does a 2-cube × 1-cube × 3-cube rectangular prism have the same volume as a 3-cube × 2-cube × 1-cube rectangular prism? Explain.

514

1. **Name** the solid figure that is 2 cubes × 2 cubes × 2 cubes. What is its volume?

Count or multiply to find the volume.

2. 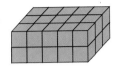   3.    4.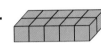

▶ **Practice and Problem Solving**

Find the volume.

5.   6.   7.

8.    9.   10.

Copy and complete the table.

	LENGTH	WIDTH	HEIGHT (number of layers)	VOLUME (cubic units)
11.	5 ft	2 ft	3 ft	▨
12.	2 in.	6 in.	4 in.	▨
13.	10 cm	3 cm	3 cm	▨
14.	3 m	7 m	2 m	▨

15. The volume of a rectangular prism is 12 cubic units. Its length is 2 units and its width is 2 units. What is its height?

16. **REASONING** Describe the different ways you can build a rectangular prism with a volume of 8 cubic units.

▲ Apartment building in Montreal, Canada

17. ✎ **Write About It** Explain to a third-grade student how to find the volume of a rectangular prism.

Mixed Review and Test Prep

18. 12 (p. 216)
   × 13

19. 51 (p. 216)
   × 32

20. 45.61 − 43.39 (p. 436)

21. 278 ÷ 43 (p. 286)

22. **TEST PREP** Hannah divided 456 pennies into 3 equal groups. How much money is in each group? (p. 264)

   A $0.52   **C** $1.52
   B $1.18   **D** $4.53

4

Problem Solving Skill
Too Much/Too Little
Information

UNDERSTAND ▶ PLAN ▶ SOLVE ▶ CHECK ▶

HOW SWEET IT IS! The Sweets Factory packs individually wrapped caramel candies in a 12-inch by 9-inch by 2-inch box. Each box sells for $2.35, and there are 24 boxes to a case. What is the volume of one box?

- **Decide what the problem asks you to find.**
  The problem asks you to find the volume of one box.

- **Decide what information you need to solve the problem.**
  The length, width, and height of the box are needed.

  volume = length × width × height

- **Read the problem again carefully. Decide if there is too much information or not enough information in the problem.**
  There is too much information. The price of each box and how many boxes in a case are not needed.

- **Solve the problem, if possible.**
  Multiply.
  length × width × height = volume
  $12 \times 9 \times 2 = 216$

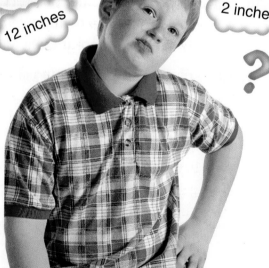

So, the volume of the box is 216 cubic inches.

Talk About It

- **What if** the width of the box was not given in the problem above? Would you have too much or too little information to solve the problem? Explain.

- What can happen if there is too much information in a problem?

516

Decide if the problem has *too much* or *too little information*.
Then solve the problem, if possible.

1. Allison has a drawer that is 33 inches long, 11 inches wide, and 8 inches high. She can fit 15 T-shirts in the drawer. What is the volume of the drawer?

8 in. · 11 in. · 33 in.

2. Rosa bought 3 CDs for $32.65. She also bought a sweater for $25.50 and a skirt for $19.95. How much did Rosa spend on clothes?

3. Kirk has a suitcase that is 40 inches long and 32 inches wide. What is the volume of his suitcase?

A 5-sided figure has one side that is 12 inches, a second side that is 29 inches, a third side that is 17 inches, and a fourth side that is 35 inches.

4. This problem contains ? to find the perimeter of the figure.
   A too much information
   B too little information
   C just enough information
   D none of the above

5. What information do you need to find the perimeter?
   F length of the fifth side
   G length of the sixth side
   H area of the figure
   J None; no information is missing.

Mixed Applications

6. Kim scored 8 goals, and Leticia scored 10 goals. Hillary scored twice as many goals as Kim. How many goals did the three girls score?

7. Together, Terrence and Mike took 24 apples to a picnic. Mike took 6 more apples than Terrence. How many apples did they each take?

8. Austin built a storage box using 4 pieces of wood. It was 2 feet high, 3 feet wide, and 5 feet long. What is the volume of his storage box?

9. The Springside Cafe has 16 tables outside on the patio and 25 tables inside. Each table seats 4 people. How many people can eat at the cafe?

10. Mrs. Fearn has $22.50. She wants to buy 4 pies that each cost $5.65. Does she have enough money? Explain.

11. ✏ **Write a problem** that contains too much information.

12. Mr. Kemp wanted to buy a horse for his farm. He drove 125 miles to Florence and then 98 miles to Camden to look at horses. When he returned home he had driven 380 miles. How far did he drive from Camden to his home?

Problem Solving Skill

Review/Test

✓ CHECK VOCABULARY AND CONCEPTS

Choose the best term from the box.

net
area
volume
cubic units

1. The measure of the space that a solid figure occupies is called _?_. (p. 514)

2. Volume is measured in _?_. (p. 514)

3. A two-dimensional pattern for a three-dimensional figure is called a _?_. (p. 512)

Write the letter of the figure that is made with each net. (pp. 512–513)

4. 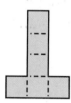   5.   6. 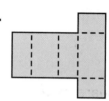   7.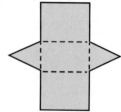

a.   b.   c.   d.

✓ CHECK SKILLS

Find the volume. (pp. 514–515)

8. 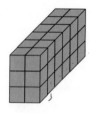   9.    10. 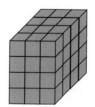   11.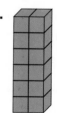

✓ CHECK PROBLEM SOLVING

For 12–13, decide if the problem has *too much* or *too little* information. Then solve the problem, if possible. (pp. 516–517)

12. Jill has $10.00. She buys a picture frame that is 8 in. long and 5 in. wide for $6.78. How much change does she receive?

13. The height of a cardboard box is 62 cm, and the width of the box is 75 cm. What is the volume of the box?

Standardized Test Prep

 **Check your work.**
See item **7.**

A room is like a rectangular prism, so you need to remember how to find volume. Draw a picture and label all the dimensions. *Also* see problem **7, p. H65.**

For 1–8, choose the best answer.

1. Which best describes the faces of a cube?

  A 4 squares   **C** 6 squares
  B 4 rectangles   **D** 6 triangles

2. Claudia drew this picture. This can be folded to be which solid figure?

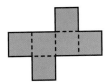

  F square pyramid
  G triangular prism
  H rectangular prism
  J cube

3. $8 \times (2 \times 2)$

  A 12   **C** 32
  B 18   **D** NOT HERE

4. What is 8,964,121 rounded to the nearest million?

  F 9,000,000   **H** 8,964,000
  G 8,900,000   **J** 8,960,000

5. Which describes the shape of the faces of this solid figure?

  A squares   **C** rectangles
  B triangles   **D** hexagons

6. The volume of a rectangular prism is 24 cubic centimeters. The height is 3 centimeters and the width is 4 centimeters. What is the length?

  F 2 cm   **H** 6 cm
  G 4 cm   **J** 12 cm

7. The height of a room is 12 feet and the width is 13 feet. What is the volume of the room?

  A 25 cubic feet
  B 156 cubic feet
  C 304 cubic feet
  D too little information

8. What is the perimeter of a square with sides 12 inches long?

  F 12 in.   **H** 48 in.
  G 24 in.   **J** 144 in.

Write What You Know

9. Lee Ann's swimming pool is 8 feet deep and 20 feet long. It has a volume of 1,600 cubic feet. Describe how to find the width of Lee Ann's pool.

10. Draw a net for the figure in Problem 5. Name the solid figure. Then make a table to show the number and type of faces, edges, and vertices it has. Tell how a net helps you find the number of faces, edges, and vertices.

PROBLEM SOLVING
MATH DETECTIVE

Case of the Unknown Weights

Case 1 — WEIGHTY WATERMELONS

At the county fair, 3 fourth-grade students entered prize winning watermelons. It is your job to use the information given to figure out how much each watermelon weighed.

John's watermelon   **Rosa's watermelon**   **Roberto's watermelon**

▧ pounds   ▧ pounds   ▧ pounds

Clue 1: Rosa's watermelon + Roberto's watermelon = 280 pounds

Clue 2: John's watermelon + Roberto's watermelon = 286 pounds

Clue 3: Rosa's watermelon − Roberto's watermelon = 14 pounds

Case 2 — POUNDS OF PUMPKINS

Three students won prizes for their pumpkins at the county fair. Can you figure out the weights of the pumpkins?

Alberto's pumpkin   **Bernie's pumpkin**   **Crystal's pumpkin**

▧ pounds   ▧ pounds   ▧ pounds

Clue 1: Alberto's pumpkin + Bernie's pumpkin = 196 pounds

Clue 2: Alberto's pumpkin − Bernie's pumpkin = 8 pounds

Clue 3: Bernie's pumpkins + Crystal's pumpkin = 177 pounds

Think It Over!

• **What if** you did not know the difference of the weight of Alberto's and Bernie's pumpkins? Write a different clue that can be used to solve the case.

• ✎ **Write About It** Explain how you solved each case.

CASE CLOSED

Challenge

Relate Benchmark Measurements

For an art project, Randy needs 5 inches of ribbon. His friend gives him 10 centimeters of ribbon. Does he have enough ribbon?

Use the table of related benchmarks to estimate conversions.

Examples

1 inch → about 2.5 centimeters

5 inches → 2.5 + 2.5 + 2.5 + 2.5 + 2.5 centimeters

2.5 + 2.5 + 2.5 + 2.5 + 2.5 = 12.5 centimeters

TABLE OF CONVERSIONS

- 1 mile is about 1,600 meters.
- 1 meter is slightly more than 1 yard.
- 1 inch is about 2.5 centimeters.
- 1 liter is a little more than 1 quart.
- 1 kilogram is a little more than 2 pounds.

So, 5 inches of ribbon is about 12.5 centimeters. Randy does not have enough ribbon for the project.

More Examples

Estimate the conversion.

A 2 yards → ■ meters

1 meter is slightly more than 1 yard.

So, 2 yards → 2 meters.

B 1 gallon → ■ liters

1 liter is a little more than 1 quart, and there are 4 quarts in a gallon.

So, 1 gallon → 4 liters.

C 50 pounds → ■ kilograms

1 kilogram is a little more than 2 pounds.

So, 50 pounds → 25 kilograms.

Try It

Estimate the conversion.

1. 7 kilograms → ■ pounds

2. 15 miles → ■ meters

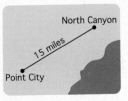

3. 2 inches → ■ centimeters
4. 9 feet → ■ meters
5. 3 kilograms → ■ pounds
6. 8 liters → ■ quarts
7. 2 meters → ■ yards
8. 12 yards → ■ meters
9. 10 pounds → ■ kilograms
10. 14 quarts → ■ liters
11. 20 miles → ■ meters

Unit 8 • Chapters 23–26 521

Study Guide and Review

VOCABULARY

Choose the best term from the box.

1. The amount of matter in an object is its _?_. (p. 478)

2. A _?_ is a closed figure with straight sides. (p. 486)

3. The measure of the space a solid figure occupies is called _?_. (p. 514)

| capacity |
| mass |
| polygon |
| volume |

STUDY AND SOLVE

Chapter 23

Use multiplication and division to change units within the customary system.

4 ft = ■ in.

feet
↓
4 × 12 = ■
↑
inches in 1 foot

4 × 12 = 48
So, 4 ft = 48 in.

• When you change larger units to smaller units, multiply.

Complete. (pp. 458–459)

4. 5 ft = ■ in.   5. 32 oz = ■ lb

6. 2 mi = ■ ft   7. 2 T = ■ lb

8. 3 gal = ■ pt   9. 8 qt = ■ pt

10. 16 c = ■ qt   11. 4 gal = ■ c

12. 4 pt = ■ qt   13. 21 ft = ■ yd

14. 48 in. = ■ ft   15. 6 qt = ■ pt

Chapter 24

Use multiplication and division to change units within the metric system.

30 dm = ■ m

decimeters
↓
30 ÷ 10 = ■
↑
decimeters in 1 meter

30 ÷ 10 = 3
So, 30 dm = 3 m.

• When you change smaller units to larger units, you divide.

Complete. (pp. 474–475)

16. 50 dm = ■ m

17. 200 cm = ■ dm

18. 7 m = ■ cm

19. 7 km = ■ dm

20. 12 L = ■ mL

21. 480 cm = ■ dm

22. 500 dm = ■ m   23. 5 kg = ■ g

24. 3 km = ■ m   25. 6 L = ■ mL

Chapter 25

Find the perimeter and the area.

Find the perimeter and the area.

Perimeter

$P = 2 + 2 + 4 + 2 + 6 + 4$

$P = 20$ cm

Area

$A = (2 \times 4) + (4 \times 2)$

$A = 8 + 8$

$A = 16$ sq cm

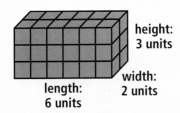

Find the perimeter and area.

(pp. 490–495)

26.

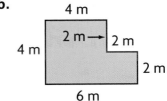

27.

5 ft

1 ft   1 ft

2 ft 2 ft

3 ft →   ← 3 ft

1 ft

Chapter 26

Find the volume of rectangular prisms.

Find the volume.

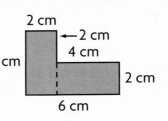

height: 3 units

width: 2 units

length: 6 units

Method 1: Count the number of cubic units in the prism: 36 cubic units.

Method 2: Multiply length × width × height.

$6 \times 2 \times 3 = 36$

Volume: 36 cubic units

For 28–29, use the figure.

(pp. 514–515)

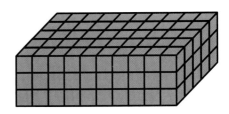

28. Name the solid figure.

29. What is the volume?

30. Mrs. Jost has a box that is 12 inches by 8 inches by 14 inches. What is the volume of her box?

PROBLEM SOLVING PRACTICE

Solve. (pp. 464–465, 480–481)

31. Margaret needs 30 cups of juice. She bought a gallon of orange juice, 2 pints of grape juice, and 3 quarts of cranberry juice. Did she buy enough? Explain.

32. Jared has a poster that is 3 feet long and 2 feet wide. He has a frame that is 3 feet long and 3 feet wide. Draw a diagram to show whether Jared's poster will fit inside his frame.

PROBLEM SOLVING ON LOCATION

with Balloons

NATIONAL BALLOON CLASSIC

Hot air balloon festivals are fun for people of all ages. At a balloon festival, you can do things like take balloon rides, listen to music, and look at interesting and unusual hot air balloons.

1. What customary unit would you use to measure the weight of a hot air balloon? the height of a hot air balloon? the volume of a hot air balloon?

2. A hot air balloon uses an average of 15 gallons of propane in an hour. On average, how many quarts of propane does a hot air balloon use in 2 hours?

3. A typical hot air balloon can be 840 inches tall. How many feet tall is this? Explain how you know.

4. Linda and Steve take a hot air balloon ride. They are in a basket that is 5 foot by 5 foot square. What is the perimeter and area of a square?

5. Daniella runs in the 5K road race at the National Balloon Classic. She knows that 1K is 1 kilometer and it is equal to about 0.62 miles. About how many miles long is a 5K road race?

6. **REASONING** Most hot air balloons fly at a height of around $\frac{1}{4}$ mile above the ground. How many feet is this? Explain how you know.

The National Balloon Classic in Indianola, Iowa takes place every summer.

Chapter 25

Find the perimeter and the area.

Find the perimeter and the area.

Perimeter

$P = 2 + 2 + 4 + 2 + 6 + 4$

$P = 20$ cm

Area

$A = (2 \times 4) + (4 \times 2)$

$A = 8 + 8$

$A = 16$ sq cm

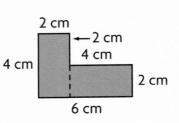

Find the perimeter and area.

(pp. 490–495)

26.

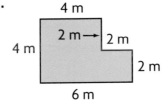

27.

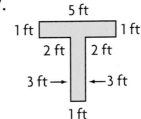

Chapter 26

Find the volume of rectangular prisms.

Find the volume.

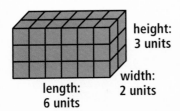

height: 3 units
width: 2 units
length: 6 units

Method 1: Count the number of cubic units in the prism: 36 cubic units.
Method 2: Multiply length × width × height.
  6 × 2 × 3 = 36
  Volume: 36 cubic units

For 28–29, use the figure.

(pp. 514–515)

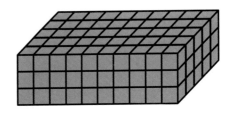

28. Name the solid figure.

29. What is the volume?

30. Mrs. Jost has a box that is 12 inches by 8 inches by 14 inches. What is the volume of her box?

PROBLEM SOLVING PRACTICE

Solve. (pp. 464–465, 480–481)

31. Margaret needs 30 cups of juice. She bought a gallon of orange juice, 2 pints of grape juice, and 3 quarts of cranberry juice. Did she buy enough? Explain.

32. Jared has a poster that is 3 feet long and 2 feet wide. He has a frame that is 3 feet long and 3 feet wide. Draw a diagram to show whether Jared's poster will fit inside his frame.

PERFORMANCE ASSESSMENT

TASK A • CLASS PARTY

You are on a committee for the class party. You need to be sure there will be enough juice for everyone at the party. There will be 24 students and 2 teachers attending the party.

The table shows the kind of juice and the amount of each kind available.

Juices	Amounts
Grapefruit	1 quart
Pineapple	4 pints
Orange	1 gallon
Kiwi	2 quarts 1 pint
Apple	1 gallon 1 quart
Grape	3 quarts
Cherry	2 pints
Strawberry	2 gallons 1 pint

a. How much juice will be needed for each person to have 2 cups?

b. How many cups can everyone have if all the juice on the list is used?

c. Decide on the amount of juice needed for the party. Tell how you made the decision. Choose at least 4 different kinds and amounts of juice you could use to make that amount.

TASK B • HOUSE ADDITION

The Millers plan to build an addition onto their home. The addition will be 12 feet by 18 feet and will include a bedroom, bath, and closet.

a. Decide how large the bedroom, bath, and closet could be. Then draw a diagram of this floor plan. Write the length and width in feet of the three rooms on your diagram.

b. The Millers will put carpet in the bedroom and in the closet. How many square feet of carpet will they need for these two rooms of your floor plan?

c. In the bedroom and the bath, they will put a wallpaper border on the walls just below the ceiling. How many feet of border will they need for your floor plan?

Technology Linkup

Mighty Math Calculating Crew • Nets for Solid Figures

Nets are patterns for solid figures. You can use Dr. Gee's Lab to identify nets for solid figures.

• Click on *Dr. Gee* to go to the *3D Lab*.

• Click  and choose Grow Slide Level C. Then click OK.

• Study the net and the solid figure shown. Click Yes or No to tell whether the net matches the solid figure shown. Continue answering the questions for Level C.

Use the arrow keys below the ▲ solid figure to turn it in order to look at all the faces.

Practice and Problem Solving

1. Choose Grow Slide Level D. Answer at least 5 questions. Draw a picture of each solid figure and its net. Write the name of the figure.

Solve.

2. What is the least number of faces a solid figure can have? Name the solid figure.

3. 📖 **Write About It** How can you use what you know about the faces of a solid figure to identify its net?

4. **STRETCH YOUR THINKING** Dr. Gee needs to make a net for a cylinder. Draw a picture of the net she needs.

Multimedia Math Glossary www.harcourtschool.com/mathglossary

5. **Vocabulary** Look up *square pyramid* in the Multimedia Math Glossary. Using the example shown in the glossary, draw a net for the square pyramid.

PROBLEM SOLVING ON LOCATION

with Balloons

NATIONAL BALLOON CLASSIC

Hot air balloon festivals are fun for people of all ages. At a balloon festival, you can do things like take balloon rides, listen to music, and look at interesting and unusual hot air balloons.

1. What customary unit would you use to measure the weight of a hot air balloon? the height of a hot air balloon? the volume of a hot air balloon?

2. A hot air balloon uses an average of 15 gallons of propane in an hour. On average, how many quarts of propane does a hot air balloon use in 2 hours?

3. A typical hot air balloon can be 840 inches tall. How many feet tall is this? Explain how you know.

4. Linda and Steve take a hot air balloon ride. They are in a basket that is 5 foot by 5 foot square. What is the perimeter and area of a square?

5. Daniella runs in the 5K road race at the National Balloon Classic. She knows that 1K is 1 kilometer and it is equal to about 0.62 miles. About how many miles long is a 5K road race?

6. **REASONING** Most hot air balloons fly at a height of around $\frac{1}{4}$ mile above the ground. How many feet is this? Explain how you know.

The National Balloon Classic in Indianola, Iowa takes place every summer.

ALBUQUERQUE INTERNATIONAL BALLOON FIESTA

The Albuquerque International Balloon Fiesta in New Mexico is the largest ballooning festival in the world. In 1999 more than 1 million people attended the event.

1. The flames from a hot air balloon burner can shoot up to 15 or 20 feet high. How many inches high can the flames shoot?

2. Ruth rides in a hot air balloon whose basket is 5 feet long, 5 feet wide, and 3 feet tall. What is the volume of the balloon basket?

3. Suppose a balloon has a rectangular piece with a perimeter of 50 feet and a width of 12 feet. What is the area of the rectangle? Explain how you know.

4. **STRETCH YOUR THINKING** Look at the shape of the balloon part of a hot air balloon. What solid figures could you combine to make this shape? Draw a picture of your answer.

5. The Balloon Fiesta has a drop competition where pilots fly their balloons over targets and drop weighted bags. If Tosha's bag lands 4 feet 9 inches from the target and Emma's bag lands 59 inches from the target, which bag is closer to the target? Explain how you know.

Hot air balloons usually have volumes of 65,000 to 105,000 cubic feet. ▼

Outcomes

The gum in gum ball machines comes in many colors.

PROBLEM SOLVING Look at the table that shows the gum ball colors in one machine. Which color is most likely to come out of the machine? Which color is least likely to come out?

GUM BALL COLORS

Color	Number of Gum Balls
Blue	114
Green	100
Red	65
Yellow	123

Use this page to help you review and remember
important skills needed for Chapter 27.

✓ VOCABULARY

Choose the best term from the box.

1. An event that is sure to happen is ?.

2. An event that will never happen is ?.

> certain
> impossible
> possible

✓ CERTAIN AND IMPOSSIBLE (For Intervention, see p. H31.)

Write *certain* or *impossible* for each event.

3. pulling a red pencil from a box of blue pencils

4. pulling a quarter from a bag full of quarters

5. tossing a number less than 7 on a number cube labeled 1 to 6

6. spinning a number less than 10 on a spinner labeled 9, 2, 8, 4, and 6

7. going to Mars for a vacation today

8. pulling a quarter from a bag of nickels and dimes

9. spinning a 3 on a spinner labeled 5, 16, 13, 9, 10

10. pulling a purple marble from a box of purple marbles

✓ IDENTIFY POSSIBLE OUTCOMES (For Intervention, see p. H31.)

Write the possible outcomes for each event.

11. tossing a coin

12. tossing a six-sided number cube, labeled 1 to 6

13. spinning the pointer of this spinner

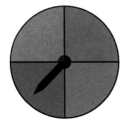

14. pulling a marble from this bag

15. pulling a marble from this bag

HANDS ON

Record Outcomes

Quick Review

1. ■ × 2 = 8 + 2
2. 6 × 6 = ■ × 4
3. 63 ÷ 9 = ■ × 7
4. 320 ÷ 8 = 10 × ■
5. 7 × 6 = ■ ÷ 2

▶ Explore

You can use a table to record the **outcomes**, or results, of an experiment. An **event** can be one outcome or a set of outcomes.

Do an experiment in which you toss a coin and spin the pointer on a spinner with four equal parts.

VOCABULARY

outcomes   event

MATERIALS coin; 4-part spinner colored red, blue, yellow, and green; 3-part spinner colored blue, yellow, and red

Activity 1

STEP 1

Make a table to show all the possible outcomes. 8 possible outcomes:

- The coin lands heads up and the pointer stops on red, blue, yellow, or green: H red, H blue, H yellow, H green.
- The coin lands tails up and the pointer stops on red, blue, yellow, or green: T red, T blue, T yellow, T green.

Technology Link

More Practice: Use E-Lab, *Organizing Data*.

www.harcourtschool.com/
elab2002

STEP 2

Toss the coin and spin the pointer. Record the outcome in your table.

Repeat for a total of 20 times, recording the outcome after each toss and spin.

Coin	Color			
	red	blue	Yellow	green
heads	l		l	l
tails		ll	l	

Try It

- Make a table to record the outcomes of tossing a coin and spinning the pointer on a spinner with three equal parts colored blue, yellow, and red. Record 20 outcomes.

The coin shows heads. Where did the pointer stop?

It is important to include all the possible outcomes in your table.

How can you organize the outcomes of an experiment in which you use a spinner with 3 equal parts colored red, blue, and green and toss a cube labeled 1 to 6?

Activity 2

MATERIALS: number cube; 3-part spinner colored red, blue, and green

STEP 1

Make a table. Use the 18 possible outcomes of the experiment to name the parts of the table. 18 possible outcomes:

- The pointer stops on red, blue, or green, and the cube shows 1, 2, 3, 4, 5, or 6.

STEP 2

Spin the pointer and toss the cube 20 times. After each spin and toss, record the outcome in the table.

EXPERIMENT RESULTS						
	Number					
Color	1	2	3	4	5	6
red						
blue						
green						

- **REASONING** Will there be more than 1 tally mark for any of the outcomes in your table? Explain how you know.

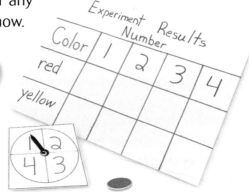

► **Practice and Problem Solving**

USE DATA For 1–3, use the table.

Bernadette and Charles organized their outcomes in this table. They tossed a counter and used a spinner labeled 1, 2, 3, and 4.

1. Name the possible outcomes for this experiment.

2. How many possible outcomes are there?

3. How many outcomes would there be if they had used a spinner with 6 different numbers?

4. A plane left at 10:02 A.M. The flight arrived 3 hr 17 min after it took off. If the flight should have arrived at 12:45 P.M., how late was the plane?

Mixed Review and Test Prep

5. 350 cm = ■ dm (p. 474)

6. 25 m = ■ cm (p. 474)

7. How many sides does a hexagon have? (p. 486)

8. Find the value of 58 − (3 + 7 + 9). (p. 54)

9. **TEST PREP** Find 32 − 1.86. (p. 436)

  A 30.14 **B** 30.24 **C** 31.24 **D** 31.86

Tree Diagrams

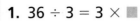
▶ **Learn**

DECISIONS, DECISIONS Betsy will choose a frozen yogurt flavor and a topping for her sundae. She can choose lemon, vanilla, or peach yogurt and cherry or nut topping. How many different sundaes can Betsy make? List the different sundaes.

A **tree diagram** shows all the possible outcomes or choices.

Yogurt	Topping	Outcomes
lemon	cherry	*lemon yogurt with cherry topping*
	nut	*lemon yogurt with nut topping*
vanilla	cherry	*vanilla yogurt with cherry topping*
	nut	*vanilla yogurt with nut topping*
peach	cherry	*peach yogurt with cherry topping*
	nut	*peach yogurt with nut topping*

So, Betsy can make 6 different sundaes.

REASONING There are 6 possible outcomes with 3 flavors of yogurt and 2 toppings. What number sentence could you write to find the number of outcomes?

MATH IDEA You can use a tree diagram to list all the possible outcomes and to find the total number of outcomes.

▶ **Check**

1. **Tell** how many different peach sundaes Betsy can make.

Find the number of possible outcomes by making a tree diagram.

2. You can plant one type of flower in one location.
   Type: tulips, roses, or irises
   Location: front yard, back yard, or side yard

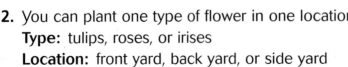

For 3–5, you are choosing one of each. Find the number of possible outcomes by making a tree diagram.

3. Event choices
  Event: play, movie, zoo, or concert
  Day: Friday, Saturday, or Sunday

4. Clothing choices
  Pants: blue, black, or tan
  Shirt: red or green

5. Sandwich choices
  Meat: ham, turkey, roast beef, or pastrami
  Cheese: American, Swiss, Jack, or Colby

6. Bill has 8 different sets of 1 shirt and 1 tie. This means he could have 1 shirt and 8 ties. List 3 other possibilities for the number of shirts and ties Bill could have.

7. Julie can choose from three types of pasta: rigatoni, linguini, and fettucine. She can make either marinara, meat, alfredo, or vegetable sauce. How many different outcomes are possible?

USE DATA For 8–10, use the table.

8. Nick bought a helmet and a snack. Joan bought 2 reflectors, a T-shirt, and a water bottle. Who spent more? How much more?

BIKE SHOP PRICE LIST			
Helmet:	$18.75	Water Bottle:	$2.55
Horn:	$4.95	Reflector:	$2.99
T-shirt:	$12.00	Snack:	$1.50

9. Josh has $17.00. He buys a T-shirt and wants to buy 2 more items. What items can he buy?

10. 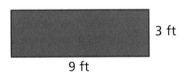 **Write a problem** using two or more items from the Bike Shop Price List.

Mixed Review and Test Prep

For 11–12, use the diagram.

3 ft
9 ft

11. Find the perimeter in feet. (p. 490)

12. Find the area in square feet. (p. 492)

13. Round 4.37 to the nearest tenth. (p. 428)

14. $4\frac{5}{8} - 2\frac{1}{8}$ (p. 392)

15. **TEST PREP** Find the elapsed time between 10:35 A.M. and 4:17 P.M. (p. 120)

  A 5 hr 18 min   **C** 6 hr 18 min
  B 5 hr 42 min   **D** 6 hr 42 min

Problem Solving Strategy
Make an Organized List

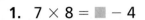
PROBLEM Winnie is playing a game at the fair using a 3-part spinner labeled 1, 2, and 3. If her total for two spins is greater than 4, she wins a prize. Show all the possible outcomes of the two spins and find the ways that Winnie can win a prize.

UNDERSTAND

- What are you asked to find?
- What information will you use?
- Is there any information you will not use? Explain.

PLAN

- What strategy can you use to solve the problem?

  Make an organized list.

SOLVE

- How can you use the strategy to solve the problem?

  Make a list of the 9 possible outcomes. Three outcomes have a total greater than 4.

  Possible Outcomes

1, 1	2, 1	3, 1
1, 2	2, 2	3, 2
1, 3	2, 3	3, 3	Spin 1 is **red**.

  So, she can win a prize by spinning 2, 3; 3, 2; or 3, 3.

CHECK

- What other strategy could you use?

▶ Problem Solving Practice

🔍 PROBLEM SOLVING STRATEGIES

Draw a Diagram or Picture
Make a Model or Act It Out
Make an Organized List
Find a Pattern
Make a Table or Graph
Predict and Test
Work Backward
Solve a Simpler Problem
Write an Equation
Use Logical Reasoning

Make an organized list to solve.

1. **What if** the spinner were labeled 3, 4, and 5? List the possible outcomes of spinning the pointer two times. Write the total number of outcomes.

2. Gwen is choosing a computer project. She can make a card, a calendar, a postcard, or a banner. She can use black or red ink. In how many ways can she do the project?

For 3–4, find the possible outcomes of spinning each pointer one time.

3. How many possible outcomes are there?
   A 6   **B** 10   **C** 12   **D** 24

4. How many of the possible outcomes include red?
   F 1   **G** 3   **H** 6   **J** 12

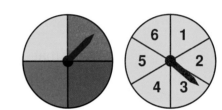

Mixed Strategy Practice

5. A furniture store sells sofas made of vinyl, leather, or cloth. Each kind of sofa comes in green, blue, or black. Make an organized list. How many sofa choices are there?

6. **REASONING** Mr. Johnson counted his change. He had 4 quarters, 8 dimes, 9 nickels and 19 pennies. What was the total amount of money?

7. Mrs. Landry wants to use string to show where the sides of her garden will be. Use the diagram to find the number of feet of string she needs.

   3 yd
   10 yd

8. Five students played basketball. Bob, Mike, and Jill scored 8 points each. Lynn scored half as many points as Chris. The total number of points scored was 54. How many points were scored by Lynn and Chris?

9. Anita put trim around 12 tables. Eight tables were 3 feet by 6 feet. Four tables were 3 feet by 8 feet. How much trim did she use for the 12 tables?

10. Each side of a square patio is 10 feet long. Starting at one corner, there is a plant every 2 feet around the border of the patio. How many plants are around the patio?

Problem Solving Strategy

Predict Outcomes of Experiments

Quick Review

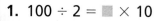

1. $100 \div 2 = \blacksquare \times 10$

2. $5 \times \blacksquare = 3 \times 10$

3. $36 \div \blacksquare = 80 \div 20$

4. $24 = 2 \times \blacksquare \times 3$

5. $3 \times 7 = 50 - \blacksquare$

VOCABULARY

predict	**likely**
unlikely	**equally likely**

▶ Learn

WHAT'S POSSIBLE?   Bag A has 28 red marbles and 2 blue marbles. Bag B has 15 red marbles and 15 blue marbles. If you pull a marble from Bag A, what color do you think you will pull?

Before doing an experiment, you can **predict** what will probably happen.

Bag A has many more red marbles than blue marbles.

So, it is **likely**, but not certain, that you will pull a red marble from Bag A.

It is **unlikely**, but not impossible, that you will pull a blue marble from Bag A.

What color do you predict you will pull from Bag B?

Since Bag B has the same number of red and blue marbles, you are **equally likely** to pull red or blue. Each color has the same chance of being pulled.

Examples

Look at the number cube labeled 1 to 6. Tell whether the events are *likely, unlikely,* or *equally likely.*

Ⓐ   tossing a 2; tossing a 3
equally likely

Ⓑ   tossing a composite number
unlikely

Ⓒ   tossing a number greater than 5
unlikely

Ⓓ   tossing a number less than 6
likely

Predicted vs. Actual Outcomes

In the activity below, there are six possible outcomes: blue, blue, blue, red, red, yellow. You can make predictions based on the likelihood of each outcome.

Activity

MATERIALS: circle; marker; blue, red, and yellow crayons; paper clip; brad

STEP 1 Make a spinner with 6 equal parts. Color 3 parts blue, 2 parts red, and 1 part yellow.

STEP 2 Copy the table. Predict the outcomes for spinning the pointer 30 times. Use tally marks to show the number of times you think the pointer will land on each color.

Color	Predicted Outcomes	Actual Outcomes
red		
blue		
yellow		

STEP 3 Now spin the pointer 30 times. Record the actual outcomes in your table. Compare your predictions from Step 2 to your actual outcomes here.

- Name an unlikely outcome of the experiment. Name an impossible outcome.

Technology Link

More Practice: Use E-Lab, *Predicting Outcomes*.

www.harcourtschool.com/ elab2002

▶ Check

1. **Explain** how you made your predictions.

Write *likely, unlikely,* or *equally likely* for the events.

2. tossing a prime number on a cube labeled 3, 5, 7, 9, 11, 13

3. tossing a 1; tossing a 3; number cube labeled 1 to 6

4. pulling a green marble from a bag that has 1 green, 6 yellow, and 5 red marbles

5. spinning an even number on a spinner with 6 equal parts labeled 1, 3, 5, 7, 9, 10

LESSON CONTINUES ▶

Write *likely, unlikely,* or *equally likely* for the events.

6. pulling a green marble

7. spinning a 5

8. spinning an odd number on a spinner that has 6 equal parts labeled 1, 3, 5, 6, 7, and 9

9. tossing a 6; tossing a 1; number cube labeled 1–6

For 10–12, look at the spinner.

10. What sum are you most likely to get if you spin the pointer 2 times?

11. Which sum for 2 spins is more likely, 2 or 3, or are they equally likely?

12. Make a spinner like the one shown. Spin the pointer twice and record the sum. Do this 11 more times. Explain how your results compare with your answers above.

For 13–14, look at the bag of marbles.

13. Which two colors are you equally likely to pull from the bag?

14. Which color marble are you unlikely to pull? Why?

15. **REASONING** Describe a number cube that would make tossing an even number unlikely.

16. **? What's the Question?** Scott had some stamps. He gave away 5 stamps and then arranged the rest in 6 rows of 12. The answer is 77 stamps.

17. **? What's the Error?** Sally has a number cube labeled 3, 5, 7, 8, 9, and 11. She says it is certain that she will toss an odd number. Describe her error.

18. **Write About It** Give an example of a likely outcome and a possible outcome, when pulling marbles out of a bag.

19. Angela has 1 nickel, 2 dimes, and 3 quarters in her pocket. Is it *certain* or *impossible* that she will pull less than $0.20 if she pulls out 3 coins at the same time?

20. Sonia has $10.00. She wants to buy a hat for $4.52, a belt for $3.99 and a pair of socks for $2.50. Does she have enough money? Explain.

Mixed Review and Test Prep

21. Write $\frac{3}{5}$ as a decimal. (p. 406)

22. $57.3 - 8.5$ (p. 434)

23. $\frac{3}{4} - \frac{1}{3}$ (p. 400)

24. 13 ft = ■ in. (p. 458)

25. $\begin{array}{r} 14 \\ \times 25 \end{array}$ (p. 224)

26. TEST PREP Which decimal is equivalent to $8\frac{7}{10}$? (p. 414)

A 0.87   **C** 87.0

B 8.7   **D** 87.7

27. TEST PREP Which number is a prime number? (p. 302)

F 4   **H** 9

G 6   **J** 11

PROBLEM SOLVING LiNKUP ... to Reading

STRATEGY · CLASSIFY AND CATEGORIZE

To *classify* information is to group information that is alike. To *categorize* information is to label the classified groups, or categories.

Daisy is choosing a clown outfit. She has a polka-dot suit, a rainbow wig, and a striped suit. She also has black boots, a pink wig, and red floppy shoes.

Category →

Classify {

WIGS	SUITS	SHOES
pink	?	?
?	?	?

1. Copy and complete the table using the information above.

2. How many outfits (1 wig, 1 suit, 1 pair of shoes) does she have?

3. Daisy is making party favors. She had blue and green balloons and gold ribbon. She bought red balloons and silver ribbon. How can you classify and categorize the information? How many kinds of favors can she make?

Review/Test

✅ CHECK VOCABULARY AND CONCEPTS

Choose the best term from the box.

> outcome
> likely
> event
> equally likely

1. You can use a table to record each _?_, or result, of an experiment. (p. 528)

2. Two events with the same chance of happening are _?_. (p. 534)

✅ CHECK SKILLS

3. Tonia tosses a coin and spins the pointer on a spinner colored red, white, blue, and gold. Name all the possible outcomes. (pp. 528–529)

Find the number of possible outcomes by making a tree diagram. (pp. 530–531)

4. Winter sports
   Sport: skiing, skating, sledding
   Day: Friday, Saturday, Sunday

5. Guitars
   Type: electric, acoustic
   Color: tan, brown, black

For 6–7, look at the spinner. Write *likely, unlikely,* **or** *equally likely* **for the events.** (pp. 534–537)

6. spinning a number greater than 10

7. spinning an odd number

✅ CHECK PROBLEM SOLVING

For 8–9, make an organized list to solve. (pp. 532–533)

8. Claire had chicken, beef, broccoli, and carrots. She bought corn, white bread, and wheat bread. How many choices of meat, vegetable, and 1 bread are there?

9. Anna is ordering a skirt. She can choose black, white, green, blue, or gray, and short or long length. How many choices are there?

10. A bag has 3 red, 20 blue, and 40 green marbles. Give an example of an unlikely event with this bag.

⭐Standardized Test Prep

Look for important words.
See item **7.**

The important words are *2 colors* and *are equally likely.* Use the drawing to find the two colors.

Also see problem **2,** p. H62.

For 1–8, choose the best answer.

1. Which expression has a value of 26?

   A $35 + (6 - 3)$ **C** $(35 - 6) + 3$
   B $35 - (6 + 3)$ **D** $35 - (6 - 3)$

For 2–4, use these spinners.

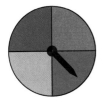 

2. How many outcomes are possible if you spin both pointers?

   F 10   **G** 12   **H** 14   **J** 24

3. How many of the possible outcomes include orange?

   A 3   **B** 4   **C** 6   **D** 8

4. Which outcome is **not** possible?

   F 6 and red   **H** 8 and blue
   G 2 and orange   **J** 5 and green

5. How is $6,000,000 + 800,000 + 50,000 + 600 + 30 + 1$ written in standard form?

   A 6,851,631   **C** 6,805,631
   B 6,850,631   **D** 850,631

For 6–7, use this bag of marbles.

6. Which color of marble is least likely to be pulled from the bag?

   F yellow   **H** green
   G red   **J** blue

7. Which 2 colors of marbles are equally likely to be pulled from the bag?

   A yellow and green
   B red and blue
   C red and yellow
   D blue and green

---

8. Mrs. Martin's class will choose one place and one day for a field trip. How many outcomes are possible?
   Place: zoo, art gallery, museum
   Day: Monday, Tuesday, Thursday

   F 3   **G** 6   **H** 9   **J** 12

Write What You Know

9. Suppose you toss a counter with a red side and a yellow side, and spin the pointer of a spinner labeled 1, 2, 3, 4, 5, and 6. List the outcomes that are possible. Show your method.

10. Todd had a nickel, dime, and quarter in his pocket. He chose two coins without looking. Show whether it is likely or unlikely that their total value was greater than 25 cents. Explain your answer.

Probability

The piñata (peen•YAH•tuh) comes from Latin America and is now enjoyed all over the world by many cultures.

PROBLEM SOLVING   The graph shows items found in some piñatas. What is the probability of finding a stuffed animal?

DATA LINK

PIÑATA TREATS

Stuffed Animals		
Toys		
Chocolate Candy		
Hard Candy		

0 10 20 30 40 50 60 70
Number of Items

CHECK WHAT YOU KNOW

Use this page to help you review and remember
important skills needed for Chapter 28.

✓ VOCABULARY

Choose the best term from the box.

1. The _?_ of a fraction tells how many of the equal parts of the whole or group are being considered.

2. The _?_ of a fraction tells the total number of equal parts or groups into which the whole or group has been divided.

3. A number that names part of a whole is a _?_ .

4. Fractions that name the same amount are _?_ .

| fraction |
| equivalent |
| denominator |
| numerator |
| equation |

✓ PARTS OF A SET (For Intervention, see p. H18.)

Write a fraction for the part of the set named.

5. green marbles

6. red marbles

7. blue or green marbles

8. circles

9. squares

10. triangles or squares

✓ PARTS OF A GROUP (For Intervention, see p. H19.)

Write a fraction for the shaded part.

11.

12.

13.

14.

15.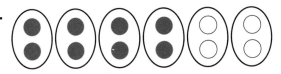

Probability as a Fraction

▶ **Learn**

COLOR WHEEL The spinner has six equally likely color outcomes: red, blue, yellow, orange, purple, and green. What is the probability of the pointer stopping on red?

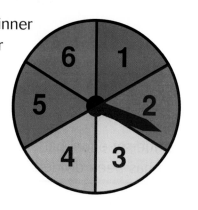

VOCABULARY

mathematical probability

Mathematical probability is a comparison of the number of favorable outcomes to the number of possible outcomes. The probability of an event can be written as a fraction.

Probability of red = $\frac{1}{6}$ ← **one favorable outcome (red)**
 ← **total possible outcomes (red, blue, yellow, orange, purple, green)**

So, the probability of spinning red is $\frac{1}{6}$, or 1 out of 6.

The probability of an event occurring is expressed as 0, 1, or a fraction between 0 and 1.

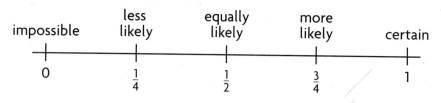

Technology Link

More Practice: Use Mighty Math Number Heroes, *Probability*, Levels F and H.

Example

Use the spinner to find the probability of the pointer stopping on 2 or 3.

Probability of 2 or 3 = $\frac{2}{6}$ ← favorable outcomes (2 or 3)
 ← total possible outcomes (1, 2, 3, 4, 5, 6)

An equivalent fraction for $\frac{2}{6}$ is $\frac{1}{3}$. So, the probability of spinning a 2 or 3 can be written as $\frac{2}{6}$, or $\frac{1}{3}$.

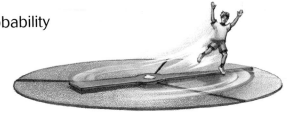

1. **Tell** which event is more likely, one with a probability of $\frac{1}{4}$ or one with a probability of $\frac{3}{4}$. Why?

Look at the spinner on page 542. Find the probability of each event.

2. the number 7

3. *not* 3 or 4

4. a number greater than 3

5. a prime number

▶ **Practice and Problem Solving**

Look at the spinner on page 542. Find the probability of each event.

6. blue

7. orange or red

8. black

9. *not* brown

Look at the bag of marbles. Write *impossible, less likely, equally likely, more likely,* **or** *certain* **for each event, and find the probability.**

10. a marble that is *not* blue

11. a green marble

12. a yellow marble

13. **? What's the Error?** Rita says that an event with a probability of $\frac{1}{2}$ is more likely to occur than an event with a probability of $\frac{3}{5}$. Explain her error.

14. **REASONING** When you toss a number cube labeled 1 to 6 and toss a coin, what is the probability of getting heads and a 5?

15. **GEOMETRY** A spinner has 6 equal sections. Two sections have a triangle, and three have a square. One section has a pentagon. What is the probability the pointer will stop on a figure with fewer than 5 sides?

Mixed Review and Test Prep

16. Write 3.6 using a fraction. (p. 414)

17. $32.02 - 4.76$ (p. 434)

18. Which has a greater measure, an acute angle or a right angle? (p. 320)

19. $739 \div 8$ (p. 266)

20. **TEST PREP** What are the perimeter and area of a rectangle with length 4 cm and width 3 cm? (p. 496)
   A 7 cm; 12 sq cm
   B 10 cm; 14 sq cm
   C 12 cm; 14 sq cm
   D 14 cm; 12 sq cm

More About Probability

▶ **Explore**

The mathematical probability of tossing each number on a number cube labeled 1 to 6 is $\frac{1}{6}$. When you do an experiment, your results may not match this probability.

Mary tossed a number cube labeled 1 to 6 twelve times. Look at the table. Mary's results show that she tossed a 5 in her experiment 3 out of 12 times. You can also write $\frac{3}{12}$, or $\frac{1}{4}$.

MARY'S EXPERIMENT						
Number	1	2	3	4	5	6
Frequency	3	2	2	1	3	1

Activity

STEP 1

Make a tally table. Toss the number cube 12 times and record the outcomes in the tally table.

STEP 2

Make a frequency table like Mary's. Use your tally table to complete the frequency table.

• Using your results, write a fraction that shows the results for getting each number. How do your results compare to the mathematical probability of getting each number?

Talk About It

• You spin the pointer of this spinner 10 times. The pointer stops on yellow 4 times and on blue 6 times. Based on this experiment, what fraction of times does the pointer stop on blue? on yellow?

Quick Review

Write each fraction in simplest form.

1. $\frac{4}{12}$ 2. $\frac{3}{9}$

3. $\frac{10}{20}$ 4. $\frac{8}{10}$

5. $\frac{6}{8}$

MATERIALS number cube labeled 1 to 6

Technology Link

To learn more about probability, watch the **Harcourt Math Newsroom Video** *Monarch Migration.*

Connect

As the number of tries increases, the fraction for each event in the experiment comes closer to the mathematical probability of the event.

Kiley combined her 10 tosses with those of nine of her classmates.

- Look at the table. Which fraction for heads is closer to the mathematical probability of $\frac{1}{2}$?

  Think: Is $\frac{40}{100}$ or $\frac{52}{100}$ closer to $\frac{50}{100}$?

COIN TOSS EXPERIMENT				
	Heads	**Fraction**	**Tails**	**Fraction**
Kiley	4	$\frac{4}{10}$, or $\frac{40}{100}$	6	$\frac{6}{10}$, or $\frac{60}{100}$
Total	52	$\frac{52}{100}$	48	$\frac{48}{100}$

Practice and Problem Solving

1. Toss a coin 50 times. Record your results. Write the fraction of tosses that showed heads and the fraction that showed tails based on your experiment.

2. **REASONING** Think of a spinner with equal parts numbered 1 to 5. Is it reasonable that the pointer will stop on 4 twice in 10 spins? Explain.

USE DATA For 3–5, use the spinner and the table.

3. What is the mathematical probability of the pointer stopping on each color on the spinner?

4. Use the table to find the fraction of spins when the pointer stopped on green. How does this compare to the mathematical probability?

SPINNER EXPERIMENT (100 SPINS)				
Outcome	**Red**	**Blue**	**Yellow**	**Green**
Tally	𝍤 𝍤 𝍤 𝍤 𝍤 𝍤	𝍤 𝍤 𝍤 𝍤 ||	𝍤 𝍤 𝍤 𝍤 𝍤 𝍤 ||||	𝍤 𝍤 𝍤 𝍤	

5. ✎ **Write About It** Based on the results, explain how to find the fraction of spins when the pointer stopped on yellow.

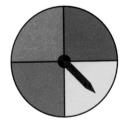

Mixed Review and Test Prep

For 6–9, write the greater measurement.

6. 38 in. or 3 ft (p. 458)

7. 4 m or 42 dm (p. 474)

8. 8 pt or 2 qt (p. 460)

9. 5,200 cm or 52 dm (p. 474)

10. **TEST PREP** Find $68 - 4.37$. (p. 436)

  A 63.63   C 64.37

  B 63.73   D 64.63

Test for Fairness

▶ Learn

FAIR PLAY   Victor and Brian are playing a game. They take turns using a spinner. Victor scores 1 point when he spins an odd number. Brian scores 1 point when he spins an even number. Which of the spinners should be chosen so that the game is fair?

Spinner A

Spinner B

Spinner C

Fairness in a game means that one player is as likely to win as another. So, each player has an equal chance of winning.

Quick Review

1. $\frac{1}{4} + \frac{2}{4}$ 2. $\frac{1}{6} + \frac{5}{6}$

3. $\frac{3}{5} + \frac{1}{5}$ 4. $\frac{3}{10} + \frac{2}{10}$

5. $\frac{5}{8} + \frac{3}{8}$

VOCABULARY

fairness

Example

STEP 1	STEP 2	STEP 3

STEP 1

Find the probability of getting an odd number on each spinner.

Spinner A: $\frac{1}{2}$

Spinner B: $\frac{1}{3}$

Spinner C: $\frac{3}{4}$

STEP 2

Find the probability of getting an even number on each spinner.

Spinner A: $\frac{1}{2}$

Spinner B: $\frac{2}{3}$

Spinner C: $\frac{1}{4}$

STEP 3

Find the spinner with equal probabilities for an odd number and an even number.

	ODD	**EVEN**
Spinner A:	$\frac{1}{2}$	$\frac{1}{2}$
Spinner B:	$\frac{1}{3}$	$\frac{2}{3}$
Spinner C:	$\frac{3}{4}$	$\frac{1}{4}$

There is the same probability of getting an odd number as an even number on Spinner A. So, in order for the game to be fair, Spinner A should be chosen.

MATH IDEA   Testing for fairness can help you decide if you have an equal chance of winning a game.

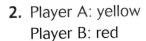
1. **Draw** a different spinner that Victor and Brian could use for a fair game.

Look at the spinner. Each player scores 1 point when the pointer stops on his or her choice. Write *yes* or *no* to tell if each game is fair. Explain.

2. Player A: yellow
   Player B: red

3. Player A: odd number
   Player B: even number

4. Player A: number that can be divided
   evenly by 2
   Player B: number that can be divided
   evenly by 3

Practice and Problem Solving

Look at the spinner. Each player scores 1 point when the pointer stops on his or her choice. Write *yes* or *no* to tell if each game is fair. Explain.

5. Player A: number between 8 and 15
   Player B: number greater than 10

6. Player A: prime number
   Player B: number that can be
   divided evenly by 4

For 7–8, use the spinner.

Each player chooses a different shape on the spinner. Each player scores 1 point when the pointer stops on his or her choice.

7. What is the probability of the pointer stopping on each shape?

8. How could you change the game to make it fair?

9. **? What's the Question?** There are 5 red marbles and 2 blue marbles. The answer is $\frac{5}{7}$.

Mixed Review and Test Prep

10. $1,982 \times 4$ (p. 200)

11. List the factors of 27. (p. 300)

12. How many sides does a pentagon have? (p. 486)

13. $2\frac{5}{8} + 4\frac{1}{8}$ (p. 392)

14. **TEST PREP** Which numbers are in order from greatest to least? (p. 418)
   A 11, 2.3, 1.47
   B 0.15, 0.51, 0.01
   C 3.51, 3.6, 8
   D 0.25, 0.52, 0.20

Problem Solving Skill
Draw Conclusions

UNDERSTAND ▸ PLAN ▸ SOLVE ▸ CHECK

MATH MYSTERY   Ken and Amy are playing a game with a bag of 6 numbers. Ken scores 1 point when he pulls an odd number. Amy scores 1 point when she pulls an even number. Use the clues to find the mystery numbers. Tell if the game is fair.

You can use probability to **draw conclusions** about the fairness of a game.

ANALYZE

CONCLUDE

CLUE 1: The probability of pulling a number greater than 10 and less than 22 is $\frac{6}{6}$, or 1.

The possible numbers are 11, 12, 13, 14, 15, 16, 17, 18, 19, 20, and 21.

CLUE 2: The probability of pulling a number greater than 13 that can be divided evenly by 3 is $\frac{3}{6}$, or $\frac{1}{2}$.

Three of the numbers are 15, 18, and 21.

CLUE 3: The probability of pulling a number that can be divided evenly by 10 is $\frac{1}{6}$.

One of the numbers is 20.

CLUE 4: The probability of pulling a prime number less than 17 is $\frac{2}{6}$, or $\frac{1}{3}$.

The last 2 numbers are 11 and 13.

The mystery numbers are 11, 13, 15, 18, 20, and 21.

The probability of pulling an odd number is $\frac{4}{6}$, or $\frac{2}{3}$.

The probability of pulling an even number is $\frac{2}{6}$, or $\frac{1}{3}$.

So, the game is not fair since $\frac{2}{3} > \frac{1}{3}$.

Talk About It

• **What if** Clue 3 were changed to a number that can be divided evenly by 8? Would the game be fair?

• Would the game be fair if Amy scored 1 point when the sum of the digits of the number is less than 5?

1. Sam and Tran are playing a game with a number cube. If an odd number is on top, Sam scores 1 point. If an even number is on top, Tran scores 1 point. Use the clues to find the numbers on the cube. Tell whether the game is fair. Explain.

NUMBER CUBE CLUES

- The probability of getting the product of 2 and 3 is $\frac{2}{6}$, or $\frac{1}{3}$.
- The probability of getting the least prime number is $\frac{1}{6}$.
- The probability of getting the number 5 is $\frac{1}{2}$.

For 2–3, use the letter tiles.

Rich and Ali made up rules for a two-player game using the letter tiles shown. Tell whether each game is fair. Explain.

2. Rich's Game
   Player 1: 1 point for E
   Player 2: 1 point for T

3. Ali's Game
   Player 1: 2 points for a vowel
   Player 2: 2 points for a consonant

M A T H
L E T T E R
G A M E

Mixed Applications

For 4–5, use the information.

A spinner has 12 equal parts. There are 2 red, 5 yellow, 3 blue, and 2 green parts.

4. On which color is the pointer most likely to land?
   A blue
   B green
   C red
   D yellow

5. Which two colors together cover exactly $\frac{1}{3}$ of the spinner?
   F blue and green
   G blue and yellow
   H red and green
   J red and blue

6. ✏ **Write a problem** about a game using a number cube labeled 1, 1, 1, 2, 2, and 3. Write the problem so that the game is fair. Rewrite the problem so that the game is unfair.

7. Nick painted for a total of 13 hours on Friday, Saturday, and Sunday. He painted 3 hours more on Saturday than on Sunday. On Friday he painted for 4 hours. How many hours did he paint each day?

Review/Test

✅ CHECK VOCABULARY

Choose the best term from the box.

> fairness
> **mathematical probability**
> impossible

1. A comparison of the number of favorable outcomes to the number of possible outcomes is _?_. (p. 542)

2. In a game, _?_ means that one player is as likely to win as another. (p. 546)

✅ CHECK SKILLS

3. What is the probability of getting a 1 or a 6 on a number cube labeled 1 to 6? (pp. 542–543)

Use the spinner to find the probability of each event. (pp. 542–543)

4. red
5. not green
6. red or blue

7. yellow
8. purple or green
9. brown

Look at the spinner. Each player scores 1 point when the pointer stops on his or her choice. Write *yes* or *no* to tell if each game is fair. Explain. (pp. 546–547)

10. Player A: purple
    Player B: yellow

11. Player A: even number
    Player B: odd number

12. Player A: number that can be divided evenly by 2
    Player B: number that can be divided evenly by 5

13. Player A: number greater than 17
    Player B: number less than 9

✅ CHECK PROBLEM SOLVING

For 14–15, use the cards. Write *yes* or *no* to tell if each game is fair. Explain. (pp. 546–549)

p r o b a b i l i t y

14. The cards are placed in a bag. When a vowel or the letter Y is pulled, Jenny scores 1 point. When a consonant other than Y is pulled, Rob scores 1 point.

15. Player 1 chooses B.
    Player 2 chooses I.
    Player 3 chooses L.
    Each player scores 2 points when his or her letter is pulled.

Standardized Test Prep

TIP! **Choose the answer.**

See item **2.**

Use the drawing to find the probability for pulling each color. Relate each answer choice to the problem one by one and choose the best answer.

Also see problem **6**, p. H64.

For 1–8, choose the best answer.

1. $\begin{array}{r} 1{,}345 \\ \times \quad 41 \\ \hline \end{array}$

    A 43,945 **C** 55,145

    B 54,145 **D** NOT HERE

For 2–4, use the bag of marbles.

2. Which two colors are equally likely to be pulled?

    F yellow and purple

    G blue and purple

    H red and yellow

    J red and blue

3. What is the probability of pulling a marble that is **not** red?

    A $\frac{5}{15}$ **B** $\frac{8}{15}$ **C** $\frac{2}{3}$ **D** $\frac{1}{3}$

4. How many outcomes are possible for tossing a cube labeled 1 through 6 and tossing a coin?

    F 6 **G** 8 **H** 12 **J** 24

For 5–6, use these spinners.

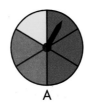 

A   B

5. What is the probability of the pointer on Spinner A stopping on a blue space?

    A $\frac{1}{3}$ **B** $\frac{1}{2}$ **C** $\frac{4}{6}$ **D** $\frac{3}{3}$

6. Diane made up a game to play with the spinners. If the pointer stops on a blue or yellow space, the person spinning earns a point. If the pointer stops on red, the other player earns a point. For which spinner would this game be fair?

    F only Spinner A

    G only Spinner B

    H Spinner A and Spinner B

    J neither Spinner A nor Spinner B

7. $\frac{3}{4} + \frac{1}{6}$

    A $\frac{4}{10}$ **B** $\frac{4}{6}$ **C** 1 **D** $\frac{11}{12}$

8. $52\overline{)1{,}196}$

    F 24 **H** 22

    G 23 **J** NOT HERE

Write What You Know

9. A spinner has 6 equal sections labeled 1 through 6. Explain how you could determine the probability of the pointer stopping on 1, 3, or 6.

10. How can you determine whether a game is fair? Describe a fair game that could be played using Spinner A from Problem 5.

Algebra: Explore Negative Numbers

Adélie penguins
in Antarctica

In Antartica, the sun shines for only six months each year.

PROBLEM SOLVING The table shows the average high and low temperatures in Antarctica for four months. Which month has the coldest average temperature?

DATA LINK

AVERAGE TEMPERATURES IN ANTARCTICA

Month	High	Low
January	⁻28°C	⁻37°C
April	⁻52°C	⁻70°C
July	⁻60°C	⁻81°C
October	⁻51°C	⁻70°C

Use this page to help you review and remember
important skills needed for Chapter 29.

✔ **COMPARE WHOLE NUMBERS** (For Intervention, see p. H3.)

Compare. Write <, >, or = for each ●.

1. 8 ● 10

2. 2,385 ● 2,356

3. 578 ● 597

4. 89 ● 100

5. 8,925,876 ● 8,792,354

6. 7,258 ● 7,892

7. 2,985,456 ● 2,955,758

Write the greater number.

8. 18; 21

9. 2,986; 2,867

10. 125; 265

11. 567,987; 565,029

12. 10,920; 1,920

13. 892; 894

14. 5,001; 5,100

15. 47,826; 47,789

✔ **READ A NUMBER LINE** (For Intervention, see p. H24.)

Use the number line to name the number each letter
represents.

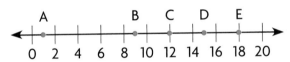

16. A

17. B

18. C

19. D

20. E

Use the number line to compare. Write <, >, or = for
each ●.

21. 100 ● 0

22. 40 ● 20

23. 55 ● 50

24. 60 ● 65

25. 50 ● 75

26. 60 ● 40

27. 52 ● 63

28. 83 ● 36

Use the number line to add or subtract.

29. 2 + 7

30. 6 + 4

31. 10 − 6

32. 8 − 7

Temperature: Fahrenheit

▶ Learn

BELOW ZERO **Degrees Fahrenheit (°F)** are customary units for measuring temperature.

Read 68°F as 68 degrees Fahrenheit.

Some temperatures are less than 0°F. These are negative temperatures. The lowest temperature marked on the thermometer at the right is ⁻10°F.

Read ⁻10°F as 10 degrees below zero.

Example

Look at the table. How much would the temperature change if it dropped from the normal high to the normal low?

MINNEAPOLIS-ST. PAUL, MINNESOTA, IN JANUARY	
Normal High: 16°F	**Normal Low:** ⁻2°F

STEP 1 First find the change in temperature from 16°F to 0°F.

The change in temperature is 16°.

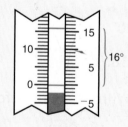

STEP 2 Then, find the change in temperature from 0°F to ⁻2°F.

The change in temperature is 2°.

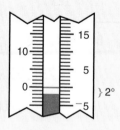

STEP 3 Add the two changes. 16° + 2° = 18°.

So, the change in temperature from 16°F to ⁻2°F is 18°.

Quick Review

Write the temperature that is 20° warmer.

1. 60°F   **2.** 15°F

3. 0°F   **4.** 49°F

5. 192°F

VOCABULARY

degrees Fahrenheit (°F)

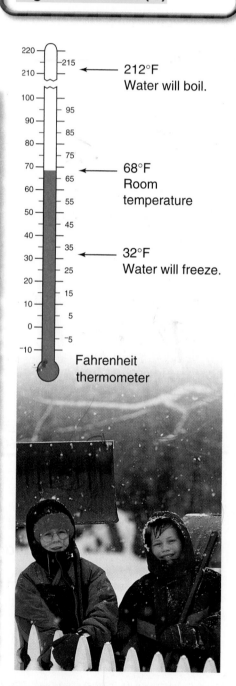

212°F
Water will boil.

68°F
Room temperature

32°F
Water will freeze.

Fahrenheit thermometer

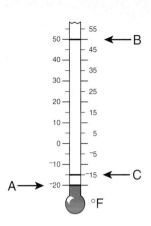

1. **REASONING** Explain how you would estimate the temperature change of a day that starts out warm and becomes cool.

Use the thermometer to find the temperature each letter represents.

2. A   **3.** B   **4.** C

► **Practice and Problem Solving**

Use the thermometer to find the temperature, in °F.

5. 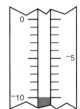   **6.** 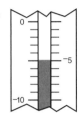   **7.**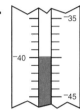

For 8–13, use a thermometer to find the change in temperature.

8. ⁻10°F to 20°F   **9.** ⁻5°F to ⁻30°F   **10.** 25°F to 82°F

11. 45°F to 90°F   **12.** ⁻5°F to 20°F   **13.** 40°F to 115°F

Choose the temperature that is a better estimate.

14. a bowl of hot soup   **15.** an ice cube   **16.** a classroom
      80°F or 180°F   30°F or 100°F   70°F or 120°F

17. Complete the pattern.
     ⁻10°F, ⁻4°F, 2°F, ▓, 14°F

18. 📖 **Write a problem** that shows a change between two temperatures.

Mixed Review and Test Prep

19. Divide. $375 \div 15$ (p. 286)

20. $28.6 + \blacksquare = 30.0$ (p. 436)

21. Multiply. $5,000 \times 12$ (p. 210)

22. $2,435 + 7,892$ (p. 40)

23. **TEST PREP** Find the median.
     36, 42, 85, 18, 24 (p. 86)
     A 24   **B** 36   **C** 41   **D** 42

Temperature: Celsius

▶ **Learn**

BRRR . . . Degrees Celsius (°C) are metric units for measuring temperature.

Read 20°C as 20 degrees Celsius.

Some temperatures are less than 0°C. The lowest temperature marked on the thermometer at the right is ⁻50°C.

Read ⁻50°C as 50 degrees below zero.

Example

In two months, the average temperature in Moline, Illinois, rises from about ⁻4°C to 10°C. How many degrees does the temperature rise?

STEP 1	First find the change in temperature from ⁻4°C to 0°C. The change in temperature is 4°.
STEP 2	Then, find the change from 0°C to 10°C. The change in temperature is 10°.
STEP 3	Add the two changes. 4° + 10° = 14°

So, the rise in temperature from ⁻4°C to 10°C is 14°.

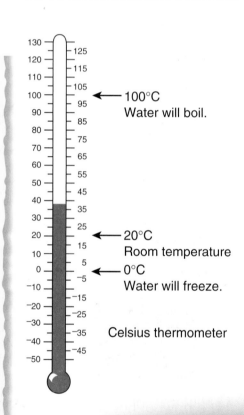

100°C
Water will boil.

20°C
Room temperature

0°C
Water will freeze.

Celsius thermometer

• What is the change in temperature from ⁻10°C to 23°C?

▶ Check

1. Compare ⁻15°C and ⁻5°C to determine which is warmer. What is the change in temperature?

Use the thermometer to find the temperature for each letter.

2. A

3. B

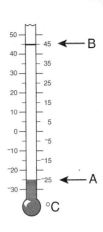

▶ Practice and Problem Solving

Use the thermometer to find the temperature, in °C.

4.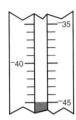

5.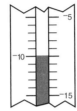

6.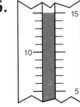

For 7–12, use a thermometer to find the change in temperature.

7. ⁻12°C to 5°C

8. 0°C to ⁻30°C

9. 5°C to 82°C

10. 40°C to 90°C

11. 5°C to ⁻2°C

12. ⁻20°C to 11°C

Choose the temperature that is a better estimate.

13. hot chocolate
90°C or 10°C

14. glass of milk
10°C or 50°C

15. inside a library
20°C or 68°C

16. Complete the pattern.
0°C, 32°C, 64°C, ■, 128°C

17. ❓ **What's the Question?** The difference in temperatures is 32°.

Mixed Review and Test Prep

18. 8 yd = ■ in. (p. 458)

19. $(5 + 2) - 3$ (p. 56)

20. Order from least to greatest. (p. 20)
29.6, 23.2, 29.5, 25.78

21. $1\frac{2}{3} + 8\frac{1}{3}$ (p. 392)

22. TEST PREP Which shows $\frac{10}{25}$ as a decimal? (p. 410)
A 0.10   **B** 0.25   **C** 0.40   **D** 0.50

Negative Numbers

Quick Review

Compare. Write <, >, or = for each ●.

1. 32 ● 42
2. 12 ● 12.0
3. $\frac{1}{5}$ ● $\frac{4}{10}$
4. 9.23 ● 9.2
5. 1,231 ● 1,230

VOCABULARY

negative numbers

positive numbers

▶ **Learn**

COUNTDOWN   At 9 minutes before a shuttle launch, or ⁻9 minutes, crews check to make sure the weather conditions are good for liftoff.

Look at the number line. All numbers to the left of 0 are called **negative numbers**. So ⁻9, read as "negative nine," is a negative number. All numbers to the right of 0 are called **positive numbers**. The number 0 is neither positive nor negative.

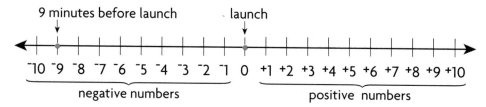

9 minutes before launch   launch

⁻10 ⁻9 ⁻8 ⁻7 ⁻6 ⁻5 ⁻4 ⁻3 ⁻2 ⁻1 0 +1 +2 +3 +4 +5 +6 +7 +8 +9 +10

        negative numbers   positive numbers

Example

Here are two ways to help you understand negative numbers.

One Way

The thermometer shows ⁻5°F. On a thermometer, negative numbers are below 0.

Another Way

⁻5 ⁻4 ⁻3 ⁻2 ⁻1 0 +1 +2 +3 +4 +5

The blue point on the number line shows ⁻5. On a number line, negative numbers are to the left of 0.

• Describe how a thermometer is like a number line.

Compare Negative Numbers

There are other real-life examples of the use of negative numbers.

Examples

A Ben owes his brother 7 dollars.

Negative number: ⁻7

B The football team lost 4 yards.

Negative number: ⁻4

MATH IDEA Just as the positive numbers decrease as you move left on a number line, the negative numbers also decrease as you move left. As you move right on the number line, the numbers increase.

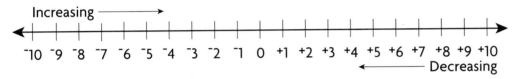

Increasing ⟶

⁻10 ⁻9 ⁻8 ⁻7 ⁻6 ⁻5 ⁻4 ⁻3 ⁻2 ⁻1 0 +1 +2 +3 +4 +5 +6 +7 +8 +9 +10

⟵ Decreasing

You can use a number line to compare numbers.

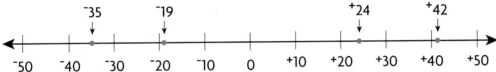

⁻35   ⁻19   +24   +42

⁻50 ⁻40 ⁻30 ⁻20 ⁻10 0 +10 +20 +30 +40 +50

More Examples

Compare. Use < and >.

C 24 ● ⁻35

Hint: 24 is the same as ⁺24, read as "positive twenty-four."

⁺24 is to the right of ⁻35 on the number line.

So, ⁺24 > ⁻35.

D ⁻35 ● ⁻19

⁻35 is to the left of ⁻19 on the number line.

So, ⁻35 < ⁻19.

- On a number line, which numbers are to the right of 0? to the left of 0?

LESSON CONTINUES

1. **Explain** how you know that a negative number is always less than a positive number.

Use the number line to name the number each letter represents.

2. A   3. B   4. C   5. D   6. E

▶ **Practice and Problem Solving**

Use the number line to name the number each letter represents.

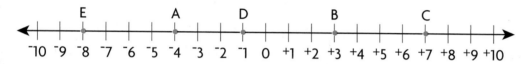

7. A   8. B   9. C   10. D   11. E

Compare. Write <, >, or = for each ●.

12. ⁻7 ● ⁺8   13. ⁻4 ● ⁻1   14. ⁻10 ● ⁻9   15. 0 ● ⁻8

16. ⁻7 ● ⁺3   17. ⁻8 ● ⁻9   18. 1 ● ⁺1   19. ⁻2 ● ⁻1

20. ⁻4 ● ⁻5   21. ⁻7 ● ⁺6   22. ⁻6 ● 0   23. ⁺8 ● 8

Order the numbers from *least* to *greatest*.

24. ⁻6, ⁻7, ⁻10   25. ⁻8, 7, ⁻10   26. 5, 8, ⁻10, ⁻9

27. After the space shuttle has lifted off, the NASA crew still monitors the shuttle's systems. What integer would represent 10 minutes after the launch?

28. List 5 numbers from greatest to least that make this statement true. ⁻5 < ■

29. ✎ **Write About It** Explain how you know that ⁻6 is less than ⁻5.

For 30–32, use the graph.

30. What was the coldest time of the day?

31. What was the difference between the temperature at 10:30 A.M. and the temperature at 3:00 P.M.?

32. What was the temperature at noon?

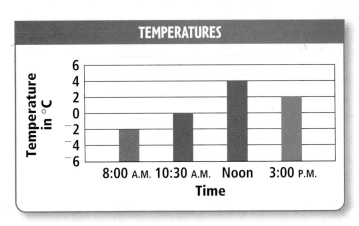

TEMPERATURES

Temperature in °C / Time
8:00 A.M. 10:30 A.M. Noon 3:00 P.M.

33. Mr. Lee's house is on land that is 10 feet below sea level. Mrs. Walker's house is on land that is 5 feet below sea level. Whose house is closer to sea level?

34. In January, the average temperature in Fairbanks, AK, is ⁻10°F. The average temperature in Bismarck, ND, for January is ⁺9°F. Which city has the higher average temperature?

Mixed Review and Test Prep

35. List the factors of 27. (p. 142)

36. Write $\frac{8}{10}$ in simplest form. (p. 368)

37. **TEST PREP** Which fraction is equivalent to $\frac{16}{24}$? (p. 368)

  A $\frac{1}{2}$ B $\frac{3}{4}$ C $\frac{2}{3}$ D $\frac{1}{4}$

38. 2 lb = ▇ oz (p. 462)

39. **TEST PREP** Brenda ran 48 miles in 3 days. She ran the same number of miles each day. How many miles did Brenda run each day? (p. 250)

  F 16 miles   H 84 miles
  G 68 miles   J 144 miles

PROBLEM SOLVING LiNKUP ... to Science

Have you ever had socks cling to your clothes after you take them out of the dryer? That's static electricity.

When there is static electricity in your clothes, there are more negative (−) charges in one place and more positive charges (+) in another place. The unlike charges attract each other, so your clothes cling together. Charges that have the same sign, such as two positive charges, repel each other, or move apart.

For 1–3, if two objects had the charges given, would they *repel* or *attract* each other?

  1. ⁺3 and ⁺3   2. ⁻7 and ⁺7   3. ⁻5 and ⁻8

Problem Solving Skill
Make Generalizations

UNDERSTAND ▶ PLAN ▶ SOLVE ▶ CHECK

WIND-BREAKER When the wind blows, the outside air feels colder than the actual temperature. The table below gives the windchill temperature. That is what the temperature feels like with the wind blowing.

Suppose that it is 30°F and the wind is blowing at 15 miles per hour. What temperature does it feel like? How do the two temperatures compare?

Quick Review

Name the next number for each pattern.

1. 3, 6, 9, 12, ■
2. 10, 20, 30, ■
3. 480, 48, 4.8, ■
4. ⁻20, ⁻15, ⁻10, ■
5. ⁻10, ⁻5, 0, ■

WIND CHILL TABLE

		Outside Temperatures (°F)				
Wind Speed (mph)	35	30	25	20	15	10
5	32	27	22	16	11	6
10	22	16	10	3	⁻3	⁻9
15	16	9	2	⁻5	⁻11	⁻18
20	12	4	⁻3	⁻10	⁻17	⁻24
25	8	1	⁻7	⁻15	⁻22	⁻29

Find the outside temperature. Find the row with the wind speed. The number shown where the column and row meet tells what the temperature feels like.

So, when the wind is blowing at 15 miles per hour and the outside temperature is 30°F, the windchill temperature is 9°F. When the wind is blowing, the temperature feels colder than the actual temperature.

Talk About It

• What is the windchill temperature when it is 20°F and the wind is blowing at 25 miles per hour?

• What generalizations can you make about what the temperature feels like when the wind is blowing?

1. Darcy wants to walk to a store 1 mile from her house. The outside temperature is 20°F and the wind is blowing at 10 mph. What is the windchill temperature?

2. Krista is getting ready to walk to school. The outside temperature is 0°C and the wind is blowing at 11 mph. Will it feel warmer or cooler than the actual temperature? Explain.

In the summer when the humidity is high, the temperature can feel hotter than the actual temperature.

USE DATA For 3–4, use the heat index table below.

HEAT INDEX TABLE

Relative Humidity	Outside Temperature (°F)						
	70	75	80	85	90	95	100
60%	70	76	82	90	100	114	132
70%	70	77	85	93	106	124	144
80%	71	78	86	97	113	136	157
90%	71	79	88	102	122	150	170

3. Find the heat index for an outside temperature of 85°F with relative humidity of 80%.

   A 86°F **B** 93°F **C** 97°F **D** 102°F

4. What would be the relative humidity if it is 90°F with a heat index of 106°F?

   F 70%   **H** 90%

   G 80%   **J** 95%

Mixed Applications

5. Claude practiced piano for 15 minutes on Monday, 20 minutes on Tuesday, 45 minutes on Wednesday, and an hour on Thursday. What was the mean number of minutes Claude practiced?

6. Mrs. Sloan bought a stereo for $280. She made a down payment of $100 and paid the rest in equal payments of $60 each month. How many months did it take to pay for the stereo?

USE DATA For 7–8, use the heat index table above.

7. **? What's the Error?** The relative humidity was 90% and the temperature 80°F. Kaley said the heat index was 113. Describe and correct her error.

8. What generalization can you make about the heat index when the outside temperature is at or above 75°F and the relative humidity is greater than 70%?

9. Casey has to find homes for her dog's 6 puppies. If she wants to make 4 flyers for each pup and post them in 4 different locations, how many flyers will Casey have to make?

Review/Test

✓ CHECK VOCABULARY AND CONCEPTS

Choose the best term from the box.

| negative numbers |
| temperature |
| degrees Celsius (°C) |
| degrees Fahrenheit (°F) |

1. The metric units for measuring temperature are _?_. (p. 556)

2. The customary units for measuring temperature are _?_. (p. 554)

3. Numbers to the left of 0 on the number line are called _?_. (p. 558)

✓ CHECK SKILLS

Use the thermometer to find the temperature. (pp. 554–557)

4. 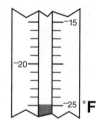 °F

5. 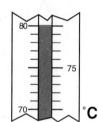 °C

6. 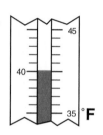 °F

7. 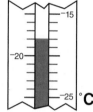 °C

Choose the temperature that is a better estimate. (pp. 554–557)

8. cup of coffee
190°F or 60°F

9. snow
23°F or 59°F

10. ice cream
⁻10°C or 20°C

Use a thermometer to find the change in temperature. (pp. 554–557)

11. 40°F to ⁻3°F

12. ⁻5°C to 13°C

13. 17°C to 32°C

Compare. Write <, >, or = for each ●. (pp. 558–561)

14. 10 ● 9   15. ⁻3 ● ⁻5   16. ⁻8 ● 2   17. ⁻5 ● ⁻8   18. ⁻17 ● ⁻13

✓ CHECK PROBLEM SOLVING

Solve. (pp. 554–555, 562–563)

19. Using the heat index table on page 563, find the relative humidity if it is 80°F with a heat index of 86°F.

20. When Fran left her house, the outside temperature was 12°F. When she arrived back home, the temperature was ⁻7°F. What was the change in temperature?

⭐Standardized Test Prep

 Get the information you need.
See item **7.**

Use the thermometer to find the mark
that is 4° less than ⁻2°C. Think about
how to write this temperature that is
below 0°C.

Also see problem **3**, p. H63.

For 1–8, choose the best answer.

1. Which numbers are in order from *least*
   to *greatest*?
   A ⁻30, ⁺20, ⁻6, ⁺2
   B ⁻30, ⁻6, ⁺2, ⁺20
   C ⁻30, ⁻6, ⁺20, ⁺2
   D ⁺2, ⁻6, ⁺20, ⁻30

2. $41\overline{)865}$
   F 20 r3   H 201 r4
   G 21 r3   J 21 r4

3. What is the value of $45 - (6 - n)$
   when $n = 4$?
   A 35   B 43   C 47   D 53

4. 16×315
   F 5,040   H 2,205
   G 4,910   J NOT HERE

5. What is $\frac{35}{60}$ written in simplest form?
   A $\frac{7}{10}$ B $\frac{7}{30}$ C $\frac{7}{12}$ D $\frac{3}{5}$

6. Which city was the warmest?

TEMPERATURES FOR FEBRUARY					
City	Temperatures for 5 days				
Baltimore	52	50	44	45	49
New York	45	50	38	39	41
Houston	79	80	76	76	77
Atlanta	68	62	62	62	65

   F Baltimore   H Houston
   G New York   J Atlanta

For 7–8, use the thermometers.

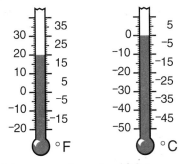

7. What temperature is 4°C less
   than ⁻2°C?
   A 6°C   B 2°C   C ⁻6°C   D ⁻8°C

8. What temperature is 6°F greater
   than ⁻15°F?
   F ⁻9°F   G 9°F   H 21°F   J ⁻21°F

Write What You Know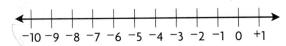

9. Use the number line to explain why
   ⁻1 > ⁻10.

   ```
   ←─┼──┼──┼──┼──┼──┼──┼──┼──┼──┼──┼──┼─→
    ⁻10 ⁻9 ⁻8 ⁻7 ⁻6 ⁻5 ⁻4 ⁻3 ⁻2 ⁻1  0  ⁺1
   ```

10. At noon, the temperature was 2°F. By 6
    P.M., the temperature had risen 4°. Then,
    by midnight, the temperature had fallen
    10°. What was the change in temperature
    from noon to midnight? Draw a number
    line to explain your answer.

Explore the Coordinate Grid

MAP OF HURRICANE ZELDA

The National Weather Service uses latitude and longitude to give the position of a hurricane. People can track the storm using a map with a coordinate grid. "Hurricane Zelda" is an imaginary storm the National Weather Service uses for tracking.

PROBLEM SOLVING Use the table and the tracking chart on this page to tell which state Hurricane Zelda comes closest to.

TRACK OF HURRICANE ZELDA

Latitude	Longitude
27.0° N	66.0° W
27.5° N	67.0° W
28.0° N	67.5° W
28.5° N	67.5° W
29.5° N	69.0° W
31.0° N	72.0° W
32.5° N	72.5° W
33.5° N	73.0° W
35.0° N	76.0° W

CHECK WHAT YOU KNOW

Use this page to help you review and remember important skills needed for Chapter 30.

✓ LOCATE POINTS ON A COORDINATE GRID (For Intervention, see p. H10.)

Look at the map of Hamilton. Name the place you find at each point.

1. (3,5)   **2.** (2,3)

3. (1,1)   **4.** (1,3)

5. (3,1)   **6.** (2,2)

Map of Hamilton

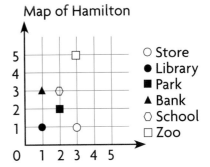

○ Store
● Library
■ Park
▲ Bank
○ School
□ Zoo

✓ USE FUNCTION TABLES (For Intervention, see p. H25.)

Use the rule to complete each table.

7. Rule: Add 2.

INPUT	OUTPUT
3	5
5	7
7	■
9	■

8. Rule: Add 9.

INPUT	OUTPUT
0	9
2	■
4	■
6	■

9. Rule: Subtract 2.

INPUT	OUTPUT
2	0
5	■
8	■
11	■

10. Rule: Subtract 1.

INPUT	OUTPUT
2	1
5	■
8	■
9	■

11. Rule: Multiply by 2.

INPUT	OUTPUT
0	0
1	■
2	■
3	■

12. Rule: Divide by 3.

INPUT	OUTPUT
3	1
6	■
9	■
12	■

13. Rule: Multiply by 3.

INPUT	OUTPUT
2	■
4	■
6	■
8	■

14. Rule: Add 5.

INPUT	OUTPUT
2	■
5	■
6	■
9	■

15. Rule: Divide by 2.

INPUT	OUTPUT
2	■
6	■
12	■
20	■

Use a Coordinate Grid

Quick Review

1. $3 + x$ for $x = 6$

2. $7 \times y$ for $y = 3$

3. $a - 4$ for $a = 12$

4. $b \div 7$ for $b = 14$

5. $(5 \times c) - 8$ for $c = 4$

▶ **Learn**

Use an **ordered pair** (x,y) to locate points on a coordinate grid. A coordinate grid has an **x-axis** (a horizontal line) and a **y-axis** (a vertical line).

In an ordered pair such as (x,y), the **x-coordinate** tells how far to move horizontally along the x-axis. The **y-coordinate** tells how far to move vertically along the y-axis.

(5,9)
x-coordinate ⌐ ⌐ y-coordinate

VOCABULARY

ordered pair

x-axis   **x-coordinate**

y-axis   **y-coordinate**

Example 1   A gardener used a coordinate grid to map where she planted each type of flower. What did she plant at (7,3)?

STEP 1

Start at 0. Count 7 units horizontally.

STEP 2

Then count 3 units vertically.
(7 units horizontally →, 3 units vertically ↑)

So, the gardener planted tulips at (7,3).

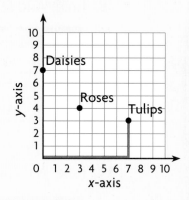

Example 2   What ordered pair tells where the roses are?

STEP 1

Start at the point labeled Roses. Look down at the x-axis. It is 3 units to the right. The x-coordinate is 3.

STEP 2

Then look across at the y-axis. It is 4 units up. The y-coordinate is 4.

So, the ordered pair (3,4) tells where the roses are.

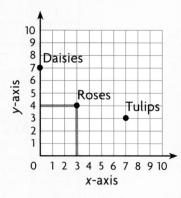

▶ Check

1. **Tell** how you would locate a point at (6,5).

2. **Name** the ordered pair that locates the daisies in the garden map.

Graph each ordered pair on a coordinate grid.

3. (2,2) 4. (6,5) 5. (3,4) 6. (8,1) 7. (10,10) 8. (1,8)

Write the ordered pair for each object on the map.

9. tree

10. playhouse

11. swings

12. wading pool

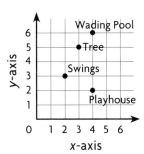

▶ Practice and Problem Solving

Graph each ordered pair on a coordinate grid.

13. (1,5) 14. (1,3) 15. (5,1) 16. (4,9) 17. (0,7) 18. (4,1)

Write the ordered pair for each point on the coordinate grid.

19. point A   20. point B

21. point C   22. point D

23. point E   24. point F

25. point G   26. point H

27. point J   28. point K

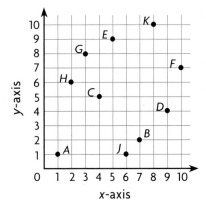

29. **? What's the Error?** Ann graphed (3,6) by starting from 0. She counted 3 units up then 6 units right. Describe and correct her error.

Mixed Review and Test Prep

30. What is the value of the blue digit? 2,390,410 (p. 8)

31. What value makes the fractions equivalent? $\frac{2}{5} = \frac{\blacksquare}{100}$ (p. 368)

32. 1,396 + ■ = 1,404 (p. 40)

33. ■ − 13 = 2,977 (p. 36)

34. **TEST PREP** What is the mean of this set of numbers? 1, 3, 5, 7, 9 (p. 276)

   A 3   **C** 5

   B 4   **D** 8

EXTRA PRACTICE page H61, Set A

Use an Equation

▶ **Learn**

VARYING VALUES Some equations have two variables.

$y = 3 \times x$ or $y = 3x$

When you multiply a variable and a number, you don't need to write the multiplication sign. So, $3 \times x = 3x$.

VOCABULARY

function table

Think of this equation as a rule that says: To find y, multiply x by 3. To find values for y, use a **function table**. In a function table, there is only one output for each input. Put *in* values for x to get *out* values for y.

Input	Multiply by 3.	Output
2	→	6

INPUT	x	2	4	5
OUTPUT	y	6	12	15

Example Use the equation $y = 3x + 5$ to find 3 pairs of x and y values that make the equation true.

STEP 1

Choose any three values for x.

INPUT	x	1	2	3
OUTPUT	y	■	■	■

STEP 2 Replace x in the equation with the values you chose.

Replace x with 1.

$y = 3x + 5$ $x = 1$
$y = (3 \times 1) + 5$
$y = 3 + 5$
$y = 8$

Replace x with 2.

$y = 3x + 5$ $x = 2$
$y = (3 \times 2) + 5$
$y = 6 + 5$
$y = 11$

Replace x with 3.

$y = 3x + 5$ $x = 3$
$y = (3 \times 3) + 5$
$y = 9 + 5$
$y = 14$

STEP 3

Record in the function table the values you found for y.

INPUT	x	1	2	3
OUTPUT	y	8	11	14

• What will y be when $x = 4$?

Example

Sarah used the equation below to complete a function table.
Are the values in her table correct?

$y = (x \div 2) + 3$ Rule: Divide x by 2 and then add 3.

INPUT	x	4	6	8
OUTPUT	y	5	6	9

The values for x and y can be written as ordered pairs.

Check each pair of values given. Replace x and y in the equation with the numbers.

To check the pair (4,5), use 4 for x and 5 for y in the equation.

$$y = (x \div 2) + 3$$
$$\downarrow \quad \downarrow$$
$$5 = (4 \div 2) + 3$$
$$5 = 2 + 3$$
$$5 = 5 \quad true$$

To check the pair (6,6), use 6 for x and 6 for y in the equation.

$$y = (x \div 2) + 3$$
$$\downarrow \quad \downarrow$$
$$6 = (6 \div 2) + 3$$
$$6 = 3 + 3$$
$$6 = 6 \quad true$$

To check the pair (8,9), use 8 for x and 9 for y in the equation.

$$y = (x \div 2) + 3$$
$$\downarrow \quad \downarrow$$
$$9 = (8 \div 2) + 3$$
$$9 = 4 + 3$$
$$9 = 7 \quad not\ true$$

So, (4,5) and (6,6) are correct. (8,9) is not correct.

- How can you find the correct value for y when $x = 8$?

MATH IDEA   An equation is like a rule, and you can use it to complete a function table.

Remember

In an ordered pair, (x,y), the first number is always x and the second number is always y. In the ordered pair (3,5), x is 3 and y is 5.

▶ Check

1. **Explain** how you would show that $x = 10$ and $y = 17$ makes the equation $y = 2x - 3$ true.

Use the equation to complete each function table.

2. $y = 4x + 1$

INPUT, x	OUTPUT, y
1	▨
2	▨
3	▨

3. $y = (x \div 2) + 7$

INPUT, x	OUTPUT, y
4	▨
6	▨
8	▨

4. $y = (x - 3) + 2$

INPUT, x	OUTPUT, y
7	▨
12	▨
15	▨

Do the values given make $y = (x + 3) \div 2$ true?
Write *yes* or *no*.

5. (5,4)   **6.** (9,6)   **7.** (11,8)   **8.** (3,3)

LESSON CONTINUES ▶

Do the values given make $y = (x + 5) \div 2$ true? Write *yes* or *no*.

9. (5,5) **10.** (9,7) **11.** (11,8) **12.** (7,4)

13. (7,12) **14.** (11,6) **15.** (13,9) **16.** (1,2)

17. (1,3) **18.** (21,13) **19.** (23,26) **20.** (27,16)

Use the equation to complete each function table.

21. $y = 5x + 4$

INPUT	x	1	2	3
OUTPUT	y	■	■	■

22. $y = (x \div 2) - 2$

INPUT	x	4	6	8
OUTPUT	y	■	■	■

23. $y = (x - 1) + 5$

INPUT	x	7	9	10
OUTPUT	y	■	■	■

24. $y = 2x + 3$

INPUT	x	3	4	8
OUTPUT	y	■	■	■

25. $y = (x \div 3) + 2$

INPUT	x	6	9	12
OUTPUT	y	■	■	■

26. $y = (x + 3) - 2$

INPUT	x	5	8	13
OUTPUT	y	■	■	■

27. $y = 2x - 1$

INPUT	x	3	6	9
OUTPUT	y	■	■	■

28. $y = (x \div 5) + 6$

INPUT	x	5	10	15
OUTPUT	y	■	■	■

29. $y = (x - 3) + 2$

INPUT	x	3	12	16
OUTPUT	y	■	■	■

30. $y = 4x - 2$

INPUT	x	■	■	■
OUTPUT	y	2	6	10

31. $y = (x \div 2) + 1$

INPUT	x	■	■	■
OUTPUT	y	3	4	5

32. $y = (x - 5) + 3$

INPUT	x	■	■	■
OUTPUT	y	5	7	8

33. Tim says that if he multiplies 3 by the number of tricycles, x, he can figure out how many wheels, y, he has. If Tim has 8 tricycles, does he have 24 wheels? Explain how you know.

34. Monique had some money in her pocket. Bernard gave her $20 more. Monique gave Brenda $5. Then Monique had $18. How much money did Monique start with?

35. ✎ **Write a problem** about using the equation $y = 12x$ to find how many cupcakes you would get if you bought more than 1 dozen.

36. **REASONING** Beth is 4 years older than 3 times her cousin's age. If her cousin is 6, how old is Beth?

37. Mr. Barrett has $600 to landscape his lawn. He plans to spend $210 on new trees. Each tree costs $42. How many trees can he buy?

38. Lyle and Miranda both bought new bikes. On Monday Lyle paid $135.75 for his bike. On Tuesday Miranda bought the same bike on sale for $125.36. How much more did Lyle pay than Miranda?

39. Miguel has 28 blank pages in his sports card album. Each page holds 6 cards. How many more cards will he be able to put in his album?

Mixed Review and Test Prep

40. $\frac{1}{4} + \frac{1}{4} = $ ▨ (p. 386) **41.** $\frac{3}{7} - \frac{2}{7} = $ ▨ (p. 390)

42. $\frac{5}{9} - \frac{1}{9} = $ ▨ (p. 390) **43.** $\frac{3}{4} + \frac{1}{2} = $ ▨ (p. 398)

44. What number is represented by the letter, *M*, on the number line? (p. 558)

M

−5 −4 −3 −2 −1 0 1 2 3 4 5

45. **TEST PREP** Which is the most reasonable unit of measure for the length of a car? (p. 471)

  A cm   **B** dm   **C** m   **D** km

46. **TEST PREP** A field has 22 rows of trees planted. Each row has 38 trees. How many trees are in the field? (p. 224)

  F 16   **G** 60   **H** 826   **J** 836

PROBLEM SOLVING LiNKUP ... to Social Studies

Foresters plant tree seedlings to replace trees that have died, have been cut down, or have been destroyed by forest fires. About half the seedlings planted survive until they are fully grown. To find half of a number, divide it by 2.

If you let *x* represent the number of seedlings planted and let *y* represent the number that become fully grown, the equation would be $y = x \div 2$.

1. Complete the function table using the equation $y = x \div 2$.

INPUT	x	10	20	30	40
OUTPUT	y	▨	▨	▨	▨

2. If a tree farmer plants 60 seedlings, would it be reasonable to expect 40 of those seedlings to survive until they are fully grown? Explain.

3. **? What's the Question?** A tree farmer planted 100 tree seedlings. The answer is 50 trees.

4. **What if** a tree farmer has 32 trees that are fully grown? About how many seedlings did he plant?

EXTRA PRACTICE page H61, Set B

Graph an Equation

▶ **Learn**

GET A GRID!   You can show on a coordinate grid some of the *x* and *y* values that make an equation true.

HANDS ON

Activity 1

Find 10 ordered pairs for the equation *y* = 2*x* + 1. Then graph the ordered pairs on a coordinate grid.

MATERIALS: coordinate grid with x-axis labeled 0–15 and y-axis labeled 0–25, ruler

- Make a function table for values of *x* and *y*. You can use any 10 values for *x*.

INPUT	**x**	0	1	2	3	4	5	6	7	8	9	10
OUTPUT	**y**	1	▧	▧	▧	▧	▧	▧	▧	▧	▧	▧

- Then use the equation to find the pairs of *x* and *y* values that make the equation true.

$y = 2x + 1$

INPUT	**x**	0	1	2	3	4	5	6	7	8	9	10
OUTPUT	**y**	1	3	5	7	9	11	13	15	17	19	21

- Write the values for *x* and *y* as ordered pairs, (*x,y*).

- Graph the ordered pairs on a coordinate grid. Connect the points with a line.

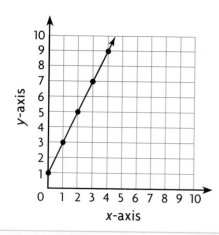

MATH IDEA   Values of *x* and *y* that make an equation true can be written as ordered pairs and then graphed.

Extend the Line

Use the graph of an equation to find other solutions to the equation by extending the line on the coordinate grid.

Activity 2

The graph shows the equation $y = x + 3$. Use the graph to find the value of y when $x = 4$.

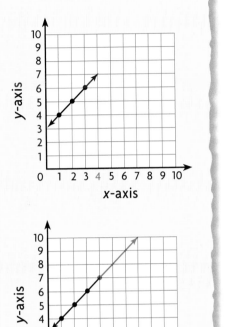

STEP 1

Use a ruler to extend the line of the graph.

STEP 2

Mark a point on the line where $x = 4$.

STEP 3

From the point, look left at the y-axis to identify the y-coordinate.

Write the x- and y-coordinates as an ordered pair. (4,7)

So, when $x = 4$, $y = 7$.

• According to the graph, what is the value of y when $x = 0$?

Check

1. **Explain** how you can extend the graph of an equation to find other x and y values that are solutions to the equation.

For 2–4, use the equation $y = x + 2$.

2. Copy and complete the function table.

INPUT	x	1	2	3	4	5	6	7	8	9	10
OUTPUT	y	3									

3. Write the input/output values as ordered pairs.

4. Graph the ordered pairs. Connect the points with a line.

LESSON CONTINUES ▶

For 5–9, use the equation $y = x + 5$.

5. Copy and complete the function table.

INPUT	x	0	1	2	3	4	5	6	7	8	9
OUTPUT	y	5	■	■	■	■	■	■	■	■	■

6. Write the input/output values as ordered pairs (x, y).

7. Graph the ordered pairs on a coordinate grid.

8. Connect the points. What pattern is formed by the connected points on the graph?

9. Extend your graph to find the y values when $x = 11$ and when $x = 0$.

Make a function table using the values 1 through 10 for x. Write the input/output values as ordered pairs. Then graph the ordered pairs.

10. $y = 2x$

11. $y = 3x - 1$.

For each graph, write the y value for the x value indicated in red.

12. $y = 2x - 2$

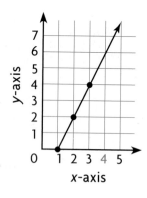

13. $y = x + 4$

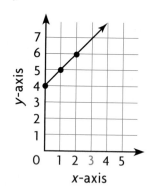

14. $y = x - 4$

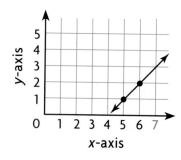

15. Ben manages the East Side Soccer League. He needs 2 soccer balls for every 3 players. Graph the data in the table at the right. Use your graph to find how many soccer balls he needs for 21 players.

PLAYERS	x	3	6	9	12
SOCCER BALLS	y	2	4	6	8

16. ✎ **Write About It** Describe the graph of x and y values that make the equation $y = 2x + 1$ true.

17. The table shows how much water and mix are needed for a certain recipe. Graph the data. Use your graph to find how many cups of mix are needed with 5 cups of water.

WATER	x	1 cup	2 cups	3 cups
MIX	y	3 cups	5 cups	7 cups

18. Neville cut a figure from paper. It had 4 equal sides and 4 right angles. What is the figure?

19. A pentagon has 5 equal sides. Each side measures 8 meters. What is the distance around the pentagon?

Mixed Review and Test Prep

20. $7\overline{)52}$ (p. 246)   **21.** $5\overline{)48}$ (p. 246)

22. ■ $- 83 = 1,405$ (p. 40)

23. $1,892 + $ ■ $= 2,000$ (p. 40)

24. $2,365 + $ ■ $= 2,405$ (p. 40)

25. Dorie bought 2 pencils and an eraser. One pencil cost twice as much as the other. The eraser cost $0.50. She spent $1.55. How much did each pencil cost? (p. 234)

26. **TEST PREP** Each floor of a 12-story building has 9 offices. How many offices does the building have? (p. 148)

A 21   **B** 108   **C** 120   **D** 180

27. **TEST PREP** At 7 P.M. the temperature was 70°F. Later the temperature was 55°F. How many degrees did the temperature change? (p. 554)

F 10°   **G** 15°   **H** 20°   **J** 25°

PROBLEM SOLVING · ThiNker's CorNer

REASONING You can write equations for other relationships, such as multiplication facts, and look at these relationships on the coordinate grid.

MATERIALS: coordinate grid with x-axis labeled 0–10 and y-axis labeled 0–25, ruler

multiply by 2: $y = 2x$ multiply by 4: $y = 4x$
multiply by 6: $y = 6x$

SECOND FACTOR	x	1	2	3	4	5	6	7	8	9	10
PRODUCT	y	■	■	■	■	■	■	■	■	■	■

1. Complete a function table for each of the equations. Then graph each set of ordered pairs and connect the points.

2. How are the three lines similar? different?

3. What rule could you write for multiplying by 12?

EXTRA PRACTICE page H61, Set C

Problem Solving Skill
Identify Relationships

UNDERSTAND ⟩ PLAN ⟩ SOLVE ⟩ CHECK

IT'S ALL RELATIVE! Pam wants to identify the relationship between the length of one side of a square and its perimeter. What relationship does the length of one side of a square have with its perimeter?

LENGTH OF ONE SIDE	**x**	1	2	3	4	5
PERIMETER	**y**	4	8	12	16	20

You can identify the relationship by writing a rule for the values of x and y in the table.

Rule: Multiply x by 4 to get y.

So, the relationship is that the perimeter is 4 times the length of one side.

You can also find relationships from points on a coordinate grid.

Look at the grid. What is the relationship between the length of one side of a square and the area of the square?

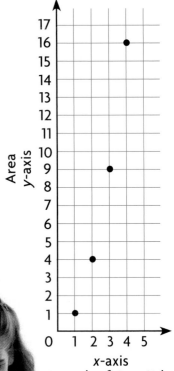

Area
y-axis

x-axis
Length of One Side

Example

> **STEP 1** Write the ordered pairs in a table.
>
> | LENGTH OF ONE SIDE | **x** | 1 | 2 | 3 | 4 |
> |---|---|---|---|---|---|
> | AREA | **y** | 1 | 4 | 9 | 16 |
>
> **STEP 2** Write a rule.
>
> Rule: Multiply the value of x by itself to get y.

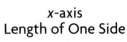

So, the relationship is that the area equals the length multiplied by itself.

► Problem Solving Practice

For 1–2, use the function tables.

1. Describe the relationship between *x* and *y*.

INPUT	x	2	3	4	5	6
OUTPUT	y	1	2	3	4	5

2. Describe the relationship between *x* and *y*.

INPUT	x	1	2	3	4	5
OUTPUT	y	2	4	6	8	10

For 3–4, use the graph that shows the length and width of squares.

3. Which rule shows the relationship between length and width?
   - **A** The length is 1 more than the width.
   - **B** The length is 1 less than the width.
   - **C** The length equals the width.
   - **D** The length is twice the width.

4. What will the length be when the width is 12?
   - **F** 11
   - **G** 12
   - **H** 13
   - **J** 24

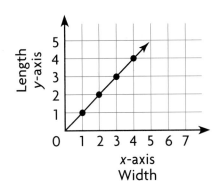

Mixed Applications

5. The temperature at 8:00 P.M. was 12°F. By 11:30 P.M., the temperature had dropped 10°. What was the temperature at 11:30 P.M.?

6. Ginny paid $12.99 each for 2 music CDs and $13.99 for a third CD. She gave the clerk two $20 bills. How much change did she receive?

7. Each student in Brad's class needs an 18-inch length of ribbon for an art project. If there are 24 students in the class, how many yards of ribbon will the class need?

8. Al has 135 baseballs to ship. He can put 12 in each carton. How many cartons does he need to ship all the baseballs?

9. Sammy is twice as old as Pete. Their ages add up to 21 years. How old is each one?

10. **? What's the Question?** There are 60×60 seconds in 1 hour. The answer is 86,400 seconds.

11. Allison's flight from New York City to Boston is scheduled to leave at 3:15 P.M. The flight will take 1 hr 10 min and is 30 minutes late. What time will the flight arrive in Boston?

Review/Test

✓ CHECK VOCABULARY AND CONCEPTS

Choose the best term from the box.

| x-axis |
| y-axis |
| x-coordinate |
| y-coordinate |
| function table |

1. The first number in an ordered pair is called the ? . (p. 568)

2. The vertical line on a coordinate grid is the ? . (p. 568)

3. When you have an equation such as $y = 5x$, you can use a ? to find y, output values, or x, input values. (p. 570)

✓ CHECK SKILLS

Graph each ordered pair on a coordinate grid. (pp. 568–569)

4. (1,2)   5. (3,7)   6. (5,5)   7. (3,9)

Write the ordered pair for each point on the coordinate grid. (pp. 568–569)

8. point A   9. point B

10. point C   11. point D

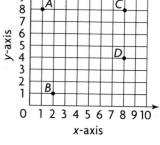

Do the values given make $y = (x - 2) + 6$ true? Write yes or no. (pp. 570–573)

12. (5,9)   13. (8,11)   14. (9,18)   15. (10,14)   16. (16,20)

For 17–18, use the equation $y = x + 3$. (pp. 574–577)

17. Copy and complete the function table.

INPUT	x	1	2	3	4	5	6	7	8
OUTPUT	y								

18. Write the input and output values as ordered pairs (x,y). Graph the ordered pairs on a coordinate grid.

✓ CHECK PROBLEM SOLVING

Use the function table at the right. Solve. (pp. 578–579)

INPUT	x	3	5	7	9	11
OUTPUT	y	1	3	5	7	9

19. Describe the relationship between x and y.

20. What will the value of y be when x is 14?

Standardized Test Prep

TIP! **Understand the problem.**

See item **1**.

To find each value of *y*, use the equation $y = 2x + 3$, replacing each *x* value with a number from the table. Choose the list of values for *y* that completes the function table.

Also see problem **1**, p. H62.

For 1–7, choose the best answer.

1. Which values complete the function table for the equation $y = 2x + 3$?

INPUT	x	1	2	3	4
OUTPUT	y	▢	▢	▢	▢

   A 5, 8, 10   **C** 5, 7, 9
   B 6, 7, 8   **D** 5, 6, 9

2. Which best describes the relationship between *x* and *y* in this table?

INPUT	x	5	6	7	8
OUTPUT	y	3	4	5	6

   F *y* is 1 more than *x*
   G *y* is 2 less than *x*
   H *x* is 3 more than *y*
   J *x* is 2 less than *y*

3. What is the value of $62 - (4 \times 3)$?
   A 50   **C** 174
   B 55   **D** NOT HERE

4. $\frac{1}{8} + \frac{3}{8} = $ ▪

   F $\frac{4}{16}$ **G** $\frac{2}{8}$ **H** $\frac{3}{8}$ **J** $\frac{1}{2}$

5. Which shows the values for this table written as ordered pairs?

INPUT	x	1	2	3	4
OUTPUT	y	3	4	5	6

   A (1,3), (4,2), (4,5), (4,6)
   B (1,3), (2,4), (3,5), (4,6)
   C (3,1), (2,4), (3,5), (4,6)
   D (2,3), (1,4), (3,5), (4,6)

For 6–7, use the coordinate grid.

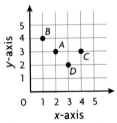

6. Which ordered pair names point *B*?
   F (4,1)   **G** (2,3)   **H** (1,3)   **J** (1,4)

7. Which point names the ordered pair (3,2)?
   A *A*   **B** *B*   **C** *C*   **D** *D*

Write What You Know

8. Graph the ordered pairs (1,0), (2,4), (3,8), and (4,12). Use the graph to name the *y* value when the *x* value is 5. Describe the method you used to find the *y* value.

9. Explain how to use the equation $y = 5x - 8$ to find *y* for any value of *x*. Then use what you wrote to find *y* when $x = 20$.

In the Bag

Use what you know about probability to determine the solution to each problem given below. Good luck!

A bag contains a total of **16 marbles**. The collection of marbles includes red marbles, blue marbles, yellow marbles, and green marbles. How many marbles of each color are likely to be in the bag if you know:

1.

you are equally likely to pull red, blue, yellow, or green?

2.

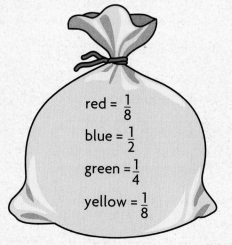

red = $\frac{1}{8}$

blue = $\frac{1}{2}$

green = $\frac{1}{4}$

yellow = $\frac{1}{8}$

3.

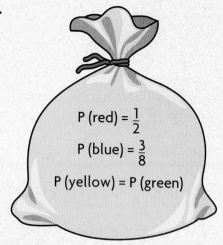

P (red) = $\frac{1}{2}$

P (blue) = $\frac{3}{8}$

P (yellow) = P (green)

4.

red = blue

red > green

green > yellow

Think It Over!

- A bag of marbles contains 4 red and 4 blue marbles. What is the probability of pulling a red marble?

- **STRETCH YOUR THINKING** Suppose a blue marble is pulled from Bag 1 and not replaced. Would the probability of then pulling a red marble still be the same? Why or why not?

582 Unit 9

Challenge

Find All Possible Ways

Using the digits 4, 6, and 8, how many 3-digit numbers can you make with the digit 6 in either the hundreds place or the tens place? Make a list to find all arrangements. For each number, use each digit only once.

STEP 1 Place 6 in the hundreds place.
6 __ __

Find all possible ways to arrange the 4 and 8.

648   684

STEP 2 Place 6 in the tens place.
__ 6 __

Find all possible ways to arrange the 4 and 8.

468   864

So, there are 4 possible numbers: 648, 684, 468, and 864.

Example

Theresa wants to take a picture of her classmates, Helen, Bob, Jim, and Karen. How many different ways can Theresa line them up if Jim is first?

Jim, Helen, Bob, Karen
Jim, Helen, Karen, Bob
Jim, Bob, Helen, Karen

Jim, Bob, Karen, Helen
Jim, Karen, Bob, Helen
Jim, Karen, Helen, Bob

So, Theresa can line up her classmates in 6 different ways.

Talk About It

• **Explain** how making a list helps you find all possible ways to order items in a problem.

Try It

1. If Theresa lines up her classmates with boys and then girls, how many ways can she line them up?

2. Mrs. Dean needs to line up three students: Alan, Mark, and Lena. In how many different ways can she line them up?

Study Guide and Review

VOCABULARY CHECK

Choose the best term from the box.

1. You can use a table to record each __?__ , or result of an experiment. (p. 528)

2. When you compare the number of favorable outcomes to the number of possible outcomes, you are finding the __?__ . (p. 542)

3. You can use an __?__ to locate points on a coordinate grid. (p. 568)

> **mathematical probability**
> **ordered pair**
> **negative number**
> **outcome**

STUDY AND SOLVE

Chapter 27

Record outcomes of an experiment in different forms.

Meat: Turkey, Roast Beef, Ham
Vegetable: Carrots, Corn, Broccoli

Meat	Vegetable	Outcomes
Turkey	Carrots	Turkey with carrots
	Corn	Turkey with corn
	Broccoli	Turkey with broccoli
Roast Beef	Carrots	Roast Beef with carrots
	Corn	Roast Beef with corn
	Broccoli	Roast Beef with broccoli
Ham	Carrots	Ham with carrots
	Corn	Ham with corn
	Broccoli	Ham with broccoli

Find the number of possible outcomes by making a tree diagram. (pp. 530–531)

4. Activity choices
   Activity: volleyball, tennis, basketball
   Day: Saturday, Sunday

5. Clothing choices
   Shirts: red, pink, or white
   Pants: blue, black, or tan pants

6. Sundae choices
   Ice Cream: vanilla, or chocolate
   Toppings: chopped nuts or pecans

Chapter 28

Write the probability of a simple event.

Find the probability of the pointer stopping on the color red.

Probability of red = $\frac{2}{6}$, or $\frac{1}{3}$

So, the probability of the pointer stopping on red is $\frac{1}{3}$.

A bag of marbles contains 7 blue, 3 green, 5 purple, and 3 red marbles. Find the probability of each event. (pp. 542–545)

7. a green marble

8. a blue marble

9. a purple marble

10. a marble that is not purple

Chapter 29

Use a number line to compare.

−5 and −3
−5 is to the left of −3 on the number line.
So, −5 < −3.

Compare. Write <, >, or = for each ⬤. (pp. 558–565)

11. −9 ⬤ +4   **12.** −6 ⬤ +8

13. −1 ⬤ −3   **14.** −3 ⬤ +9

15. −8 ⬤ +8   **16.** +7 ⬤ −2

17. −2 ⬤ +2   **18.** −4 ⬤ −10

19. −8 ⬤ −6   **20.** 0 ⬤ −1

Chapter 30

Identify ordered pairs that make an equation true.

$$y = 25 - (x \times 2)$$

$y = 25 - (x \times 2)$ • Check values for (4, 17)

$17 = 25 - (4 \times 2)$ • Replace x and y in the
$17 = 25 \ - \ 8$ equation. Use 4 and 17 for
$17 = 17$ true   x and y.

Do the values given make $y = (x - 3) ÷ 2$ true? Write yes or no. (pp. 570–573)

21. (17, 5)   **22.** (27, 12)

23. (39, 18)   **24.** (21, 10)

25. (23, 10)   **26.** (25, 9)

27. (18, 6)   **28.** (31, 14)

PROBLEM SOLVING PRACTICE

Solve. (pp. 532–533)

29. Tracy made up rules for a 2-player game using cards for each letter in the word *mathematics*. Use probability to tell whether the game is *fair* or *not fair*. Explain.

Player 1: scores 1 point for T
Player 2: scores 1 point for S or C

30. When Aaron went to school the temperature was 10°C. When he returned home it was −2°C. What was the change in temperature?

PERFORMANCE ASSESSMENT

TASK A • VISIT TO GRANDMOTHER

Robyn is planning a one-week visit with her grandmother. She wants to pack enough shorts and shirts so that she has enough different outfits to wear during the week.

a. Robyn wants to take more shirts than shorts. How many of each should she pack? Make a tree diagram to show that your answer is correct.

b. Suppose Robyn packs her favorite red shirt. If she reaches into her suitcase and pulls out a shirt without looking, is it likely or unlikely that she will pull out the red shirt?

c. Robyn buys one more shirt during her visit. How many different outfits will she be able to make now?

TASK B • FUN AND GAMES

The Fun-for-All Toy Company asked some students to create some new games.

a. In one game a spinner will show how many spaces to move on a board. Decide how many different numbers of spaces a player can move. Include a "lose a turn" space. There are 6 sections on the spinner. Draw a spinner for this game, and tell how many possible outcomes there are.

b. Choose one of the outcomes for the spinner you drew. Write a fraction that describes the probability for that outcome. Explain how you found the probability.

c. In another game a regular number cube with the numbers 1–6 on it is used. To begin, each player tosses the number cube. The player with the greatest number plays first. Explain how this is or is not a fair way to begin the game.

Technology Linkup

E-Lab • Predicting Outcomes

Linda has a spinner with 5 equal sections labeled 1, 2, 3, 4, 5. Predict how many times the pointer will stop on each number if Linda spins the pointer 20 times.

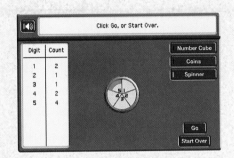

You can use E-Lab to test predicted outcomes.

- Click on *Predicting Outcomes*.
- Click on *Spinner*.
- Type 5. Press *Enter*.
- List the possible outcomes. Predict the outcomes of spinning the pointer 20 times.
- Click on *Go* 20 times to spin the pointer.
- Copy the results from the frequency table.

How do your predictions compare to the actual outcomes?

Practice and Problem Solving

Use E-Lab. Click on *Start Over*. Click on *Number Cube*.

1. List the possible outcomes for the number cube. Predict the outcomes for tossing the number cube 30 times.

2. Click on *Go* 30 times. Copy the results from the frequency table. How do your predictions compare to the actual outcomes?

Use E-Lab. Click on *Start Over*. Click on *Coins*. Type 2 and press *Enter* to set the number of coins.

3. List the possible outcomes for the number of heads. Predict the outcomes for tossing the coins 30 times.

4. Click on *Go* 30 times. Copy the results from the frequency table. How do your predictions compare to the actual outcomes?

Multimedia Math Glossary www.harcourtschool.com/mathglossary

5. **Vocabulary** Look up *probability* in the Multimedia Math Glossary. Using the spinner in the example, find the probability of not spinning green.

PROBLEM SOLVING ON LOCATION
with
Monarch Butterflies

MONARCH MIGRATION

Every year, monarch butterflies migrate south for the winter—up to 3,000 miles. The monarch butterflies west of the Rocky Mountains migrate to California, and the monarch butterflies east of the Rockies migrate to Mexico.

The table shows the relationship between the number of children who enter the Butterfly Pavilion in Colorado and the cost of entering.

For 1–2, use the table.

NUMBER OF CHILDREN	x	1	2	3	4	5
COST IN DOLLARS	y	4	8	12	16	20

1. Write a rule for the values of x and y in the table. How much would it cost for 6 children to enter the Pavilion?

2. Write the values as ordered pairs. Graph the ordered pairs. Connect the points with a line.

3. An individual monarch butterfly usually does not migrate south and then north in its lifetime. Knowing this, which statement below is true?

a. It is *likely* for a monarch butterfly to migrate south and then north.

b. It is *unlikely* for a monarch butterfly to migrate south and then north.

Every September the people of Kelley's Island, Ohio celebrate the return of the monarch butterflies.

MARVELOUS MONARCHS

Monarch butterflies can be found throughout the world, but they are mainly found in North America.

1. Monarch butterflies migrate south in the winter because they can't survive in cold temperatures. The lowest temperature ever recorded in Cleveland, Ohio was ⁻19°F. Show ⁻19 on a number line.

2. Dolores is making a butterfly garden. She can use plants to attract monarch, clipper, or owl butterflies. She can make her garden in the front or back yard. How many different outcomes are possible? Make a tree diagram to show your answer.

For 3–5, use this information.

Scott bought 30 postcards to send to his friends and family. There were 10 lacewing, 3 paper kite, 7 monarch, and 10 zebra longwing butterfly pictures on the postcards.

3. Which types of butterfly pictures are *equally likely* to be on a postcard Scott sends to his cousin?

4. What is the probability that a monarch butterfly will be on the postcard he sends to his grandmother?

5. What kind of butterfly picture is *least likely* to be on the postcard he sends to a friend?

▲ Tagging butterflies helps scientists study where and how far the butterflies migrate.

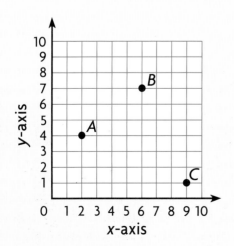

For 6–7, use the coordinate grid.

6. Mark graphed 3 points on the coordinate grid to show where he found butterflies. Write an ordered pair for the location where each butterfly was found.

7. Suppose Mark located another monarch butterfly at (5,1). Explain how you would graph the position of this butterfly on the coordinate grid above.

STUDENT HANDBOOK

Troubleshooting H2

PREREQUISITE SKILLS REVIEW Do you have the math skills needed to start a new chapter? Use this list of skills to review and remember your skills from last year.

Skill	Page
Benchmark Numbers to 100 . . . . H2	
Read and Write Numbers to Thousands. . H2	
Place Value to Thousands. H3	
Compare Numbers to Thousands. H3	
Order Numbers to Thousands H4	
Round to Tens and Hundreds H4	
2-Digit Addition and Subtraction. H5	
3-Digit Addition and Subtraction. H5	
Column Addition H6	
Missing Addends. H6	
Addition and Subtraction Fact Families. . H7	
Number Patterns H7	
Read Pictographs. H8	
Tallies to Frequency Tables H8	
Parts of a Graph H9	
Read Bar Graphs H9	
Identify Points on a Grid H10	
Time to the Half and Quarter Hour . . . H10	
Time to the Minute H11	
Use a Calendar H11	
Meaning of Multiplication and Division. . H12	
Multiplication and Division Facts. H12	
Missing Factors H13	
Model Multiplication H13	
Record Multiplication H14	
Multiply by Tens, Hundreds, and Thousands H14	
Multiply by 1-Digit Numbers. H15	
Estimate Products H15	
Multiplication and Division Fact Families H16	
Use Multiplication Facts and Patterns. . H16	

Skill	Page
Place Value H17	
Divide with Remainders H17	
Use Compatible Numbers H18	
Parts of a Whole H18	
Parts of a Group H19	
Compare Parts of a Whole H19	
Model Decimals H20	
Relate Decimals to Money H20	
Fractions with Denominators of 10 and 100. . . . . H21	
Round Money Amounts H21	
Add and Subtract Money H22	
Measure to the Nearest Inch and Half-Inch H22	
Multiplication. H23	
Measure to the Nearest Centimeter . . . H23	
Multiply by 10, 100, and 1,000 . . . . . . H24	
Read a Number Line. H24	
Use Function Tables H25	
Identify Angles. H25	
Compare Figures. H26	
Identify Symmetric Figures H26	
Sort Polygons. H27	
Identify Plane Figures H27	
Find Perimeter. H28	
Find Area. H28	
Identify Solid Figures. H29	
Draw Plane Figures H29	
Multiply 3 Factors H30	
Classify Angles H30	
Certain and Impossible. H31	
Identify Possible Outcomes. H31	

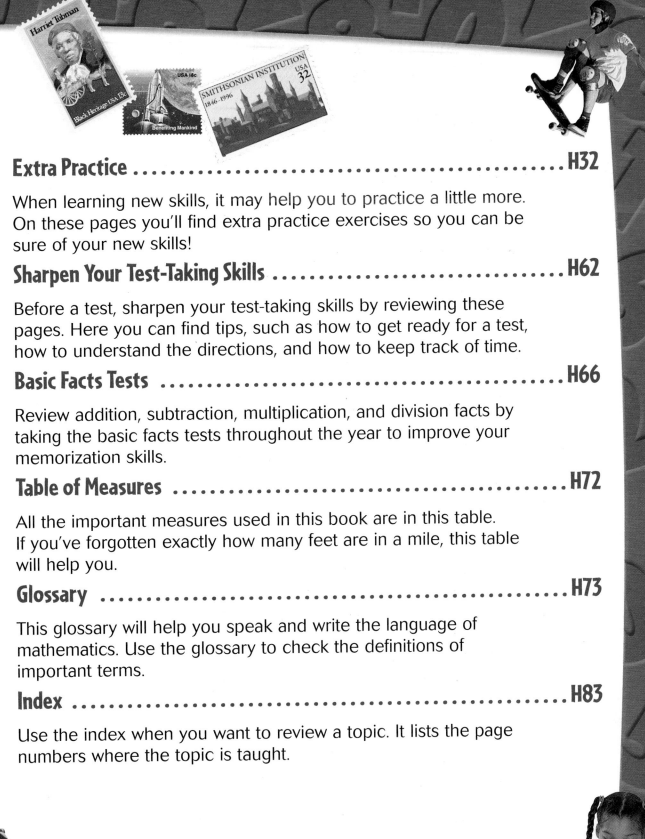

Extra Practice H32

When learning new skills, it may help you to practice a little more. On these pages you'll find extra practice exercises so you can be sure of your new skills!

Sharpen Your Test-Taking Skills H62

Before a test, sharpen your test-taking skills by reviewing these pages. Here you can find tips, such as how to get ready for a test, how to understand the directions, and how to keep track of time.

Basic Facts Tests H66

Review addition, subtraction, multiplication, and division facts by taking the basic facts tests throughout the year to improve your memorization skills.

Table of Measures H72

All the important measures used in this book are in this table. If you've forgotten exactly how many feet are in a mile, this table will help you.

Glossary H73

This glossary will help you speak and write the language of mathematics. Use the glossary to check the definitions of important terms.

Index H83

Use the index when you want to review a topic. It lists the page numbers where the topic is taught.

TROUBLESHOOTING

BENCHMARK NUMBERS TO 100

Numbers that help you estimate a number of objects without counting are **benchmark numbers**. Any useful number can be a benchmark.

Example

The first box contains 100 straws. Choose the best estimate for the number of straws in the second box.

Think: The number of straws in the second box is a little less than in the first box.

So, a good estimate is 60.

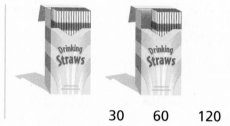

30   60   120

▶ Practice

Choose the best estimate for the second set of objects.

1.
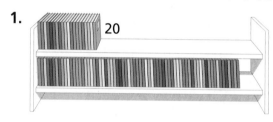
20

10   40   70

2.
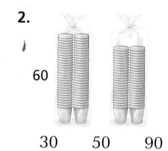
60

30   50   90

READ AND WRITE NUMBERS TO THOUSANDS

Use a place-value chart to read and write numbers to thousands. Numbers can be written in word form and in standard form.

Example

Write nine thousand, forty-eight in standard form.

Thousands	Hundreds	Tens	Ones
9,	0	4	8

- Write 9 in the thousands place.
- Write a zero in the hundreds place.
- Write 4 in the tens and 8 in the ones place.

So, the number is 9,048 in standard form.

▶ Practice

Write each number in standard form.

1. six thousand, one hundred forty-four

2. eight hundred three

3. seven thousand, fifty-one

4. nine hundred ninety-nine

H2 Prerequisite Skills Review

PLACE VALUE TO THOUSANDS

The value of a digit depends on its place-value position in a number. A place-value chart can help you to find the value.

Example

Tell the value of the digits 4 and 5 in 4,325.

Thousands	Hundreds	Tens	Ones
4,	3	2	5

• The 4 is in the thousands place. So, it has a value of 4 × 1,000, or 4,000.

• The 5 is in the ones place. So, it has a value of 5 × 1, or 5.

▶ Practice

Write the value of the blue digit.

1. 2,317   **2.** 8,002   **3.** 5,681   **4.** 1,947

5. 2,508   **6.** 7,095   **7.** 9,462   **8.** 4,723

COMPARE NUMBERS TO THOUSANDS

You can compare two numbers by comparing their digits from left to right.

Compare 7,613 and 7,631.

STEP 1

Compare digits in each place-value position from left to right.

7,613	7,631	The thousands digits are the same.
7,613	7,631	The hundreds digits are the same.
7,613	7,631	The tens digits are not the same.

STEP 2

Determine which number is greater or less. Compare the digits in the place-value position in which the digits are different.

Since 3 > 1,
7,631 > 7,613.
    OR
Since 1 < 3,
7,613 < 7,631.

▶ Practice

Compare. Write < or > for each ●.

1. 1,863 ● 3,899   **2.** 4,302 ● 4,203   **3.** 9,801 ● 8,976   **4.** 2,305 ● 9,740

5. 2,578 ● 5,207   **6.** 2,611 ● 2,161   **7.** 3,574 ● 3,754   **8.** 5,623 ● 6,101

9. 897 ● 1,010   **10.** 9,477 ● 7,974   **11.** 1,624 ● 1,642   **12.** 1,066 ● 977

TROUBLESHOOTING

ORDER NUMBERS TO THOUSANDS

You order numbers by comparing digits in the same place-value position.

Write the numbers in order from *least* to *greatest*. 3,962; 3,978; 3,899

STEP 1	STEP 2	STEP 3
Compare the thousands.	Compare the hundreds.	Compare the tens.
3,962	3,962	3,962
3,978	3,978	3,978
3,899	3,899	
same number of thousands	8 < 9, so 3,899 is least.	6 < 7, so 3,978 is greatest.

So, the order from *least* to *greatest* is 3,899; 3,962; 3,978.

▶ Practice
Write the numbers in order from *least* to *greatest*.

1. 5,478; 5,576; 4,589   **2.** 3,275; 2,854; 3,189   **3.** 1,746; 978; 1,066

Write the numbers in order from *greatest* to *least*.

4. 6,734; 6,546; 6,874   **5.** 743; 7,341; 7,431   **6.** 3,601; 3,610; 3,106

ROUND TO TENS AND HUNDREDS

To round to the tens or hundreds place, look at the digit to the right of the rounding place. If the digit is less than 5, the digit in the rounding place stays the same. If the digit is 5 or more, the digit in the rounding place increases by 1.

Examples

> **A**   Round 724 to the nearest ten.
>
> **Think:** The digit in the rounding place is 2. The digit to its right is 4. Since 4 < 5, the tens digit stays the same.

So, 724 rounds to 720.

> **B**   Round 2,851 to the nearest hundred.
>
> **Think:** The digit in the rounding place is 8. The digit to its right is 5. Since 5 = 5, the hundreds digit increases by 1.

So, 2,851 rounds to 2,900.

▶ Practice
Round each number to the nearest ten.

1. 694   **2.** 126   **3.** 439   **4.** 322

Round each number to the nearest hundred.

5. 9,127   **6.** 1,035   **7.** 2,665   **8.** 7,251

✓ 2-DIGIT ADDITION AND SUBTRACTION

The **sum** is the total of two or more addends. The **difference** is the amount left after one number is subtracted from another.

Find the sum. 87 + 46 = ■

STEP 1	**STEP 2**
Add the ones. 7 + 6 = 13	Add the tens. 1 + 8 + 4 = 13
Regroup 13 ones as 1 ten 3 ones.	Regroup 13 tens as 1 hundred 3 tens.

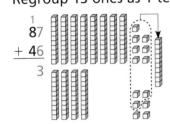
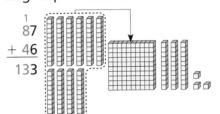

▶ Practice

Find the sum or difference.

1.	**2.**	**3.**	**4.**	**5.**
31	92	55	40	73
+ 79	− 24	+ 85	+ 16	− 37

✓ 3-DIGIT ADDITION AND SUBTRACTION

You can add and subtract 3-digit numbers as you do with 2-digit numbers.

Find the difference. 423 − 156 = ■

STEP 1	**STEP 2**
Since 6 > 3, regroup 2 tens 3 ones as 1 ten 13 ones. Subtract the ones.	Regroup 4 hundreds 1 ten as 3 hundreds 11 tens. Subtract the tens. Subtract the hundreds.

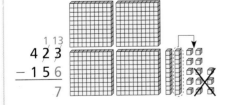
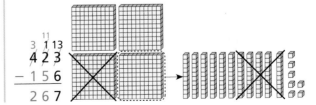

▶ Practice

Find the sum or difference.

1.	**2.**	**3.**	**4.**	**5.**
243	527	766	346	877
+ 168	+ 385	− 488	+ 359	− 788

TROUBLESHOOTING

✓ COLUMN ADDITION

More than two numbers can be added together. Look for ways
to group the addends in each column of numbers.

Find the sum.

STEP 1	STEP 2	STEP 3
Add the ones. $9 + 8 + 2 =$ $9 + (8 + 2) =$ $9 + 10 = 19$	Regroup 19 ones as 1 ten 9 ones.	Add the tens. $1 + 4 + 6 = 11$. Regroup 11 tens as 1 hundred 1 ten. So, the sum is 119.

STEP 1:
```
   9
  48
+ 62
```

STEP 2:
```
  1
   9
  48
+ 62
   9
```

STEP 3:
```
  1
   9
  48
+ 62
 119
```

▶ Practice

Find the sum.

1.
```
   52
   25
+ 17
```

2.
```
    8
   13
+ 49
```

3.
```
   24
   76
+ 53
```

4.
```
   64
   46
+ 55
```

5.
```
   75
   81
+ 12
```

6.
```
   19
   32
+ 47
```

7.
```
   37
   15
+ 28
```

8.
```
   92
   11
+  7
```

9.
```
   41
   58
+ 29
```

10.
```
   17
   14
+ 23
```

✓ MISSING ADDENDS

A missing addend can be found when one addend and the sum are known.

Examples

Find the missing addend.

A $9 + ■ = 16$

Think: $9 + \boxed{\text{What number?}} = 16$

So, the missing addend is 7.

B $■ + 6 = 14$

Think: $\boxed{\text{What number?}} + 6 = 14$

So, the missing addend is 8.

▶ Practice

Find the missing addend.

1. $7 + ■ = 11$ **2.** $■ + 5 = 12$ **3.** $■ + 8 = 13$ **4.** $2 + ■ = 8$

5. $2 + ■ = 20$ **6.** $■ + 17 = 23$ **7.** $■ + 13 = 22$ **8.** $5 + ■ = 18$

✓ ADDITION AND SUBTRACTION FACT FAMILIES

The fact families for addition and subtraction can have either two or four number sentences.

Examples

A An addition family with two different addends has four related facts. This fact family is for 3, 4, and 7.

$$3 + 4 = 7 \qquad 7 - 3 = 4$$
$$4 + 3 = 7 \qquad 7 - 4 = 3$$

B An addition family with the same addends has only two related facts. This fact family is for 5, 5, and 10.

$$5 + 5 = 10$$
$$10 - 5 = 5$$

▶ Practice

Find the missing numbers.

1. $2 + 6 = 8$
$6 + 2 = \blacksquare$
$8 - 2 = \blacksquare$
$8 - \blacksquare = 2$

2. $5 + 7 = 12$
$7 + 5 = \blacksquare$
$12 - 7 = \blacksquare$
$\blacksquare - 5 = 7$

3. $8 + 8 = \blacksquare$
$16 - 8 = \blacksquare$

4. $9 + 4 = 13$
$4 + \blacksquare = 13$
$13 - 4 = \blacksquare$
$13 - 9 = \blacksquare$

5. $5 + 9 = 14$
$14 - 9 = \blacksquare$

6. $3 + 8 = 11$
$11 - 3 = \blacksquare$

7. $6 + 8 = 14$
$14 - 6 = \blacksquare$

8. $9 + 9 = 18$
$18 - \blacksquare = \blacksquare$

✓ NUMBER PATTERNS

A list of numbers can show a number pattern.

Examples

Write the next three numbers in the pattern.

A 14, 24, 34, 44, ■, ■, ■

Think: Skip-count by tens.

$44 + 10 = 54 \quad 54 + 10 = 64 \quad 64 + 10 = 74$

So, the next three numbers are 54, 64, and 74.

B 20, 18, 16, 14, ■, ■, ■

Think: Count back by twos.

$14 - 2 = 12 \quad 12 - 2 = 10 \quad 10 - 2 = 8$

So, the next three numbers are 12, 10, and 8.

▶ Practice

Write the next three numbers in the pattern.

1. 17, 20, 23, 26, ■, ■, ■

2. 14, 21, 28, 35, ■, ■, ■

3. 15, 21, 27, 33, ■, ■, ■

4. 78, 74, 70, 66, ■, ■, ■

5. 37, 32, 27, 22, ■, ■, ■

6. 34, 37, 40, 43, ■, ■, ■

TROUBLESHOOTING

✓ READ PICTOGRAPHS

Pictographs show data by using pictures. The **key** at the bottom of a pictograph tells how many items each picture stands for. This pictograph shows the numbers of bushels of apples picked.

Example

How many more bushels of Granny Smith apples were picked than of Delicious apples?

Think: Skip-count the bushels to find the number of Granny Smith apples and Delicious apples. Then subtract.

$35 - 25 = 10$

So, 10 more bushels of Granny Smith apples were picked.

BUSHELS OF APPLES	
Cortland	🍎🍎🍎
Delicious	🍎🍎🍎🍎🍎
Granny Smith	🍎🍎🍎🍎🍎🍎🍎

Key: Each 🍎 **= 5 bushels.**

▶ Practice

For 1–2, use the pictograph.

1. In all, how many bushels of Cortland and Granny Smith apples were picked?

2. Were there two more bushels of Delicious apples picked than Cortland apples? Explain.

✓ TALLIES TO FREQUENCY TABLES

Tally marks show numbers of things counted in groups of five. You can skip-count to find a total.

KINDS OF SHIRTS	
Kind	**Number**
Striped	ЖЖ II
Solid	ЖЖЖ I
Flowered	Ж I

A **frequency table** uses numbers to show how often something happens.

KINDS OF SHIRTS	
Kind	**Number**
Striped	12
Solid	16
Flowered	6

▶ Practice

Use the tally table to complete the frequency table at the right.

	FLOWER SALES	
	Type	**Number**
1.	Irises	ЖЖЖЖ II
2.	Lilies	ЖЖЖ III
3.	Roses	ЖЖЖЖЖЖЖЖ III

FLOWER SALES	
Type	**Number**
Irises	▪
Lilies	▪
Roses	▪

4. Were more roses sold than irises and lilies together? Explain.

PARTS OF A GRAPH

A **graph** is used to compare information. A bar graph should have a title, a label for the scale, and a label for the data.

Example

What is the label for the scale? Look at the bottom of the graph.

The label for the scale is Number of Videos.

What is the label for the data? Look along the left side of the graph.

The label for the data is Types of Videos.

► Practice

For 1–2, use the bar graph. It shows the number of fourth graders who went on each field trip.

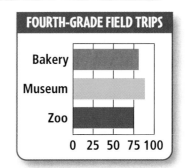

1. What would you label the scale?

2. What would you label the data?

READ BAR GRAPHS

Bar graphs help to compare information, or data. The graph below shows the number of instruments in each section of the orchestra.

Example

Which section has the fewest instruments?

Think: The section that has the shortest bar has the fewest instruments.

So, the percussion section has the fewest instruments.

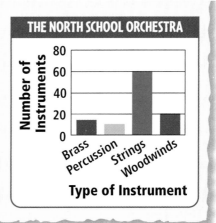

► Practice

For 1–2, use the bar graph.

1. Which section has the most instruments?

2. Which section has more instruments than the other three sections combined?

IDENTIFY POINTS ON A GRID

An **ordered pair** of numbers locates a point on a grid. The first number in an ordered pair tells how many units to the right of zero the point is. The second number tells how many units above zero it is.

Example

What is the ordered pair for point *M*?

Think: Point *M* is 4 units to the right of zero and 1 unit up from zero.

So, the ordered pair is (4,1).

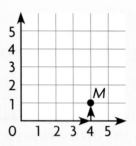

▶ Practice

For 1–6, use the grid. Write the ordered pair for each fruit or vegetable.

1. Tomatoes
2. Pears
3. Carrots

4. Peaches
5. Beans
6. Corn

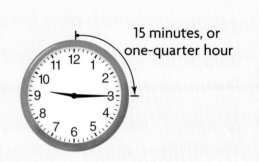

TIME TO THE HALF AND QUARTER HOUR

One hour is 60 minutes, so one-half hour is 30 minutes. One-quarter hour is 15 minutes, and three-quarters of an hour is 45 minutes.

Example

Write the time shown on the clock.

Think: The short hand is the hour hand. It is a little past 9. The long hand is the minute hand. It shows 15 minutes after the hour.

So, the time is 9:15, or quarter past nine.

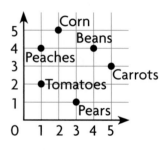

▶ Practice

Write the time.

1.

2.

3.

4.

✓ TIME TO THE MINUTE

The minute hand on an analog clock moves 60 minutes each hour. A
digital clock shows time as the hour and the number of minutes after the hour.

Examples

A Write the time after the hour.

Think: Skip-count by
fives to the number 2.
Then count one more
minute.

So, the time is 4:11.

B Write the time before the hour, or
as it is shown on a digital clock.

Think: Count back
by fives starting
with 12. Then count
by ones to the mark
where the minute
hand is pointing.

So, the time is nine minutes before eight,
or 7:51.

▶ Practice

Write the time.

1.
2.
3.
4.

✓ USE A CALENDAR

Calendars show the months and days in a year.

Example

Use the calendar. What is the date of the
second Sunday in February?

Think: The first column is Sunday.
The second number in the column is 10.

So, the second Sunday is February 10.

FEBRUARY						
Sun	Mon	Tue	Wed	Thu	Fri	Sat
					1	2
3	4	5	6	7	8	9
10	11	12	13	14	15	16
17	18	19	20	21	22	23
24	25	26	27	28		

▶ Practice

1. Presidents' Day is the third Monday in
   February. What is the date?

2. What day of the week is February 28?

3. The fourth graders have music on Wednesdays and Fridays.
   How many days in February will they have music?

TROUBLESHOOTING

✓ MEANING OF MULTIPLICATION AND DIVISION

When you **multiply**, you put equal groups together. When you **divide**, you separate into equal groups.

Examples

A Find the product.
$3 \times 4 = \blacksquare$

This array shows 3 rows of 4, or 12. So, $3 \times 4 = 12$.

B Find the quotient.
$12 \div 3 = \blacksquare$

12 counters in all
3 groups
4 in each group
So, $12 \div 3 = 4$.

▶ Practice

Find the product or quotient. You may wish to draw arrays or groups.

1. $2 \times 5 = \blacksquare$ **2.** $21 \div 3 = \blacksquare$ **3.** $3 \times 9 = \blacksquare$ **4.** $3 \times 5 = \blacksquare$

5. $45 \div 5 = \blacksquare$ **6.** $4 \times 7 = \blacksquare$ **7.** $2 \times 10 = \blacksquare$ **8.** $40 \div 5 = \blacksquare$

9. $24 \div 3 = \blacksquare$ **10.** $4 \times 6 = \blacksquare$ **11.** $50 \div 5 = \blacksquare$ **12.** $15 \div 3 = \blacksquare$

✓ MULTIPLICATION AND DIVISION FACTS

The answer to a multiplication fact is the **product**.
The answer to a division fact is the **quotient**.

Examples

A Find the product. $5 \times 7 = \blacksquare$

You can skip-count to find a product. Starting with 0, count out 7 fives. So, the product is 35.

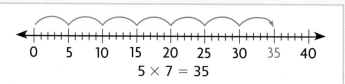

$5 \times 7 = 35$

B Find the quotient. $35 \div 5 = \blacksquare$

You can count back to find a quotient. Starting with 35, count back by fives. So, the quotient is 7.

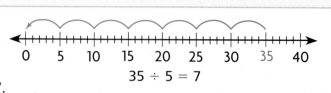

$35 \div 5 = 7$

▶ Practice

Find the product or quotient. You may wish to use number lines.

1. $6 \times 8 = \blacksquare$ **2.** $28 \div 7 = \blacksquare$ **3.** $8 \times 9 = \blacksquare$ **4.** $6 \times 6 = \blacksquare$

5. $56 \div 8 = \blacksquare$ **6.** $7 \times 7 = \blacksquare$ **7.** $9 \times 10 = \blacksquare$ **8.** $35 \div 7 = \blacksquare$

H12 Prerequisite Skills Review

✓ MISSING FACTORS

You can find a missing factor by recalling basic facts.

Examples

Ⓐ $7 \times \blacksquare = 28$

Think: $7 \times \boxed{\text{What number?}} = 28$

So, the missing factor is 4.

Ⓑ $\blacksquare \times 5 = 30$

Think: $\boxed{\text{What number?}} \times 5 = 30$

So, the missing factor is 6.

▶ Practice

Find the missing factor.

1. $6 \times \blacksquare = 24$ **2.** $\blacksquare \times 7 = 28$ **3.** $9 \times \blacksquare = 27$ **4.** $\blacksquare \times 8 = 64$

5. $\blacksquare \times 5 = 35$ **6.** $7 \times \blacksquare = 14$ **7.** $4 \times \blacksquare = 32$ **8.** $\blacksquare \times 5 = 0$

9. $\blacksquare \times 6 = 30$ **10.** $10 \times \blacksquare = 30$ **11.** $\blacksquare \times 7 = 63$ **12.** $6 \times \blacksquare = 18$

13. $9 \times \blacksquare = 63$ **14.** $8 \times \blacksquare = 32$ **15.** $\blacksquare \times 5 = 45$ **16.** $\blacksquare \times 4 = 48$

✓ MODEL MULTIPLICATION

Use base-ten blocks to find the product. $3 \times 16 = \blacksquare$

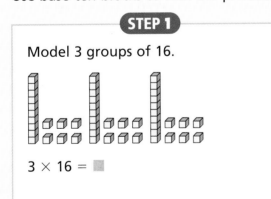

STEP 1

Model 3 groups of 16.

$3 \times 16 = \blacksquare$

STEP 2

Combine the tens and ones. Regroup 10 ones as 1 ten.

So, $3 \times 16 = 48$.

▶ Practice

Use base-ten blocks to find each product.

1. $\begin{array}{r} 23 \\ \times\ 8 \\ \hline \end{array}$ **2.** $\begin{array}{r} 17 \\ \times\ 5 \\ \hline \end{array}$ **3.** $\begin{array}{r} 26 \\ \times\ 3 \\ \hline \end{array}$ **4.** $\begin{array}{r} 32 \\ \times\ 4 \\ \hline \end{array}$ **5.** $\begin{array}{r} 56 \\ \times\ 2 \\ \hline \end{array}$

6. $\begin{array}{r} 13 \\ \times\ 7 \\ \hline \end{array}$ **7.** $\begin{array}{r} 43 \\ \times\ 6 \\ \hline \end{array}$ **8.** $\begin{array}{r} 21 \\ \times\ 9 \\ \hline \end{array}$ **9.** $\begin{array}{r} 44 \\ \times\ 6 \\ \hline \end{array}$ **10.** $\begin{array}{r} 65 \\ \times\ 3 \\ \hline \end{array}$

✔ RECORD MULTIPLICATION

Use place value to find and record the product.

Find the product. $4 \times 27 = \blacksquare$

STEP 1

Multiply the ones.
Record the product.

Hundreds	Tens	Ones
	2	7
$\times$		4
	2	8	← $4 \times 7 = 28$

STEP 2

Multiply the tens. Record the product.
Add the products.

Hundreds	Tens	Ones
	2	7
$\times$		4
	2	8
$+$	8	0	← $4 \times 20 = 80$
1	0	8

▶ Practice

Find the product.

1. 22
   $\times\ 8$

2. 57
   $\times\ 6$

3. 25
   $\times\ 9$

4. 54
   $\times\ 7$

5. 4×55

6. 5×33

7. 8×19

8. 7×49

✔ MULTIPLY BY TENS, HUNDREDS, AND THOUSANDS

To multiply by tens, hundreds, and thousands, look for a basic fact and a pattern.

Examples

A
$3 \times 30 = 90$
$3 \times 300 = 900$
$3 \times 3,000 = 9,000$

B
$2 \times 60 = 120$
$2 \times 600 = 1,200$
$2 \times 6,000 = 12,000$

C
$10 \times 10 = 100$
$10 \times 100 = 1,000$
$10 \times 1,000 = 10,000$

▶ Practice

Multiply. Use a basic fact and a pattern.

1. 10×8

2. 10×80

3. 10×800

4. $10 \times 8,000$

5. 10
   $\times\ 4$

6. 10
   $\times\ 40$

7. 100
   $\times\ 40$

8. 1,000
   $\times\ 400$

✓ MULTIPLY BY 1-DIGIT NUMBERS

An understanding of place value can help you find products.

Find the product. $8 \times 43 = $ ▩

STEP 1

Multiply the ones.

8×3 ones $= 24$ ones

Regroup 24 ones
as 2 tens 4 ones.

$$\begin{array}{r} 2 \\ 43 \\ \times\ 8 \\ \hline 4 \end{array}$$

STEP 2

Multiply the tens.

8×4 tens $= 32$ tens

Add the regrouped tens.

2 tens $+ 32$ tens $= 34$ tens

Record the product. So, $8 \times 43 = 344$.

$$\begin{array}{r} 2 \\ 43 \\ \times\ 8 \\ \hline 344 \end{array}$$

▶ Practice

Multiply.

1. $\begin{array}{r} 15 \\ \times\ 6 \\ \hline \end{array}$
2. $\begin{array}{r} 27 \\ \times\ 3 \\ \hline \end{array}$
3. $\begin{array}{r} 59 \\ \times\ 2 \\ \hline \end{array}$
4. $\begin{array}{r} 86 \\ \times\ 5 \\ \hline \end{array}$
5. $\begin{array}{r} 63 \\ \times\ 8 \\ \hline \end{array}$

6. $7 \times 48 = $ ▩
7. $4 \times 143 = $ ▩
8. $5 \times 483 = $ ▩

✓ ESTIMATE PRODUCTS

You can round one factor to estimate products.

Examples

Ⓐ $6 \times 87 = $ ▩

6×87
↓
$6 \times 90 = 540$

Round 87 to the nearest 10.

Use a basic fact and patterns.

So, $6 \times 90 = 540$.

Ⓑ $4 \times 329 = $ ▩

4×329
↓
$4 \times 300 = 1,200$

Round 329 to the nearest 100.

Use a basic fact and patterns.

So, $4 \times 300 = 1,200$.

▶ Practice

Round the blue number. Estimate the product.

1. $\begin{array}{r} 23 \\ \times\ 9 \\ \hline \end{array}$
2. $\begin{array}{r} 46 \\ \times\ 6 \\ \hline \end{array}$
3. $\begin{array}{r} 73 \\ \times\ 8 \\ \hline \end{array}$
4. $\begin{array}{r} 91 \\ \times\ 7 \\ \hline \end{array}$
5. $\begin{array}{r} 68 \\ \times\ 5 \\ \hline \end{array}$

6. $4 \times 145 = $ ▩
7. $9 \times 589 = $ ▩
8. $6 \times 358 = $ ▩

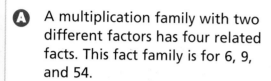

TROUBLESHOOTING

✓ MULTIPLICATION AND DIVISION FACT FAMILIES

The fact families for multiplication and division can have either two or four number sentences.

Examples

A A multiplication family with two different factors has four related facts. This fact family is for 6, 9, and 54.

$9 \times 6 = 54$ $54 \div 6 = 9$
$6 \times 9 = 54$ $54 \div 9 = 6$

B A multiplication family with the same factors has only two related facts. This fact family is for 12, 12, and 144.

$12 \times 12 = 144$
$144 \div 12 = 12$

▶ Practice

Find the missing numbers.

1. $3 \times 7 = 21$
   $7 \times 3 = \blacksquare$
   $21 \div \blacksquare = 7$
   $21 \div \blacksquare = 3$

2. $8 \times 9 = 72$
   $9 \times 8 = \blacksquare$
   $\blacksquare \div 8 = 9$
   $72 \div \blacksquare = 8$

3. $4 \times 5 = 20$
   $5 \times 4 = \blacksquare$
   $20 \div 5 = \blacksquare$
   $20 \div \blacksquare = 5$

4. $6 \times 12 = 72$
   $12 \times 6 = \blacksquare$
   $72 \div \blacksquare = 12$
   $72 \div 12 = \blacksquare$

5. $10 \times 10 = 100$
   $\blacksquare \div 10 = 10$

6. $7 \times 7 = 49$
   $\blacksquare \div 7 = 7$

7. $8 \times 8 = 64$
   $\blacksquare \div 8 = 8$

8. $6 \times 6 = 36$
   $36 \div \blacksquare = \blacksquare$

✓ USE MULTIPLICATION FACTS AND PATTERNS

You can use basic facts and place-value patterns to multiply numbers greater than 10.

Place-Value Pattern
$1 \times 5 = 5$
1 ten $\times 5 = 5$ tens, or 50
1 hundred $\times 5 = 5$ hundreds, or 500

Example

Complete the number sentences.

$\blacksquare \times 3 = 9$ **Think:** What number times 3 equals 9? 3

$\blacksquare \times 30 = 90$ **Think:** What number times 3 tens equals 9 tens? 3

$\blacksquare \times 300 = 900$ **Think:** What number times 3 hundreds equals 9 hundreds? 3

So, the missing factor in each sentence is 3.

▶ Practice

Copy and complete.

1. $\blacksquare \times 1 = 7$
   $\blacksquare \times 10 = 70$
   $\blacksquare \times 100 = 700$

2. $\blacksquare \times 4 = 8$
   $\blacksquare \times 40 = 80$
   $\blacksquare \times 400 = 800$

3. $\blacksquare \times 5 = 15$
   $\blacksquare \times 50 = 150$
   $3 \times \blacksquare = 1,500$

4. $8 \times 4 = \blacksquare$
   $8 \times \blacksquare = 320$
   $8 \times \blacksquare = 3,200$

✔ PLACE VALUE

A place-value chart can help you understand greater numbers.
The position of each digit tells you its value.

Example

In the chart, the value of 6 is 6 ten thousands, or
60,000. The value of 5 is 5 hundreds, or 500.

THOUSANDS			ONES		
Hundreds	Tens	Ones	Hundreds	Tens	Ones
	6	7,	5	1	3

Standard form: 67,513

Word form: sixty-seven thousand,
five hundred thirteen

▶ Practice

Write the value of the blue digit.

1. 12,673   **2.** 109,531   **3.** 26,285   **4.** 314,018

Write the number in standard form.

5. forty-two thousand, fifty-six   **6.** three hundred thousand,
five hundred eighteen

✔ DIVIDE WITH REMAINDERS

When you divide into equal groups, you may have some left over.
The leftover amount is called a **remainder**.

Example

Find the quotient and remainder. 23 ÷ 4

20 ÷ 4 = 5   Find the greatest number of 4s in 23.

23 − 20 = 3   Subtract to find how many are left over.

23 ÷ 4 = 5 r3   Write the quotient and its remainder.

So, 23 ÷ 4 = 5 r3.

▶ Practice

Write the quotient and the remainder.

1. 27 ÷ 4   **2.** 82 ÷ 9   **3.** 60 ÷ 8   **4.** 22 ÷ 3   **5.** 26 ÷ 6

6. 68 ÷ 8   **7.** 37 ÷ 5   **8.** 46 ÷ 8   **9.** 45 ÷ 7   **10.** 70 ÷ 9

11. 49 ÷ 5   **12.** 58 ÷ 12   **13.** 81 ÷ 4   **14.** 94 ÷ 7   **15.** 83 ÷ 6

TROUBLESHOOTING

✓ USE COMPATIBLE NUMBERS

Compatible numbers can be used to estimate quotients mentally.
Compatible numbers are in a fact family close to the numbers
you need to divide.

Estimate the quotient. 50 ÷ 6

STEP 1	**STEP 2**
Name the fact families close to 50 ÷ 6. 50 ÷ 10 48 ÷ 6 54 ÷ 6	Determine which fact family is closest to the actual dividend and divisor. The fact family with the closest numbers is 48 ÷ 6.

So, a good estimate is 50 ÷ 6 = 8.

▶ Practice
Estimate the quotient. Tell the compatible numbers you used.

1. $5\overline{)27}$ 2. $7\overline{)65}$ 3. $6\overline{)29}$ 4. $8\overline{)61}$ 5. $12\overline{)45}$

6. 73 ÷ 9   7. 58 ÷ 11   8. 67 ÷ 7   9. 87 ÷ 9   10. 59 ÷ 7

✓ PARTS OF A WHOLE

Fractions can describe parts of a whole.

The **numerator**, or top number, tells how
many of the parts are being used.

The **denominator**, or bottom number, tells
the total number of equal parts in the whole.

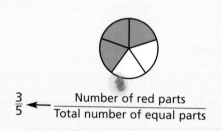

$$\frac{3}{5} \leftarrow \frac{\text{Number of red parts}}{\text{Total number of equal parts}}$$

▶ Practice
Write a fraction for each shaded part.

1.   2.   3.   4.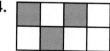

Draw a picture for each description.

5. A rectangle with three parts shaded out of four parts.

6. A square with one part shaded out of two parts.

7. A triangle with two parts shaded out of three parts.

✔ PARTS OF A GROUP

Fractions can describe parts of a group.

The **numerator**, or top number, tells how many of the parts are being used.

The **denominator**, or bottom number, tells the total number of equal parts in the group.

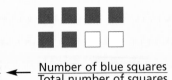

$\dfrac{6}{8}$ ← $\dfrac{\text{Number of blue squares}}{\text{Total number of squares}}$

▶ Practice

Write a fraction for the shaded part.

1. ● ○ ○ ○ ○

2. ● ● ● ● ● ●
● ○ ○ ○ ○ ○

3.

Draw a picture for each description.

4. Two out of three triangles are shaded.

5. Two out of five squares are shaded.

6. Seven out of ten circles are shaded.

✔ COMPARE PARTS OF A WHOLE

You can compare fractions using models. The model with the greater part shaded represents the greater fraction.

Example

Compare using a model. Write <, >, or = for each ●.

The first rectangle shows $\frac{1}{4}$. The second rectangle shows $\frac{2}{4}$ has more shaded parts.

So, $\frac{1}{4} < \frac{2}{4}$.

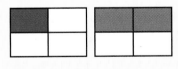

$\frac{1}{4}$ ● $\frac{2}{4}$

▶ Practice

Compare the shaded areas of the rectangles.
Write <, >, or = for each ●.

1.

$\dfrac{4}{5}$ ● $\dfrac{3}{5}$

2.

$\dfrac{5}{8}$ ● $\dfrac{6}{8}$

3.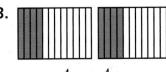

$\dfrac{4}{12}$ ● $\dfrac{4}{12}$

✓ MODEL DECIMALS

You can write a decimal for a model that shows tenths or hundredths.

Examples

Write a decimal for each shaded part.

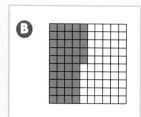

So, 0.2 is shaded.   So, 0.45 is shaded.

▶ Practice

Write the decimal for the shaded part.

1.    **2.**   **3.**

✓ RELATE DECIMALS TO MONEY

The number before a decimal point shows the number of dollars. The number after the decimal point shows hundredths of dollars, or cents.

Example

Write three dollars and fifty-eight cents as a decimal.

• Write the whole number of dollars. $3
• Write the cents as hundredths. $0.58

So, write $3.58.

▶ Practice

Write a decimal for each money amount.

1. six dollars and forty-five cents   **2.** ninety-eight cents

3. ten dollars and two cents   **4.** one dollar and fifty-four cents

5. twenty dollars and ten cents   **6.** two dollars and fifty-one cents

✅ FRACTIONS WITH DENOMINATORS OF 10 AND 100

You can use decimal models to write equivalent fractions.

Each whole square has the same amount shaded. So, the models represent equivalent fractions.

Example

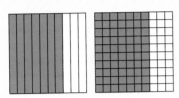

- The first square shows 7 out of 10 parts shaded.
- The second square shows 70 out of 100 parts shaded. Both have the same amount shaded.

So, $\frac{7}{10} = \frac{70}{100}$.

▶ Practice

Complete to show equivalent fractions.

1.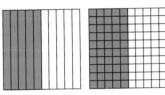

$$\frac{5}{10} = \frac{\blacksquare}{100}$$

2.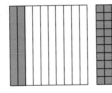

$$\frac{2}{10} = \frac{\blacksquare}{100}$$

3.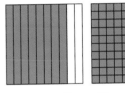

$$\frac{8}{10} = \frac{\blacksquare}{\blacksquare}$$

✅ ROUND MONEY AMOUNTS

You can round money amounts to estimate values.

Examples

A Round $2.63 to the nearest dollar.

Think: Look at the digit to the right of the ones place. If the digit is 5 or more, the digit in the ones place increases by 1.

The digit to the right is 6. $6 > 5$.

So, $2.63 rounds to $3.00.

B Round $36.14 to the nearest dollar.

Think: Look at the digit to the right of the ones place. If the digit is less than 5, the ones digit stays the same.

The digit to the right is 1. $1 < 5$.

So, $36.14 rounds to $36.00.

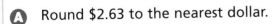

▶ Practice

Round each to the nearest dollar.

1. $7.82   **2.** $7.14   **3.** $9.58   **4.** $7.39   **5.** $22.98

6. $64.85   **7.** $82.45   **8.** $12.09   **9.** $38.11   **10.** $99.99

TROUBLESHOOTING

✓ ADD AND SUBTRACT MONEY

Knowing how to add and subtract whole numbers can help you add and subtract money amounts.

Examples

Add or subtract the money amounts.

<table>
<tr><td>

A

$3.12
+ $1.56

Think: 312
    + 156
     468
</td><td>

B

$4.17
− $2.02

Think: 417
    − 202
     215
</td></tr>
</table>

So, $3.12 + $1.56 = $4.68.

So, $4.17 − $2.02 = $2.15.

▶ Practice

Find the sum or difference.

1. $17.52
   + $ 9.84

2. $109.76
   − $105.50

3. $55.02
   + $36.12

4. $9.81
   − $2.76

5. $22.25
   + $ 7.40

6. $8.06
   + $2.94

7. $19.00
   − $13.50

8. $4.80
   − $3.17

9. $43.99
   − $20.00

10. $37.05
   + $27.20

MEASURE TO THE NEAREST INCH AND HALF-INCH

Linear units are used to measure length, width, height, and distance.

Example

Measure the length to the nearest inch.

- Find the inch mark that is nearest the eraser's length.
- The length is closer to the 2-inch mark.

So, the eraser is about 2 inches long.

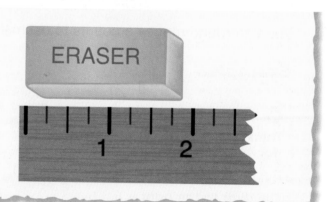

▶ Practice

Measure the length of each to the nearest inch.

1.

2.

Measure the length to the nearest half-inch.

3.

✓ MULTIPLICATION

You can use patterns to skip-count and to multiply.

Example

Describe the pattern. Find the next number in the pattern.

12, 24, 36, 48, ▨

You can multiply to find the next number.

This pattern shows multiples of 12.

$12 \times \mathbf{1} = 12$, $12 \times \mathbf{2} = 24$, $12 \times \mathbf{3} = 36$, $12 \times \mathbf{4} = 48$

So, the next number is 12×5, or 60.

▶ Practice

Describe each pattern. Write the next number in each pattern.

1. 6, 12, 18, 24, ▨   **2.** 8, 16, 24, 32, ▨   **3.** 5, 10, 15, 20, 25, ▨

Find the product.

4. $8 \times 12 = $ ▨   **5.** $3 \times 48 = $ ▨   **6.** $3 \times 5,280 = $ ▨

✓ MEASURE TO THE NEAREST CENTIMETER

You can use centimeters to measure length and distance.

Example

Measure to the nearest centimeter.

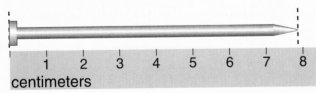

- Find the centimeter mark that is nearest to the nail's length.
- The nail's length is closer to 8 cm.

So, to the nearest centimeter, the nail is 8 cm long.

▶ Practice

Measure the length of each object to the nearest centimeter.

1.

2.

3.

TROUBLESHOOTING

✓ MULTIPLY BY 10, 100, AND 1,000

You can multiply by multiples of 10, 100, and 1,000 to find products.

Example

Find the value of *n*. $15 \times 100 = n$

$15 \times 1 = 15$ 15×1 ten = 15 tens, or 150 15×1 hundred = 15 hundreds, or 1,500

So, $n = 1,500$.

▶ Practice

Find the value of *n*.

1. $17 \times 100 = n$

2. $8 \times n = 8,000$

3. $n \times 100 = 2,400$

4. $130 \times n = 1,300$

5. $350 \times 10 = n$

6. $n \times 1,000 = 40,000$

✓ READ A NUMBER LINE

As you move from left to right on a number line, the numbers increase.

Examples

Ⓐ Use the number line to compare.

124 ● 122

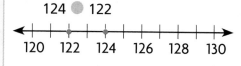

Ⓑ Use the number line to subtract.

$14 - 5 = \blacksquare$

Since 124 is to the right of 122, 124 > 122.

So, $14 - 5 = 9$.

▶ Practice

Use the number line to compare. Write < or >.

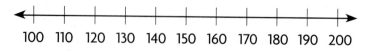

100 110 120 130 140 150 160 170 180 190 200

1. 120 ● 150

2. 175 ● 135

3. 181 ● 189

4. 135 ● 110

Use the number line to add and subtract.

0 1 2 3 4 5 6 7 8 . 9 10 11 12 13 14 15

5. $9 + 4$

6. $15 - 8$

7. $12 - 9$

8. $3 + 11$

USE FUNCTION TABLES

You can use a **function table** and a rule to find output numbers.

Example

Use the rule to complete the table.

INPUT	2	5	8
OUTPUT	6	9	■

Rule: add 4

Think: The rule is to add 4. That means you add 4 to each input number.

Input   Rule   Output
8   + 4 =   12

So, the output value that completes the table is 12.

▶ Practice

Use the rule to complete each table.

1. Rule: Subtract 5.

INPUT	5	9	13
OUTPUT	0	4	■

2. Rule: Add 9.

INPUT	4	7	9
OUTPUT	13	■	■

3. Rule: Multiply by 4.

INPUT	1	3	5
OUTPUT	4	■	■

IDENTIFY ANGLES

You can use the size of a right angle to determine whether another angle is less than or greater than a right angle.

 right angle

Examples

Tell if the angle is a *right* angle, *greater than* a right angle, or *less than* a right angle.

A   The dashed line shows where the ray would be if the angle were a right angle.

B   The dashed line shows where the ray would be if the angle were a right angle.

So, the angle is *less than* a right angle.   So, the angle is *greater than* a right angle.

▶ Practice

Tell if each angle is a *right* angle, *greater than* a right angle, or *less than* a right angle.

1.   **2.**   **3.**

✓ COMPARE FIGURES

Congruent figures have the same size and shape.

Examples

Are the figures congruent?

 A

Think: The figures are the same shape but not the same size.

So, the figures are not congruent.

B

Think: The figures are the same shape and the same size.

So, the figures are congruent.

▶ Practice

Are the figures congruent? Write *yes* or *no*.

1. 　　**2.** 　　**3.**

✓ IDENTIFY SYMMETRIC FIGURES

A figure has **line symmetry** if you can fold it on the line and the two parts match.

Examples

Is the blue line a line of symmetry?

 A

If you fold the circle on the line, the two parts are identical.

B

If you fold the circle on the line, the two parts are not identical.

So, the blue line is a line of symmetry.　　So, the blue line is not a line of symmetry.

▶ Practice

Is the blue line a line of symmetry? Write *yes* or *no*.

1. 　　**2.** 　　**3.** 　　**4.**

✔️ SORT POLYGONS

A **polygon** is a closed plane figure with straight sides. A corner is where two sides meet. Square corners form right angles.

Example

The figure at the right has six sides. Two right angles are formed at *A* and *B*.

A   B

▶ Practice

For 1–4, use the polygons.

1. Which have 3 sides?

2. Which have 4 sides?

3. Which have 5 sides?

4. Which have right angles?

A   B   C

D   E   F

✔️ IDENTIFY PLANE FIGURES

A **plane figure** is flat and is all in one plane. It can be closed or open.

These are types of closed plane figures.

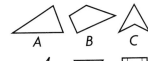

circle   triangle   quadrilateral   pentagon   hexagon   octagon

▶ Practice

Write a name for each figure.

1.

2.

3.

4.

5.

6.

7.

8.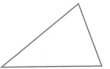

✔ FIND PERIMETER

Perimeter is the distance around a figure. You can find the perimeter of a polygon by counting or adding the lengths of its sides.

Count to find the perimeter of the figure.

STEP 1	**STEP 2**
Count to find the length of each side. The sides are each 3 units in length.	Add the lengths of the sides. 3 + 3 + 3 + 3 = 12 So, the figure has a perimeter of 12 units.

▶ Practice

Count to find the perimeter of each figure.

1.   **2.**   **3.**   **4.**

✔ FIND AREA

Area is the number of square units that cover a closed figure. To find the area of the figure, you can count the number of square units.

Example

Count to find the area of the blue figure.

The blue figure covers exactly 12 squares.

So, the area is 12 square units.

▶ Practice

Count to find the area of each blue figure.

1. 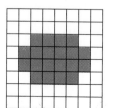 **2.** 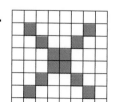 **3.** 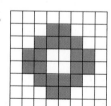 **4.**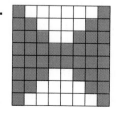

✔ IDENTIFY SOLID FIGURES

Solid figures have three dimensions: length, width, and height.

These figures are called solid figures.

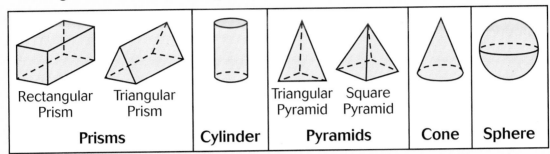

Rectangular Prism Triangular Prism	Cylinder	Triangular Pyramid Square Pyramid	Cone	Sphere
Prisms	**Cylinder**	**Pyramids**	**Cone**	**Sphere**

▶ Practice

Name the solid figure that each object looks like.

1.

2.

3.

Name each solid figure.

4.

5.

6.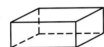

✔ DRAW PLANE FIGURES

When you draw a plane figure, think about the properties of the figure.

Example

Draw a triangle.

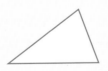

Think: A triangle has 3 sides.

▶ Practice

Draw the plane figure named.

1. pentagon   **2.** rectangle   **3.** octagon   **4.** hexagon   **5.** circle

TROUBLESHOOTING

✓ MULTIPLY 3 FACTORS

When you multiply three factors, you can use the **Grouping Property** to group them so that it is easy to multiply mentally.

Example

Find the product. $6 \times 8 \times 5$

One Way

Multiply the first two factors.

$(6 \times 8) \times 5$

48×5

240

Another Way

Multiply the last two factors.

$6 \times (8 \times 5)$

6×40

240

Either way you group the factors, the product is the same.
So, the product is 240.

▶ Practice

Group the factors two different ways. Find the product.

1. $6 \times 4 \times 2 =$ ▪

2. $5 \times 3 \times 4 =$ ▪

3. $5 \times 6 \times 7 =$ ▪

4. $10 \times 3 \times 5 =$ ▪

5. $27 \times 9 \times 0 =$ ▪

6. $14 \times 1 \times 7 =$ ▪

7. $12 \times 3 \times 4 =$ ▪

8. $20 \times 5 \times 5 =$ ▪

9. $6 \times 2 \times 30 =$ ▪

✓ CLASSIFY ANGLES

An angle is formed by two rays that have the same endpoint.

A **right angle** forms a square corner.

Shows right-angle measure.

right angle

An **acute angle** is an angle that measures less than a right angle.

acute angle

An **obtuse angle** is an angle that measures greater than a right angle.

obtuse angle

▶ Practice

Tell if each angle is acute, right, or obtuse.

1.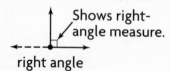

2.

3.

✓ CERTAIN AND IMPOSSIBLE

An event that is sure to happen is **certain**. An event that will never happen is **impossible**.

Examples

Write *certain* or *impossible* for each event.

> **A** ice melting on a warm summer day
>
> Ice will always melt at temperatures above freezing.

> **B** snow from a cloudless day
>
> It cannot snow without clouds.

So, this event is certain.

So, this event is impossible.

▶ Practice

Write *certain* or *impossible* for each event.

1. tossing a number less than 7 with a number cube labeled 1 to 6

2. making change for a $5 bill when you only have three $1 bills

3. drawing a 4-sided triangle

4. drawing a 4-sided square

✓ IDENTIFY POSSIBLE OUTCOMES

Some events can happen in only a few ways. You can list the ways, or **possible outcomes**, of the event.

Examples

Write the possible outcomes for each event.

> **A** tossing a counter with one red side and one blue side
>
> The 2 possible outcomes are red side up and blue side up.

> **B** pulling a coin from a bag of one dime, one nickel, and one penny
>
> The 3 possible outcomes are pulling a dime, pulling a nickel, and pulling a penny.

▶ Practice

Write the possible outcomes for each event.

1. pulling a balloon out of the bag

2. spinning the pointer of this spinner

3. tossing a number cube labeled 1–6

4. tossing two coins

Set A (pp. 2–3)

Use the benchmark to decide which is the more reasonable number.

1. table tennis balls in bucket B

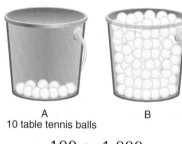

A
10 table tennis balls
B

100 or 1,000

2. sunflower seeds in bird feeder B

A
300 sunflower seeds
B

3,000 or 30,000

Set B (pp. 4–5)

Write the value of the digit 6 in each number.

1. 7,056

2. 15,608

3. 60,789

4. 16,340

Set C (pp. 6–7)

Write each number in word form.

1. 482,907   **2.** 500,128   **3.** 961,542   **4.** 271,964

Write the value of the blue digit.

5. 137,568   **6.** 739,062   **7.** 561,342   **8.** 906,723

9. 354,781   **10.** 827,910   **11.** 269,415   **12.** 478,536

Set D (pp. 8–11)

Write each number in word form.

1. 68,247,311   **2.** 8,601,824   **3.** 3,714,069   **4.** 742,093,758

Write the value of the blue digit.

5. 13,749,568   **6.** 4,739,062   **7.** 397,561,342   **8.** 458,906,723

9. Write the word form of the number that is 10,000,000 greater than 6,021,849.

Set A (pp. 18–19)

Compare. Write <, >, or = for each ●.

1. 2,310 ● 2,340
2. 25,050 ● 23,050
3. 22,790 ● 22,790

4. 130,290 ● 130,230
5. 365,280 ● 361,792
6. 12,941 ● 115,226

7. 47,569 ● 47,650
8. 594,031 ● 594,010
9. 731,598 ● 703,892

10. Casey has 3,500 stickers, Joanie has 3,573 stickers, and Emil has 3,432 stickers. Who has the most stickers?

USE DATA For 11–12, use the table.

11. Which plane traveled the greatest distance?

12. A plane called Streak flew a distance of 6,156 miles. Is this greater than or less than Silver's distance?

FLIGHT DISTANCES	
Name of Plane	**Distance Traveled (in miles)**
Flying Eagle	5,987
Max 7	6,251
Silver	6,076
Flight Machine	5,872

Set B (pp. 20–23)

Write the numbers in order from _least_ to _greatest_.

1. 13,069; 13,960; 13,609
2. 76,214; 74,612; 76,421

3. 160,502; 160,402; 163,500
4. 7,450,343; 7,429,203; 7,492,393

Write the numbers in order from _greatest_ to _least_.

5. 37,456; 34,567; 37,654
6. 49,325; 49,852; 49,538

7. 560,898; 560,908; 560,890
8. 2,864,305; 2,648,509; 2,986,413

Set C (pp. 26–29)

Round each number to the place value of the blue digit.

1. 512,399
2. 238,299
3. 942,310

4. 251,003
5. 499,210
6. 9,449,390

7. 3,215,007
8. 1,924,308
9. 4,957,021

10. Which number rounds to three million, nine hundred ninety thousand: 3,985,762 or 3,958,617?

Set A (pp. 34–35)

Estimate the sum or difference.

1. $16,453$
   $-11,019$

2. $678,401$
   $+259,345$

3. $27,645$
   $-10,067$

4. $85,103$
   $-76,495$

5. $405,501$
   $+570,201$

6. $854,633$
   $+941,053$

Set B (pp. 36–39)

Add or subtract mentally. Write the strategy you used.

1. $98 + 46$

2. $109 - 94$

3. $246 + 176$

4. $64 - 23$

5. $346 + 79$

6. $169 - 104$

Set C (pp. 40–41)

Find the sum or difference. Estimate to check.

1. $7,625$
   $+4,092$

2. $9,004$
   $-5,762$

3. $2,435$
   $+2,576$

4. $8,941 - 6,701$

5. $1,241 + 1,367 + 495$

6. $9,204 - 4,352$

Set D (pp. 42–43)

Find the difference. Estimate to check.

1. $8,000$
   $-4,934$

2. $6,001$
   -638

3. $9,040$
   $-3,401$

4. $2,000$
   $-1,764$

5. $1,600$
   $-1,041$

6. $4,006$
   $-1,706$

7. $3,900$
   $-1,461$

8. $7,060$
   $-1,515$

Set E (pp. 44–47)

Find the sum or difference. Estimate to check.

1. $311,420$
   $+592,780$

2. $496,140$
   $+328,628$

3. $406,709$
   $-123,456$

4. $950,275$
   $-450,550$

5. $37,810$
   $26,510$
   $+34,670$

6. $78,540$
   $+31,204$

7. $95,384$
   $-59,412$

8. $65,041$
   $-20,159$

Set A (pp. 54–55)

Tell what you would do first. Then find the value of each expression.

1. $(186 + 45) - 10$ 2. $(96 - 47) + 3$ 3. $76 + (34 - 10)$ 4. $100 - (75 + 2)$

5. $467 - (103 + 246)$ 6. $(4,765 - 3,041) + 34$ 7. $6,741 - (2,105 + 1,231)$

Set B (pp. 56–57)

Choose the expression that shows the given value.
Write a or b.

1. 33

   a. $90 - (22 + 35)$

   b. $(90 - 22) + 35$

2. 376

   a. $1,653 - (75 + 1,202)$

   b. $(1,653 - 75) + 1,202$

3. 615

   a. $651 - (90 + 54)$

   b. $(651 - 90) + 54$

Set C (pp. 58–59, 60–63)

1. There are 10 people in a group. 7 people leave and 3 new people join. Write an expression. How many people are left?

2. There are 17 book bags in a room. Some bags are picked up and now 7 are left. Write an equation. Choose a variable for the unknown.

Set D (pp. 64–65)

Find the rule. Write the rule as an equation.

1.

INPUT	OUTPUT
r	s
12	9
20	17
31	28
45	42

2.

INPUT	OUTPUT
w	v
18	25
42	49
65	72
86	93

3.

INPUT	OUTPUT
x	y
54	66
41	53
29	41
15	27

4.

INPUT	OUTPUT
d	f
87	73
63	49
45	31
27	13

Set E (pp. 66–69)

Complete to make the equation true.

1. 2 dimes + 2 nickels + 3 pennies = 3 dimes + _?_

2. 1 quarter + 4 dimes = 2 quarters + 1 dime + _?_

3. $5 + 7 = 7 + \blacksquare$

4. $26 + \blacksquare = 15 + 15$

Set A (pp. 82–85)

Marta is bringing juice for the class party. She took a survey to find the favorite juice of the class. The table shows the results of her survey.

FAVORITE JUICE		
Juice	Frequency	Cumulative Frequency
Apple	8	8
Orange	9	17
Grapefruit	4	21
Grape	7	28

1. What type of juice did the greatest number of students choose as the favorite?

2. How many students were surveyed?

3. How many more students chose apple juice than grapefruit juice?

Set B (pp. 88–89)

Use the line plot.

1. How many total prizes were won?

2. Find the range, the mode(s), and the median of the data.

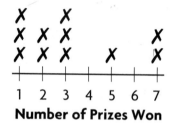

Number of Prizes Won

Set C (pp. 90–91)

Use the stem-and-leaf plot.

1. The plot shows the students' spelling test scores in Mr. Berry's class. Which digits are stems?

2. What is the mode? What is the median?

3. Find the lowest score and the highest score. What is the range?

Spelling Test Scores

Stem	Leaves				
7	1	4	4	6	6
8	2	2	3	5	8	8
9	2	2	4	4	4	4

Set D (pp. 92–93)

Use the graph.

1. What is the scale of the graph? What is the interval?

2. How would the bars change if the interval were 10? if it were 3?

3. Which activities received more votes than reading?

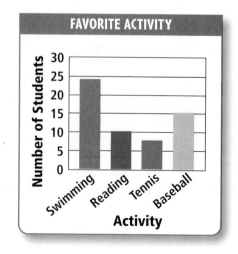

Set A (pp. 102–103)

For 1–6, use the graph.

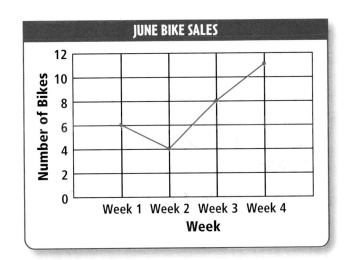

1. During which week were the most bikes sold? the fewest?

2. During which weeks were more than 6 bikes sold?

3. How many bikes were sold during the third week?

4. Between which two weeks was there the greatest increase in the number of bikes sold?

5. If 12 more bikes were sold in July than in June, how many bikes were sold in July?

6. The bike shop's goal is to sell 100 bikes during June, July, and August. How many bikes does it need to sell in August?

Set B (pp. 106–109)

For 1–4, write the kind of graph or plot you would choose.

1. to show your weight each year since birth

2. to compare boys' and girls' favorite ice cream flavors

3. to show this week's science test scores of your classmates

4. to show the kinds of movies watched by your classmates

5. Mr. Cason surveyed his students to find out their favorite activities. He organized the data in a table. What is the best graph or plot to show the data? Make the graph or plot.

6. Mrs. Varga recorded the attendance of the Parkview bowling league for 5 weeks. She organized the data in a table. What is the best graph or plot to show the data? Make the graph or plot.

FAVORITE ACTIVITIES

	Board Game	Playground	Movie	Puzzle
Boys	7	9	4	2
Girls	5	7	2	3

LEAGUE ATTENDANCE

Week	1	2	3	4	5
Members	29	21	27	30	24

Set A (pp. 116–117)

Write the time shown on the clock in two different ways.

1.

2.

3.

Write the time as shown on a digital clock.

4. 11 minutes before one   **5.** 4 minutes after eight   **6.** 37 minutes before two

Set B (pp. 118–119)

Write A.M. or P.M.

1. Camera shop opens at 10:00.

2. Lunch is served at 11:30.

3. Band practice starts at 4:00.

Set C (pp. 120–123)

Copy and complete the tables.

	START TIME	END TIME	ELAPSED TIME
1.	9:05 A.M.	11:35 A.M.	■
3.	■	2:23 P.M.	46 minutes
5.	3:35 P.M.	■	5 hr 17 min

	START TIME	END TIME	ELAPSED TIME
2.	■	5:35 P.M.	1 hr 45 min
4.	10:40 A.M.	■	1 hr 50 min
6.	9:20 A.M.	12:35 P.M.	■

Set D (pp. 126–129)

For 1–2, use the calendars.

1. Today is October 19. Janice has 3 weeks to train for the marathon. What is the date of the marathon?

2. A road race was scheduled 16 days after the 1st Thursday in November. What is the date of the road race?

October						
Sun	Mon	Tue	Wed	Thu	Fri	Sat
		1	2	3	4	5
6	7	8	9	10	11	12
13	14	15	16	17	18	19
20	21	22	23	24	25	26
27	28	29	30	31		

November						
Sun	Mon	Tue	Wed	Thu	Fri	Sat
					1	2
3	4	5	6	7	8	9
10	11	12	13	14	15	16
17	18	19	20	21	22	23
24	25	26	27	28	29	30

3. What year was 3 centuries 8 decades before 1999?

Set A (pp. 140–141)

Find the value of the variable. Write a related equation.

1. $15 \div 3 = c$
2. $7 \times 4 = b$
3. $4 \times 5 = k$
4. $27 \div 9 = y$

5. $24 \div 3 = a$
6. $2 \times 6 = n$
7. $8 \times 4 = d$
8. $25 \div 5 = z$

Write the fact family for each set of numbers.

9. 3, 4, 12
10. 4, 6, 24
11. 2, 9, 18
12. 3, 7, 21

Set B (pp. 142–143)

Find the product or quotient.

1. 7×3
2. 8×4
3. 5×6
4. 4×4

5. $5 \div 5$
6. $36 \div 4$
7. $27 \div 3$
8. $12 \div 4$

9. 9×2
10. 5×4
11. 7×5
12. 5×8

13. $20 \div 5$
14. $16 \div 8$
15. $24 \div 4$
16. $45 \div 9$

Set C (pp. 144–147)

1. For 7×6, draw one array to show the expression. Next, use the break-apart strategy and draw two arrays that could be used to solve the problem.

Find the product or quotient. Show the strategy you used.

2. 7×4
3. 8×9
4. 6×6
5. 9×4

6. $55 \div 5$
7. $36 \div 6$
8. $27 \div 9$
9. $60 \div 5$

10. 9×11
11. 12×4
12. 7×7
13. 7×8

14. $54 \div 6$
15. $48 \div 8$
16. $63 \div 9$
17. $64 \div 8$

Set D (pp. 150–151)

Find the product.

1. $(2 \times 4) \times 3$
2. $(4 \times 3) \times 3$
3. $5 \times (2 \times 6)$
4. $(2 \times 3) \times 6$

5. $5 \times (2 \times 2)$
6. $3 \times (3 \times 2)$
7. $(4 \times 2) \times 9$
8. $(9 \times 0) \times 4$

Set A (pp. 158–161)

Find the value of the expression.

1. $(4 \times 5) + 2$
2. $100 - (8 \times 5)$
3. $(56 \div 7) + 20$
4. $13 + (5 \times 6)$
5. $(12 + 8) \div 4$
6. $(27 \div 3) \times 7$
7. $85 - (3 \times 7)$
8. $50 + (72 \div 9)$

Set B (pp. 162–163)

Choose the expression that matches the words. Write *a* or *b*.

1. There were 3 shelves with 7 footballs each. Lin took 4 footballs out to the field.

   a. $3 \times (4 - 7)$ **b.** $(3 \times 7) - 4$

2. The 5 cases of puppy food each contain 10 bags. Kate placed 6 extra bags next to the cases.

   a. $(5 \times 10) + 6$ **b.** $10 \times (5 + 6)$

Set C (pp. 164–167)

Multiply both sides by the given number. Find the new value.

1. $(3 + 1) = (2 \times 2)$; multiply both sides by 5.
2. $(5 + 7) = (4 \times 3)$; multiply both sides by 4.

Set D (pp. 168–169)

Find the value of the expression.

1. $12 \times h$ if $h = 3$
2. $12 \times h$ if $h = 7$
3. $36 \div s$ if $s = 12$
4. $8 \times y$ if $y = 11$
5. $36 \div s$ if $s = 9$

Set E (pp. 170–171)

Write an equation for each. Choose a variable for the unknown. Tell what the variable represents.

1. 4 boxes with 3 games in each box is the total number of games.
2. 25 people in some cars is 5 people in each car.

Set F (pp. 172–173)

Find the rule. Write the rule as an equation.

1.

INPUT	OUTPUT
b	c
18	9
14	7
10	5
6	3

2.

INPUT	OUTPUT
d	f
2	6
4	12
6	18
8	24

3.

INPUT	OUTPUT
w	z
6	54
8	72
10	90
12	108

4.

INPUT	OUTPUT
r	s
33	11
27	9
24	8
21	7

Set A (pp. 186–187)

Use a basic fact and patterns to write each product.

1. a. 3×50
   b. 3×500

2. a. 4×80
   b. 4×800

3. a. 2×900
   b. $2 \times 9,000$

4. a. $5 \times 4,000$
   b. $5 \times 40,000$

5. Ruiz has four $20 bills and three $5 bills. How much money does he have?

Set B (pp. 188–189)

Round one factor. Estimate the product.

1. 728×7

2. 219×8

3. 704×4

4. 237×4

5. 590×5

6. About how many weeks are in 4 years 4 weeks?

Set C (pp. 190–193)

Find the product. Estimate to check.

1. 18×3

2. 27×5

3. 58×4

4. 83×7

5. 64×9

Set D (pp. 196–199)

Find the product. Estimate to check.

1. 403×7

2. 783×2

3. 197×5

4. 816×9

5. Mr. Jones bought 4 boxes of brushes for his students to use. Each box has 125 brushes. How many brushes did he buy?

Set E (pp. 200–203)

Find the product. Estimate to check.

1. $2 \times 4,232$

2. $3 \times \$4.34$

3. $4 \times 3,991$

4. $9 \times \$8.61$

5. $7 \times \$78.07$

6. $5 \times 1,054$

7. $6 \times \$69.03$

8. $8 \times 2,007$

9. Amanda earns $12.75 each week baby-sitting for her cousin. How much does she earn in 4 weeks?

Set A (pp. 210–211)

Use a basic fact and patterns to find the product.

1. 90×500 **2.** $40 \times 8,000$ **3.** 90×30 **4.** 700×200

5. $\begin{array}{r} 80 \\ \times 60 \\ \hline \end{array}$ **6.** $\begin{array}{r} 7,000 \\ \times\ \ 200 \\ \hline \end{array}$ **7.** $\begin{array}{r} 800 \\ \times\ 50 \\ \hline \end{array}$ **8.** $\begin{array}{r} 300 \\ \times\ 10 \\ \hline \end{array}$

Find the value of n.

9. $10,000 \times n = 500,000$ **10.** $7,000 \times 7,000 = n$ **11.** $n \times 50 = 10,000$

12. $900 \times 12,000 = n$ **13.** $n \times 250,000 = 6,250,000$ **14.** $80 \times n = 320,000$

Set B (pp. 212–213)

Find the product.

1. $\begin{array}{r} 11 \\ \times 20 \\ \hline \end{array}$ **2.** $\begin{array}{r} 45 \\ \times 30 \\ \hline \end{array}$ **3.** $\begin{array}{r} 18 \\ \times 40 \\ \hline \end{array}$ **4.** $\begin{array}{r} 25 \\ \times 30 \\ \hline \end{array}$ **5.** $\begin{array}{r} 62 \\ \times 50 \\ \hline \end{array}$

6. $\begin{array}{r} 43 \\ \times 20 \\ \hline \end{array}$ **7.** $\begin{array}{r} 32 \\ \times 60 \\ \hline \end{array}$ **8.** $\begin{array}{r} 17 \\ \times 50 \\ \hline \end{array}$ **9.** $\begin{array}{r} 65 \\ \times 30 \\ \hline \end{array}$ **10.** $\begin{array}{r} 84 \\ \times 40 \\ \hline \end{array}$

11. 73×40 **12.** 21×70 **13.** 19×20 **14.** 55×30 **15.** 47×80

16. 24×20 **17.** 64×40 **18.** 16×30 **19.** 12×80 **20.** 38×60

21. Erin swims 25 laps a day. How many laps is this in 30 days?

Set C (pp. 214–215)

Round each factor. Estimate the product.

1. $\begin{array}{r} 13 \\ \times 22 \\ \hline \end{array}$ **2.** $\begin{array}{r} 479 \\ \times\ 48 \\ \hline \end{array}$ **3.** $\begin{array}{r} \$73 \\ \times\ 18 \\ \hline \end{array}$ **4.** $\begin{array}{r} 529 \\ \times\ 11 \\ \hline \end{array}$ **5.** $\begin{array}{r} 91 \\ \times 27 \\ \hline \end{array}$

6. $\begin{array}{r} 87 \\ \times 61 \\ \hline \end{array}$ **7.** $\begin{array}{r} 52 \\ \times 48 \\ \hline \end{array}$ **8.** $\begin{array}{r} 734 \\ \times\ 52 \\ \hline \end{array}$ **9.** $\begin{array}{r} 618 \\ \times\ 37 \\ \hline \end{array}$ **10.** $\begin{array}{r} 129 \\ \times\ 23 \\ \hline \end{array}$

11. 32×79 **12.** 43×57 **13.** 89×531 **14.** 93×271 **15.** 32×197

16. 18×39 **17.** 58×129 **18.** 72×489 **19.** $21 \times \$625$ **20.** 42×385

Set A (pp. 224–227)

Choose a method to find the product. Estimate to check.

1. 71
    ×64

2. $29
    × 23

3. 92
    ×33

4. 89
    ×16

5. 46
    ×96

6. 14 × 36   7. 22 × 81   8. $45 × 15   9. 93 × 39   10. 45 × 67

11. 13 × 19   12. 79 × 99   13. 81 × 74   14. 24 × 25   15. 79 × 35

Set B (pp. 228–229)

Find the product. Estimate to check.

1. $226
    × 65

2. 547
    × 53

3. 924
    × 38

4. 839
    × 22

5. 409
    × 28

6. 91 × 682   7. 59 × $474   8. 47 × 567   9. 38 × 106   10. 35 × 750

11. 56 × 590   12. 23 × 492   13. 11 × 711   14. 72 × 219   15. 79 × 35

Set C (pp. 230–231)

Find the product. Estimate to check.

1. 325
    ×   4

2. 9,234
    ×   9

3. 805
    × 66

4. $3,782
    ×   21

5. 7,760
    ×   48

6. 76 × 6,432   7. 4 × 1,210   8. 26 × 7,031   9. 37 × 108   10. 18 × 5,160

11. 56 × $3,804   12. 33 × 9,968   13. 8 × 2,002   14. 78 × 9,542   15. 26 × $4,853

Set D (pp. 232–233)

Find the product. Estimate to check.

1. 19
    ×66

2. 74
    ×76

3. 805
    × 45

4. 1,249
    ×   64

5. $3.95
    ×   25

6. $57
    × 20

7. 8,019
    ×   50

8. $454
    ×   18

9. $7.68
    ×   90

10. $56.10
     ×   32

11. 68 × 79   12. 57 × $9.80   13. 13 × 492   14. 44 × 3,013   15. 50 × 1,008

Set A (pp. 246–247)

Divide. You may wish to use counters.

1. $28 \div 4$ 2. $84 \div 9$ 3. $37 \div 5$ 4. $39 \div 5$

5. $3\overline{)29}$ 6. $4\overline{)19}$ 7. $5\overline{)42}$ 8. $6\overline{)59}$

Set B (pp. 250–253)

Divide and check.

1. $5\overline{)84}$ 2. $6\overline{)68}$ 3. $2\overline{)47}$ 4. $4\overline{)75}$ 5. $3\overline{)31}$

6. $95 \div 3$ 7. $69 \div 4$ 8. $77 \div 6$ 9. $58 \div 5$ 10. $76 \div 9$

Write the check step for each division problem.

11. $73 \div 9 = 8$ r1   12. $64 \div 5 = 12$ r4   13. $99 \div 6 = 16$ r3   14. $87 \div 7 = 12$ r3

15. Patty brought 30 cookies to a party. Each of 7 children took the same number of cookies. How many cookies were left over?

Set C (pp. 256–257)

Use a basic division fact and patterns to write each quotient.

1. $480 \div 6$ 2. $210 \div 7$ 3. $320 \div 4$ 4. $360 \div 6$

   $4{,}800 \div 6$ $2{,}100 \div 7$ $3{,}200 \div 4$ $3{,}600 \div 6$

Write the value of n.

5. $36{,}000 \div 6 = n$ 6. $n \div 6 = 200$ 7. $2{,}800 \div n = 40$ 8. $2{,}400 \div 8 = n$

9. $4{,}200 \div 7 = n$ 10. $n \div 3 = 900$ 11. $560 \div 8 = n$ 12. $n \div 7 = 900$

Set D (pp. 258–261)

Estimate by using compatible numbers.

1. $481 \div 9$ 2. $516 \div 7$ 3. $298 \div 6$ 4. $359 \div 5$

5. $602 \div 8$ 6. $183 \div 5$ 7. $282 \div 4$ 8. $709 \div 8$

9. $341 \div 6$ 10. $675 \div 9$ 11. $193 \div 3$ 12. $244 \div 9$

13. $207 \div 8$ 14. $413 \div 7$ 15. $312 \div 4$ 16. $429 \div 5$

Set A (pp. 264–265)

Tell where to place the first digit. Then divide.

1. $3\overline{)242}$ 2. $8\overline{)679}$ 3. $2\overline{)469}$ 4. $5\overline{)412}$ 5. $7\overline{)887}$

6. $9\overline{)851}$ 7. $3\overline{)297}$ 8. $5\overline{)751}$ 9. $7\overline{)460}$ 10. $2\overline{)263}$

Set B (pp. 266–267)

Divide and check.

1. $4\overline{)672}$ 2. $9\overline{)121}$ 3. $2\overline{)496}$ 4. $5\overline{)453}$ 5. $3\overline{)587}$

6. There are 8 swimmers on a team. The swim club has 152 swimmers. How many teams does the swim club have?

Set C (pp. 268–271)

Write the number of digits in each quotient.

1. $5\overline{)709}$ 2. $8\overline{)650}$ 3. $4\overline{)407}$ 4. $5\overline{)360}$ 5. $7\overline{)620}$

Divide and check.

6. $6\overline{)700}$ 7. $4\overline{)828}$ 8. $3\overline{)921}$ 9. $5\overline{)740}$ 10. $7\overline{)370}$

11. $570 \div 4$ 12. $754 \div 8$ 13. $310 \div 4$ 14. $890 \div 7$ 15. $820 \div 4$

Set D (pp. 272–273)

Divide.

1. $9\overline{)2,899}$ 2. $6\overline{)4,827}$ 3. $6\overline{)\$31.50}$ 4. $8\overline{)7,609}$

5. Rashad has 1,386 baseball cards. If he puts an equal number in each of 7 boxes, how many cards are in each box?

Set E (pp. 276–277)

Find the mean.

1. 65; 73; 126; 81; 60
2. 1,371; 1,656; 1,899
3. $42.48; $35.97; $56.58; $13.89
4. 376; 408; 523; 293
5. 415; 86; 137; 294
6. $13.89; $4.76; $12.52; $8.99

Set A (pp. 282–283)

Write the numbers you would use to estimate the quotient. Then estimate.

1. $832 \div 21$ **2.** $394 \div 19$ **3.** $542 \div 38$ **4.** $407 \div 72$

5. $163 \div 43$ **6.** $269 \div 56$ **7.** $717 \div 65$ **8.** $808 \div 91$

9. Jim wants to make a scrapbook. He can fit 6 baseball cards on each page. He has 123 baseball cards. About how many pages will he need?

10. The 12 members of a scout troop sold 230 buckets of popcorn. About how many buckets of popcorn did each member sell?

Set B (pp. 286–287)

Divide.

1. $12\overline{)645}$ **2.** $24\overline{)721}$ **3.** $18\overline{)212}$ **4.** $34\overline{)310}$

5. $15\overline{)927}$ **6.** $13\overline{)172}$ **7.** $26\overline{)525}$ **8.** $37\overline{)454}$

9. A pen manufacturer packed 718 pens. Each package holds 12 pens. How many packages is this? How many pens are left over?

10. The bicycle club cycled 375 miles in 15 days. If they cycled the same number of miles each day, how many miles did they cycle each day?

Set C (pp. 288–291)

Write *too high, too low,* or *just right* for each estimate. Then divide.

1. $78\overline{)612}^{\;7}$ **2.** $34\overline{)312}^{\;8}$ **3.** $26\overline{)156}^{\;6}$ **4.** $56\overline{)448}^{\;9}$

Divide.

5. $49\overline{)872}$ **6.** $63\overline{)917}$ **7.** $14\overline{)275}$ **8.** $32\overline{)564}$

9. $25\overline{)743}$ **10.** $52\overline{)816}$ **11.** $17\overline{)194}$ **12.** $21\overline{)836}$

13. Lawrence School won 215 trophies. A trophy case holds 22 trophies. How many trophy cases does the school need to buy?

14. The bookstore has 329 new books. Each shelf holds 28 books. How many shelves will be used?

Set A (pp. 298–299)

Use what you know about multiplication. Find as many factors as you can for each product.

1. 15   **2.** 16   **3.** 20   **4.** 44

5. 45   **6.** 46   **7.** 50   **8.** 60

9. 18   **10.** 28   **11.** 32   **12.** 56

Set B (pp. 300–301)

Write an equation for the arrays shown.

1.

2.

3.

4.

5.

6.

Set C (pp. 302–305)

Write *prime* or *composite* for each number.

1. 8   **2.** 25   **3.** 39   **4.** 17

5. 2   **6.** 84   **7.** 31   **8.** 43

9. 72   **10.** 21   **11.** 41   **12.** 16

Set D (pp. 306–307)

Write each as a product of prime factors.

1. 66   **2.** 75   **3.** 80   **4.** 100   **5.** 150

Write the missing factor.

6. $16 = 2 \times 2 \times 2 \times \blacksquare$ **7.** $56 = 2 \times \blacksquare \times 7$ **8.** $63 = 3 \times \blacksquare \times \blacksquare$

Set A (pp. 320–323)

Draw and label an example of each.

1. line segment *DC*   **2.** point *J*   **3.** ray *JK*

4. obtuse angle *RST*   **5.** right angle *B*   **6.** line *GH*

Set B (pp. 324–325)

Name any line relationship you see in each figure.
Write *intersecting, parallel,* **or** *perpendicular lines.*

1.   **2.**    **3.**

Set C (pp. 326–329)

Tell how each figure was moved. Write *slide, flip,* **or** *turn.*

1.   **2.** 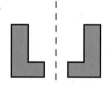   **3.** 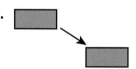   **4.**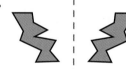

Tell whether the two figures are *congruent, similar,*
or *neither.*

5. 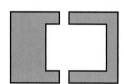   **6.**   **7.**

Set D (pp. 330–331)

Tell whether the figure has *rotational symmetry, line symmetry,*
or *both.*

1.   **2.** 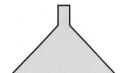   **3.**   **4.**

Set A (pp. 342–343)

For 1–4, use the drawing and a ruler marked in inches.

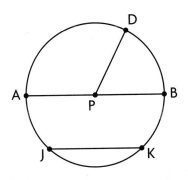

1. The line segment *AB* is a _?_. It measures _?_ inches.

2. The line segment *JK* is a _?_.

3. Three points on the circle are _?_, _?_, and _?_.

4. The line segment *PD* is a _?_. It measures _?_ inch.

5. The radius of one circle is 5 inches longer than the radius of another circle. Together they measure 29 inches. How long is each radius?

Set B (pp. 346–347)

Classify each triangle. Write *isosceles, scalene,* or *equilateral.*

1. 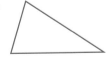   2.   3.   4.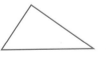

Classify each triangle by the lengths of its sides. Write *isosceles, scalene,* or *equilateral*.

5. 4 feet, 4 feet, 4 feet

6. 8 yards, 10 yards, 7 yards

7. 7 inches, 4 inches, 7 inches

8. I have 3 sides and 3 angles. All 3 of my angles are acute. What figures could I be?

Set C (pp. 348–351)

Classify each figure in as many ways as possible. Write *quadrilateral, parallelogram, rhombus, square, rectangle,* or *trapezoid*.

1.   2.   3.    4.

5. Elena cut a figure from paper. It has 4 sides, 2 of which are parallel. There are no right angles. What is the figure?

Set A (pp. 364–365)

Write the fraction for the shaded part.

1.

2.

3.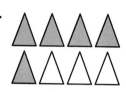

4.

Draw the picture and divide it into the number of parts given.
Then shade one part and write a fraction for the shaded part.

5. 15 circles
   5 parts

6. 18 squares
   2 parts

7. 24 stars
   6 parts

Set B (pp. 368–371)

Use fraction bars to write each fraction in simplest form.

1. $\frac{2}{4}$ 2. $\frac{4}{4}$ 3. $\frac{6}{10}$ 4. $\frac{4}{12}$ 5. $\frac{6}{9}$

Is the fraction in simplest form? Write *yes* or *no*.

6. $\frac{1}{5}$ 7. $\frac{2}{12}$ 8. $\frac{2}{5}$ 9. $\frac{6}{8}$ 10. $\frac{5}{7}$

Set C (pp. 372–375)

Order the fractions from *greatest* to *least*.

1. $\frac{2}{3}, \frac{2}{5}, \frac{1}{2}$ 2. $\frac{3}{4}, \frac{1}{2}, \frac{2}{3}$ 3. $\frac{1}{6}, \frac{3}{4}, \frac{2}{5}$ 4. $\frac{1}{5}, \frac{1}{3}, \frac{1}{2}$

5. $\frac{4}{12}, \frac{1}{10}, \frac{3}{5}$ 6. $\frac{5}{6}, \frac{1}{2}, \frac{2}{3}$ 7. $\frac{1}{8}, \frac{3}{5}, \frac{1}{2}$ 8. $\frac{3}{8}, \frac{1}{4}, \frac{3}{4}$

Set D (pp. 378–381)

Write a mixed number for each picture.

1.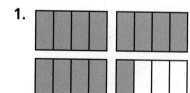

2.

3.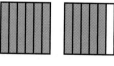

Rename each fraction as a mixed number.

4. $\frac{7}{2}$ 5. $\frac{13}{8}$ 6. $\frac{11}{5}$ 7. $\frac{14}{6}$

Set A (pp. 386–389)

Find the sum.

1. $\frac{1}{8} + \frac{5}{8}$

2. $\frac{2}{6} + \frac{1}{6}$

3. $\frac{4}{12} + \frac{5}{12}$

4. $\frac{3}{10} + \frac{4}{10}$

5. $\frac{1}{4} + \frac{2}{4}$

6. $\frac{2}{7} + \frac{3}{7}$

7. $\frac{2}{10} + \frac{5}{10}$

8. $\frac{4}{10} + \frac{6}{10}$

9. $\frac{4}{8} + \frac{2}{8}$

10. $\frac{7}{12} + \frac{5}{12}$

11. $\frac{4}{5} + \frac{3}{5}$

12. $\frac{8}{12} + \frac{2}{12}$

13. $\frac{4}{5} + \frac{2}{5}$

14. $\frac{6}{7} + \frac{1}{7}$

15. $\frac{5}{8} + \frac{1}{8}$

16. $\frac{3}{4} + \frac{3}{4}$

17. $\frac{3}{10} + \frac{2}{10}$

18. $\frac{5}{12} + \frac{8}{12}$

19. $\frac{5}{6} + \frac{5}{6}$

20. $\frac{5}{7} + \frac{3}{7}$

21. There was $\frac{2}{3}$ cup of milk left in the refrigerator after Robbie used $\frac{1}{3}$ cup of milk in his oatmeal. How much milk was in the refrigerator before Robbie ate his oatmeal?

22. On Monday $\frac{3}{10}$ inch of snow fell in Leigh's yard. On Tuesday $\frac{7}{10}$ inch of snow fell. How much snow fell in all?

Set B (pp. 392–395)

Find the sum or difference.

1. $4\frac{2}{5}$
$+3\frac{1}{5}$

2. $3\frac{2}{4}$
$+3\frac{3}{4}$

3. $2\frac{1}{3}$
$+4\frac{1}{3}$

4. $3\frac{5}{9}$
$+4\frac{4}{9}$

5. $3\frac{2}{10}$
$+6\frac{3}{10}$

6. $8\frac{1}{2}$
$+7\frac{1}{2}$

7. $5\frac{2}{3}$
$-4\frac{1}{3}$

8. $8\frac{8}{12}$
$-5\frac{3}{12}$

9. $5\frac{4}{6}$
$-1\frac{3}{6}$

10. $7\frac{4}{5}$
$-3\frac{2}{5}$

11. $4\frac{5}{10}$
$-3\frac{2}{10}$

12. $3\frac{3}{4}$
$-1\frac{1}{4}$

13. Joseph fed his dogs. Brewster got $\frac{1}{4}$ cup of dry dog food and Beatrice got $\frac{1}{4}$ cup of dry dog food. How much dry dog food was used altogether?

14. Sam's book report is $3\frac{4}{8}$ pages long. Mary's book report is $2\frac{2}{8}$ pages long. Whose report is longer? How many pages longer?

Set A (pp. 406–409)

Write the decimal and fraction shown by each model or number line.

1.   **2.**   **3.**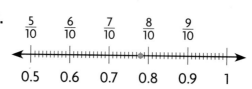

Write each fraction as a decimal.

4. $\frac{2}{10}$ **5.** $\frac{33}{100}$ **6.** $\frac{8}{100}$ **7.** $\frac{9}{10}$ **8.** $\frac{84}{100}$

9. Patty has $\frac{3}{4}$ of a dollar. How much money does she have?

Set B (pp. 410–411)

Write each decimal as a fraction.

1. 0.372   **2.** 0.025   **3.** 0.009   **4.** 0.461   **5.** 0.073

Write each decimal in two other forms.

6. 0.548   **7.** 0.029   **8.** 0.004   **9.** 0.473   **10.** 0.081

11. eighty-three thousandths **12.** $0.7 + 0.01 + 0.006$ **13.** 8 thousandths

Set C (pp. 414–417)

Write an equivalent mixed number or decimal.

1. 6.5   **2.** $7\frac{1}{4}$ **3.** $\frac{6}{4}$ **4.** 3.75   **5.** 8.25

6. 9.75   **7.** $5\frac{1}{2}$ **8.** $\frac{8}{3}$ **9.** $\frac{61}{7}$ **10.** $\frac{73}{9}$

Set D (pp. 418–421)

Compare. Write <, >, or = for each ●.

1. 0.30 ● 0.3   **2.** 5.67 ● 6.75   **3.** 3.60 ● 3.06   **4.** 1.2 ● 1.20

Order the decimals from *least* to *greatest.*

5. 0.19; 0.21; 0.91; 0.12; 1.69   **6.** 4.35; 3.45; 5.43; 4.53; 3.54

Set A (pp. 428–429)

Round each number to the place of the blue digit.

1. 9.301   2. $5.79   3. $8.65   4. 4.531   5. $3.49

6. $8.93   7. 4.57   8. 16.895   9. 39.472   10. $7.65

Set B (pp. 430–431)

Estimate the sum or difference.

1. $2.80
   +$2.30

2. 2.356
   +1.192

3. 2.356
   −0.918

4. $23.47
   −$14.96

5. 18.92
   +39.45

Set C (pp. 432–433)

Find the sum. Estimate to check.

1. 10.29
   +33.46

2. $15.98
   +$12.04

3. 6.419
   +7.234

4. 7.98
   +2.31

5. 6.29
   +8.88

6. 0.31 + 4.57   7. 4.875 + 8.136   8. 7.5 + 8.3   9. 5.76 + 2.18

10. Bob went to the store. He bought oranges for $1.32, juice for $1.48, and eggs for $1.10. How much money did he spend?

Set D (pp. 434–435)

Find the difference. Estimate to check.

1. 2.56
   −1.38

2. 13.287
   − 3.534

3. $11.27
   −$ 7.55

4. $75.43
   −$18.63

5. 67.30
   −31.72

6. $0.78 − $0.51   7. 5.94 − 5.76   8. $9.49 − $6.23   9. $7.65 − $1.85

Set E (pp. 436–439)

Find the sum or difference. Estimate to check.

1. $13.95
   +$21.76

2. 22.49
   +14.3

3. $26.00
   −$18.94

4. 74.3
   − 6.794

5. 14.28
   + 8.5

6. 16.3 + 1.094   7. $8 − $3.78   8. 3.7 − 1.99   9. 46 − 3.059

Set A (pp. 452–453)

Choose the most reasonable unit of measure.
Write *in., ft, yd,* or *mi*.

1. The length of my foot is about 8 _?_.

2. The height of my dog is about 3 _?_.

3. The width of my room is 4 _?_.

Name the greater measurement.

4. 15 ft or 15 yd

5. 300 mi or 300 ft

6. 39 in. or 39 yd

7. 72 in. or 72 ft

8. 4 ft or 4 mi

9. 93 yd or 93 ft

Set B (pp. 454–457)

Estimate to the nearest inch. Then measure to the nearest $\frac{1}{4}$ inch.

1.

2.

Estimate to the nearest inch. Then measure to the nearest $\frac{1}{8}$ inch.

3.

4.

Order the measurements from *least* to *greatest*.

5. $7\frac{1}{2}$ in., $7\frac{5}{8}$ in., $7\frac{1}{4}$ in.

6. $4\frac{1}{8}$ in., $4\frac{3}{4}$ in., $3\frac{7}{8}$ in.

7. $2\frac{1}{2}$ in., $2\frac{3}{8}$ in., $2\frac{5}{8}$ in.

Set C (pp. 458–459)

Complete. Tell whether you multiply or divide.

1. 60 in. = ■ ft

2. 12 yd = ■ in.

3. 5 mi = ■ yd

4. 900 ft = ■ yd

Compare. Write <, >, or = for each ●.

5. 7 ft ● 80 in.

6. 4 mi ● 2,200 ft

7. 108 in. ● 2 yd

8. 36 in. ● 1 yd

9. 210 ft ● 75 yd

10. 3 ft ● 36 in.

11. 144 in. ● 3 yd

12. 3,520 yd ● 3 mi

13. 20 ft ● 60 yd

Set A (pp. 470–473)

**Choose the most reasonable unit of measure.
Write *cm, dm, m,* or *km*.**

1. The width of a book is 22 _?_ .

2. The length of a marker is 13 _?_ .

3. The height of a child's bike is 1 _?_ .

4. The height of a desk is 10 _?_ .

5. distance from Chicago to Las Vegas

6. width of a CD

7. length of a calculator

8. length of a peanut

9. length of a shoe

10. height of a building

Compare. Write < or > for each ●.

11. 750 m ● 75 dm

12. 20 cm ● 20 dm

13. 2 cm ● 2 m

14. 70 dm ● 70 m

15. 500 m ● 500 km

16. 200 m ● 200 cm

Set B (pp. 474–475)

Complete.

1. 3 m = ▦ dm

2. 7 m = ▦ cm

3. 5 dm = ▦ cm

4. 150 cm = ▦ dm

5. 28 dm = ▦ cm

6. 3 km = ▦ m

Write the correct unit.

7. 5 _?_ = 50 cm

8. 40 dm = 4 _?_

9. 3 _?_ = 300 cm

10. 45 _?_ = 4,500 cm

11. 7 m = 70 _?_

12. 500 cm = 5 _?_

Compare. Write <, >, or = for each ●.

13. 45 cm ● 4 dm

14. 60 cm ● 3 dm

15. 30 dm ● 3 m

16. 72 m ● 95 cm

17. 9,000 cm ● 6,400 dm

18. 70 km ● 700 m

19. 51 dm ● 15 m

20. 8,000 m ● 8 km

21. 18 m ● 180 cm

22. Four people rode their bikes 3,000 meters each in a relay race. John rode his bike 15 kilometers. Who rode farther, the whole relay team or John? How much farther?

23. Josh ran 1,600 meters on Saturday and 3,200 meters on Sunday. How many meters did he run in all? Is that more than or less than 5 kilometers?

Set A (pp. 486–491)

Find the perimeter.

1.

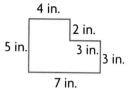

4 in.
2 in.
5 in.
3 in.
3 in.
7 in.

2.

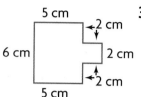

5 cm
2 cm
6 cm
2 cm
2 cm
5 cm

3.

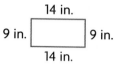

14 in.
9 in.
9 in.
14 in.

4.

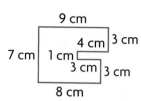

9 cm
4 cm
3 cm
7 cm
1 cm
3 cm
3 cm
8 cm

Set B (pp. 492–495)

Find the area.

1.

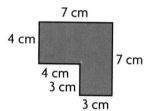

7 cm
4 cm
7 cm
4 cm
3 cm
3 cm

2.

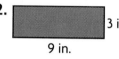

3 in.
9 in.

3.

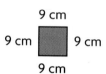

9 cm
9 cm
9 cm
9 cm

4.

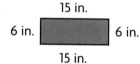

15 in.
6 in.
6 in.
15 in.

5. Mandy's class is painting a mural on a wall. The wall is 6 feet high and 4 feet wide. What is the area of the wall?

Set C (pp. 496–497)

Find the area and perimeter of each figure. Then draw another figure that has the same perimeter but a different area.

1.

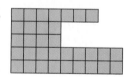

2.

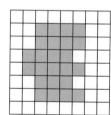

3.

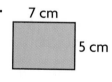

7 cm
5 cm

4.

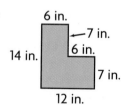

6 in.
7 in.
6 in.
14 in.
7 in.
12 in.

5. The floor of Ben's tree house is a square measuring 6 feet on each side. What are the area and perimeter of the floor?

Set D (pp. 498–499)

Complete for each rectangle.

1. Area = 121 sq in.
Length = 11 in.
Width = ▇ in.

2. Area = 96 sq ft
Width = 4 ft
Length = ▇ ft

3. Area = 72 sq mi
Length = 9 mi
Width = ▇ mi

Set A (pp. 508–511)

Which solid figure do you see in each?

1.

2.

3.

4.

Write the names of the plane figures that are the faces of each solid figure.

5.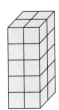

6.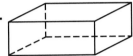

7.

8.

Set B (pp. 512–513)

Write the letter of the figure that is made with each net.

1.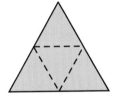

2.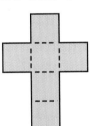

3.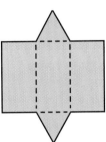

4.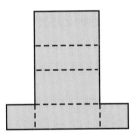

a.

b.

c.

d.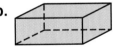

5. Draw a net that when cut and folded will form a square pyramid.

Set C (pp. 514–515)

Find the volume.

1.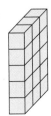

2.

3.

4.

Set A (pp. 530–531)

For 1–6, find the number of possible outcomes by making a tree diagram.

1. Decorations
   Type: balloons, streamers
   Color: blue, gold, silver, white, purple

2. Activities
   Sport: soccer, tennis, football
   Day: Monday, Tuesday, Wednesday, Thursday

3. Outfits
   Tops: sweater, blouse, T-shirt
   Bottoms: jeans, skirt, shorts

4. Lunch
   Main: meatloaf, sandwich, taco
   Side: salad, fruit

5. Sundaes
   Ice cream: chocolate, vanilla, strawberry, coconut
   Toppings: hot fudge, caramel, chocolate syrup, nuts

6. Pizzas
   Crust: thin, Sicilian, hand-tossed, garlic-flavored
   Toppings: beef, mushrooms, chicken

7. The Smiths are taking a vacation. They can go to the beach, mountains, or lake. They can go in June, July, or August. How many choices do they have?

8. Sam can buy either milk or orange juice. Each comes in a pint, quart, or gallon. How many choices does Sam have?

Set B (pp. 534–537)

For 1–3, look at the bag of marbles and the spinner at the right.

1. Which color are you unlikely to pull from the bag?

2. Is it likely or unlikely the pointer will land on blue?

3. Which two colors are you equally likely to spin?

Write *likely, unlikely,* or *equally likely* for the events.

4. tossing a 4; tossing a 6; cube labeled 1 to 6

5. spinning an odd number on a 6-part spinner labeled 1, 3, 6, 9, 12, 15

6. pulling a quarter from a bag of 10 quarters and 4 dimes

7. pulling a white sock from a drawer with 10 black socks and 2 white socks

8. tossing an even number; tossing an odd number; cube labeled 1, 2, 3, 4, 5, 6

9. spinning a number greater than 7 on a 6-part spinner labeled 2, 4, 5, 6, 8, 10

Set A (pp. 542–543)

Look at the spinner. Find the probability of each event.

1. yellow

2. an odd number

3. 6

4. an even number

5. yellow or even

6. a number less than 6

One tile is drawn from the bag of tiles.
Find the probability of each event.

7. red

8. green or blue

9. green or not green

10. A spinner has 8 equal sections. Two are purple, $\frac{1}{2}$ are green, and the remaining sections are blue. What is the probability of getting blue?

11. On a cube, two sides have an 8, $\frac{1}{3}$ of the sides have a 5, $\frac{1}{6}$ have a 3, and $\frac{1}{6}$ have a 4. What is the probability of tossing a 5?

Set B (pp. 546–547)

Look at the spinner. Each player scores 1 point when the pointer stops on his or her choice. Write *yes* or *no* to tell if each game is fair. Explain.

1. Player A: odd number
Player B: number that can be divided evenly by 3

2. Player A: even number
Player B: number between 14 and 30

3. Player A: number less than 20
Player B: number greater than 20

4. Player A: prime number
Player B: number that can be divided evenly by 4

In a game with these cards, Player 1 scores a point when a prime number is pulled. Player 2 scores a point when a composite number is pulled.

1	5	9	13
15	19	23	27

5. What is the probability of pulling a prime number?

6. Is the game fair? Explain.

7. In a game, players take turns tossing a cube labeled 1 to 6. Player A scores a point when an even number is tossed. Player B scores a point when an odd number is tossed. Is the game fair? Explain.

Set A (pp. 554–555)

Use the thermometer to find the temperature for each letter.

1. A   **2.** B   **3.** C

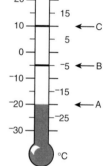

Use the thermometer to find the change in temperature.

4. ⁻20°F to 15°F   **5.** ⁻8°F to ⁻22°F

Choose the temperature that is a better estimate.

6. ice cream   **7.** glass of orange juice   **8.** banana
31°F or 55°F   45°F or 100°F   34°F or 68°F

Set B (pp. 556–557)

Use the thermometer to find the temperature for each letter.

1. A   **2.** B   **3.** C

Use the thermometer to find the change in temperature.

4. ⁻20°C to 5°C   **5.** ⁻11°C to ⁻26°C

Choose the temperature that is a better estimate.

6. yogurt   **7.** bath water   **8.** Florida in the summer
4°C or 35°C   10°C or 37°C   10°C or 34°C

Set C (pp. 558–561)

Use the number line to name the number for each letter.

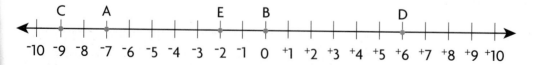

1. A   **2.** B   **3.** C   **4.** D   **5.** E

Compare. Write <, >, or = for each ⬤.

6. 5 ⬤ ⁺5   **7.** 7 ⬤ ⁻9   **8.** ⁻3 ⬤ 0   **9.** ⁻2 ⬤ ⁻7

10. ⁻39 ⬤ ⁺28   **11.** ⁺14 ⬤ ⁻17   **12.** ⁺42 ⬤ ⁻24   **13.** 56 ⬤ ⁺56

Set A (pp. 568–569)

Plot each ordered pair on a coordinate grid.

1. (3,3) **2.** (8,0) **3.** (4,2) **4.** (1,7)

Write the ordered pair for the point on the coordinate grid.

5. point A **6.** point B

7. point C **8.** point D

9. point E **10.** point F

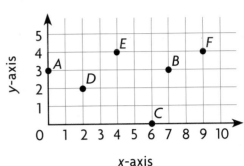

Set B (pp. 570–573)

Use the equation to complete each function table.

1. $y = (x \div 2) + 1$

INPUT	x	4	6	8	10
OUTPUT	y				

2. $y = (x - 2) + 4$

INPUT	x	5	9	15	18
OUTPUT	y				

Do the values given make $y = (x + 1) \div 3$ true? Write *yes* or *no*.

3. (2,1) **4.** (5,3) **5.** (11,4)

Do the values make $y = (5 + x) \times 2$ true? Write *yes* or *no*.

6. (3,16) **7.** (6,22) **8.** (7,18) **9.** (2,8)

10. (5,15) **11.** (4,9) **12.** (0,10) **13.** (1,12)

Set C (pp. 574–577)

For 1–3, use the equation $y = x - 2$.

1. Complete the function table.

2. Write the input/output values as ordered pairs (x,y).

INPUT	x	2	3	4	5	6	7	8	9	10	11
OUTPUT	y										

3. Graph the ordered pairs on a coordinate grid.

Tips for Taking Math Tests

Being a good test-taker is like being a good problem solver. When you answer test questions, you are solving problems. Remember to **UNDERSTAND**, PLAN, **SOLVE**, and **CHECK**.

UNDERSTAND

Read the problem.

- Look for math terms and recall their meanings.
- Reread the problem and think about the question.
- Use the details in the problem and the question.

1. Twenty students signed up for a new club. Six more girls than boys signed up. How many boys signed up?

   A 16 boys   **C** 7 boys

   B 13 boys   **D** 4 boys

TIP! Understand the problem.

Six more girls than boys is the difference between the number of girls and boys. So find two numbers with a difference of 6 whose sum is 20. The number of boys will be the lesser number. The answer is **C**.

- Each word is important. Missing a word or reading it incorrectly could cause you to get the wrong answer.
- Pay attention to words that are in **bold** type, all CAPITAL letters, or *italics* and words like *round, best,* or *least to greatest.*

2. Skates cost $219. Kent rounded the price to the nearest hundred dollars. Abby rounded to the nearest ten dollars. Which statement is true?

   F Kent's amount and Abby's amount are the same.

   G Kent's amount is $20 less than Abby's amount.

   H Kent's amount is $10 less than Abby's amount.

   J Kent's amount is $10 more than Abby's amount.

TIP! Look for important words.

The words *rounded* and *true* are important words. Kent and Abby each rounded the price to a different place value. Round the price as Kent and Abby did. Then compare your rounded amounts to each answer choice to determine which one is true. The answer is **G**.

Think about how you can solve the problem.

- Can you solve the problem with the information given?
- Pictures, charts, tables, and graphs may have the information you need.
- You may need to recall information not given.
- The answer choices may have information you need.

3. Trains run from New York to Washington, D.C., every 30 minutes. Which train takes the least amount of time to reach Washington, D.C.?

Train Number	Leaves New York	Arrives in Washington
15	9:00 A.M.	11:54 A.M.
26	9:30 A.M.	12:30 P.M.
86	10:30 A.M.	1:03 P.M.
127	10:00 A.M.	12:44 P.M.

A Train 15   **C** Train 86

B Train 26   **D** Train 127

TIP! Get the information you need.
Use the schedule to find how much time it takes each train to reach Washington, D.C. Then compare the times to find the one that is less than the others. The answer is **C.**

- You may need to write a number sentence and solve it.
- Some problems have two steps or more.
- You may need to look at relationships rather than compute.
- If the path to the solution isn't clear, choose a problem solving strategy and use it to solve the problem.

4. Roberto has $38, which is $2 more than twice the amount Janet has. How much money does Janet have?

F $14   **H** $40

G $18   **J** $78

TIP! Decide on a plan.
Use the strategy *work backward.* Start with Roberto's $38. When you work backward, each operation will be opposite to what is in the problem. *$2 more* means add $2, so you would subtract $2. *Twice the amount* means multiply by 2, so you would divide by 2. The answer is **G.**

Follow your plan, working logically and carefully.

- Estimate your answer. Are any answer choices unreasonable?

- Use reasoning to find the most likely choices.

- Make sure you solved all steps needed to answer the problem.

- If your answer does not match any of the answer choices, check the numbers you used. Then check your computation.

5.

Which figure has the same area as the one above but a different perimeter?

A

B

C

D

TIP! Eliminate choices.
You can eliminate choices B and C because they do not have the same area. Only answer choices A and D have an area of 8. Since D is congruent to the figure but in a different position, its perimeter will be the same. The answer is **A**.

- If your answer still does not match, look for another form of the number such as a decimal instead of a fraction.

- If answer choices are given as pictures, look at each one by itself while you cover the other three.

- If the answer choices include NOT HERE and your answer is not given, make sure your work is correct and then mark NOT HERE.

- Read answer choices that are statements and relate them to the problem one by one.

- Change your plan if it isn't working. Try a different strategy.

6. Which statement is true?

   F All circles are congruent.

   G All squares are similar.

   H All rectangles are similar.

   J All squares are congruent.

TIP! Choose the answer.
Read each statement to decide if it is true. If you aren't sure which is true, think about the properties of circles, squares, and rectangles. The answer is **G**.

Take time to catch your mistakes.

- Be sure you answered the question asked.
- Check that your answer fits the information in the problem.
- Check for important words you might have missed.
- Be sure you used all the information you needed.
- Check your computation by using a different method.
- Draw a picture when you are unsure of your answer.

7. What number is inside the triangle, inside the square, and is an even number?

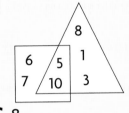

A 5

B 6

C 8

D 10

TIP! **Check your work.**
Look at your answer choice. Does it match all the descriptions given in the problem? If not, look for important words you might have missed. The answer is **D**.

Don't Forget!

Before the test

- Listen to the teacher's directions and read the instructions.
- Write down the ending time if the test is timed.
- Know where and how to mark your answers.
- Know whether you should write on the test page or use scratch paper.
- Ask any questions you may have before the test begins.

During the test

- Work quickly but carefully. If you are unsure how to answer a question, leave it blank and return to it later.
- If you cannot finish on time, look over the questions that are left. Answer the easiest ones first. Then go back to answer the others.
- Fill in each answer space carefully. Erase completely if you change an answer. Erase any stray marks.
- Check that the answer number matches the question number, especially if you skip a question.

ADDITION FACTS TEST

	K	L	M	N	O	P	Q	R
A	6 + 7	9 + 6	3 + 5	8 + 9	0 + 7	2 + 8	6 + 4	7 + 7
B	1 + 6	8 + 4	5 + 1	2 + 7	3 + 3	8 + 2	4 + 5	2 + 6
C	6 + 6	3 + 7	7 + 8	4 + 6	9 + 0	4 + 2	10 + 4	3 + 8
D	6 + 1	5 + 9	10 + 6	5 + 7	3 + 9	9 + 8	8 + 7	8 + 1
E	7 + 6	7 + 1	6 + 9	4 + 3	5 + 5	8 + 0	9 + 5	2 + 9
F	9 + 1	8 + 5	7 + 0	8 + 3	7 + 2	4 + 7	10 + 5	4 + 8
G	5 + 3	9 + 9	3 + 6	7 + 4	0 + 8	4 + 4	7 + 10	6 + 8
H	8 + 6	10 + 7	0 + 9	7 + 9	5 + 6	8 + 10	6 + 5	9 + 4
I	9 + 7	8 + 8	1 + 9	5 + 8	10 + 9	6 + 3	6 + 2	9 + 10
J	9 + 2	7 + 5	6 + 0	10 + 8	5 + 4	4 + 9	9 + 3	10 + 10

	K	L	M	N	O	P	Q	R
A	13 − 4	7 − 1	9 − 7	9 − 9	11 − 5	6 − 3	12 − 7	8 − 5
B	8 − 8	16 − 8	15 − 6	10 − 2	6 − 5	8 − 7	14 − 4	11 − 9
C	9 − 5	12 − 8	15 − 8	11 − 7	14 − 8	18 − 9	15 − 5	8 − 1
D	10 − 4	16 − 10	13 − 9	9 − 1	7 − 2	7 − 0	13 − 8	6 − 4
E	10 − 9	9 − 6	17 − 9	7 − 3	6 − 0	11 − 8	8 − 6	9 − 4
F	8 − 4	13 − 6	11 − 2	15 − 7	19 − 10	12 − 3	17 − 8	7 − 5
G	9 − 8	13 − 7	7 − 4	15 − 9	8 − 2	10 − 6	14 − 7	12 − 5
H	10 − 7	6 − 6	8 − 0	12 − 4	14 − 6	11 − 4	6 − 2	17 − 7
I	13 − 5	12 − 9	16 − 7	7 − 6	10 − 5	11 − 3	12 − 6	14 − 9
J	10 − 8	11 − 6	14 − 5	16 − 9	9 − 3	5 − 4	18 − 10	20 − 10

MULTIPLICATION FACTS TEST

	K	L	M	N	O	P	Q	R
A	5 × 6	5 × 9	7 × 7	9 × 10	7 × 5	12 × 2	10 × 6	6 × 7
B	6 × 6	0 × 6	2 × 7	12 × 8	9 × 2	3 × 5	5 × 8	8 × 3
C	7 × 0	5 × 1	4 × 5	9 × 9	6 × 8	8 × 11	11 × 7	10 × 5
D	1 × 7	9 × 4	0 × 7	2 × 5	9 × 7	10 × 9	3 × 3	12 × 7
E	5 × 7	1 × 9	4 × 3	7 × 6	11 × 3	3 × 8	4 × 2	10 × 10
F	10 × 12	5 × 5	6 × 4	9 × 8	0 × 8	9 × 6	11 × 2	12 × 6
G	5 × 3	4 × 6	6 × 3	7 × 9	12 × 5	0 × 9	5 × 4	12 × 11
H	7 × 1	6 × 9	1 × 6	4 × 4	3 × 7	11 × 11	4 × 8	12 × 9
I	7 × 4	2 × 4	8 × 6	3 × 4	11 × 5	2 × 9	8 × 9	7 × 8
J	8 × 0	3 × 9	12 × 12	8 × 5	4 × 7	6 × 2	9 × 5	8 × 8

	K	L	M	N	O	P	Q	R
A	$7\overline{)56}$	$5\overline{)40}$	$6\overline{)24}$	$6\overline{)30}$	$6\overline{)18}$	$7\overline{)42}$	$8\overline{)16}$	$9\overline{)45}$
B	$3\overline{)9}$	$10\overline{)90}$	$1\overline{)1}$	$1\overline{)6}$	$10\overline{)100}$	$3\overline{)12}$	$10\overline{)70}$	$8\overline{)56}$
C	$6\overline{)48}$	$12\overline{)60}$	$4\overline{)32}$	$6\overline{)54}$	$7\overline{)0}$	$3\overline{)18}$	$9\overline{)90}$	$11\overline{)55}$
D	$2\overline{)16}$	$3\overline{)21}$	$5\overline{)30}$	$3\overline{)15}$	$11\overline{)110}$	$9\overline{)9}$	$8\overline{)64}$	$9\overline{)63}$
E	$4\overline{)28}$	$2\overline{)10}$	$9\overline{)18}$	$1\overline{)5}$	$7\overline{)63}$	$8\overline{)32}$	$2\overline{)8}$	$9\overline{)108}$
F	$8\overline{)24}$	$4\overline{)4}$	$2\overline{)14}$	$11\overline{)66}$	$8\overline{)72}$	$4\overline{)12}$	$7\overline{)21}$	$6\overline{)36}$
G	$12\overline{)36}$	$5\overline{)20}$	$7\overline{)28}$	$7\overline{)14}$	$4\overline{)24}$	$11\overline{)121}$	$9\overline{)36}$	$11\overline{)132}$
H	$9\overline{)27}$	$3\overline{)27}$	$7\overline{)49}$	$4\overline{)20}$	$9\overline{)72}$	$5\overline{)60}$	$8\overline{)88}$	$10\overline{)80}$
I	$4\overline{)44}$	$8\overline{)48}$	$5\overline{)35}$	$8\overline{)40}$	$5\overline{)10}$	$2\overline{)12}$	$10\overline{)60}$	$9\overline{)54}$
J	$10\overline{)120}$	$12\overline{)72}$	$9\overline{)81}$	$4\overline{)16}$	$1\overline{)7}$	$12\overline{)60}$	$12\overline{)96}$	$12\overline{)144}$

ADDITION AND SUBTRACTION FACTS TEST

	K	L	M	N	O	P	Q	R
A	15 + 4	9 − 3	6 + 5	10 − 7	13 − 9	5 + 8	11 − 4	0 + 9
B	12 − 4	9 + 9	16 − 9	8 − 6	12 + 5	11 − 6	5 + 3	16 − 9
C	8 + 6	0 + 8	9 − 6	7 + 7	19 − 6	3 + 6	11 − 8	7 + 4
D	6 + 3	14 − 6	8 + 8	11 − 11	16 − 8	7 + 9	9 + 7	12 − 3
E	6 + 8	17 − 9	7 − 3	6 + 6	8 − 4	1 + 9	8 + 7	12 − 9
F	8 + 5	3 + 9	11 − 3	3 + 7	10 − 2	9 + 0	12 − 8	7 + 2
G	5 + 7	13 − 4	4 + 6	20 − 10	7 − 0	6 + 9	14 − 7	11 + 8
H	6 + 7	7 − 4	9 − 9	9 + 8	13 − 5	8 + 10	17 − 10	10 − 5
I	8 + 4	14 − 9	3 + 3	9 + 10	10 − 6	4 + 7	9 − 7	7 + 6
J	11 − 2	9 + 5	15 − 7	12 + 12	18 − 9	7 + 8	11 − 5	13 − 7

	K	L	M	N	O	P	Q	R
A	3)‾21‾	9 × 6	6)‾30‾	10 × 7	10 × 1	4)‾32‾	7)‾63‾	5 × 4
B	8 × 3	7)‾56‾	8)‾88‾	12 × 2	8 × 11	1)‾8‾	9)‾90‾	10 × 5
C	12)‾36‾	7 × 8	5 × 7	10)‾90‾	5)‾45‾	12 × 4	8)‾16‾	5 × 5
D	7 × 7	4)‾48‾	9)‾99‾	11 × 3	12 × 3	9)‾108‾	11)‾88‾	10 × 2
E	7 × 10	12 × 9	12)‾84‾	2)‾20‾	12 × 0	10 × 11	3)‾36‾	10)‾110‾
F	4)‾44‾	12)‾72‾	7 × 11	12 × 6	7)‾56‾	9)‾45‾	10 × 3	9 × 7
G	12)‾108‾	6)‾60‾	9 × 2	8 × 12	12)‾144‾	6)‾24‾	4 × 4	7)‾84‾
H	4 × 9	8 × 8	11)‾44‾	7)‾77‾	6 × 11	12 × 5	11)‾132‾	6)‾42‾
I	8)‾96‾	8)‾56‾	12 × 11	12 × 12	5)‾60‾	3 × 6	10 × 10	4)‾40‾
J	11 × 5	9 × 9	10)‾80‾	7)‾35‾	11 × 4	10)‾120‾	9 × 8	10 × 9

TABLE OF MEASURES

METRIC

Length

1 centimeter (cm) = 10 millimeters (mm)
1 decimeter (dm) = 10 centimeters
1 meter (m) = 10 decimeters
1 meter = 100 centimeters
1 kilometer (km) = 1,000 meters

Mass/Weight

1 gram (g) = 1,000 milligrams (mg)
1 kilogram (kg) = 1,000 grams

Capacity

1 liter (L) = 1,000 milliliters (mL)
1 metric cup = 250 milliliters

CUSTOMARY

Length

1 foot (ft) = 12 inches (in.)
1 yard (yd) = 3 feet, or 36 inches
1 mile (mi) = 1,760 yards, or 5,280 feet

Mass/Weight

1 pound (lb) = 16 ounces (oz)
1 ton (T) = 2,000 pounds

Capacity

1 tablespoon (tbsp) = 3 teaspoons (tsp)
1 cup (c) = 8 fluid ounces (fl oz)
1 pint (pt) = 2 cups
1 quart (qt) = 2 pints
1 gallon (gal) = 4 quarts

TIME

1 minute (min) = 60 seconds (sec)
1 hour (hr) = 60 minutes
1 day = 24 hours
1 week (wk) = 7 days

1 year (yr) = 12 months (mo), or about 52 weeks
1 year = 365 days
1 leap year = 366 days

MONEY

1 penny = 1 cent (¢)
1 nickel = 5 cents
1 dime = 10 cents

1 quarter = 25 cents
1 half dollar = 50 cents
1 dollar ($) = 100 cents

SYMBOLS

⊥	is perpendicular to	∠ABC	angle ABC	=	is equal to
‖	is parallel to	△ABC	triangle ABC	°F	degrees Fahrenheit
$\overleftrightarrow{AB}$	line AB	<	is less than	°C	degrees Celsius
$\overrightarrow{AB}$	ray AB	>	is greater than	(2,3)	ordered pair (x,y)
$\overline{AB}$	line segment AB				

A.M. [ā•em′] Between midnight and noon (p. 118)

acute angle [ə•kyōōt′ an′gəl] An angle that has a measure less than a right angle (less than 90°) (p. 321)
Example:

addend [a′dend] A number that is added to another in an addition problem
Example: 2 + 4 = 6;
    2 and 4 are addends.

analog clock [a′nəl•ôg kläk] A device for measuring time by moving hands around a circle for showing hours, minutes, and sometimes seconds.
Example:

angle [an′gəl] The figure formed by two line segments or rays that share the same endpoint (p. 321)
Example:

area [âr′ē•ə] The number of square units needed to cover a surface (p. 492)
Example:

area = 9 square units

array [ə•rā′] An arrangement of objects in rows and columns

Associative Property of Addition [ə•sō′shē•ə•tiv prä′pər•tē əv ə•di′shən] See *Grouping Property of Addition.*

Associative Property of Multiplication [ə•sō′shē•ə•tiv prä′pər•tē əv mul•tə•plə•kā′shən] See *Grouping Property of Multiplication.*

average [av′rij] See *mean.*

B

bar graph [bär′graf] A way to show information that uses bars to stand for data.

benchmark [bench′märk] A known number of things that helps you understand the size or amount of a different number of things (p. 2)

C

calendar [ka′lən•dər] A table that shows the days, weeks, and months of a year (p. 126)

capacity [kə•pa′sə•tē] The amount a container can hold when it is filled (p. 460)

center [sen′tər] A point inside a circle from which all points on the circle are the same distance (p. 342)
Example:

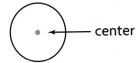
center

centimeter (cm) [sən′tə•mē•tər] A unit of length in the metric system. 100 centimeters = 1 meter (p. 470)

century [sen′chə•rē] A measure of time on a calendar, equal to 100 years (p. 127)

certain [sər′tən] An event that is sure to happen

chord [kôrd] A line segment with endpoints on a circle (p. 342)
Example:

$\overline{AB}$ is a chord.

circle [sər′kəl] A closed figure made up of points that are the same distance from the center point (p. 342)
Example:

circle C

circle graph [sər′kəl graf] A graph in the shape of a circle that shows data as a whole made up of different parts
Example:

VEGETABLES

Carrots — — Lettuce

Radishes

circumference [sər•kum′fər•əns] The distance around a circle (p. 344)

closed figure [klōzd fi′gyər] A shape that begins and ends at the same point
Examples:

compass [kəm′pəs] A tool used to construct circles (p. 342)

compatible numbers [kəm•pa′tə•bəl num′bərz] Nearby numbers that are easy to compute mentally (p. 258)

composite number [kəm•pä′zət num′bər] A number that has more than two factors (p. 302)
Example: 9 is composite since its factors are 1, 3, and 9

cone [kōn] A solid, pointed figure that has a flat, round base. (p. 508)
Example:

congruent [kən•grōo′ənt] Having the same size and shape (p. 326)

corner [kôr′nər] See *vertex.*

cube [kyōob] A solid figure with six congruent square faces
Example:

cubic unit [kyōo′bik yōo′nət] A unit of volume with dimensions of 1 unit × 1 unit × 1 unit (p. 514)

cumulative frequency [kyōo′myə•lə•tiv frē′kwən•sē] The sum of the frequency of data as they are collected. A running total of items being counted (p. 83)

cup (c) [kup] A customary unit for measuring capacity. 8 ounces = 1 cup (p. 460)

cylinder [si′lən•dər] A solid or hollow figure that is shaped like a can (p. 508)
Example:

D

data [dā′tə] Information collected about people or things from which conclusions may be drawn

decade [de′kād] A measure of time, equal to 10 years (p. 127)

decimal [de′sə•məl] A number with one or more digits to the right of the decimal point (p. 406)

decimal point [de′sə•məl point] The mark in a decimal number that separates the ones and the tenths places
Example: 6.4

↑ decimal point

decimeter (dm) [de′sə•mē•tər] A unit of length in the metric system. 10 decimeters = 1 meter (p. 470)

degree [di•grē′] The unit used for measuring angles or temperatures (p. 338)

degrees Celsius (°C) [di•grēz′ sel′sē•əs] A standard unit for measuring temperature in the metric system (p. 556)

degrees Fahrenheit (°F) [di•grēz′ fâr′ən•hīt] A standard unit for measuring temperature in the customary system (p. 554)

denominator [di•nä′mə•nā•tər] The part of a fraction that tells how many equal parts are in the whole
Example: $\frac{3}{4}$ ←denominator

diameter [di•am′ə•tər] A chord passing through the center of a circle (p. 342)
Example:

— diameter

difference [di′fər•əns] The answer to a subtraction problem (p. 33)

digit [di′jət] Any one of the ten symbols 0, 1, 2, 3, 4, 5, 6, 7, 8, or 9 used to write numbers (p. 1)

digital clock [di′jə•təl kläk] A clock that shows time to the minute using digits
Example:

dimension [də•men′shən] A measure in one direction

Distributive Property of Multiplication [di•stri′byə•tiv prä′pər•tē əv mul•tə•plə•kā′shən] The property that states that multiplying a sum by a number is the same as multiplying each addend by the number and then adding the products (p. 216)

dividend [di′və•dend] The number that is to be divided in a division problem (p. 198)
Example: 36 ÷ 6; 6)36; the dividend is 36.

divisible [də•vi′zə•bəl] Capable of being divided without a remainder (p. 313)
Example: 21 is divisible by 3.

division [də•vi′zhən] The process of sharing a number of items to find how many groups can be made or how many items will be in a group; the opposite operation of multiplication

divisor [də•vi′zər] The number that divides the dividend (p. 140)
Example: 15 ÷ 3; 3)15; the divisor is 3.

double-bar graph [du′bəl bär graf] A graph used to compare similar kinds of data (p. 100)
Example:

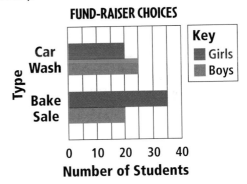

doubles [du′ bəlz] Two addends that are the same number

E

edge [ej] The line made where two or more faces of a solid figure meet (p. 509)
Example:

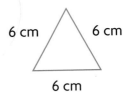
— edge

elapsed time [i•lapst′ tīm] The time that passes from the start of an activity to the end of that activity (p. 120)

equally likely [ē′kwə•lē li′klē] Having the same chance of happening as something else (p. 534)

equation [i•kwā′zhən] A number sentence which states that two amounts are equal (p. 61)
Example: 4 + 5 = 9

equilateral triangle [ē•kwə•la′tə•rəl trī′an•gəl] A triangle with 3 congruent sides (p. 346)
Example:

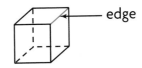
6 cm   6 cm
6 cm

equivalent [ē•kwiv′ə•lənt] Having the same value or naming the same amount

equivalent decimals [ē•kwiv′ə•lənt de′sə•məlz] Two or more decimals that name the same amount (p. 412)

equivalent fractions [ē•kwiv′ə•lənt frak′shənz] Two or more fractions that name the same amount (p. 366)
Example: $\frac{2}{4}$ and $\frac{1}{2}$ name the same amount.

estimate [es′tə•māt] To find an answer that is close to the exact answer (p. 50)

event [i•vent′] One outcome or a combination of outcomes in an experiment (p. 528)

expanded form [ik•span′dəd fôrm] A way to write numbers by showing the value of each digit (p. 1)
Example: $253 = 200 + 50 + 3$

expression [ik•spre′shən] A part of a number sentence that has numbers and operation signs but does not have an equal sign (p. 54)

face [fās] A polygon that is a flat surface of a solid figure (p. 509)
Example:

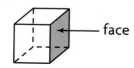

fact family [fakt fam′ə•lē] A set of related multiplication and division, or addition and subtraction, equations using the same numbers (p. 140)
Example: $7 \times 8 = 56$; $8 \times 7 = 56$;
$56 \div 7 = 8$; $56 \div 8 = 7$

factor [fak′tər] A number that is multiplied by another number to find a product (p. 140)

factor tree [fak′tər trē] A diagram that shows the prime factors of a number (p. 306)
Example:
```
        30
       /  \
      5 × 6
     /   / \
    5 × 2 × 3
```

fairness [fâr′nəs] Fairness in a game means that one player is as likely to win as another. Each player has an equal chance of winning (p. 546)

flip [flip] A movement of a figure to a new position by flipping the figure over a line (p. 326)

foot (ft) [foot] A customary unit used for measuring length or distance (p. 452)

formula [fôr′myə•lə] A set of symbols that expresses a mathematical rule (p. 486)
Example: $A = l \times w$

fraction [frak′shən] A number that names a part of a whole (p. 364)

frequency [frē′kwen•sē] The number of times a response occurs (p. 83)

frequency table [frē′kwen•sē tā′bəl] A table that uses numbers to record data about how often something happens (p. 83)

function table [funk′shən tā′bəl] A table that matches each input value with an output value. The output values are determined by the function (p. 570)

gallon (gal) [ga′lən] A customary unit for measuring capacity. 4 quarts = 1 gallon (p. 460)

gram (g) [gram] A unit of mass in the metric system. 1,000 grams = 1 kilogram (p. 478)

greater than (>) [grā′tər than] A symbol used to compare two numbers, with the greater number given first (p. 20)
Example: $6 > 4$

grid [grid] Evenly divided and equally spaced squares on a figure or flat surface

Grouping Property of Addition [grōo′ping prä′pər•tē əv ə•di′shən] The property stating that you can group addends in different ways and still get the same sum (p. 33)
Example: $3 + (8 + 5) = (3 + 8) + 5 = 16$

Grouping Property of Multiplication [grōo′ping prä′pər•tē əv mul•tə•plə•kā′shən] The property stating that you can group factors in different ways and still get the same product (p. 150)
Example: $3 \times (4 \times 2) = (3 \times 4) \times 2 = 24$

hexagon [hek′sə•gän] A polygon with six sides and six angles (p. 486)

hour (hr) [our] A standard measure of time, equal to 60 minutes

hour hand [our hand] The short hand on an analog clock or watch that shows the hour

hundredth [hən′drədth] One of one hundred equal parts (p. 406)
Example:

hundredth

I

impossible [im·pä'sə·bəl] Never able to happen (p. 528)

inch (in.) [inch] A customary unit used for measuring length (p. 452)
Example:

intersecting lines [in·tər·sek'ting linz] Lines that cross each other at exactly one point (p. 324)
Example:

interval [in'tər·vəl] The distance between two numbers on the scale of a graph (p. 92)

inverse operations [in'vərs ä·pə·rā'shənz] Operations that undo each other, like addition and subtraction, or multiplication and division (p. 140)
Example: $6 \times 8 = 48$ and $48 \div 6 = 8$

isosceles triangle [ī·sä'sə·lēz trī'an·gəl] A triangle with only two congruent sides (p. 346)
Example:

10 in. 10 in.
7 in.

K

key [kē] The part of a map or graph that explains the symbols

kilogram (kg) [ki'lə·gram] A unit of mass in the metric system.
1 kilogram = 1,000 grams (p. 478)

kilometer (km) [kə·lä'mə·tər] A unit of length in the metric system.
1,000 meters = 1 kilometer (p. 470)

L

leaf [lēf] A ones digit in a stem-and-leaf plot (p. 90)

less than (<) [les than] A symbol used to compare two numbers, with the lesser number given first (p. 20)
Example: $3 < 7$

like fractions [līk frak'shənz] Fractions with the same denominator (p. 386)

likely [līk'lē] Having a greater than even chance of happening (p. 534)

line [līn] A straight path in a plane, extending in both directions with no endpoints (p. 320)
Example:

S   T

line graph [līn graf] A graph that uses a line to show how data change over a period of time (p. 102)
Example:

SAVINGS

line plot [līn plät] A graph that shows the frequency of data along a number line (p. 88)
Example:

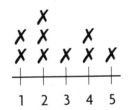

Cookies Eaten

line segment [līn seg'mənt] A part of a line that has two endpoints (p. 320)
Example:

A   B

line symmetry [līn si'mə·trē] What a figure has if it can be folded about a line so that its two parts match exactly (p. 330)
Example:

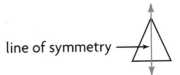
line of symmetry

linear units [li'nē·ər yoo'nəts] Units which measure length, width, height, or distance (p. 452)

liter (L) [lē'tər] A unit of capacity in the metric system. 1 liter = 1,000 milliliters (p. 476)

mass [mas] The amount of matter in an object (p. 478)

mathematical probability [math·ma′ti·kəl prä·bə·bi′lə·tē] A comparison of the number of favorable outcomes to the number of possible outcomes of an event (p. 542)

mean [mēn] The average of a set of numbers, found by dividing the sum of the set by the number of addends (p. 276)

median [mē′dē·ən] The middle number in an ordered set of data (p. 86)

meter (m) [mē′tər] A unit of length in the metric system. 100 centimeters = 1 meter (p. 470)

mile (mi) [mīl] A customary unit for measuring length. 5,280 feet = 1 mile (p. 452)

milliliter (mL) [mi′lə·lē·tər] A unit of capacity in the metric system. 1,000 milliliters = 1 liter (p. 476)

millions [mil′yənz] The period after thousands, equal to one thousand thousands (p. 8)

minute (min) [mi′nət] Unit to measure short amounts of time. There are 60 minutes in one hour.

minute hand [mi′nət hand] The long hand on an analog clock or watch that shows the minute

mixed number [mikst nəm′bər] An amount given as a whole number and a fraction (p. 378)

mode [mōd] The number or item that occurs most often in a set of data (p. 86)

multiple [mul′tə·pəl] The product of a given whole number and another whole number (p. 298)

multiplication [mul·tə·plə·kā′shən] A process to find the total number of items in equal-sized groups, or to find the total number of items in a given number of groups when each group contains the same number of items. Multiplication is the inverse of division.

multistep problem [mul′ti·step prä′bləm] A problem requiring more than one step to solve (p. 234)

negative numbers [ne′gə·tiv num′bərz] All the numbers to the left of zero on the number line (p. 558)

net [net] A two-dimensional pattern of a three-dimensional figure (p. 512)
Example:

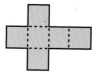

number line [num′bər līn] A line with equally spaced tick marks named by numbers. A number line does not always start at 0.
Example:

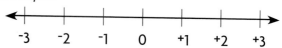

numerator [noo′mə·rā·tər] The part of a fraction that tells how many parts of the whole are being considered

obtuse angle [əb·toos′ an′gəl] An angle that has a measure greater than a right angle (greater than 90°) (p. 321)
Example:

octagon [äk′tə·gän] A polygon with eight sides and eight angles (p. 486)

open figure [ō′pən fi′gyər] A figure that does not begin and end at the same point
Examples:

Order Property of Addition [ôr′dər prä′pər·tē əv ə·di′shən] The property stating that you can add two numbers in any order and get the same sum (p. 36)
Example: 4 + 2 = 2 + 4 = 6

Order Property of Multiplication [ôr′dər prä′pər·tē əv′ mul·tə·plə·kā′shən] The property stating that you can multiply two factors in any order and get the same product (p. 145)
Example: 4 × 2 = 2 × 4 = 8

ordered pair [ôr′dərd pâr] A pair of numbers used to locate a point on a coordinate grid. The first number tells how far to move horizontally, and the second number tells how far to move vertically. (p. 568)

ounce (oz) [ouns] A customary unit used for measuring weight. 16 ounces = 1 pound (p.462)

outcomes [out′kəmz] The possible results of an experiment (p. 528)

outlier [out′li•ər] A value separated from the rest of the data (p. 88)

P.M. [pē•em′] After noon (p. 118)

parallel lines [pâr′ə•lel linz] Lines that never intersect (p. 324)
Example:

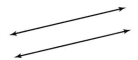

parallelogram [pâr•ə•lel′ə•gram] A quadrilateral whose opposite sides are parallel and congruent (p. 349)
Example:

parentheses [pə•ren′thə•sēz] The symbols used to show which operations in an expression should be done first (p. 54)

partial product [pär′shəl prä′dəkt] A method of multiplying in which the ones, tens, hundreds, and so on, are multiplied separately and then the products are added together (p. 216)

pentagon [pen′tə•gän] A polygon with five sides and five angles (p. 486)
Example:

perimeter [pə•ri′mə•tər] The distance around a figure (p. 486)

period [pir′ē•əd] Each group of three digits in a number (p. 6)
Example: 85,643,900 has three periods.

perpendicular lines [pər•pən•di′kyə•lər linz] Lines that intersect to form four right angles (p. 324)
Example:

pictograph [pik′tə•graf] A graph that uses pictures to show and compare information
Example:

HOW WE GET TO SCHOOL

Walk	✸ ✸ ✸
Ride a Bike	✸ ✸ ✸ ✸
Ride a Bus	✸ ✸ ✸ ✸ ✸ ✸
Ride in a Car	✸ ✸

Key: Each ✸ = 10 students.

pint (pt) [pint] A customary unit for measuring capacity. 2 cups = 1 pint (p. 460)

place value [plās val′yoo] The value of a place, such as ones or tens, in a number

plane [plān] A flat surface that extends without end in all directions (p. 320)
Example:

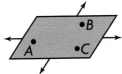

plane figure [plān fi′gyər] A closed figure that lies on a flat surface and is all in one plane. Points, lines, rays, and polygons are plane figures.

point [point] A location on an object or in space (p. 320)

polygon [pä′lē•gän] A closed plane figure with straight sides. Each side is a line segment. (p. 486)
Examples:

positive numbers [pä′zə•tiv num′bərz] All the numbers to the right of zero on the number line (p. 558)

pound (lb) [pound] A customary unit used for measuring weight. 16 ounces = 1 pound (p. 462)

predict [pri•dikt′] To tell what might happen (p. 534)

prime factor [prīm fak′tər] A factor that is a prime number (p. 306)

prime number [prīm num′bər] A number that has only two factors: 1 and itself (p. 302)
Examples: 5, 7, 11, 13, 17, and 19 are prime numbers.

probability [prä·bə·bi′lə·tē] The likelihood that an event will happen. See also *mathematical probability*. (p. 542)

product [prä′dəkt] The result of multiplication. (p. 140)

Property of One [prä′pər·tē əv wən] The property that states that the product of any number and 1 is that number.

protractor [prō′trak·tər] A tool for measuring the size of an angle (p. 340)

pyramid [pir′ə·mid] A solid figure with a polygon base and triangular sides which meet at a single point.
Example:

quadrilateral [kwä·drə·la′tə·rəl] A polygon with four sides and four angles (p. 486)

quart (qt) [kwôrt] A customary unit for measuring capacity. 2 pints = 1 quart (p. 460)

quotient [kwō′shənt] The number, not including the remainder, that results from dividing
Example: 8 ÷ 4 = 2; 2 is the quotient. (p. 140)

radius [rā′dē·əs] A line segment with one endpoint at the center of a circle and the other endpoint on the circle (p. 342)
Example:

range [rānj] The difference between the greatest and the least values in a set of data (p. 88)

ray [rā] A part of a line; begins at one endpoint and goes on forever in one direction (p. 320)
Example:

rectangle [rek′tan·gəl] A plane figure with opposite sides that are equal and parallel, and with four right angles (p. 349)
Example:

rectangular prism [rek·tan′gyə·lər pri′zəm] A solid figure in which all six faces are rectangles (p. 508)
Example:

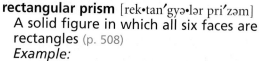

regroup [rē·groop′] To exchange amounts of equal value when renaming a number
Example: 5 + 8 = 13 ones or 1 ten 3 ones

regular polygon [reg′yə·lər pä′lē·gän] a polygon that has sides that are the same length (p. 486)

remainder [ri·mān′dər] The amount left over after you find a quotient (p. 246)

rhombus [räm′bəs] A parallelogram with four congruent sides and with opposite angles that are congruent (p. 349)
Example:

right angle [rīt an′gəl] An angle that forms a square corner and has a measure of 90°
(p. 321)
Example:

right triangle [rīt trī′an·gəl] A triangle that has one right angle
Example:

rotational symmetry [rō·tā′shən·əl si′mə·trē] What a figure has if it can be turned about a central point and still look the same in at least two positions (p. 330)

rounding [roun′ding] Replacing a number with another number that tells about how many or how much (p. 26)

scale [skāl] A series of numbers placed at fixed distances on a graph to help label the graph (p. 92)

scalene triangle [skā′ lēn trī′an·gəl] A type of triangle with no congruent sides (p. 346)
Example:

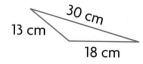

schedule [skeʹjo͞ol] A table that lists activities or events and the times they happen

seconds (sec) [seʹkəndz] A small unit of time. 60 seconds = 1 minute (p. 116)

similar [siʹmə•lər] Having the same shape as something else but possibly different in size (p. 326)
Example:

simplest form [simʹpləst fôrm] A fraction is in simplest form when 1 is the only number that can divide evenly into the numerator and the denominator. (p. 368)

slide [slīd] A movement of a figure to a new position without turning or flipping it (p. 326)

sphere [sfer] A round object whose curved surface is the same distance from the center to all its points (p. 508)
Example:

square [skwâr] A polygon with 4 equal sides and 4 right angles (p. 349)
Example:

square number [skwâr numʹbər] The product of a number and itself
Example: 2 × 2 = 4, so 4 is a square number.

square pyramid [skwâr pirʹə•mid] A solid figure with a square base and with four triangular faces that have a common point (p. 508)
Example:

square unit [skwâr yo͞oʹnət] A square with a side length of one unit; used to measure area

standard form [stanʹdərd fôrm] A way to write numbers by using digits (p. 9)
Example: 3,540 ←standard form

stem [stem] A tens digit in a stem-and-leaf plot (p. 90)

stem-and-leaf plot [stem ənd lēf plät] A table that shows groups of data arranged by place value (p. 90)

Example:

Number of Sit-Ups

Stem	Leaves
1	1 3 4
2	0 1 2 3
3	3 4 6
4	0 5 9

sum [sum] The answer to an addition problem (p. 33)

survey [sûrʹvā] A method of gathering information by asking questions and recording people's answers (p. 82)

tally table [taʹlē tāʹbəl] A table that uses tally marks to show how often something happens (p. 82)

tenth [tenth] One of ten equal parts (p. 406)
Example:

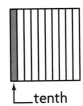
└─tenth

thousandth [thouʹzəndth] one of one thousand equal parts (p. 410)

three-dimensional [thrē•də•menʹshən•əl] Measured in three directions, such as length, width, and height (p. 508)
Example:

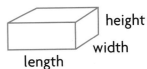
height
width
length

time line [tīm līn] A schedule of events or an ordered list of historic moments

ton (T) [tun] A customary unit used for measuring weight. 2,000 pounds = 1 ton (p. 462)

transformation [trans•fər•māʹshən] The movement of a figure by a slide, flip, or turn (p. 326)

trapezoid [traʹpə•zoid] A quadrilateral with only one pair of parallel sides (p. 349)
Example:

tree diagram [trē dī'ə•gram] An organized list that shows all possible outcomes for an event (p. 530)

trends [trendz] On a graph, areas where the data increase, decrease, or stay the same over time (p. 105)

triangle [trī'an•gəl] A polygon with three sides and three angles (p. 486)
Example:

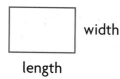

turn [tûrn] A movement of a figure to a new position by rotating the figure around a point (p. 326)

two-dimensional [tōō•də•men'shən•əl] Measured in two directions, such as length and width (p. 508)
Example:

width

length

unlike fractions [un'lik frak'shənz] Fractions with different denominators (p. 398)

unlikely [un•li'klē] Having a less than even chance of happening (p. 534)

variable [vâr'ē•ə•bəl] A letter or symbol that stands for any number (p. 60)

Venn diagram [ven dī'ə•gram] A diagram that shows relationships among sets of things (p. 352)
Example:

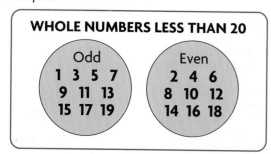

WHOLE NUMBERS LESS THAN 20

Odd
1 3 5 7
9 11 13
15 17 19

Even
2 4 6
8 10 12
14 16 18

vertex [vûr'teks] The point at which two rays of an angle or two or more line segments meet in a plane figure, or where three or more sides meet in a solid figure (p. 321)

Example:

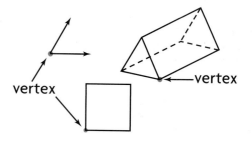

vertex

vertex

volume [väl'yəm] The measure of the space a solid figure occupies (p. 514)

weight [wāt] How heavy an object is (p. 462)

whole number [hōl num'bər] One of the numbers 0, 1, 2, 3, 4, ... The set of whole numbers goes on without end.

word form [wûrd fôrm] A way to write numbers by using words (p. 9)
Example: Sixty-two million, four hundred fifty-three thousand, two hundred twelve

x-axis [eks'ak'səs] The horizontal line on a coordinate grid (p. 568)

x-coordinate [eks'kō•ôrd'ə•nət] The first number in an ordered pair. It tells the distance to move horizontally. (p. 568)

y-axis [wī'ak'səs] The vertical line on a coordinate grid (p. 568)

y-coordinate [wī'kō•ôrd'ə•nət] The second number in an ordered pair. It tells the distance to move up or down. (p. 568)

yard (yd) [yärd] A customary unit for measuring length. 3 feet = 1 yard (p. 452)

Zero Property of Multiplication [zē'rō prä'pər•tē əv mul•tə•plə•kā'shən] The property that states that the product of any number and 0 is 0

A

Act It Out strategy, 70–71
Acute angles, 321–323, H30
Acute triangles, 346–347
Addition
  column, H6
  decimals, 432–433, 436–439
  estimating sums, 34–35, 430–431
  four-digit numbers, 40–41
  fractions
    like fractions, 386–389
    mixed numbers, 392–395
    unlike fractions, 398–399
  greater numbers, 40, 44–47
  mental math and, 36–39
  missing addends, H6
  of money, 437–438, H22
  Order/Commutative Property, 39
  three-digit numbers, 55, H5
  two-digit numbers, 58, H5
Algebra, 13, 149, 211, 267, 374, 388, 394, 465, 474–475, 479
  coordinate grid, 568–569, 574–577, H10
  equations, solving, 61–63, 66–71, 165–167, 170–171
    equals added to equals are equal, 66–68
    equals multiplied by equals are equal, 164–167
    graphing simple linear, 574–577
  expressions, 54–60, 158–163, 168–169
    evaluating, 54–57
  finding a rule, 64–65, 172–173, 578–579, H25
  formulas, using, 486
  inequalities. See Comparing
  input/output patterns, 64–65, 172–173, 570–580, H25
  inverse operations, multiplication and division, 140, 142, 145
  parentheses, 54–59, 150–151, 158–163
  recognizing a linear pattern, 129, 145, 309
  See also Function table, Properties, and Variables
Analyze data. *See* Data
Analyzing information, 203
Angles
  acute, 321–323, H30
  classifying, 321–323, H25, H30
  in congruent figures, 326
  measuring, 340–341
  obtuse, 321–323, H30
  in polygons, 486
  in quadrilaterals, 486
  right, 321–323, H25, H27, H30
  in triangles, 486
  turns and degrees, 338–339
Arabic numerals, 11
Area, 492, H28
  of complex and irregular figures, 493–494, H28
  relating to perimeter, 496–497
  using appropriate units for, 492
  using formulas to find
    for rectangles, 492–495, 498
    for squares, 493
Arrays
  in multiplying, 142–146, 300–303
  in relating multiplication and division, 142, 143
Assessment
  Chapter Review/Test, 14, 30, 50, 72, 96, 112, 130, 154, 176, 206, 220, 236, 260, 278, 294, 310, 334, 354, 382, 402, 424, 442, 466, 482, 504, 518, 538, 550, 564, 580

Cumulative Review. See Mixed Applications, Mixed Review and Test Prep, Mixed Strategy Practice, and Standardized Test Prep
Mixed Review and Test Prep, 3, 5, 7, 11, 19, 23, 29, 35, 39, 41, 43, 47, 55, 57, 59, 63, 65, 69, 85, 87, 89, 91, 93, 101, 103, 105, 109, 117, 119, 123, 129, 141, 143, 147, 149, 151, 161, 163, 167, 169, 171, 173, 187, 189, 199, 203, 211, 213, 217, 227, 229, 231, 233, 235, 247, 249, 253, 257, 259, 265, 267, 271, 273, 277, 283, 285, 287, 291, 299, 301, 305, 307, 323, 325, 329, 331, 339, 341, 343, 345, 347, 351, 365, 367, 371, 375, 381, 389, 391, 395, 399, 401, 409, 413, 417, 421, 429, 431, 433, 435, 439, 453, 457, 459, 461, 463, 473, 475, 477, 479, 487, 491, 495, 497, 501, 511, 513, 515, 529, 531, 537, 543, 545, 547, 555, 557, 561, 569, 573, 577
Standardized Test Prep, 15, 31, 51, 73, 97, 113, 131, 155, 177, 207, 221, 237, 261, 279, 295, 311, 335, 355, 383, 403, 425, 443, 467, 483, 505, 519, 539, 551, 565, 581
Study Guide and Review, 76–77, 134–135, 180–181, 240–241, 314–315, 358–359, 446–447, 522–523, 584–585
Associative Property of Addition. *See* Grouping/Associative Property of Addition
Associative Property of Multiplication. *See* Grouping/Associative Property of Multiplication
Average, 276–277

B

Bar graphs, 13, 100–101, 106–110, H9
Basic facts
  dividing by multiples of ten, 282–283
  dividing multiples of 10, 100, and 1,000, 257
  multiplication table through ten times ten, 145, 148–149, 298
  multiplying multiples of 10, 100, and 1,000, 186–187, H14, H16, H24
  multiplying multiples of ten, 186–187, 212–213, H14, H16, H24
  related addition and subtraction facts, H7
  related multiplication and division facts, 140
  tests, H66–H69
Benchmark, 2, 52, H2
  in estimation, 2–3

C

Calculator, 44–45, 230, 243, 272, 317, 458
Calendars, 126–130, 307, H11
Capacity, 460
  customary, 460
  estimating, 476–477
  metric, 476–477
Categorizing, 351
Cause and effect, 167
Celsius temperature, 556–557
Centimeter (cm), 470–475, 481
Century, 127–129
Challenge, 75, 133, 179, 239, 313, 357, 445, 521, 587
Chapter Review/Test, 14, 30, 50, 72, 96, 112, 130, 154, 176, 206, 220, 236, 260, 278, 294, 310, 334, 354, 382, 402, 424, 442, 466, 482, 504, 518, 538, 550, 564, 580

Check What You Know, 1, 17, 33, 53, 81, 99, 115, 139, 157, 185, 209, 223, 245, 263, 281, 297, 319, 337, 363, 385, 405, 427, 451, 469, 485, 507, 527, 541, 553, 567

Choose a Method, 44–47, 230–231, 272–273

Chord, 342, 343

Circle graphs, 445

Circles, 342
  center, 342, 343
  chord, 342, 343
  circumference, 344–345
  diameter, 342, 343
  drawing, 342–343
  measuring, 342–343
  radius, 342, 343

Circumference, 344–345

Classifying, 351
  polygons, 486
  quadrilaterals, 348–350
  solid figures, 508–510
  triangles, 346–347

Clocks
  analog, 116–117, 118, 120, 121, 130, 339, H10–H11
  digital, 116–117

Clustering, 47

Column addition, H6

Commutative Property. *See* Order/Commutative Property

Compare strategies, 464–465

Comparing
  customary units, 452–453, 454–457, 458–459, 462–463
  decimals, 418–421, 431
  fractions, 372–375, 380, 388
  graphs, 92–93
  metric units, 470–473, 474–475
  negative numbers, 558–561
  on a number line, 18, 418, 559, 560, H24
  weight/mass, 463
  whole numbers, 18–19, 24, 43, 46, 48, 56, H3
  See also Ordering

Compass, 342

Compatible numbers, 39, 258–259, H18

Composite numbers, 302–305

Computer software
  E-Lab, 21, 42, 101, 121, 149, 225, 276, 303, 343, 367, 391, 413, 480, 493, 529, 535
  Mighty Math Calculating Crew, 67, 201, 250, 289, 393, 415, 432, 508, 512
  Mighty Math Number Heroes, 45, 146, 197, 286, 308, 348, 373, 542
  See also Technology

Conclusions, drawing, 110–111, 548–549

Cones, 508–510

Congruent figures, 326–329, H26

Coordinate grid, 568
  graphing simple lines on, 574–577
  representing points on, 568–569, H10

Corner. *See* Vertex

Counting, skip, H23

Cross-curricular connections
  art, 323, 389
  health, 85, 147
  history, 305, 381
  music, 371
  social studies, 23, 129, 271, ?457, 511, 561, 573

Cubes, 508–510, 512–515, 534, 544, 549

Cubic units, 514–515

Cumulative frequency, 83–85

Cups (c), 460–461

Curved surface, 508–510

Customary system
  capacity, 460–461
  changing units, 458–459, 460–461, 462–463
  Fahrenheit temperature, 554–555
  length, 452–453, H22
  weight, 462–463
  See also Measurement

Cylinders, 508–510

D

Data
  analyzing charts and tables, 110, 111, 122–123, 125, 141, 160, 166, 171–173, 176, 235, 270, 273, 277, 299, 374, 394
  categorical, 87
  collecting, 82–85
  frequency tables, 82–85, 235, H8
  graphs
    bar, 92–93, 94, 106–110, H9
    circle, 445
    coordinate. *See* Coordinate grid
    double-bar, 100–101
    line, 18, 102–109, 111, 202, 227, 252
    pictographs, 12–13, 257, H8
  interpreting, 110–111, 120–123
  intervals and, 92–93
  line plots, 88–89, 107, 108
  making tables, 24–25, 109, 464
  mean, 276–277
  median, 86–87
  mode
    for categorical data, 87
    for numerical data, 86–87
  numerical, 122–123
  organized lists, 532–533
  organizing, 82–85, 528–529
  outcomes, 528–533
  outliers, 88
  range and, 88
  scale and, 92
  schedules, 116, 117, 122–125, 130
  stem-and-leaf plots, 90–91, 106
  surveys, 82
  tally tables, 82, 83, H8
  tree diagrams, 530–531
  Venn diagrams, 352–353

Decade, 127–129

Decimal point, 406

Decimals, 406, 417
  adding, 432–433, 436–439
  comparing and ordering, 418–421, 431
  equivalent, 412–413, 415, 416
  estimate sums and differences, 430–431
  fractions
    modeling equivalencies, 406–409, 412, 415, 416, 473
    relating on a number line, 406–409, 414–417
    writing as, 406–409, H21
  hundredths, 406–409
  metric measures, 473
  mixed numbers
    modeling equivalencies, 414–417
    relating on a number line, 414–417
  modeling, 406–407, 412–413, 415, 416, 473, H20
  money and, 429, 437–438, H20
  number line and, 407, 408, 414–416, 418–420, 429

place value in, 406–407, 409
reading and writing, 406–409
rounding, 428–431
subtracting, 434–439
tenths, 406–407
thousandths, 410–411
See also Money
Decimeters (dm), 470–475
Degrees
in angles, 340–341
Celsius, 556–557
in circles, 338–339
Fahrenheit, 554–555
Denominator, 364, H18, H19
Diagrams. *See also* Data
drawing, 249, 330, 352–353, 480–481, 493, 496, 530–531
Venn, 352–353
Diameter, 342–343
Differences. *See* Subtraction
Digits. *See also* Place value
missing, 213
Distance, 227, H22, H23
Dividends, 140
three-digit, 266–267
two-digit, 140–149
Divisibility rules, 313
Division, H12
averages and, 276–277
basic facts, H12
choosing the operation, 152, 153, 292–293
concept of, 140
connecting to multiplication, 140–143, 145, 286–287
dividends, 140
divisors, 140
estimation in, 258–259, 282–283, 286–291
fact families and, 140, 141, H16
of four-digit dividends, 272–273
interpreting remainders, 274–275
as inverse of multiplication, 140, 142, 143, 145
mental math, 256–257
modeling, 142, 246–249, 284–285
by multiples of ten, 282–283
on multiplication table, 145
by one-digit divisors, 140–146
patterns in, 256, 282–283
placing first digit in quotients, 264–265
procedures in, 250–253, 286–287
quotient, 140, H12
recording, 249, 285
remainders in, 246–247, 250–253, 254–255, 274–275, H17
of three-digit dividends, 266–267
by two-digit divisors, 282–291
zeros in, 256, 268–271, 282
Divisor. *See* Division, divisors
Double-bar graphs, 100–101
Draw a Diagram or Picture strategy, 464–465, 480–481
Drawing conclusions, 110–111, 548–549

E

Edge, 509
E-Lab. *See* Technology Links
Elapsed time, 120–123, 129, 130
Equally likely events, 534–536, 542–543, 546–549
Equations, 164
equals added to equals are equal, 66–68
equals multiplied by equals are equal, 164–167
graphing, 574–577
manipulate, 66–69, 164–167
using, 570–573
with parentheses, 165–167, 174–175
with variables, 61–62, 170–171
writing, 170–173, 204–205, 301
Equilateral triangles, 330, 346–347
Equivalent decimals, 412–413, 415, 416
Equivalent fractions, 366–371
Equivalent measures in metric system, 474, H70
Error analysis. *See* What's the Error?
Estimates, checking, 288–290
Estimation, 34
of area, 492–495
benchmarks, 2–3
to check reasonableness of answer, 197, 201–202, 225–226, 229, 231, 233, 433, 435, 438
of differences, 34–35, 430–431
of fractions, 380
of measurements, 455–456, 477, 555, 557
or exact answer, 48–49
of perimeters, 490–491
of products, 188–189, 214–215, H15
of quotients, 258–259, 282–283, 288–291
reasonableness, 2, 40–44, 200–205, 224–233, 258–259, 266–267
and rounding, 188–189, 214–215, 430–431, H15
of sums, 34–35
of volume, 514–515
Events, 528–529. *See also* Probability
Expanded form of whole numbers, 6, 7, 9, 10
Explain, 56, 145, 149, 196, 197, 201, 210, 214, 228, 230, 232, 246, 265, 269, 273, 286, 298, 366, 373, 455, 471, 575
Expressions, 54
evaluating, 54–57, 158–163, 169
interpret, 54–57, 158–163
matching words and, 58–59, 162–163
with parentheses, 54–59, 158–163
with variables, 168–169
Extra Practice, H32–H61

F

Faces, 508–511
Fact families
addition and subtraction, H7
multiplication and division, 140, 141, H16
Factor tree, 306
Factors, 140
finding, 298–301
missing, 307, H13
prime, 306–307
Fahrenheit temperature, 102–103, 554–555
Fairness, in probability, 546–549
Figures. *See* Plane figures
Find a Pattern strategy, xxv, 308–309, 502–503
Finding a rule, 64–65, 172–173, 578–579
Flips. *See* Reflections
Foot (ft), 452–453, 458
Formulas, 486
for area of rectangle, 498
for area of square, 493
for perimeter of rectangle, 490
for perimeter of square, 490

Fractions, 364, H18
  adding
    like denominators, 386–389
    unlike denominators, 398–399
  comparing and ordering, 372–375, 380, 388, 456–457
  concepts of, 364
  decimals and, 406–409, H21
    modeling equivalencies, 412–413, 414–417
    relating on a number line, 406–409, 414–417
    writing as, 406–409
  denominator, 364, H18, H19
  estimating, 395
  equivalent, 366–371
  improper, 379–380
  like, 386–392
  in measurements, 454–457, 473
  modeling
    as division of whole numbers by whole numbers, 368–369
    with fraction bars, 366–369, 372–375, 379, 386, 388–393, 398–401
    on a number line, 376–377
    parts of a group, H19
    parts of a set. *See* parts of a group
    parts of a whole, 412, 415, 416, H18, H19
    with shapes, 365
  numerator, 364, H18, H19
  probability as, 542–543, 544–545
  reading and writing, 364–365
  simplest form of, 369–371
  subtracting
    like denominators, 390–391
    mixed numbers, 393–395
    unlike denominators, 400–401
  written as percents, 417
  See also Mixed numbers
Frequency table, 82–85, H8
Function table, 64–65, 172–173, 570–579, H25

G

Gallons (gal), 460–461
Generalizations, making, 562–563
Geometry, 543
  angles, 321–323
    acute, 321–323, H30
    classifying and identifying, 321–323, H25, H30
    obtuse, 321–323, H30
    in polygons, 486
    in quadrilaterals, 486
    right, 321–323, H25, H27, H30
    turns, 338–339
  area
    of complex figures, 493–494, H28
    of rectangles, 492–495, 498
    of squares, 493
  classifying, 351
    angles, 321–323, H25, H30
    figures by number of dimensions, 508
    polygons, 486
    quadrilaterals, 348–350, 486
    solids, 508–510
  congruent figures, 326–329, H26
  coordinate grids
    graphing simple lines on, 574–577
    representing points on, 568–569, H10

  faces in solids, 508–511
  flips. *See* Reflections
  lines, 320
    angle relationships and, 321–325
    parallel, 324–325
    perpendicular, 324–325
    ray, defined, 320
    segment, defined, 320
  motion
    reflections (flips), 326–328
    rotations (turns), 326–328
    translations, 326–328
  patterns with, 512–513
  perimeter, 486–489, 490–491, 496–497, H28
  plane, defined, 320
  point, defined, 320
  polygons, 486
  quadrilaterals, 486
    parallelograms, 348–350
    rectangles, 490–498, 502–503, 508–511
    rhombuses, 348–350
    squares, 490, 493
    trapezoids, 348–350
  rotations
    angles of, 338–339
    motion geometry, 326–329
  similar figures, 326–328
  slides. *See* Translations
  solid figures, H29
    cones, 508–510
    cubes, 508–510, 512–513
    cylinders, 508–510
    edges of, 508–511
    faces of, 509–511
    prisms, 508–510
    pyramids, 508, 510
    rectangular prisms, 508–510
    spheres, 508–510
  symmetry
    line, 330, H26
    rotational, 330–331
  three-dimensional figures, 508–511, 512–513, 514–515
  transformations. *See* Motion geometry
  triangles
    acute, 346–347
    attributes of, 486
    equilateral, 330, 346–347
    as face of solid figure, 508–509, 513. *See also* Pyramids
    isosceles, 346–347
    obtuse, 346–347
    scalene, 346–347
  turns
    angles of rotation, 338–339
    motion geometry, 326–329
  two-dimensional figures, 486–503, 508
    volume, 514–515
    See also Plane figures
Glossary, H79–H88
Gram, 478–479
Graphic aids, defined, 123
Graphs
  analyzing, 106–111
  choosing, 106–109
  comparing, 92–93
  data labels on, H9
  identifying parts of, H9
  interpreting, 102–103, 110–111

intervals and, 92–93
kinds of
    bar, 18, 107–110, 152, H9
    circle, 445
    double-bar, 100–101
    line, 18, 102–109, 111
    line plots, 88–89, 107, 108
    pictographs, 12–13, 257, H8
    stem-and-leaf plot, 90–91, 107, 108
    tree diagram, 530–531
making, 94–95, 104–105
ordered pairs, 568–569, H10
scale and, 93
"Greater than" symbol, 18, 20
Grouping/Associative Property of Addition, 33
Grouping/Associative Property of Multiplication, 150–151
Guess and Check Strategy. *See* Predict and Test strategy

H

Hands On, 86–87, 100–101, 104–105, 148–149, 194–195, 216–217, 248–249, 284–285, 338–341, 344–345, 366–367, 390–391, 398–401, 412–413, 460–463, 476–479, 528–529, 544–545
Harcourt Math Newsroom Videos, 10, 127, 144, 214, 283, 327, 407, 454, 544
Hexagons, 486
Horizontal
    defined, 568
Hour, 118–119
Hundred thousands, 6–7, 8–10, 20
Hundredths, 406–409

I

Improper fractions, 378–381
Inches (in.), 452–453, 458–459
Inequalities, 18, 20. *See also* Comparing
Input/output tables, 64–65, 172–173, 570–579, H25
Integers
    comparing and ordering, 558–561
    concept of, 558
    on a number line, 558–560
Internet activities. *See* Computer software
Intersecting lines, 324–325
Intervals, 92–93
Inverse operations, 140, 142, 143, 145
Isosceles triangle, 346–347

K

Key, of graph, 100–101, 229, 293
Kilogram (kg), 478–480
Kilometer (km), 470–475

L

Leaf. *See* Stem-and-leaf plots
Length. *See* Measurement
"Less than" symbol, 18, 20
Like fractions, 386–391
Likely events, 534–536

Line graphs, 18, 102–109, 111
Line plots, 88–89, 107, 108
Line segments, 320
Line symmetry, 330, H26
Linear units, 452, H22. *See also* Measurement *and* Perimeter
    changing, 458–459, 474–475, 479
    choosing, 452–453
    customary units, 452–453, 458
    fractions with, 454–457
    metric, 470–475, 480–481
Lines, 320
    intersecting, 324–325
    parallel, 324–325
    perpendicular, 324–325
    of symmetry, 330, H26
Linkup
    architecture, 495
    art, 147, 323, 389
    health, 63
    history, 129, 305, 381
    music, 371
    reading, 29, 123, 167, 203, 291, 351, 457, 501, 537
    science, 129, 409, 561
    social studies, 271, 511
    technology, 79, 137, 183, 243, 317, 361, 449, 525, 587
Liters (L), 476–477
Logical Reasoning, 422–423

M

Make a Graph strategy, 94–95
Make a Model strategy, 70–71, 332–333, 376–377
Make a Table strategy, xxv, 24–25, 464–465
Manipulatives and visual aids, 123
    arrays
        multiplication, 142–146, 300–303
    base-ten blocks
        division, 248–249, 284–285
        multiplication, 194–195, H13
        place value, 248–249
    compasses, 342
    counters, 70, 247
    decimal squares, 434
    dot paper
        congruence, 326–329
        symmetry, 331
        transformations, 326–329
    fraction bars, 366–369, 372–375, 379, 386, 388–393, 398–401
    geometric solids
        number cubes, 534, 544, 549
        unit cubes, 514–515
    grid paper
        area, 492–495, 496–499
        multiplication, 144, 300
        perimeter, 496–497
    multiplication tables, 142, 145, 148–149, 298
    number lines
        decimals, 414–416, 418–420, 429, 455
        fractions, 376–377, 407, 408, 414–416, 455
        subtraction, H24
        whole numbers, 18, 26, 558–560, H24
    rulers, 326, 327, 343, 470–472, 490, 494
    spinners, 532, 533, 536, 542–547
Mass, 478–479
Math Detective. *See* Reasoning

Mathematical expressions. *See* Expressions
Mathematical probability, 542–543, 544–545, 546–547
Mathematical symbols. *See* Symbols
Mean, 276–277
Measurement
  of angles, 340–341
  area, 492–495
    square feet, 492–503
    square units, 492–503
  area/perimeter relationship, 496–497
  capacity/liquid volume, 460–461, 476–477
  changing units, 458–459, 474–475, 477, 479
  choosing a reasonable unit, 452–453, 463, 470–472, 476, 477, 478–479
  comparing, 452, 459, 472, 475
  customary system, 452–453, 458, 460–463, 492–503
  estimating
    length, 455–456
    liquid volume, 477
    surface area of solids, 492
    temperature, 555, 557
    volume of solids, 514–515
    weight/mass, 462–463
  fractions in, 454–457
  length/distance, 454–457, H22–H23
    centimeters, 470–475, 481
    decimeters, 470–475
    feet, 452–453, 458
    inches, 452–453, 458
    kilometers, 470–475
    meters, 470–475
    meterstick, 470–473
    miles, 452–453, 458
    using rulers, 326, 327, 343, 470–472, 490, 494
    yards, 452–453, 458
  metric system, 470–481, H70
  perimeter of a polygon, 486–491
  table of measures, 458, 461, 463, 474, 479, H70
  temperature, 554–557
    Celsius degrees, 556–557
    Fahrenheit degrees, 554–555
  three-dimensional figures, 508, 514–517
  time
    days, 126–129
    hours, 118–119
    minutes, H10–H11
    seconds, 116
  two-dimensional figures, 486–503
  unit conversions, 458–460, 464–465, 474–475
  weight/mass, 462–463, 478–479
    grams, 478–479
    kilograms, 478–480
    ounces, 462–463
    pounds, 462–463
    tons, 462–463
Median, 86–87
Mental math, 413
  addition, 36–39, 160
  division, 256–257
  multiplication, 160, 186–187, 210–211, 226, 305
  subtraction, 36–39, 305
Meter (m), 470–475
Metric system
  capacity
    liters, 476–477
    milliliters, 476–477
  changing units in, 477, 479
  equivalent measures in, 474, H70
  length
    centimeters, 470–475, 481, H23
    decimeters, 470–475
    kilometers, 471–475
    meters, 470–475
  mass
    grams, 478–479
    kilograms, 478–480
  relating units in, H70
Mighty Math Software. *See* Computer software
Mile (mi), 452–453, 458
Milliliters (mL), 476–477
Millions, 8–11, 21
Minutes, H10–H11
Mixed Applications, 13, 49, 111, 125, 153, 235, 275, 293, 397, 441, 517, 549, 563, 579
Mixed numbers, 378–381
  adding
    with decimals, 430–433, 436–439
    with fractions, 392–395
  decimals, 414–417
  improper fractions, 378–381
  modeling, 415, 416
  subtracting
    with decimals, 434–439
    with fractions, 393–395
Mixed Review and Test Prep, 3, 5, 7, 11, 19, 23, 29, 35, 39, 41, 43, 47, 55, 57, 59, 63, 65, 69, 85, 87, 89, 91, 93, 101, 103, 105, 109, 117, 119, 123, 129, 141, 143, 147, 149, 151, 161, 163, 167, 169, 171, 173, 187, 189, 199, 203, 211, 213, 215, 217, 227, 229, 231, 233, 235, 247, 249, 253, 257, 259, 265, 267, 271, 273, 277, 283, 285, 287, 291, 299, 301, 305, 307, 323, 325, 329, 331, 339, 341, 343, 345, 347, 351, 365, 367, 371, 375, 381, 389, 391, 395, 399, 401, 409, 413, 417, 421, 429, 431, 433, 435, 439, 453, 457, 459, 461, 463, 473, 475, 477, 479, 487, 491, 495, 497, 501, 511, 513, 515, 529, 531, 537, 543, 545, 547, 555, 557, 561, 569, 573, 577
Mixed Strategy Practice, 25, 71, 95, 175, 205, 219, 255, 309, 333, 353, 377, 423, 465, 481, 503, 533
Mode, 86
  of categorical data, 87
  of numerical data, 86–87
Models. *See also* Manipulatives and visual aids
  addition, 70–71
  decimals, ?362–412, 415, 416, 473, H20
  division, 142, 246–249, 284–285
  equivalency of decimals and fractions, 406, 415, 416, 473
  fractions, 376–377, 415, 416
  mixed numbers, 378–381, 415, 416
  multiplication, 142, 194–195, 216–217, H13
  negative numbers, 558–561
  place value, 4–11
Money
  adding, 437–438, H22
  decimal points with, 437–438, H20
  dividing, 276–277
  estimating with, 430–431
  multiplying, 201, 232
  rounding, 429, H21
  subtracting, 437–438, H22
Motion geometry (transformations)
  reflections (flips), 326–328
  rotations (turns), 326–328
  translations (slides), 326–328

Multiples, 298
estimating products with, H15
finding with multiplication table, 298–299
of hundred, H14
patterns with, 186, 210, H16, H23
of ten
dividing by, 282–283
multiplying by, 186–187, 212–213, H14, H16, H24
of thousand, H14
Multiplication, 140–141, 158–163, 168–175, 232–233, H12.
    See also Measurement changing units
basic facts, 142–149, H12
choosing the operation, 152–153
division as inverse operation, 140, 142, 143, 145, 287
estimating products, 188–189, 214–215, 229, H15
factors in, 298–299
mental math and, 160, 186–187, 210–211, 226
modeling, 142, 194–195, 216–217, 224, H13
by multiples of ten, 186–187, 212–213, H14, H16, H24
one-digit by four-digit, 200–203
one-digit by three-digit, 196–199, H15
one-digit by two-digit, 144–146, 148–149, 190–193, H15
partial products, 216–219
patterns in, H16, H23
products of, H12
properties of
Grouping/Associative Property, 150–151
Order/Commutative Property, 145
recording, H14
skip-counting in, H23
of three factors, 150–151, H30
two-digit by four-digit, 230–231
two-digit by three-digit, 228–229
two-digit by two-digit, 224–227
using place value and, 197
Multiplication table, division on, 145
Multistep problems, 3, 5, 7, 11, 13, 22, 25, 28, 35, 38, 43, 47, 49, 55, 57, 63, 65, 68, 69, 71, 87, 89, 91, 93, 95, 103, 105, 109, 111, 117, 122, 125, 128, 141, 143, 149, 151, 153, 160, 161, 166, 167, 169, 171, 173, 187, 189, 193, 195, 198, 202, 203, 205, 211, 213, 217, 219, 227, 229, 231, 234–235, 247, 249, 252, 253, 257, 259, 265, 267, 270, 273, 275, 277, 283, 285, 290, 293, 299, 301, 304, 307, 309, 322, 323, 325, 331, 333, 339, 341, 343, 350, 351, 353, 365, 367, 375, 377, 380, 388, 391, 395, 396, 399, 401, 408, 413, 416, 420, 423, 429, 433, 435, 438, 441, 453, 457, 459, 461, 463, 465, 475, 477, 479, 481, 488, 491, 494, 495, 497, 501, 511, 513, 515, 517, 529, 533, 537, 543, 545, 547, 555, 557, 560, 561, 563, 572, 576, 577, 579

N

Negative numbers, 558–561
Nets, 512–513
Networks, 227
Number lines
comparing on, 18, 418–421, 558–561, H24
decimals on, 407–408, 428–429
fractions on, 376–377
mixed numbers on, 414–417
negative numbers on, 558–561
ordering on, 419, 420
rounding on, 26, 429
Number patterns, 129, 146, H7

Number sense, 5, 299, 306–307
adding
decimals, 432–433, 436–439
fractions, 386–389, 392–395, 398–399
multidigit numbers, 40–41, H5
checking division with multiplication, 251–252, 265, 267, 269, 270
checking subtraction with addition, 393
comparing and ordering whole numbers, 20–23
dividing by multiples of ten, 282–283
dividing multidigit numbers by a single digit, 140–146, 246–259
dividing multiples of 10, 100, and 1000, 186–187
equivalency of fractions and decimals, 406
estimating sums and differences, 34–35
Grouping/Associative Property of multiplication, 150–151
identifying place value of digits. *See* Place value
modeling simple fractions to show equivalency, 406, 473
multiplication table through ten times ten, 145, 148–149, 298
multiplying multidigit numbers by a double-digit number, 228–231
multiplying multidigit numbers by a single-digit number, 148–149, 196–203, H15
multiplying multiples of 10, 100, and 1,000, 186–187, H14, H16, H24
multiplying multiples of ten, 186–187, 212–213, H14, H16, H24
Order/Commutative Property
of addition, 39
of multiplication, 145
reading, writing decimals to thousandths, 409
reading, writing whole numbers to 1,000,000, 8–11
related addition and subtraction facts, H7
related multiplication and division facts, 140–141
rounding off to nearest ten, hundred, thousand, ten thousand, or hundred thousand, 26–29, 34, H4, H15
subtracting
decimals, 434–439
multidigit numbers, 40–41, H5
using concepts of negative numbers, 558
Number sentence. *See* Equations
Numbers
average of, 276–277
benchmark, 2–3, H2
comparing, 18–25, 46, 48, 56, 372–375, 380, 388, 418–421, 431, 453, 456–457, 459, 463, 472, 475, 557, 559, 560, H3, H24
composite, 302–305
expanded form, 6, 7, 9, 10
factoring, 298–307
mean of, 276–277
millions, 8–11, 21
mixed, 378–381, 392–395, 414–417
negative, 558–561
ordered pairs of, 568–569, H10
ordering, 20–23, 373–375, 419–420, 456–457, 560, H4
place value in, 4–11, 18
prime, 302–307
reading and writing, 6–11, 364–365, 406–409
rounding, 26–29, 34, 188–189, 428–431, H4, H15
standard form, 6, 7, 9, 10
thousandths, 409
in words, 6, 9, 10
Numeration system, 11
Numerator, 364, 387, H18, H19

O

Obtuse angles, 321–323, H30
Obtuse triangles, 346–347
Octagons, 486
One-digit numbers
    dividing by, 140–146
    multiplying by, H15
Operations, choosing, 152, 153, 396–397, 459
Order/Commutative Property
    of addition, 39
    of multiplication, 145
Ordered pairs, 568–569, H10
Ordering
    decimals, 418–421
    fractions, 372–375, 454–457
    greater numbers, H4
    measurements, 475
    negative numbers, 558–561
    on number line, 418–420
Organized lists, 532–533
Ounce (oz), 462–463
Outcomes, 528–533, 534–537
Outlier, 88

P

Parallel lines, 324–325
Parallelograms, 349–350
Parentheses, 54–59, 150–151, 158–161
Patterns
    in division, 257, 282–283
    extending, 187, 309
    finding, xxv, 64, 129, 146, 308–309, 502–503
    input/output, 64–65, 172–173, 570–579, H25
    linear, 129
    with multiples, 146, 210, H16, H23
    for solid figures, 512–513
Pentagonal prism, 486
Pentagons, 486
Percent, 417
Performance Assessment, 78, 136, 182, 242, 316, 360, 448, 524, 586
Perimeter
    of plane figures, 486–491, 498–501, H28
    relating to area, 496–497
    See also Formulas
Periods, 6, 8
Perpendicular lines, 324–325
Pictographs, 12–13, 257, H8
Pint (pt), 460–461
Place value, 4, 18, 197, H14, H17
    hundred thousands, 6–7, 8–10
    hundreds, 2, 4–5, 8
    hundredths, 406–409
    in large numbers, 2
    millions, 8–11, 21
    modeling, 4, 6, 8, 9
    ones, 4–5, 6, 8
    periods, 6, 8
    placing first digit in quotient and, 264–265
    tens, 2, 4–5, 6, 8
    tenths, 406–407

thousands, 2, 4–5, 6, 8, H2
thousandths, 409
Plane, defined, 320
Plane figures, 486, H27. *See also* Quadrilaterals
    circles, 342–345
    drawing, 342, H29
    hexagons, 486
    octagons, 486
    parallelograms, 348–351
    pentagons, 486
    polygons, 486
    rectangles, 490–498, 502–503, 508–511
    rhombuses, 348–351
    squares, 490, 493
    trapezoids, 348–351
    triangles, 327, 330, 346–347, 350, 351, 486, 508–509
    vertices of, 321
Plots
    line, 88–89, 107, 108
    stem-and-leaf, 90–91, 107, 108
Points, with ordered pairs, 568–569, H10
Polygons. *See* Plane figures
Possible outcomes, 528–537, 542–543
Pound (lb), 462–463
Predict and Test strategy, 254–255
Predictions, 254–255, 534
    of certain events, H31
    of likely and equally likely events, 534–537
    of outcomes, 534–537
    of possible outcomes, 534–537
    of unlikely events, 534–536
Prime factorization, 306–307
Prime numbers, 302–305
Prisms
    modeling volume, 514–515
    pentagonal, 511
    rectangular, 508–510
    triangular, 508–510
Probability, 542–545
    certain events, 542, 543, H31
    equally likely outcomes, 534–536, 542–543, 546–549
    fairness, 546–549
    impossible events, 542–543, H31
    likely events, 534–536, 542–543
    organizing results of experiments, 544–545
    performing simple experiments, 544–545
    possible outcomes, 528–533, 542–543
    predictions, 534–537
    unlikely events, 534–536
Problem solving, *See* Problem solving Linkup, Problem
        solving skills, Problem solving strategies, *and*
        Problem solving Thinker's Corner, xxiv–xxv
    analyzing, 203
    choosing a strategy, 24, 25, 464–465
    choosing the operation, 152, 153, 292–293, 396–397
    determining reasonableness, 2, 440–441
    solving multistep problems, 234–235
Problem solving applications, 12–13, 48–49, 110–111,
        124–125, 152–153, 234–235, 274–275, 292–293,
        396–397, 440–441, 516–517, 548–549, 562–563, 578–579
Problem Solving on Location, 79A–79B, 137A–137B,
        183A–183B, 243A–243B, 317A–317B, 361A–361B,
        449A–449B, 525A–525B, 587A–587B
Problem solving skills
    choose the operation, 152–153, 292–293, 396–397
    draw conclusions, 110–111, 548–549

estimate or exact answers, 48–49
evaluate reasonableness of answers, 440–441
identify relationships, 578–579
interpret the remainder, 274–275
make generalizations, 562–563
multistep problems, 234–235
sequence information, 124–125
too much/too little information, 516–517
use a graph, 12–13
Problem solving strategies
  act it out, 70–71
  compare strategies, 464–465
  draw a diagram or picture, 352–353, 464–465, 480–481
  find a pattern, xxiv – xxv, 308–309, 502–503
  make a graph, 94–95
  make a model, 70–71, 332–333, 376–377
  make a table, 24–25, 464–465
  make an organized list, 532–533
  predict and test, 254–255
  solve a simpler problem, 218–219
  use logical reasoning, 422–423
  work backward, 174–175
  write an equation, 204–205
Products. *See* Multiplication
Properties
  Grouping/Associative Property, xxvi, 150–151
  Identity, xxvi
  Order/Commutative Property, xxvi, 39, 145
Protractor, 340–341
Pyramids, 439
  square, 508–510
  triangular, 508–510

Quadrilaterals, 486
  parallelograms, 348–350
  rectangles, 490–498, 502–503, 508–510
  rhombuses, 348–350
  squares, 490, 493
  trapezoids, 348–350
Quart (qt), 460–461
Quotients, 140, H12
  correcting, 288–291
  estimating, 258–259, 282–283, 288–291
  placing first digit in, 264–265
  three-digit, 266–277
  zeros in, 268–271
  See also Division

Radius, 342–343
Random sampling, 85
Range, 88
Ratio, 381
Ray, 320
Reading Strategies
  analyze information, 203
  cause and effect, 167
  choose relevant information, 457
  classify and categorize, 351, 537
  compare and contrast, 29

observe relationships, 501
sequence information, 395
synthesize information, 291
use graphic aids, 123
Reasoning, 48–49, 218–219, 422–423, 440–441
  applying strategies from a simpler problem to solve a more complex problem, 36
  breaking a problem into simpler parts, 36
  in Challenge, 75, 133, 179, 239, 313, 357, 445, 521, 583
  choosing a problem solving strategy, 464–465
  estimating, 34–35, 188–189, 214–215, 258–259, 282–283, 430–431, 490, 492, 514
    using estimation to verify reasonableness of an answer, 40–44, 196–205, 224–233, 254–255, 258–259, 266–267
  generalizing beyond a particular problem to other situations, 562–563
  identifying relationships among numbers, 397, 578–579
  logical, 422–423
  in Math Detective, 74, 132, 179, 238, 312, 356, 444, 520, 582
  observing patterns, 11, 397
  Opportunities to explain reasoning are contained in every exercise set. Some examples are: 13, 35, 87, 89, 93, 101, 102, 117, 122, 129, 149, 160, 187, 198, 203, 226, 229, 231, 253, 277, 285, 301, 304, 309, 325, 329, 339, 343, 350, 367, 371, 374, 377, 380, 388, 397, 400, 408, 420, 421, 429, 431, 459, 463, 472, 473, 487, 497, 515, 533, 536, 543, 545, 577
  recognizing relevant and irrelevant information, 516–517
  sequencing and prioritizing information, 124–125
  in Thinker's Corner, 11, 39, 47, 69, 109, 161, 199, 227, 253, 329, 375, 409, 417, 421, 439, 473, 577
Recording division, 249, 285
Recording multiplication, H14
Rectangles
  area of, 502–503
  as face of solid figure, 508
  perimeter of, 486–491
  relationship between area and perimeter in, 496–497
Rectangular prism, 508–510
Reflections (flips), 326–328
Remainders, in division, 246–247, 274–275, H17
Rhombuses, 348–350
Right angles, 321–323, H25, H27, H30
Roman numerals, 11
Rotational symmetry, 330–331
Rotations (turns), 326–328
Rounding
  deciding when to round, 428
  decimals, 428–431, H21
  in estimation, 188–189, 214–215, 430–431, H15
  whole numbers, 26–29, 34, H4

Scale, 92–93
Scalene triangles, 346–347
Schedules, 116, 117, 120–125, 130
Seconds, 116
Self Similarity, 329
Shapes. *See* Geometry
Sharpen Your Test-Taking Skills, H62–H65
Similar figures, 326–328
Simplest form of fraction, 368–371
Skip-counting, H23

Slides. *See* Translations

Solid figures, H28
  cone, 508–510
  cylinder, 508–510
  prism, 508–510
  pyramid, 508–510
  volume of, 514–515

Solve a Simpler Problem strategy, 218–219

Sorting. *See* Ordering

Spheres, 508–510

Spinners, 532, 533, 536, 542–547

Square numbers, 179

Square pyramids, 508, 510

Square unit, 492–503

Squares, 490, 493
  area, 493
  as face of cube, 508–511
  perimeter of, 490–491

Standard form, 6, 7, 9, 10

Standardized Test Prep, 15, 31, 51, 73, 97, 113, 131, 155,
    177, 207, 221, 237, 261, 279, 295, 311, 335, 355,
    383, 403, 425, 443, 467, 483, 505, 519, 539, 551,
    565, 581

Statistics. *See also* Data
  averages, 276–277
  bar graph, 100–101, 107–110, H9
  collecting data, 82–85
  designing survey questions, 82
  interpreting data, 110–111, 120–123
  line graph, 102–109, 111
  making predictions from data, 534–537
  mean, 276–277
  median, 86–87
  mode, 86–87
  outliers, 88
  pictograph, 12–13, 257, H8
  range, 88
  survey, 82

Stem, 90

Stem-and-leaf plots, 90–91, 106

Study Guide and Review, 76–77, 134–135, 180–181, 240–241,
    314–315, 358–359, 446–447, 522–523, 584–585

Subtraction
  across zeros, 42–43
  addition and, 40–41, 44–47
  choosing, 396–397
  of decimals, 434–439
  differences, 40–41, 42–43, 45, 46
  estimation and, 34–35, 430–431
  of like fractions, 390–391
  mental math, 36–39
  of mixed numbers, 392–395
  with money, 437–438, H22
  of three-digit numbers, H5
  time, 120
  of two-digit numbers, H5

Sum, 34–35

Surveys, 82

Symbols
  angles, 320–323
  line relationships
    intersecting, 324
    parallel, 324
    perpendicular, 324
  lines, 320–323
  rays, 320–323

Symmetry
  line, 330–331, H26
  rotational, 330–331
  verifying, H26

T

Tables and charts
  analyzing data from, 110–111, 122–123, 125, 141, 160, 166,
    171–173, 235, 270, 273, 277, 299, 374, 394
  completing, 109, 283
  frequency, 82–85, H8
  making, 24–25, 109, 422
  multiplication, 142, 145, 148–149, 298
  organizing data in, 82–85, 528–529
  tally tables, 82, 83, H8

Tally tables, 82, 83, H8

Technology Linkup, 79, 137, 183, 243, 317, 361, 449, 525,
    587

Technology Links
  E-Lab Activities, 21, 42, 101, 121, 149, 225, 276, 303, 343,
    367, 391, 413, 480, 493, 529, 535
  Harcourt Math Newsroom Video, 10, 127, 144, 214, 283,
    327, 407, 454, 544
  Mighty Math Software, 45, 67, 146, 197, 201, 250, 286, 289,
    308, 348, 373, 393, 415, 432, 508, 512, 542
  See also Computer software

Temperature, 102–103, 554–555, 556–557
  measuring, 554–555, 556–557
  negative numbers, 554–555, 556–557

Tens
  adding, 36–39, H5
  dividing, 256, 282–283
  multiples of, 186–187, 212–213, 282–283, H14, H16, H24
  place value, 4–5, 6, 8
  rounding, 26, H4, H15

Tenths, 406–407

Tessellations, 357

Thermometers, 554–557

Thinker's Corner, 11, 23, 39, 47, 69, 85, 109, 161, 193, 199,
    253, 329, 375, 381, 395, 417, 421, 439, 473, 489, 577

Thousands, H2–H3
  compare numbers to, H3
  multiples of, H14
  place value, H2
  rounding, 188–189, 214–215, H4

Thousandths, 409

Three-digit numbers
  adding, H5
  dividing, 266–267
  multiplying by two-digit numbers, 224–233
  quotients, 266–277
  subtracting, H5

Three-dimensional figures, 508

Time
  A.M., 118–119
  calendars, 126–128, 307, H11
  clocks, 117, 118, 120, 121, 339, H10–H11
  elapsed, 120–123, 129
  as fraction, H10
  leap years, 127
  midnight, 118, 119
  noon, 118, 119
  P.M., 118, 119

schedules, 116, 117, 124–125
units of, 116, 126–127, H70
  century, 127–129
  decade, 127–129
  hours, 118–119
  minutes, H10–H11
  seconds, 116
  years, 127–130
  zones, 133
Time lines, 129
Tons (T), 462–463
Transformations, 326–328
Translations (slides), 326–328
Trapezoids, 349–350
Tree diagrams, 530–531
Trends, 105
Triangles
  acute, 346–347
  attributes of, 486
  equilateral, 327, 330, 346–347
  as face of solid figure, 508–509, 513. *See also* Pyramids
  finding length of side of, 499
  isosceles, 346–347
  obtuse, 346–347
  scalene, 346–347
Triangular prism, 508–510
Triangular pyramid, 508–510
Troubleshooting, H2–H31
Turns. *See* Rotations
Two-digit numbers
  adding, H5
  dividing by, 282–291
  multiplying by, 224–231
  subtracting, H5
Two-dimensional figures, 508

Unfair games, 546–549
Units of measurement. *See* Measurement
Unlikely events, 534–536
Use Logical Reasoning strategy, 422–423

Variables
  defined, 60
  equations with, 60–62, 170–171
  expressions with, 168–169
Venn diagram, 352–353
Vertex, 321
Vertical
  define, 568
Volume, 514–515

Weeks, 126–129
Weight
  ounces, 462–463
  pounds, 462–463
  tons, 462–463

What's the Error?, 10, 22, 43, 57, 68, 71, 87, 108, 122, 146, 163, 166, 187, 202, 211, 226, 231, 247, 249, 259, 270, 285, 299, 304, 307, 322, 325, 345, 367, 370, 393, 399, 433, 435, 438, 456, 463, 475, 491, 513, 536, 543, 563, 569
What's the Question?, 10, 25, 38, 65, 198, 255, 270, 287, 304, 329, 339, 394, 453, 494, 510, 536, 547, 573, 579
Whole numbers
  average of, 276–277
  comparing, 18–19, 24, 39, 46, 48, 56
  expanded form of, 6, 7, 9, 10
  mean of, 276–277
  ordering, 20–23, 560, H4
  place value of, 4–11
  reading and writing, 6–11
  rounding, 26–29, 34, H4
  standard form of, 6, 7, 9, 10
Word form, 6, 7, 9, 10
Work Backward strategy, 174–175
World Wide Web. *See* Computer software
Write a Problem, 10, 19, 22, 46, 49, 59, 93, 103, 125, 153, 160, 171, 175, 198, 213, 233, 235, 252, 257, 267, 275, 290, 347, 350, 389, 395, 409, 438, 453, 472, 481, 497, 510, 517, 531, 549, 555, 572
Write About It, 3, 5, 28, 55, 62, 84, 91, 105, 107, 108, 128, 141, 143, 169, 198, 215, 255, 265, 270, 283, 299, 307, 322, 343, 353, 370, 374, 380, 397, 408, 413, 441, 456, 477, 487, 494, 503, 515, 536, 545, 560, 576
Write an Equation strategy, 204–205
Write What You Know, 15, 31, 51, 73, 97, 113, 131, 155, 177, 207, 221, 237, 261, 279, 295, 311, 335, 355, 383, 403, 425, 443, 467, 483, 505, 519, 539, 551, 565, 585
Writing in math, 3, 19, 22, 28, 46, 55, 61, 62, 105, 108, 141, 143, 153, 160, 175, 198, 213, 215, 235, 252, 255, 257, 265, 270, 283, 299, 307, 322, 370, 374, 380, 389, 395, 397, 409, 413, 438, 441, 456, 472, 477, 487, 494, 497, 503, 515, 517, 536, 545, 560, 572, 576
  See also What's the Question?, Write About It, Write a Problem, and Write What You Know

x-axis, 568
x-coordinate, 568

Yard (yd), 452–453, 458
y-axis, 568
y-coordinate, 568
Year, 127–130

Zero
  in division, 256, 268–271, 282
  in multiplication, 210, 228
  in subtraction, 42–43

PHOTO CREDITS

Photography Credits: Page Placement key: (t) top, (b) bottom, (c) center, (l) left, (r) right, (bg) background, (i) inset.

Page: v S.I. Epperson/Index Stock; xi David Stoecklein/The Stock Market; xvii (t) NASA; xvii (b) Jeff Schultz/Alaska Stock Images; Chip Henderson/Picturesque Stock Photo; xxi David B. Fleetham/Natural Selection; xxvii (t) Barbara Reed/Animals Animals; xxviii Russell D. Curtis/Photo Researchers; xxviii (i) Elsa Peterson/The Stock Market; 4 Kyodo News International; 6 Index Stock; 7 NASA; 8 Mark E. Gibson; 9 www.roadsideamerica.com, Kirby, Smith, Wilkins; 10 NASA; 11 Ann Purcell/Words and Pictures; 12 Ariel Skelley/The Stock Market; 16 Craig Tuttle/The Stock Market; 21 Lee Snider/Corbis; 24 S.I. Epperson/Index Stock; 26 Dembinsky Photo Associates; 28 Elliott Varner Smith/International Stock; 29 J.C. Carton/Bruce Coleman, Inc.; 32 Charles Sleicher/Stone; 34 (t) Ron Kimball; 34 (b) G.K. & Vikki Hart/The Image Bank; 40 Corbis; 42 David B. Fleetham/Natural Selection; 43 Bill Hickey/The Image Bank; 45 David Muench/Corbis; 46 Wyman Meinzer/Texas Panhandle Heritage Foundation; 48 Superstock; 54 David Young Wolff/PhotoEdit; 79A,79B (t) John Elk III; 79B (b) Richard Hamilton Smith; 80 Fred Bavendam/Minden Pictures; 88 Index Stock; 89 Richard Hutchings/PhotoEdit; 91 Chuck Savage/The Stock Market; 94 Alan Thornton/Stone; 98 Jim Nilsen/Stone; 102 Gerard Fritz/International Stock; 103 Roy Engelbrecht; 106 (b) Leonard Lee Rue III/Animals Animals; 106 (t) Johnny Johnson/Animals Animals; 109 (l to r) #1 Wolfgang Kaehler/Corbis; 109 (l to r) #2 Mitsuaki Iwago/Minden Pictures; 109 (l to r) #3 Roland Seitre/Peter Arnold, Inc.; 109 (l to r) #4 Manfred Danegger/Peter Arnold, Inc.; 109 (l to r) #5 Trevor Bararett/Animals Animals; 109 (l to r) #6 Rick Edwards/Animals Animals; 110 Biological Photo Service; 111 Bradley Simmons/Bruce Coleman, Inc.; 114 Richard Cummins/Corbis; 116 Ernest H. Robl; 118 Todd Gipstein/National Geographic Image Collection; 121 A. Ramey/Stock, Boston; 122 (t) Charles Thatcher/Stone; 122 (b) Sports Illustrated; 124 (b) Superstock; 125 The Purcell Team/Corbis; 127 John Skowronski; 129 NASA; 137A Mark E. Gibson; 137B (t) Phil Shermeister/Corbis; 137B (b) Dean Dunson/Corbis; 138 Rafael Macia/Photo Researchers; 147 George Holton/Photo Researchers; 150 Underwood & Underwood/Corbis; 151 Paul Barton/The Stock Market; 152 Bob Daemmrich/Stock, Boston; 153 Owen Franken/Stock, Boston; 156 Christies Images; 166 Steven Needham/Envision; 168 Brian Bailey/Stone; 182 (t) Daemmrich/Stock, Boston; 183A (t) LaFOTO/H. Armstrong Roberts, Inc.; 183A (br) E.R. Degginger/Bruce Coleman, Inc.; 183A (bl) Thom Lang/The Stock Market; 183A (bc) Eastcott/Momatiuk/Animals Animals; 183B Doug Perrine/Innerspace Visions; 184 Dale O'Dell/The Stock Market; 188 Andre Jenny/International Stock; 208 Index Stock; 210 Index Stock; 212 AGStock USA; 222 David R. Frazier/Photo Researchers; 224 Lowell Georgia/Corbis; 225 Flip Chalfant/The Image Bank; 227 J. C. Carton/Bruce Coleman, Inc.; 223 Larry Lefever/Grant Heilman Photography; 228 Ed Young/AGStock USA; 232 Michael Newman/PhotoEdit; 243A (t) P. Degginger/H. Armstrong Roberts, Inc.; 243A (b) Annie Griffiths Belt/Corbis; 243B (b) Jack Olson; 244 Dave Fleetham/Tom Stack & Associates; 250 Jeff Simon/Bruce Coleman, Inc.; 251 Jose Luis G. Grande/Photo Researchers; 254 (t) Rex A. Butcher/Stone; 256 David Madison/Bruce Coleman, Inc.; 258 Renee Lynn/Stone; 262 Phil Kramer/The Stock Market; 266 Philip Rosenburg/Pacific Stock; 268, 269 United States Postal Service; 272 Jon Feingersh/The Stock Market; 277 Paul Barton/The Stock Market; 280 Myrleen Ferguson/PictureQuest; 288 Alan Schein/The Stock Market; 291 Johnny Stockshooter/International Stock; 293 Carl Purcell/Words and Pictures; 296 Peter Dean/Grant Heilman Photography; 299 John M. Roberts/The Stock Market; 317A (t), 317A (b) Chuck Dresner/Saint Louis Zoo; 317B (bg) Philip Roullard; 317B Chuck Keeler/The Stock Market; 318 Cliff Hollenbeck/International Stock; 320 Mark E. Gibson; 330 Adam Peiperl/The Stock Market; 331 (tr) John Watkins/Photo Researchers; 331 (trc) Peter Steiner/The Stock Market; 331 (bl) D and J Heaton/Stock, Boston; 333 Gregory G. Dimijian/Photo Researchers; 336 Joseph Sohm/ChromoSohm/Corbis; 348, 349 Christie's Images; 361A Richard Cummins/The Viesti Collection; 361B (i) Dembinsky Photo Associates; 361B R. Krubner/H. Armstrong Roberts; 362 (t) Manuel Denner/Internatonal Stock; 362 (c), 362 (b) Bob Firth/International Stock; 364 Noble Stock/International Stock; 368 (r) Barbara Reed/Animals Animals/Earth Scenes; 371 David Young-Wolff/PhotoEdit; 372 C.C.Lockwood/Animals Animals; 374 Laurie Campbell/Stone; 378 Robert Maier/Animals Animals; 381 Mark Downey/Lucid Images/PictureQuest; 389 David Muench Photography; 409 (t) NASA; 409 (b) Zeva Oelbaum/Envision; 410 Associated Press File/Wide World Photos; 414 Gary Braasch Photography; 415 Len Rue, Jr./Bruce Coleman, Inc.; 421 (b) Ed Wheeler/The Stock Market; 422 Mark E. Gibson; 426 Ken Graham/Bruce Coleman, Inc.; 428 Bruce Curtis/Peter Arnold, Inc.; 430 Hubertus Kanus/Photo Researchers; 431 Rose McNulty; 432 D. Boone/Corbis; 433 George Disario/The Stock Market; 434 J.C. Carton/Bruce Coleman, Inc.; 436 Richard Cummins/Corbis; 440 David Muench/Corbis; 441 Craig Jones/Allsport; 449A Carr Clifton/Minden Pictures; 449B (l) Raymond Gehman/Corbis; 449B (r) Tim Fitzharris/Minden Pictures; 456 E. R. Degginger/Bruce Coleman, Inc.; 468 (bg) Mark J. Thoms/Dembinsky Photo Associates; 468 (ti) Johnny Johnson/Stone; 468 (ci) Stephen J. Krasemann/DRK; 468 (bi) Marty Cordano/DRK; 470 (t) Ray Coleman/Photo Researchers; 472 D. Young-Wolff/PhotoEdit; 474 Index Stock; 475 J. Swedberg/Bruce Coleman, Inc.; 480 Index Stock; 484 Larry West/FPG International; 487 Tom Edwards/Animals Animals; 501 Federal Clip Art/One Mile Up; 506 Kevin Hart/International Stock; 508 Jean Kugler/FPG International; 509 Tom McHugh/Photo Researchers; 510 (cl) Peter Aaron/Esto; 510 Frederica Georgia/Picturesque; 510 Ezra Stoller/Esto; 511 Alex S. Maclean/Landslides; 515 Richard Barnet/Omni-Photo Communications; 521 Spencer Grant/PhotoEdit; 524 Superstock; 525A, 525B H. Abernathy/H. Armstrong Roberts; 525B (br) Ty Smedes Nature Photography; 540 Marc Romanelli/The Image Bank; 552 Roland Seitre/Peter Arnold, Inc.; 554 Bob Krist/Stone; 555 (l) Rita Maas/The Image Bank; 555 (r) Jose L. Pelaez/The Stock Market; 556 Jeff Schultz/Alaska Stock Images; 558, 559, 560 NASA; 562 Color Day/The Image Bank; 563 Ken Levinson/International Stock; 564 (c) Mark E. Gibson; 564 (l) Envision(c)Overseas; 566 NASA/Science Photo Library/Photo Researchers; 566 (i) American Red Cross/NOAA; 568 T & D McCarthy/The Stock Market; 573 David R. Frazier; 587A Michael Evans/The Image Finders; 587A (t) (bi) 587B (t) John Gillmoure/The Stock Market; 587B (b) Lance Beeny; H1 David Madison/Bruce Coleman, Inc.; H58 (r) John Margolies/Esto; H58 (cl) William Johnson/Stock, Boston; H58 (r) Ezra Stoller/Esto; H58 (cr) Frederica Georgia/Picturesque.

All other photography by Harcourt photographers listed below, © Harcourt: Weronica Ankarorn; Victoria Bowen; Richard Haynes; Ken Kinzie; Ron Kunzman; Ed McDonald; Sheri O'Neal; Terry Sinclair, Sonny Spenser.